EARTHQUAKE RESISTANT DESIGN AND RISK REDUCTION

EARTHQUAKE RESISTANT DESIGN AND RISK REDUCTION

Second Edition

David Dowrick
Tauranga, New Zealand

A John Wiley and Sons, Ltd., Publication

This edition first published 2009
© 2009, John Wiley & Sons, Ltd.

Registered office
John Wiley & Sons Ltd, The Atrium, Southern Gate, Chichester, West Sussex, PO19 8SQ, United Kingdom

For details of our global editorial offices, for customer services and for information about how to apply for permission to reuse the copyright material in this book please see our website at www.wiley.com.

The right of the author to be identified as the author of this work has been asserted in accordance with the Copyright, Designs and Patents Act 1988.

All rights reserved. No part of this publication may be reproduced, stored in a retrieval system, or transmitted, in any form or by any means, electronic, mechanical, photocopying, recording or otherwise, except as permitted by the UK Copyright, Designs and Patents Act 1988, without the prior permission of the publisher.

Wiley also publishes its books in a variety of electronic formats. Some content that appears in print may not be available in electronic books.

Designations used by companies to distinguish their products are often claimed as trademarks. All brand names and product names used in this book are trade names, service marks, trademarks or registered trademarks of their respective owners. The publisher is not associated with any product or vendor mentioned in this book. This publication is designed to provide accurate and authoritative information in regard to the subject matter covered. It is sold on the understanding that the publisher is not engaged in rendering professional services. If professional advice or other expert assistance is required, the services of a competent professional should be sought.

Library of Congress Cataloging-in-Publication Data

Dowrick, David J.
   Earthquake resistant design and risk reduction / David Dowrick. – 2nd ed.
        p. cm.
   Rev. ed. of: Earthquake risk reduction. 2003.
   Includes bibliographical references and index.
   ISBN 978-0-470-77815-9 (cloth)
   1. Earthquake engineering. 2. Earthquake resistant design. 3. Earthquake hazard analysis. I. Dowrick, David J. Earthquake risk reduction.   II. Title.
   TA654.6.D69 2009
   624.1′762—dc22
                                                                          2009008862

A catalogue record for this book is available from the British Library.

ISBN: 978-0-470-77815-9

Set in 10/12pt Times by Laserwords Private Limited, Chennai, India
Printed and bound in Singapore by Markono Print Media Pte Ltd.

*Dedicated to my wife Gulielma
and all people who live in earthquake areas.*

Contents

Preface		xv
About the Author		xvii
1	**Earthquake Risk Reduction**	1
1.1	Introduction	1
1.2	Earthquake Risk and Hazard	1
1.3	The Social and Economic Consequences of Earthquakes	3
	1.3.1 Earthquake consequences and their acceptability	3
	1.3.2 Economic consequences of earthquakes	7
1.4	Earthquake Risk Reduction Actions	9
	References	14
2	**The Nature of Earthquakes**	15
2.1	Introduction	15
2.2	Global Seismotectonics	16
2.3	The Strength of Earthquakes – Magnitude and Intensity	20
	References	24
3	**Determination of Site Characteristics**	27
3.1	Introduction	27
3.2	Local Geology and Soil Conditions	27
3.3	Ground Classes and Microzones	31
3.4	Site Investigations and Soil Tests	34
	3.4.1 Introduction	34
	3.4.2 Field determination and tests of soil characteristics	35
	3.4.3 Laboratory tests relating to dynamic behaviour of soils	39
	References	43
4	**Seismic Hazard Assessment**	45
4.1	Introduction	45
4.2	Crustal Strain and Moment Release	45
4.3	Regional Seismotectonics	47
4.4	Faulting	49

	4.4.1	*Location of active faults*	49
	4.4.2	*Types of fault*	50
	4.4.3	*Degree of fault activity*	51
	4.4.4	*Faults and earthquake magnitudes*	57
4.5	Earthquake Distribution in Space, Size and Time	59
	4.5.1	*Introduction*	59
	4.5.2	*Spatial distribution of earthquakes – maps*	60
	4.5.3	*Earthquake distribution in time and size*	63
	4.5.4	*Models of the earthquake process*	66
4.6	The Nature and Attenuation of Ground Motions	70
	4.6.1	*Earthquake source models*	70
	4.6.2	*The characteristics of strong ground motion*	72
	4.6.3	*Spatial patterns of ground motions*	84
	4.6.4	*Attenuation of ground motions, spectral response and intensity*	91
	4.6.5	*Attenuation of displacement*	102
	4.6.6	*Other conditions that influence ground motions*	102
4.7	Design Earthquakes	105
	4.7.1	*Introduction*	105
	4.7.2	*Defining design events*	109
	4.7.3	*Sources of accelerograms and response spectra*	111
	4.7.4	*Response spectra as design earthquakes*	111
	4.7.5	*Accelerograms as design earthquakes*	114
4.8	Faults – Hazard and Design Considerations	116
	4.8.1	*Introduction*	116
	4.8.2	*Probability of occurrence of fault displacements*	116
	4.8.3	*Designing for fault movements*	117
4.9	Probabilistic Seismic Hazard Assessment	118
4.10	Probabilistic vs. Deterministic Seismic Hazard Assessment	122
	References	123
5	**Seismic Response of Soils and Structures**	**131**
5.1	Introduction	131
5.2	Seismic Response of Soils	131
	5.2.1	*Dynamic properties of soils*	131
	5.2.2	*Site response to earthquakes*	138
5.3	Seismic Response of Soil–Structure Systems	149
	5.3.1	*Introduction*	149
	5.3.2	*Dynamic analysis of soil–structure systems*	151
	5.3.3	*Soil models for dynamic analysis*	151
	5.3.4	*Useful results from soil–structure interaction studies*	163
5.4	Seismic Response of Structures	169
	5.4.1	*Elastic seismic response of structures*	169
	5.4.2	*Non-linear seismic response of structures*	172
	5.4.3	*Mathematical models of non-linear seismic behaviour*	177
	5.4.4	*Level of damping in different structures*	177
	5.4.5	*Periods of vibration of structures*	179

	5.4.6	Interaction of frames and infill panels	180
	5.4.7	Methods of seismic analysis for structures	183
	References		197

6 Earthquake Vulnerability of the Built Environment — 203
6.1 Introduction — 203
6.2 Qualitative Measures of Vulnerability — 203
6.3 Quantitative Measures of Vulnerability — 206
    6.3.1 Introduction — 206
    6.3.2 Vulnerability of different classes of buildings — 213
    6.3.3 Vulnerability of contents of buildings — 214
    6.3.4 Damage models as functions of ground-motion measures — 220
    6.3.5 Microzoning effects on vulnerability functions — 225
    6.3.6 Upper and lower bounds on vulnerability — 236
    6.3.7 Earthquake risk reduction potential — 237
    6.3.8 Human vulnerability to casualties — 237
    6.3.9 Inter-earthquake effects — 240
    6.3.10 Damage due to liquefaction-induced differential ground deformations — 241
    References — 242

7 Earthquake Risk Modelling and Management — 245
7.1 Earthquake Risk Modelling — 245
7.2 Material Damage Costs — 245
    7.2.1 Damage costs directly due to ground shaking using empirical damage ratios — 246
    7.2.2 Damage costs due to earthquake induced fires — 247
    7.2.3 Damage cost estimation using structural response parameters — 248
7.3 Estimating Casualties — 249
7.4 Business Interruption — 253
7.5 Reduction of Business Interruption — 255
7.6 Management of and Planning for Earthquakes — 256
7.7 Earthquake Insurance — 259
7.8 Earthquake Risk Management in Developing Countries — 259
7.9 Impediments to Earthquake Risk Reduction — 261
7.10 Further Reading and Software — 262
    References — 262

8 The Design and Construction Process – Choice of Form and Materials — 265
8.1 The Design and Construction Process – Performance-Based Seismic Design — 265
8.2 Criteria for Earthquake Resistant Design — 267
    8.2.1 Performance-based seismic design — 267
    8.2.2 Function, cost and reliability — 270
    8.2.3 Criteria for reliability of performance — 270

8.3	\multicolumn{2}{l}{Principles of Reliable Seismic Behaviour – Form, Material and Failure Modes}	275

8.3 Principles of Reliable Seismic Behaviour – Form, Material and Failure Modes — 275
- 8.3.1 Introduction — 275
- 8.3.2 Simplicity and symmetry — 276
- 8.3.3 Length in plan — 277
- 8.3.4 Shape in elevation — 278
- 8.3.5 Uniform and continuous distribution of strength, stiffness and mass — 278
- 8.3.6 Appropriate stiffness — 280
- 8.3.7 Choice of construction materials — 282
- 8.3.8 Failure mode control — 284

8.4 Specific Structural Forms for Earthquake Resistance — 289
- 8.4.1 Moment resisting frames — 289
- 8.4.2 Framed tube structures — 289
- 8.4.3 Structural walls (shear walls) — 289
- 8.4.4 Concentrically braced frames — 290
- 8.4.5 Eccentrically braced frames — 290
- 8.4.6 Hybrid structural systems — 291

8.5 Passive Control of Structures – Seismic Isolation and Energy-Dissipating Devices — 291
- 8.5.1 Introduction — 291
- 8.5.2 Isolation from seismic motion — 293
- 8.5.3 Seismic isolation using flexible bearings — 295
- 8.5.4 Seismic isolation using flexible piles and energy dissipators — 298
- 8.5.5 Rocking structures — 298
- 8.5.6 Energy dissipators for seismically isolated structures — 301
- 8.5.7 Energy dissipators for non-isolated structures — 302

8.6 Low-Damage Structures – Damage Avoidance Design — 304
8.7 Construction and the Enforcement of Standards — 306
8.8 Developing Countries — 306
References — 307

9 Seismic Design of Foundations and Soil-Retaining Structures — 311

9.1 Foundations — 311
- 9.1.1 Introduction — 311
- 9.1.2 Shallow foundations — 312
- 9.1.3 Deep box foundations — 313
- 9.1.4 Caissons — 316
- 9.1.5 Piled foundations — 316
- 9.1.6 Foundations in liquefiable ground — 325
- 9.1.7 Further reading — 328

9.2 Soil-Retaining Structures — 328
- 9.2.1 Introduction — 328
- 9.2.2 Seismic soil pressures — 329
References — 334

10 Design and Detailing of New Structures for Earthquake Ground Shaking 337

- 10.1 Introduction 337
  - 10.1.1 Strength-based vs. displacement-based design 338
- 10.2 Steel Structures 343
  - 10.2.1 Introduction 343
  - 10.2.2 Seismic response of steel structures 343
  - 10.2.3 Reliable seismic behaviour of steel structures 344
  - 10.2.4 Steel beams 347
  - 10.2.5 Steel columns 352
  - 10.2.6 Steel frames with diagonal braces 356
  - 10.2.7 Steel connections 361
  - 10.2.8 Composite construction 366
  - 10.2.9 Further reading 366
- 10.3 Concrete Structures 367
  - 10.3.1 Introduction 367
  - 10.3.2 Seismic response of reinforced concrete 367
  - 10.3.3 Reliable seismic behaviour of concrete structures 367
  - 10.3.4 Reinforced concrete structural walls 377
  - 10.3.5 In situ concrete design and detailing: general requirements 387
  - 10.3.6 Foundations 392
  - 10.3.7 Walls 392
  - 10.3.8 Columns 393
  - 10.3.9 Beams 397
  - 10.3.10 Beam–column joints in moment resisting frames 397
  - 10.3.11 Structural precast concrete detail 399
  - 10.3.12 Precast concrete cladding detail 406
  - 10.3.13 Prestressed concrete design and detail 410
- 10.4 Masonry Structures 416
  - 10.4.1 Introduction 416
  - 10.4.2 Seismic response of masonry 417
  - 10.4.3 Reliable seismic behaviour of masonry structures 419
  - 10.4.4 Design and construction details for reinforced masonry 421
  - 10.4.5 Construction details for structural infill walls 424
  - 10.4.6 Masonry structures in regions of low and moderate seismic hazard 427
- 10.5 Timber Structures 427
  - 10.5.1 Introduction 427
  - 10.5.2 Seismic response of timber structures 429
  - 10.5.3 Reliable seismic behaviour of timber structures 432
  - 10.5.4 Foundations of timber structures 433
  - 10.5.5 Timber-sheathed walls (shear walls) 435
  - 10.5.6 Timber horizontal diaphragms 437
  - 10.5.7 Timber moment resisting frames and braced frames 438
  - 10.5.8 Connections in timber construction 441
  - 10.5.9 Fire resistance of timber construction 443

10.6	Design of New Structures in Developing Countries	444
	10.6.1 Introduction	444
	10.6.2 Some aspects of design and detailing of buildings	444
	10.6.3 Further sources of design information	444
	References	445

11	**Earthquake Resistance of Services, Equipment and Plant**	**451**
11.1	Seismic Response and Design Criteria	451
	11.1.1 Introduction	451
	11.1.2 Earthquake motion – accelerograms	452
	11.1.3 Design earthquakes	452
	11.1.4 The response spectrum design method	452
	11.1.5 Comparison of design requirements for buildings and equipment	453
	11.1.6 Equipment mounted in buildings	454
	11.1.7 Material behaviour	455
	11.1.8 Cost of providing earthquake resistance of equipment	456
11.2	Seismic Analysis and Design Procedures for Equipment	456
	11.2.1 Design procedures using dynamic analysis	456
	11.2.2 Design procedures using equivalent-static analysis	457
11.3	Seismic Protection of Equipment	462
	11.3.1 Introduction	462
	11.3.2 Rigidly mounted equipment	464
	11.3.3 Equipment mounted on isolating or energy-dissipating devices	466
	11.3.4 Light fittings	468
	11.3.5 Ductwork	469
	11.3.6 Pipework	471
	References	473

12	**Architectural Design and Detailing for Earthquake Resistance**	**475**
12.1	Introduction	475
12.2	Non-structural Infill Panels and Partitions	476
	12.2.1 Introduction	476
	12.2.2 Integrating infill panels with the structure	477
	12.2.3 Separating infill panels from the structure	477
	12.2.4 Separating infill panels from intersecting services	478
12.3	Cladding, Wall Finishes, Windows and Doors	479
	12.3.1 Introduction	479
	12.3.2 Cladding and curtain walls	480
	12.3.3 Weather seals	480
	12.3.4 Wall finishes	480
	12.3.5 Windows and architectural glass panels	481
	12.3.6 Doors	481
12.4	Miscellaneous Architectural Details	482
	12.4.1 Exit requirements	482
	12.4.2 Suspended ceilings	482
	References	484

13	**Retrofitting**	**485**
13.1	Introduction	485
13.2	To Retrofit or Not?	486
13.3	Benefit-Cost of Retrofitting	493
13.4	Retrofitting Lifelines	493
13.5	Retrofitting Structures	496
	13.5.1 Strategies for improving structural performance	496
	13.5.2 Examples of retrofitting structures	499
13.6	Retrofitting Equipment and Plant	503
13.7	Retrofitting in Developing Countries	504
13.8	Performance of Retrofitted Property in Earthquakes	506
	13.8.1 Introduction	506
	13.8.2 Earthquake performance of retrofitted unreinforced masonry buildings	506
	13.8.3 Earthquake performance of retrofitted reinforced concrete buildings	506
	References	507

Appendix A	**Modified Mercalli Intensity Scale (NZ 2007)**	**511**
Appendix B	**Structural Steel Standards for Earthquake Resistant Structures**	**517**
Index		**519**

Preface

This book is intended to help professionals of a wide range of disciplines in their attempts to reduce the social and economic risks of earthquakes. Earthquake risk reduction involves so many issues in planning, design, regulation, quality control and risk assessment, that it is difficult for any individual to gain a full perspective on the issue, or for any society to move forward in the quest at the desired speed.

The principal objectives are to:

- discuss the chief aspects of earthquake risk and their evaluation;
- present methods of reducing or managing a range of earthquake risks;
- give guidance on topics where no generally accepted method is currently available;
- suggest procedures to be adopted in earthquake regions having no official zoning or lateral force regulations; and
- indicate some of the more important specialist literature.

The general principles of this book apply to the whole built environment, while the more detailed sections relate to selected aspects of it. Whereas an attempt has been made to provide guidance on most of the more important issues, the coverage cannot be exhaustive in a single book.

The author published the predecessor to this book, under the title *Earthquake Risk Reduction*, in 2003. However because of the popularity of the previous title *Earthquake Resistant Design*, a new title combining the titles of my 1977, 1987 and 2003 books has been chosen for the present edition. In the six years or so since writing the 2003 edition, much progress has continued to be made in understanding earthquakes and in how to build more safely. In some areas of study great developments have occurred, such as in various aspects of seismic hazard, hazard analysis, Damage Avoidance Design and retrofitting techniques. However, one of the great difficulties for the designers of earthquake resistant property arises simply from the enormous volume of literature being produced on each of the many specialisms within the overall subject area. Hopefully, this book will help some of us to find our way better through this maze.

This book was written mainly, but not only, from the standpoint of designers trying to keep a broad perspective on the total process, starting from the nature of the loading through to the details of construction. To this end, the successful overall format of my previous book has been retained, with the introduction of some new topics. I have attempted to give the book as international a flavour as possible, although I have inevitably drawn more heavily on information from the literature that I know best.

To reduce earthquake risk worldwide our greatest needs are (1) retrofitting of much existing infrastructure, (2) to speed up the currently promising development of the new generation of economical low damage infrastructure (Damage Avoidance Design), (3) to avoid building in high hazard zones, (4) to improve quality control of construction, (5) to improve collaboration between engineers and architects, and (6) develop simpler methods of analysis and detailing rules.

Finally, I must again express my gratitude to all those fellow workers upon whose works I have drawn, and I also acknowledge all the valuable work to which I have not been able to refer, through the sheer enormity of the task. Special thanks are owed to my former employers, GNS Science, Lower Hutt (the Institute of Geological and Nuclear Sciences), for again supporting my writing with drafting, carried out expertly by Carolyn Hume. I also wish to thank the following earthquake professionals for recent advice: Dr John Beavan of GNS Science; Dr Charles Clifton of the University of Auckland; Dr Richard Fenwick and Dr Stefano Pampanin of the University of Canterbury, Christchurch; Graham Hancox, Nick Perrin and Grant Dellow of GNS Science whose engineering geology work was a great contribution to our revision of the New Zealand version of the Modified Mercalli Scale, given here in Appendix A and Professor Eduardo Miranda of Stanford University, California. Finally I particularly wish to thank my statistician workmate Dr David Rhoades of GNS Science for his fruitful collaboration in our research over many years (many of our papers are referred to in this book), and his contribution of the section on earthquake processes in this book. My collaboration with GNS continues with my ongoing research with David Rhoades.

David Dowrick

About the Author

DAVID DOWRICK   BE (Civil), DEng, FIPENZ, FNZSEE

David Dowrick graduated as a Civil Engineer from Auckland University in 1958, and was awarded a Doctorate of Engineering in 2003. He has worked both in consulting engineering and in earthquake research, based in London and New Zealand. He now works part-time on earthquake research from his home in Tauranga, where he is also involved in four non-engineering committees.

Major engineering design projects include the Sydney Opera House roof, restoration of York Minster and the Ohaaki power station cooling tower in New Zealand. In addition, he has advised the New Zealand Historic Places Trust on a variety of conservation projects. Dr Dowrick is the author of over 70 publications, including four editions of books on earthquake engineering. He has received five awards, four for earthquake engineering papers plus the Institution of Civil Engineers' Telford Gold Medal for his paper on York Minster. He has been made a life member of the NZ Society for Earthquake Engineering.

1

Earthquake Risk Reduction

1.1 Introduction

Earthquake risk reduction is a complex affair involving many people of many vocations, much information, many opinions and many decisions and actions. The relationships between the contributing sets of information and people are illustrated schematically by the flowchart given in Figure 1.1. Considering that this diagram is necessarily simplified, it is clear that managing the changes needed to reduce earthquake risk is a challenging task in which all of the people in any given region are explicitly or implicitly involved. The largest component of earthquake risk reduction has traditionally been known as *earthquake resistant design*, the subject of Chapters 8–13.

1.2 Earthquake Risk and Hazard

In normal English usage the work *risk* means *exposure to the chance of injury or loss*. It is noted that the word *hazard* is almost synonymous with *risk*, and the two words are used in the risk literature with subtle variations which can be confusing.

Fortunately, an authoritative attempt has been made to overcome this difficulty through the publication by the Earthquake Engineering Research Institute's glossary of standard terms for use in this subject (EERI Committee on Seismic Risk, 1984). Their terminology will be used in this book.

Thus, the definition of *seismic risk* is *the probability that social or economic consequences of earthquakes will equal or exceed specified values at a site, at several sites, or in an area, during a specified exposure time*. Risk statements are thus given in quantitative terms.

Seismic hazard, on the other hand, is *any physical phenomenon (e.g. ground shaking, ground failure) associated with an earthquake that may produce adverse effects on human activities*. Thus, hazards may be either purely descriptive terms or quantitatively evaluated, depending on the needs of the situation. In practice, seismic hazard is often evaluated for given probabilities of occurrence, for example as for ground motions in Figure 4.48.

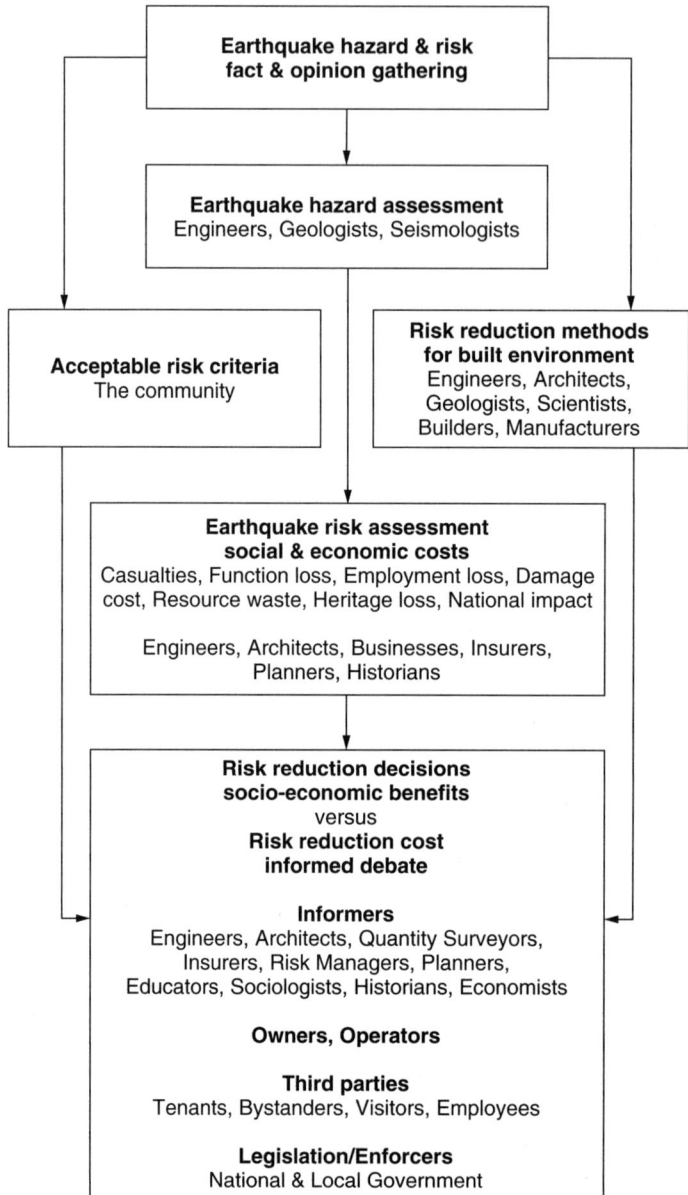

Figure 1.1 Information flow and those involved in the earthquake risk reduction process

It follows that seismic risk is an outcome of seismic hazard as described by relationships of the form

$$\text{Seismic risk} = (\text{Seismic hazard}) \times (\text{Vulnerability}) \times (\text{Value}) \qquad (1.1)$$

where *Vulnerability* is the amount of damage, induced by a given degree of hazard, and expressed as a fraction of the *Value* of the damaged item under consideration. Referring to Figure 6.7(a), the monetary seismic risk to a building could be evaluated by taking the seismic hazard to be the Modified Mercalli intensity of the appropriate probability of occurrence, the vulnerability would then be taken as the damage ratio on the appropriate curve for that intensity, and the value would be the replacement cost.

For design or risk assessment purposes the assessment of seismic hazard consists of the following basic steps:

(1) definition of the nature and locations of earthquake sources;
(2) magnitude–frequency relationships for the sources;
(3) attenuation of ground motion with distance from source;
(4) determination of ground motions at the site having the required probability of exceedance.

Because seismic risk and hazard statements are essentially forecasts of future situations, they are inherently *uncertain*. Seismic hazard assessments are attempts to forecast the likely future seismic activity rates and strengths, based on knowledge of the past and present, and significant uncertainties arise partly because the processes involved are not fully understood and partly because relevant data are generally scarce and variable in quality. For reasonable credibility considerable knowledge of both *historical seismicity* and *geology* need to be used, together with an appropriate analysis of the uncertainties. Seismicity is defined as the frequency of occurrence of earthquakes per unit area in a given region, and is illustrated in non-numerical terms by the seismicity map of the world presented in Chapter 2 (Figure 2.1). Where available, other geophysical or seismological knowledge, such as crustal strain studies, may also be helpful, particularly in evaluating regional seismic activity patterns. Once both the estimated future seismic-activity rates and the acceptable risks are known, appropriate earthquake loadings for the proposed structure may be determined, for example, loadings with mean recurrence intervals of 100 to more than 10,000 years, depending on the consequences of failure.

Because of the difficulties involved in seismic hazard evaluation, earthquake design criteria in different areas of the world vary, from well codified to inadequate or non-existent. Hence, depending on the location and nature of the project concerned, seismic risk evaluation ranging from none through arbitrary to thoroughgoing may be required.

The whole of this book is essentially to do with the explicit or implicit management of seismic risk, and hence the foregoing brief introduction to risk and hazard will be expanded upon in the subsequent text.

1.3 The Social and Economic Consequences of Earthquakes

1.3.1 Earthquake consequences and their acceptability

The primary consequence of concern in earthquakes is of course human casualties, i.e. deaths and injuries. According to Steinbrugge (1982), the greatest known number of deaths that have occurred in a single event is 830,000, in the Shaanxi, China, earthquake

Table 1.1 Numbers of deaths caused by a selection of larger twentieth-century earthquakes in various countries (from Steinbrugge, 1982, and NEIC web page)

Date	Location	Magnitude	Deaths
1906 Apr 18	USA, San Francisco	7.8	800
1908 Dec 28	Italy, Messina	7.5	83,000
1923 Sep 1	Japan, Tokyo	7.9	142,807
1927 May 22	China, Nan-Shan	8.3	200,000
1935 May 31	India, Quetta	7.5	30,000–60,000
1939 Jan 24	Chile, Chillan	8.3	28,000
1939 Dec 26	Turkey, Erzincan	7.9	30,000
1949 Aug 5	Ecuador, Pelileo	6.8	6,000
1956 Jun 10–17	Northern Afghanistan	7.7	2,000
1957 Dec 4	Outer Mongolia, Gobi-Altai	8.6	1,200
1960 Feb 29	Morocco, Agadir	5.6	12,000
1962 Sep 1	Northwestern Iran	7.1	12,230
1963 Jul 26	Yugoslavia, Skopje	6.0	1,100
1970 May 31	Northern Peru	7.8	66,794
1972 Dec 23	Nicaragua	6.2	5,000
1974 Dec 28	Pakistan	6.2	5,300
1976 Feb 4	Guatemala	7.5	23,000
1976 Jul 28	China, Tangshan	7.9	245,000–655,000
1976 Aug 17	Philippines, Mindanao	7.9	8,000
1977 Mar 4	Romania, Bucharest	7.2	1,500
1978 Sep 16	Northeast Iran	7.7	25,000
1980 Oct 10	Algeria	7.2	3,000
1985 Sep 19	Mexico	8.1	9,500–30,000
1995 Jan 10	Japan, Kobe	6.9	5,500
1999 Aug 17	Turkey, Koeceli	7.4	17,439
1999 Sep 20	Taiwan, Chi-Chi	7.6	2,400

of January 24, 1556. Thus the number of casualties in any given event varies enormously, depending on the magnitude, location and era of the earthquake. This is illustrated by a selection of 26 of the more important earthquakes of the twentieth century (mostly drawn from Steinbrugge, 1982) as listed here in Table 1.1. These earthquakes occurred in 24 countries in most parts of the world, and range in magnitude from 6.0 to 8.6. Many of the higher casualty counts have been caused by the collapse of buildings made of heavy, weak materials such as unreinforced masonry or earth. Safety in houses in developing countries remains our biggest challenge (Comartin et al., 2004).

In Figure 1.2 are plotted the approximate total numbers of deaths in earthquakes that occurred worldwide in each decade of the twentieth century. This histogram highlights the randomness of the size and location of the earthquake occurrence process, as well as the appalling societal cost, and implied economic cost, of earthquakes. The totals were found by summing the deaths in major earthquakes listed by Steinbrugge (1982) and the NEIC. The totals for each decade do not include deaths from events with less than 1000 casualties, one of the larger omissions being the 1931 Hawke's Bay, New Zealand, earthquake in which about 260 people died (Dowrick and Rhoades, 2005).

Social and Economic Consequences of Earthquakes

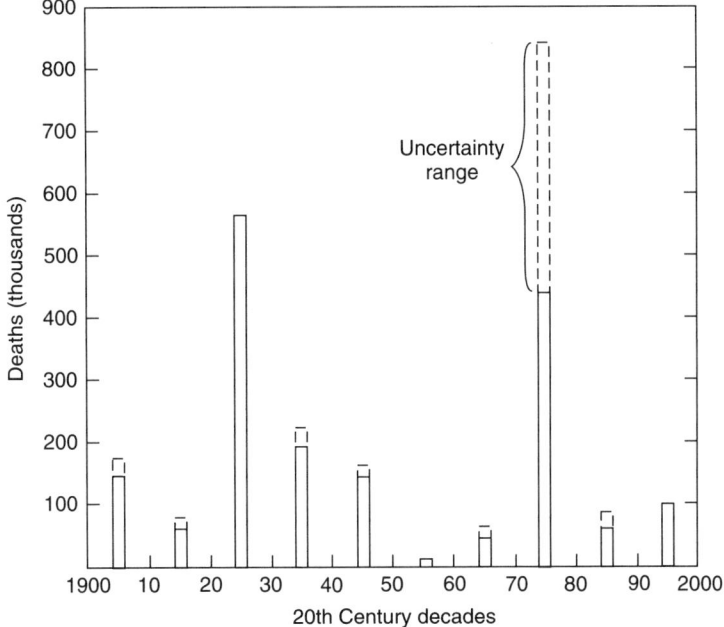

Figure 1.2 Numbers of deaths worldwide caused by large earthquakes in each decade of the twentieth century

The physical consequence of earthquakes for human beings are generally viewed under two headings:

(A) death and injury to human beings;
(B) damage to the built and natural environments.

These physical effects in turn are considered as to their social and economic consequences:

(1) numbers of casualties;
(2) trauma and bereavement;
(3) loss of employment;
(4) loss of employees/skills;
(5) loss of heritage;
(6) material damage cost;
(7) business interruption;
(8) consumption of materials and energy (sustaining resources);
(9) macro-economic impacts (negative and positive).

The above physical and socio-economic consequences should all be taken into account when the acceptable consequences are being decided (i.e. the acceptable earthquake risk).

Both financially and technically, it is possible only to *reduce* these consequences for strong earthquake shaking. The basic *planning aims* are to minimize the use of land subject to the worst shaking or ground damage effects, such as fault rupture, landslides or liquefaction. The basic *design aims* are therefore confined (a) to the reduction of loss of life in any earthquake, either through collapse or through secondary damage such as falling debris or earthquake induced fire, and (b) to the reduction of damage and loss of use of the built environment. (See also Section 6.3.7.)

Obviously, some facilities demand greater earthquake resistance than others, because of their greater social and/or financial significance. It is important to determine in the design brief not only the more obvious intrinsic value of the structure, its contents, and function or any special parts thereof, but also the survival value placed upon it by the owner.

In some countries the greater importance to the community of some types of facility is recognized by regulatory requirements, such as in New Zealand, where various public buildings are designed for higher earthquake forces than other buildings. Some of the most vital facilities to remain functional after destructive earthquakes are dams, hospitals, fire and police stations, government offices, bridges, radio and telephone services, schools, energy sources, or, in short, anything vitally concerned with preventing major loss of life in the first instance and with the operation of emergency services afterwards. In some cases, the owner may be aware of the consequences of damage to his property but may do nothing about it. It is worth noting that, even in earthquake conscious California, it is only since the destruction of three hospitals and some important bridges in the San Fernando earthquake of 1971 that there have been statutory requirements for extra protection of various vital structures.

The consequences of damage to structures housing intrinsically dangerous goods or processes is another category of consideration, and concerns the potential hazards of fire, explosion, toxicity, or pollution represented by installations such as liquid petroleum gas storage facilities or nuclear power or nuclear weapons plants. These types of consequences often become difficult to consider objectively, as strong emotions are provoked by the thought of them. Acknowledging the general public concern about the integrity of nuclear power plants, the authorities in the United Kingdom decided in the 1970s that future plants should be designed against earthquakes, although that country is one of low seismicity and seismic design is not generally required.

Since the 1960s, with the growing awareness of the high seismic risks associated with certain classes of older buildings, programmes for strengthening or replacement of such property have been introduced in various parts of the world, notably for pre-earthquake code buildings of lightly reinforced or unreinforced masonry construction. While the substantial economic consequences of the loss of many such buildings in earthquakes are, of course, apparent, the main motivating force behind these risk-reduction programmes has been social, i.e. the general attempt to reduce loss of life and injuries to people, plus the desire to save buildings or monuments of historical and cultural importance.

While individual owners, designers, and third parties are naturally concerned specifically about the consequences of damage to their own proposed or existing property, the overall effects of a given earthquake are also receiving increasing attention. Government departments, emergency services, and insurance firms all have critical interests in the physical and financial overall effects of large earthquakes on specific areas. In the case of insurance companies, they need to have a good estimate of their likely losses in any single large catastrophe event so that they can arrange sufficient reinsurance if they are over-exposed to seismic risk. Disruption of lifelines such as transport, water, and power systems obviously greatly hampers rescue and rehabilitation programmes.

1.3.2 Economic consequences of earthquakes

Figure 1.3 plots the costs of earthquake material damage worldwide per decade in the twentieth century, where known. The data for the second half of the century comes from Smolka (2000) of Munich Reinsurance. The first half of the century is incomplete, only the material damage costs for the 1906 San Francisco and the 1923 Kanto earthquakes being readily found. As with the twentieth century deaths sequence plotted in Figure 1.2, the costs sequence is seen to be random. However, there is no correlation between the deaths and costs sequences. It appears that if the costs were normalized to a constant population, and if the 1995 Kobe earthquake were not included, there would be no trend to increase

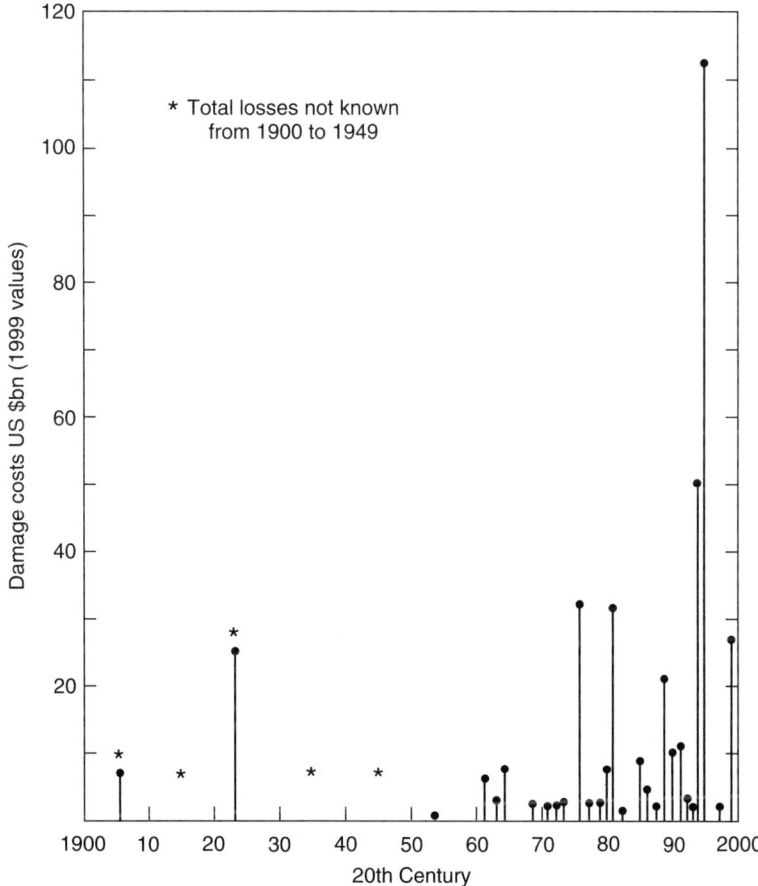

Figure 1.3 Total costs of earthquake material damage worldwide for each decade of the twentieth century (adapted from Smolka, 2000). Reproduced by permission of the Munich Reinsurance Company

with time. However, the global seriousness of earthquake damage losses is undisputed. The economic consequences of earthquakes occur both before and after the event. Those arising before the event include protection provisions such as earthquake resistance of new and existing facilities, insurance premiums, and provision of earthquake emergency services. Insurance companies themselves need to reinsure against large earthquake losses, as mentioned in the previous section.

Post-earthquake economic consequences include:

(1) cost of death and injury;
(2) cost of damage;
(3) losses of production and markets;
(4) insurance claims.

The direct cost of damage depends upon the nature of the building or other type of facility, its individual vulnerability, and the strength of shaking or other seismic hazard to which it is subjected.

During the briefing and budgeting stages of a design, the cost of providing earthquake resistance will have to be considered, at least implicitly, and sometimes explicitly, such as for the retrofitting of older structures. The cost will depend upon such things as the type of project, site conditions, the form of the structure, the seismic activity of the region, and statutory design requirements. The capital outlay actually made may in the end be determined by the wealth of the client and his or her attitude to the consequences of earthquakes, and insurance to cover losses.

Unfortunately it is not possible to give simple guides on costs, although it would not be misleading to say that most engineering projects designed to the fairly rigorous Californian or New Zealand regulations would spend a maximum of 10% of the total cost on earthquake provisions, with 5% as an average figure.

The cost of seismic upgrading of older buildings varies from as little as 10% to more than 100% of the replacement cost, depending on the nature of the building, the level of earthquake loadings used, and the amount of non-structural upgrading that is done at the same time as the strengthening. It is sad to record that many fine old buildings have been replaced rather than strengthened, despite it often being much cheaper to strengthen than to replace.

Where the client simply wants the minimum total cost satisfying local regulations, the usual cost-effectiveness studies comparing different forms and materials will apply. For this a knowledge of good earthquake resistant forms will, of course, hasten the determination of an economical design, whatever the material chosen.

In some cases, however, a broader economic study of the cost involved in prevention and cure of earthquake damage may be fruitful. These costs can be estimated on a probabilistic basis and a cost-effectiveness analysis can be made to find the relationship between capital expenditure on earthquake resistance on the one hand, and the cost of repairs and loss of income together with insurance premiums on the other.

For example, Elms and Silvester (1978) found that in communal terms the capital cost savings of neglecting seismic design and detailing would be more than offset by the increased economic losses in earthquakes over a period of time in any part of New Zealand. It is not clear just how low the seismic activity rate needs to be for it to be cheaper in the long term for any given community to omit specific seismic resistance provisions. The availability or not of private sector earthquake insurance in such circumstances would be part of the economic equation.

Hollings (1971) discussed the earthquake economics of several engineering projects. In the case of a 16-storey block of flats with a reinforced concrete ductile frame it was estimated that the cost of incorporating earthquake resistance against collapse and subsequent loss of life was 1.4% of the capital cost of building, while the cost of preventing other earthquake damage was reckoned as a further 5.0%, a total of 6.4%. The costs of insurance for the same building were estimated as 4.5% against deaths and 0.7% against damage, a total of 5.2%. Clearly, a cost-conscious client would be interested in putting up a little more capital against danger from collapse, thus reducing the life insurance premiums, and he or she might well consider offsetting the danger of damage mainly with insurance.

Loss of income due to the building being out of service was not considered in the preceding example. In a hypothetical study of a railway bridge, Hollings showed that up to 18% of the capital cost of the bridge could be spent in preventing the bridge going out of service, before this equalled the cost of complete insurance cover.

In a study by Whitman *et al.* (1974), an estimate was made of the costs of providing various levels of earthquake resistance for typical concrete apartment buildings of different heights, as illustrated on Figure 1.4. Until further studies of this type have been done, results such as those shown in the figure should be used qualitatively rather than quantitatively.

It is most important that at an early stage the owner should be advised of the relationship between strength and risk so that he can agree to what he is buying. Where stringent earthquake regulations must be followed the question of insurance versus earthquake resistance may not be a design consideration: but it can still be important, for example for designing non-structural partitions to be expendable or if a 'fail-safe' mechanism is proposed for the structure. Where there are loose earthquake regulations or none at all, insurance can be a much more important factor, and the client may wish to spend little on earthquake resistance and more on insurance.

However, in some cases insurance may be more expensive, or unavailable, for facilities of high seismic vulnerability. For example, the latter is often the case for older unreinforced masonry buildings in some high seismic risk areas of New Zealand, i.e. those built prior to the introduction of that country's earthquake loadings code in 1935. The costs of earthquake damage are discussed further in Chapter 7.

1.4 Earthquake Risk Reduction Actions

To reduce earthquake risk, each country needs to examine its strengths and weaknesses, build on the strengths, and systematically take actions which reduce or eliminate the weaknesses. An example of such an approach comes from New Zealand where a list of weaknesses was identified (Dowrick, 2003).

Over a score of weaknesses were identified there in a preliminary list of weaknesses of a wide range of types. The weaknesses have been initially divided into two main categories, named *strategic* and *tactical*, as listed in Tables 1.2(a) and 1.2(b), respectively. This division in some cases is somewhat arbitrary, but it helps in comprehending

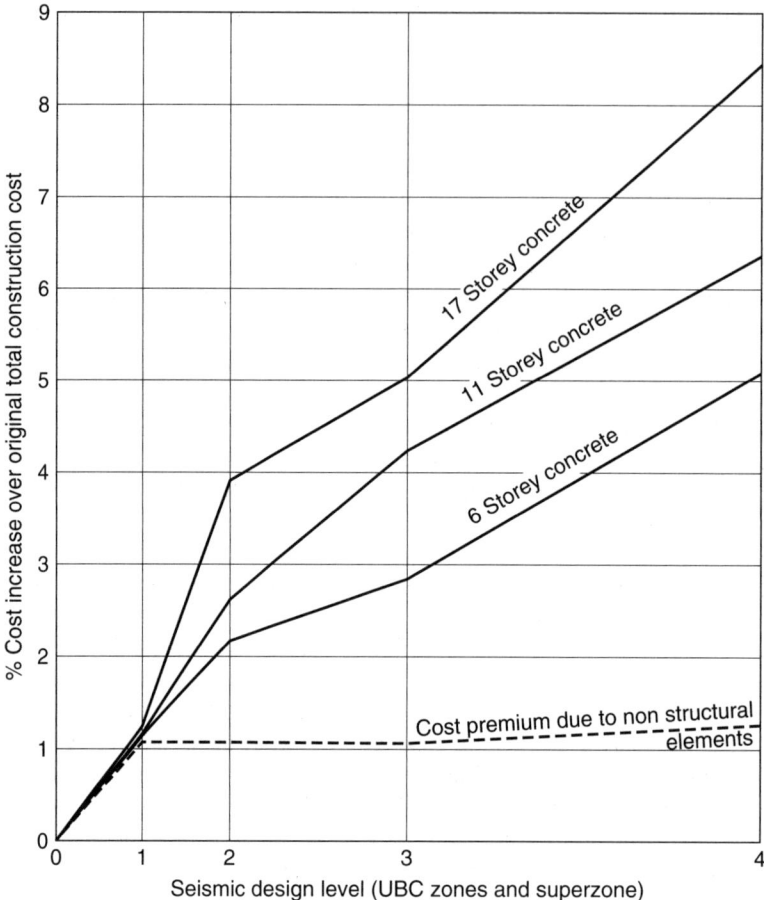

Figure 1.4 Effect on cost of earthquake resistant design of typical concrete apartment buildings in Boston (after Whitman *et al.*, 1974)

the considerable detail implied by the abbreviated descriptions given to the tabulated weaknesses.

Consider the 11 *strategic* weaknesses listed in Table 1.2(a). The first of these is clearly strategic, noting that New Zealand has no national strategy for managed progressive reduction of earthquake risk. What was needed were monitored goals of target risk reductions in a series of (say) five-year plans, with priorities assigned at both a national and a local level.

As well as listing weaknesses, Tables 1.2(a) and 1.2(b) attempt to list all parties who contribute to remedying each of the weaknesses. The first of these is *advocacy* by earthquake professionals (engineers, geologists, seismologists, architects, economists, planners, risk managers and others), and another is *funding* (rather than people). The remaining nine entities, ranging from engineers to central government, illustrate the complexity of the workings of modern society, which by fragmentation constitutes a considerable difficulty

Table 1.2 (a) Part 1 of the list of New Zealand's weaknesses in earthquake risk reduction (from Dowrick, 2003)

A	Undesirable situations – strategic	Remedial action by whom										
		A	E	a	I	M	P	G	g	L	F	O
A1	No national strategy and targets for managed incremental risk reduction with time	A	E	–	–	M	–	G	g	L	–	–
A2	Too much national vulnerability to a 'king-hit' earthquake on Wellington	A	–	–	–	M	–	G	–	L	–	–
A3	Fragmentation of the many endeavours contributing to earthquake risk reduction	A	–	–	–	–	P	G	g	L	–	–
A4	Underfunding of production of design codes and standards	A	–	–	–	–	–	G	–	–	F	–
A5	Systematic reduction of the numbers of hospitals/beds nationwide	A	–	–	–	–	P	G	g	–	F	–
A6	Too little management/modelling of business interruption losses	A	–	–	I	M	P	G	g	L	–	O
A7	Slow uptake of some new research findings	A	–	–	–	–	P	G	g	L	F	O
A8	As yet no official process for retrofitting of non-URM earthquake risk buildings	A	E	–	–	–	–	G	g	L	–	O
A9	Too much emphasis on life safety at the expense of high damage (e.g. EBFs)	A	E	–	–	–	–	–	–	–	–	O
A10	Over-design in New Zealand's lowest seismic hazard regions	–	E	–	–	–	P	–	–	L	–	–
A11	Architects who don't collaborate with engineers structural form needs	A	–	a	–	–	–	–	–	–	–	O

Notes: A = Advocacy by earthquake professions; a = Architects; E = Engineers; F = Funding needed; G = Central government; g = Government department; I = Insurance industry; L = Local government; M = Economists; O = Owners of property; P = Planners. EBF = Eccentrically braced frame; URM = Unreinforced masonry.

(i.e. a weakness) as listed in item A3. As given in Table 1.2(a), central government (G), government departments (g), local government (L) and planners (P) are all needed to address this problem, in addition to the advocacy role of earthquake professionals.

Item A10, over-design in New Zealand's lowest seismic hazard regions, results from the historical excessive conservatism of design loadings for northern regions of the North Island, a situation which was expected to be resolved in the then proposed revision of the loadings standard. This is listed as a weakness in order to illustrate the need to spend

Table 1.2 (b) Part 2 of the list of New Zealand's weaknesses in earthquake risk reduction (from Dowrick, 2003)

B	Undesirable situations – tactical	A	E	a	I	M	P	G	g	L	F	O
B1	No earthquake regulations for most equipment and plant	A	E	–	–	–	–	G	g	–	–	–
B2	Inadequate earthquake regulations for building services in buildings	A	E	–	–	–	–	G	–	L	–	O
B3	Inadequate earthquake regulations for storage of stock in shops and warehouses	A	E	–	–	–	–	G	g	L	–	O
B4	No adequate regulatory framework for existing high-risk concrete and steel buildings	A	E	–	–	–	–	G	g	–	–	–
B5	Weak powers and weak action for pre-emptive land-use planning (f,l,l,m)[1]	A	–	–	–	–	P	G	–	L	–	–
B6	Buildings astride active faults	A	EG^2	–	I	–	P	–	g	L	–	O
B7	Modern buildings built without measures for liquefiable ground	A	E	–	–	–	P	–	–	L	–	O
B8	Inadequate enforcement of some regulations	A	E	–	I	–	P	G	–	L	–	O
B9	Incomplete and/or inadequate microzoning maps nationwide	A	EG	–	–	–	P	–	–	L	–	–
B10	Some councils renting out or using earthquake risk buildings	A	E	–	I	–	P	–	–	L	–	–
B11	Are all new materials and techniques adequately researched before use? (e.g. 'chilly bins')	A	E	–	I	–	–	–	g	L	–	–
B12	No regular checks on seismic movement gaps for seismically isolated structures	A	E	–	I	–	–	–	–	L	–	O

Notes:
[1] (f, l, l, m) = faults, landslides, liquefaction, microzoning.
[2] EG = Engineers + geologists. For explanation of other abbreviations A, E, etc. see Table 1.2(a).

New Zealand's limited national financial resources wisely, and emphasize the need for national priorities for risk reduction as discussed above for item A1.

Let us now turn to the 12 *tactical* weaknesses, listed in Table 1.2(b), which generally involves more technical detail than the *strategic* weaknesses of Table 1.2(a). This is illustrated by the fact that in the *actions by whom* lists, engineers (E) appear in 11 items of Table 1.2(b) and only four of Table 1.2(a). As indicated by items B1–B4, many components of the built environment are inadequately regulated for earthquake risk purposes. The lack of mandatory regulations for earthquake protection of most built or manufactured items other than buildings is a historical situation (common worldwide) which strongly merits rectification in the interests of earthquake risk reduction. The case of stored goods (stock) in shops (item B3) is a curious and alarming example. Consider the way that goods are stacked in some shops. Lethally heavy goods are stacked needlessly high overhead in the most dangerous fashion to anyone below. The fact that loose goods or contents of buildings fall to the floor in moderate or strong shaking is common knowledge.

These situations are, in fact, a breach of the New Zealand law regarding the safety of the shop employees, and it was surprising and disappointing that the government agency, Occupational Safety and Health (OSH), had not stamped out this practice. The deaths and injuries of workers and public alike would be on the slate of the owners, OSH staff and the government, if this situation is not eliminated before the next damaging earthquake. Oddly, the New Zealand public had no statutory protection from this source of danger at the time of this study, and still had none in 2007.

In the more seismic parts of New Zealand two types of older buildings, of unreinforced masonry (URM) and some concrete buildings (item B4), pose a serious threat. While many brick buildings have been demolished or strengthened in some parts of the country, the process has been somewhat erratic, such that in 2008 there were some 5000 unstrengthened URM buildings countrywide. Even in Wellington where the City Council has been a leader in this field since about 1980, many old unreinforced brick buildings were still in use in 2008, death traps to occupants and passers-by. (However, recent changes to the Building Act should see a steady improvement in the rate of dealing with such buildings.) We might also ask why long-vacated brick buildings should not be demolished forthwith. They pose a great threat to passers-by.

The older concrete buildings that are at risk of serious earthquake damage (item B4), comprise mainly pre-1976 multi-storey buildings, which have beam and column frames rather than structural walls. In the past decade or two much work has been done in various countries such as the USA and New Zealand. In the latter country the outcome has been the publication of recommendations which cover initial evaluation, detailed assessment and improvement of structural performance (if required) of all existing buildings (New Zealand Society for Earthquake Engineering, 2006). Similar recommendations for the USA are given in FEMA-356 (American Society of Civil Engineers, 2000). The issue of what to do about substandard buildings is rightly contentious as the costs of strengthening will be considerable in many cases. Details of strengthening are discussed in Chapter 13.

An important aspect of Tables 1.2(a) and 1.2(b) is the influence of *duty of care* on who could be involved in remedial actions. Duty of care is the common law responsibility of a person or body to do something, such as warning others about a situation that they know to be dangerous, even if they are not involved, or if there is no statutory requirement. For example, building on an active fault (item B6) is known by most people to be dangerous, so that in addition to geologists, those who could act on this danger to people and property include engineers, architects, insurers, planners, government departments, local government and the owner of the building.

As the duty of care is surprisingly pervasive, Tables 1.2(a) and 1.2(b) should be widely distributed to all concerned.

References

American Society of Civil Engineers (2000) *Prestandard and Commentary for the Seismic Rehabilitation of Buildings*, FEMA-356. Federal Emergency Management Agency, Washington, DC.

Comartin C, Brzev S, Naeim F *et al.* (2004) A challenge to earthquake engineering professionals. *Earthq Spectra* **20**(4): 1049–1056.

Dowrick DJ (2003) Earthquake risk reduction actions for New Zealand. *Bull NZ Soc Earthq Eng* **36**(4): 249–259.

Dowrick DJ and Rhoades DA (2005) Risk of casualties in New Zealand earthquakes. *Bull NZ Soc Earthq Eng* **38**(4): 53–72.

EERI Committee on Seismic Risk (1984) Glossary of terms for probabilistic seismic-risk and hazard analysis. *Earthq Spectra* **1**: 33–40.

Elms DG and Silvester D (1978) Cost effectiveness of code base shear requirements for reinforced concrete frame structures. *Bull NZ Nat Soc Earthq Eng* **11**(2): 85–93.

Hollings JP (1971) The economics of earthquake engineering. *Bull NZ Soc Earthq Eng* **4**(2): 205–221.

New Zealand Society for Earthquake Engineering (2006) *Assessment and Improvement of the Structural Performance of Buildings in Earthquakes*. NZSEE, Wellington. http://www.nzsee.org.nz/PUBS/2006AISBEGUIDELINES_Corr_06a.pdf.

Smolka A (2000) Earthquake research and the insurance industry. *NZ Geophys Soc Newsletter* No 55: 22–25.

Steinbrugge KV (1982) *Earthquakes, Volcanoes, and Tsunamis: An Anatomy of Hazards*. Skandia America Group, New York.

Whitman RV, Biggs JM, Brennan J, Cornell CA, de Neufville R and Vanmarcke EH (1974) Seismic design analysis, Structures Publication No. 381, Massachusetts Institute of Technology.

2

The Nature of Earthquakes

2.1 Introduction

An earthquake is a spasm of ground shaking caused by a sudden release of energy in the earth's lithosphere (i.e. the crust plus part of the upper mantle). This energy arises mainly from stresses built up during tectonic processes, which consist of interaction between the crust and the interior of the earth. In some parts of the world, earthquakes are associated with volcanic activity. For example, in Guatemala such earthquakes occur in swarms, with an average duration of 3–4 months, the largest having a magnitude normally under 6.5. These events are of shallow focus and cause considerable damage within a radius of about 30 km from the epicentre. Human activity also sometimes modifies crustal stresses enough to trigger small or even moderate earthquakes, such as the swarms of minor tremors resulting from mining in the Midlands of England, or the sometimes larger events induced by the impounding of large amounts of water behind dams, such as the earthquakes associated with the construction of the Koyna dam in central India in 1967 (Chopra and Chakrabarti, 1973).

While the design provisions of this book apply to all earthquakes regardless of origin, any discussion of earthquakes themselves is generally confined to events derived from the main cause of seismicity, i.e. tectonic activity.

As most earthquakes arise from stress build-up due to deformation of the earth's crust, understanding of seismicity depends heavily on aspects of geology, which is the science of the earth's crust, and also calls upon knowledge of the physics of the earth as a whole, i.e. geophysics. The particular aspect of geology which sheds most light on the source of earthquakes is *tectonics*, which concerns the structure and deformations of the crust and the processes which accompany it; the relevant aspect of tectonics is now often referred to as *seismotectonics*.

Geology tells us the overall underlying level of seismic hazard which may differ from the available evidence of historical seismicity, notably in areas experiencing present-day quiescent periods.

Earthquake Resistant Design and Risk Reduction   D. Dowrick
© 2009, John Wlley & Sons, Ltd

2.2 Global Seismotectonics

On a global scale, the present-day seismicity pattern of the world is illustrated in general terms by the seismic events plotted in Figure 2.1. Most of these events can be seen to follow clearly defined belts which form a map of the boundaries of segments of the earth's crust known as *tectonic plates*. This may be seen by comparing Figure 2.1 with Figure 2.2, which is a world map of the main tectonic plates taken from the highly understandable book on the theory of continental drift by Stevens (1980). According to the latter, the earth's crust is composed of at least 15 virtually undistorted plates of lithosphere. The lithosphere moves differentially on the weaker asthenosphere which starts at the low-velocity layer in the upper mantle at a depth of about 50 km. Boundaries of plates are of four principal types:

(1) divergent zones, where new plate material is added from the interior of the earth;
(2) subduction zones, where plates converge and the under-thrusting one is consumed;
(3) collision zones, former subduction zones where continents riding on plates are colliding;
(4) transform faults, where two plates are simply gliding past one another, with no addition or destruction of plate material.

Almost all the earthquake, volcanic and mountain-building activity which marks the active zones of the earth's crust closely follows the plate boundaries, and is related to movements between them.

Divergent boundaries are found at the oceanic sea-floor ridges, affecting scattered islands of volcanic origin, such as Iceland and Tristan da Cunha, which are located on these ridges. As these zones involve lower stress levels, they generate somewhat smaller earthquakes than the other types of plate boundary.

As can be seen in Figure 2.2, subduction zones occur in various highly populated regions, notably Japan and the western side of Central and South America. Figure 2.3 shows the cross-section of the likely structure of the subduction zone formed by the Pacific plate thrusting under the Indian-Australian plate beneath the North Island, New Zealand. The seismic cross-section corresponding to Figure 2.3 is in Figure 2.4, and gives earthquakes located under the shaded region of the key map (during a period of time when no events shallower than c. 40 km occurred). The zone of diffuse seismic activity which exists down to a depth of over 300 km is believed to be related both to *volcanic activity* (movement of magma in the crust and upper mantle and related expansion and contraction), and to *faulting* within the volcanic belt and the 'New Zealand Shear Belt'. The latter is a continuation of the major Alpine Fault of the Southern Alps (Figure 2.5). Below 100 km and down to about 250 km, the pattern of earthquakes tends to lie on a well-defined plane known as a Benioff Zone, dipping 50 degrees to the north-west. This is the contact plane between the Indian-Australian plate and the Pacific plate. The isolated group of earthquakes about 600 km deep in Figure 2.4 have been conjectured to be caused by a piece of lithosphere that has become detached and has moved deeper into the mantle, as illustrated in Figure 2.3.

The progressive movement of the Pacific plate, subducting under the overlying plate, caused shear stresses to develop, as illustrated by Walcott's (1981) geophysical study (Figure 2.5), which relates the shear strains of the shear belt referred to above to large

Global Seismotectonics 17

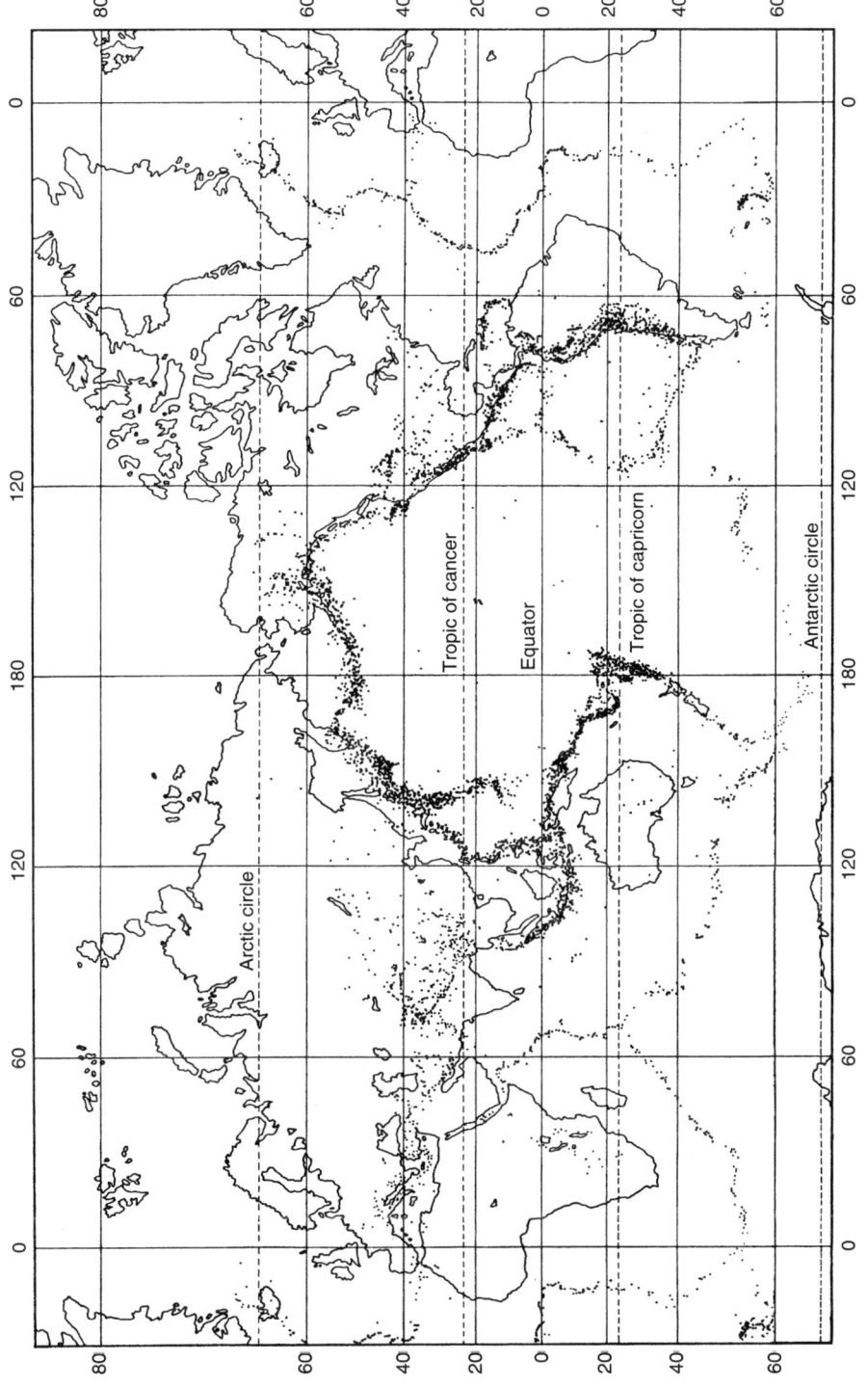

Figure 2.1 Seismicity map of the world. The dots indicate the distribution of seismic events in the mid-twentieth century (after Barazangi and Dorman, 1969)

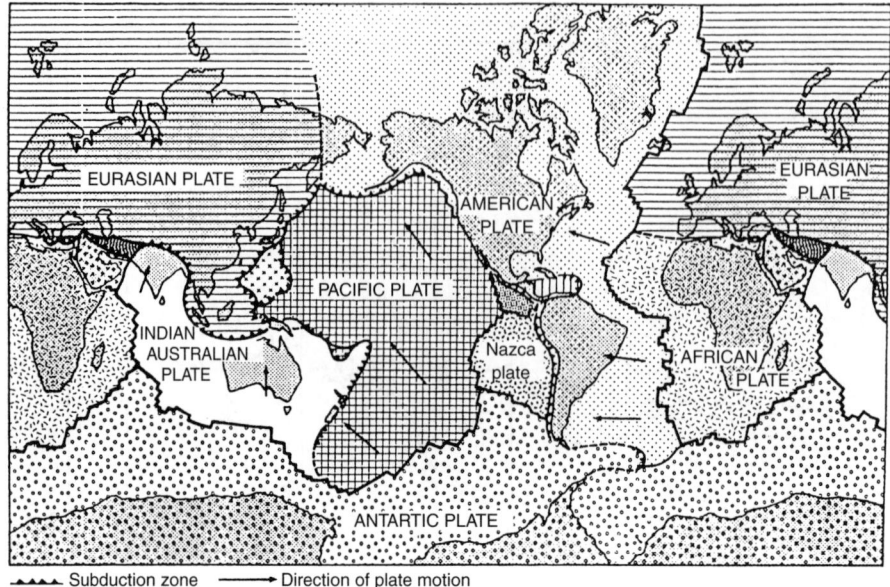

Figure 2.2 Tectonic plate map of the world, showing names of the seven largest plates and indicating subduction zones and the directions of plate movement (reproduced with permission from G.R. Stevens, 1980)

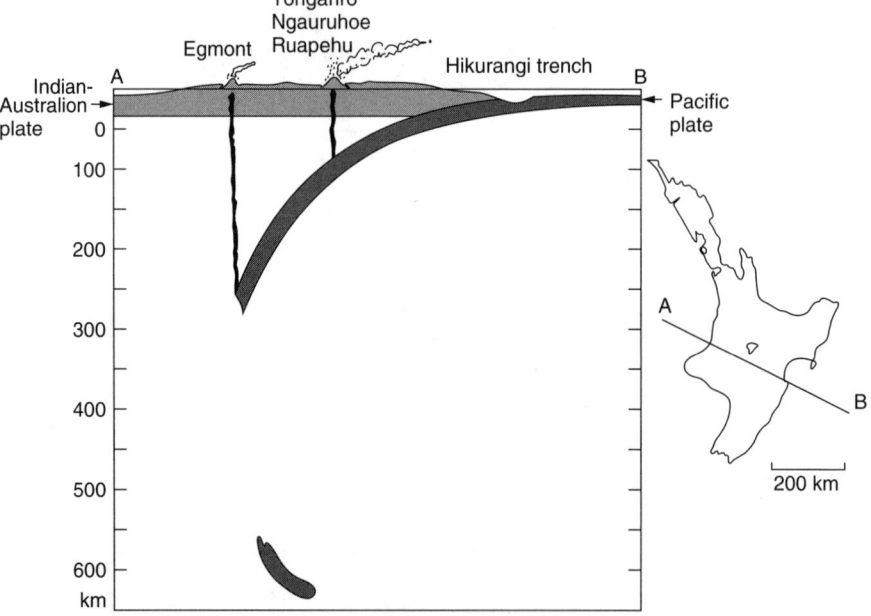

Figure 2.3 The likely structure of the subduction zone beneath the North Island, New Zealand, inferred from Figure 2.4 (reproduced with permission from G.R. Stevens, 1980)

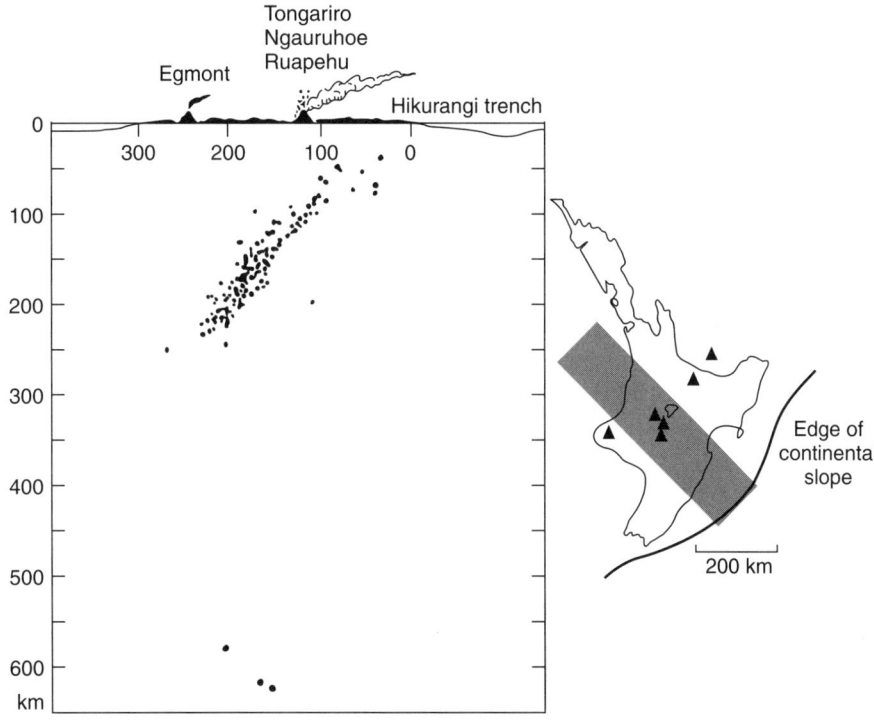

Figure 2.4 Seismic cross-section through the North Island, New Zealand, showing locations of earthquake foci (reproduced with permission from G.R. Stevens, 1980)

historical earthquakes. It is believed that the fault forming the plate boundary periodically locks together, and this leads to an accumulation of shear and compressional strain until it is in part relieved by a large thrust type of earthquake. The sudden release of strain (when the shear resistance is overcome) signals the recommencement of movement of the subducting plate in a further cycle of seismic slip, then another locking of the fault leads to the next plate interface earthquake.

As well as the 15 or so main plates shown in Figure 2.2, studies of seismic activity need to consider the smaller buffer plates or sub-plates which in certain areas tend to ease the relative movements of the world's giant plates. Buffer plates have been recognized in Tibet and China, the western USA, and at the complex junction of the African, Arabian, Iranian, and Eurasian plates, where eight Mediterranean buffer plates have been identified (Stevens, 1980).

In the foregoing discussion, tectonic plates have been described as rigid, virtually undistorted plates and the world's principal zones of seismicity have been shown to be associated with the interaction between the plates. However, occasional damaging intra-plate earthquakes also occur, well within the interior of the plates that clearly are not associated with plate boundary conditions, and so far their origins are poorly understood. The uncertainties associated with intra-plate seismicity are much greater than is the case for interplate regions of high seismicity.

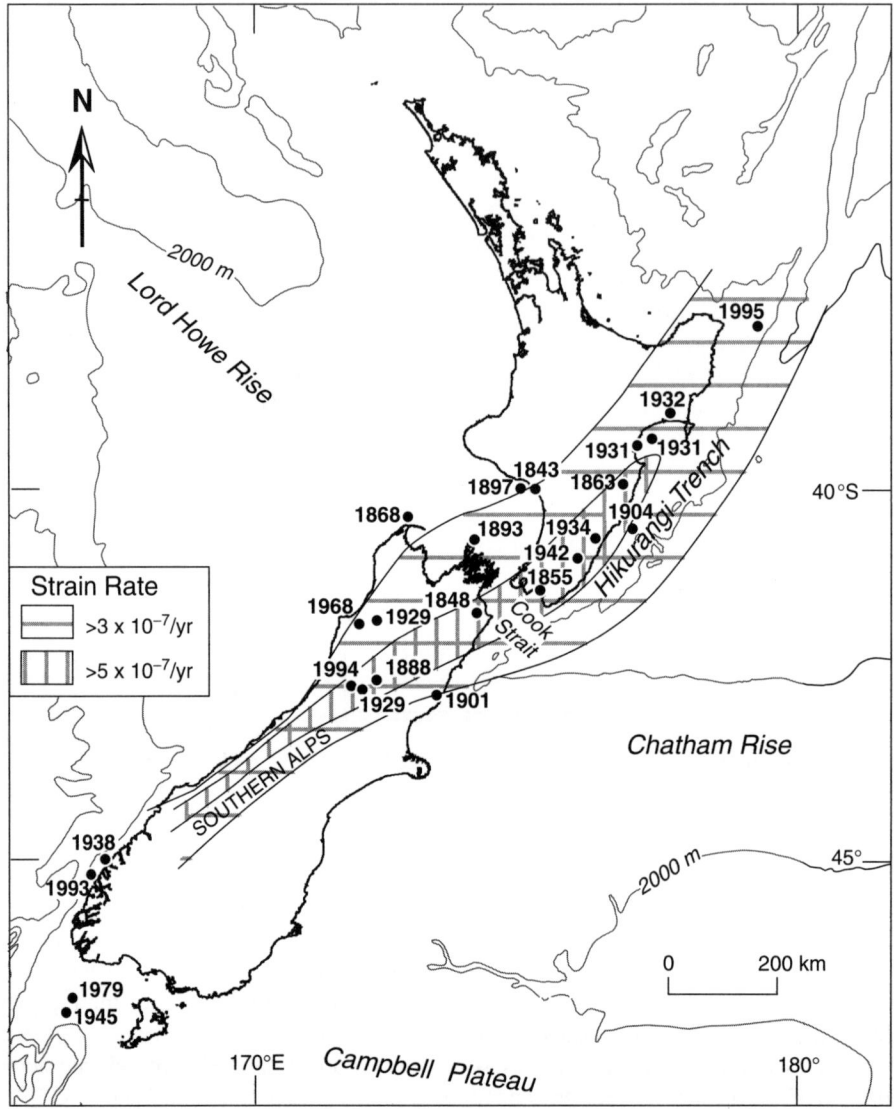

Figure 2.5 Generalized shear strain rates and large ($M_W \geq 6.8$) shallow ($h_C < 50$ km) New Zealand earthquakes, 1840–2000 (adapted from Walcott, 1981)

2.3 The Strength of Earthquakes – Magnitude and Intensity

During earthquakes the release of crustal stresses is believed generally to involve the fracturing of the rock along a plane which passes through the point of origin (the *hypocentre* or *focus*) of the event (Figure 4.25). Sometimes, especially in larger shallower earthquakes, this rupture plane, called a *fault*, breaks through to the ground surface, where it is known as a *fault trace* (Figure 4.41).

Magnitude and Intensity

The cause and nature of earthquakes is the subject of study of the science of *seismology*, and further background may be obtained from the books by Richter (1958), Bolt (2003) and Lay and Wallace (1995).

Unfortunately, for non-seismologists at least, understanding the general literature related to earthquakes is impeded by the difficulty of finding precise definitions of fundamental seismological terms. For assistance in the use of this book, definitions of some basic terms are set out below. Further definitions may be found elsewhere in this book or in the references given above.

The *strength* of an earthquake is not an official technical term, but is used in the normal language sense of 'How strong was that earthquake?' Earthquake strength is defined in two ways: first the strength of shaking at any given place (called the *intensity*) and second, the total strength (or size) of the event itself (called *magnitude*, seismic *moment*, or *moment magnitude*). These entities are described below.

Intensity is a qualitative or quantitative measure of the severity of seismic ground motion at a specific site. Over the years, various subjective scales of what is often called *felt intensity* have been devised, notably the European Macroseismic and the Mercalli scale, which are very similar. The most widely used in the English speaking world is the Modified Mercalli scale (commonly denoted MM), which has 12 grades denoted by Roman numerals I–XII. A detailed description of this intensity scale is given in Appendix A, taken from Dowrick *et al*. (2008).

Quantitative instrumental measures of intensity include engineering parameters such as peak ground acceleration, peak ground velocity, the Housner spectral intensity, and response spectra in general. Because of the high variability of both subjective and instrumental scales, the correlation between these two approaches to describing intensity is inherently weak (Figure 4.23).

Magnitude is a quantitative measure of the size of an earthquake, related indirectly to the energy released, which is independent of the place of observation. It is calculated from amplitude measurements on seismograms, and is on a logarithmic scale expressed in ordinary numbers and decimals. Unfortunately several magnitude scales exist, of which the four most common ones are described here (M_L, M_S, m_b and M_W).

The most commonly used magnitude scale is that devised by and named after Richter, and is denoted M or M_L. It is defined as

$$M_L = \log A - \log A_0, \tag{2.1}$$

where A is the maximum recorded trace amplitude for a given earthquake at a given distance as written by a Wood–Anderson instrument, and A_0 is that for a particular earthquake selected as standard.

The Wood–Anderson seismograph ceases to be useful for shocks at distances beyond about 1000 km, and hence Richter magnitude is now more precisely called *local magnitude* (M_L) to distinguish it from magnitude measured in the same way but from recordings on long-period instruments, which are suitable for more distant events. When these latter magnitudes are measured from surface wave impulses they are denoted by M_S. Gutenburg proposed what he called 'unified magnitude', denoted m or m_b, *which is dependent on body waves, and is now generally named* body wave magnitude (m_b). This magnitude scale is particularly appropriate for events with a focal depth greater than c. 45 km. All three scales M_L, m_b and M_S suffer from saturation at higher values.

The most reliable and generally preferred magnitude scale is moment magnitude, M_W. This is derived from seismic moment, M_0, which measures the size of an earthquake directly from the energy released (Wyss and Brune, 1968), through the expression

$$M_0 = \mu A D, \qquad (2.2)$$

where μ is the shear modulus of the medium (and is usually taken as 3×10^{10} Nm), A is the area of the dislocation or fault surface, and D is the average displacement or slip on that surface. Seismic moment is a modern alternative to magnitude, which avoids the shortcomings of the latter but is not so readily determined. Up to 1985, seismic moment had generally only been used by seismologists.

Moment magnitude is a relatively recent magnitude scale from Kanamori (1977) and Hanks and Kanamori (1979), which overcomes the above-mentioned saturation problem of other magnitude scales by incorporating seismic moment into its definition, such that moment magnitude is given by

$$M_W = \frac{2}{3} \log M_0 - 6.03 \qquad (M_0 \text{ in Nm}). \qquad (2.3)$$

Local magnitude M_L is inherently a poor magnitude scale, as shown by the plot in Figure 2.6 of New Zealand data from Dowrick and Rhoades (1998), who found that the best fit relationship for estimating M_W from M_L and depth, h_c, was

$$M_W = 0.96[\pm 0.49] + 0.84[\pm 0.08]M_L - 0.0055[\pm 0.0015](h_c - 25). \qquad (2.4)$$

The regression explains only 59% of the variance and has a residual standard error of 0.31. The M_L scale as estimated in other parts of the world, as well as New Zealand, is similarly unreliable.

The relation between moment magnitude M_W, surface-wave magnitude M_S and centroid depth h_c, using data restricted to modern M_W determinations (i.e. from March 8, 1964 onwards), is shown in Figure 2.7. For earthquakes of $h_c \leq 30$ km, M_S and M_W are close to being equal above magnitude 6.5. At lower magnitudes M_S is consistently smaller than M_W, and is as much as a quarter-unit smaller between magnitude 5.0 and 5.5. Depth also influences the discrepancy between M_S and M_W; for deep New Zealand earthquakes ($h_c > 50$ km) M_S is about a half-unit smaller than M_W between magnitude 5.0 and 5.5. This results from the tendency for M_S to decrease with depth for earthquakes of a given seismic moment. Karnik (1969) first dealt with this effect by proposing a focal depth correction term for M_S in relation to m_b for various parts of Europe, while Ambraseys and Free (1997) more recently estimated a focal depth correction term in relation to $\log M_0$ for European earthquakes. Considering the New Zealand data (Figure 2.7), Dowrick and Rhoades (1998) found the best fit for finding M_W in terms of M_S and h_c was the quadratic expression

$$M_W = 1.27[\pm 0.16] + 0.80[\pm 0.03]M_S + 0.087[\pm 0.031](M_S - 6)2 + 0.0031$$
$$\times [\pm 0.0006](h_c - 25). \qquad (2.5)$$

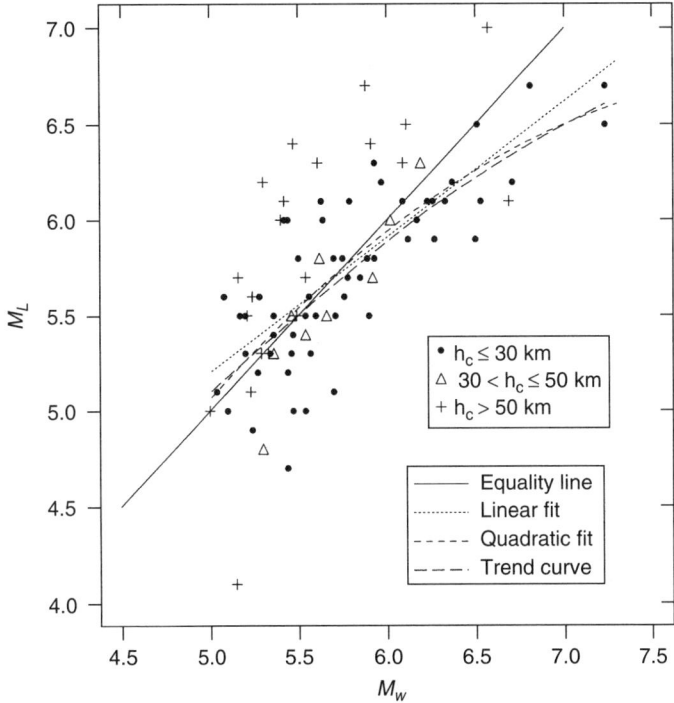

Figure 2.6 Scatter plot of local magnitude M_L against moment magnitude M_W for New Zealand earthquakes distinguishing events in different classes of centroid depth h_C. Also shown are the linear and quadratic regression fits for M_L evaluated at $h_C = 25$ km and a local regression trend curve of M_L on M_S for events with $h_C \leq 50$ km (from Dowrick and Rhoades, 1998)

The above expression explains 93% of the variance. In equation (2.5) it can be seen that the quadratic term contributes significantly to the regression because the coefficient of this term is more than twice its standard error. It is of interest to note that, although their expression is different from ours, Ambraseys and Free obtained a coefficient for their depth term of 0.0036, which is very similar to the coefficient of 0.003 in equation (2.5).

Also shown in Figure 2.7 is the relation of Ekström and Dziewonski (1988), derived from global data, between log M_O and M_S for events with $h < 50$ km. In terms of M_W, this relation is

$$M_W = \begin{cases} 2.13 + \frac{2}{3}M_S, & M_S < 5.3, \\ 9.40 - \sqrt{41.09 - 5.07M_S}, & 5.3 \leq M_S \leq 6.8, \\ 0.03 + M_S, & M_S > 6.8. \end{cases} \quad (2.6)$$

As seen in Figure 2.7, there is no great difference between this relation and the linear and quadratic fits for shallow New Zealand events over the magnitude range of the data, but the latter also describe the effect of depth.

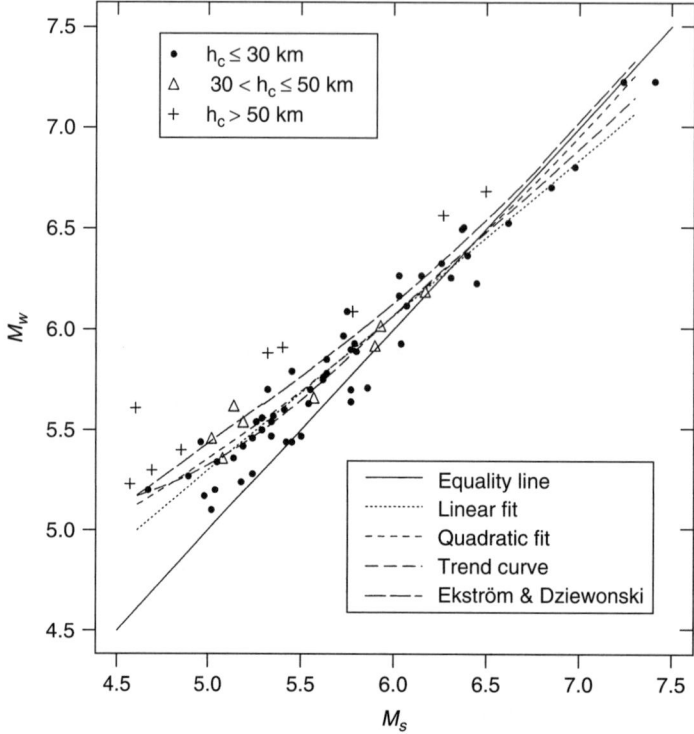

Figure 2.7 Scatter plot of moment magnitude M_W against surface wave magnitude M_S for earthquakes distinguishing events in different classes of centroid depth h_C. Also shown are the linear and quadratic (equation (2.5)) regression fits for M_W evaluated at $h_C = 25$ km and a local regression trend curve of M_W on M_S for events with $h_C \leq 50$ km, and the relation of Ekström and Dziewonski (equation (2.4)) (from Dowrick and Rhoades, 1998)

References

Ambraseys NN and Free MW (1997) Surface-wave magnitude calibration for European region earthquakes. *J Earthq Eng* **1**(1): 1–22.

Barazangi M and Dorman J (1969) World seismicity map of ESSA coast and geodetic survey epicentre data for 1961–67. *Bull Seism Soc Amer* **59**: 369–380.

Bolt B (2003) *Earthquakes*, 5th edn. WH Freeman, New York.

Chopra AK and Chakrabarti P (1973) The Koyna earthquake and damage to Koyna dam. *Bull Seism Soc Amer* **63**(2): 381–397.

Dowrick DJ and Rhoades DA (1998) Magnitudes of New Zealand earthquakes, 1901–1993. *Bull NZ Soc Earthq Eng* **31**(4): 260–280.

Dowrick DJ, Hancox GT, Perrin ND and Dellow GD (2008) The Modified Mercalli earthquake intensity scale – revisions arising from New Zealand experience. *Bull NZ Nat Soc Earthq Eng* **41**(3): 193–206.

Ekström G and Dziewonski AM (1988) Evidence of bias in estimations of earthquake size. *Nature* **332**: 319–323.

References

Hanks TC and Kanamori H (1979) A moment magnitude scale. *J Geophys Res B* **84**: 2348–2350.

Kanamori H (1977) The energy release in great earthquakes. *J Geophys Res* **82**: 2981–2987.

Karnik V (1969) *Seismicity of the European Area, Part 1*. Reidel, Dordrecht.

Lay T and Wallace TC (1995) *Modern Global Seismology*. Academic Press, San Diego, CA.

Richter CF (1958) *Elementary Seismology*. Freeman, San Francisco.

Stevens GR (1980) *New Zealand Adrift*. AH & AW Reed, Wellington.

Walcott RI (1981) The gates of stress and strain. In: *Large Earthquakes in New Zealand*. Royal Society of New Zealand, Miscellaneous Series No 5.

Wyss M and Brune J (1968) Seismic moment, stress and source dimensions. *J Geophys Res* **73**: 4681–4694.

3

Determination of Site Characteristics

3.1 Introduction

In seismic regions, geotechnical site investigations should obviously include the gathering of information about the physical nature of the site and its environs that will allow an adequate evaluation of seismic hazard to be made. The scope of the investigation will be a matter of professional judgement, depending on the seismicity of the area and the nature of the site, as well as of the proposed or existing construction. In addition to the effects of local soil conditions upon the severity of ground motion, the investigation should cover possible earthquake danger from geological or other consequential hazards such as:

- fault displacement;
- subsidence (flooding and/or differential settlement);
- liquefaction of cohesionless soils;
- failure of sensitive or quick clays;
- landslides;
- mudflows;
- dam failures;
- water waves (tsunamis, seiches);
- groundwater discharge changes.

The seismic characteristics of local geology and soil conditions described briefly in the following section provide an introduction to the site investigations, and to the determination of design ground motions and soil response analyses described in Chapters 4 and 5.

3.2 Local Geology and Soil Conditions

In many earthquakes the local geology and soil conditions have had a profound influence on site response. The term 'local' is a somewhat vague one, generally to be understood in comparison to the total terrain transversed between the earthquake source and the site. On the assumption that the gross bedrock vibration will be similar at two adjacent sites, local differences in geology and soil produce different surface ground motions at the two sites. Factors influencing the local modifications to the underlying motion are the topography and nature of the bedrock and the nature and geometry of the depositional soils. Thus, the term 'local' may involve a depth of a kilometre or more, and an area within a horizontal distance of several kilometres from the site.

Soil conditions and local geological features affecting site response are numerous, and some of the more important are now discussed with reference to Figure 3.1.

(1) The greater the *horizontal extent* (L_1 or L_2) of the softer soils, the less the boundary effects of the bedrock on the site response. Mathematical modelling is influenced by this, as discussed in Section 5.2.2.

(2) The *depth* (H_1 or H_2) *of soil overlying bedrock* affects the dynamic response, the natural period of vibration of the ground increasing with increasing depth. This helps to determine the frequency of the waves amplified or filtered out by the soils and is also related to the amount of soil–structure interaction that will occur in an earthquake (Sections 5.2 and 5.3). The Mexico earthquakes of 1957 and 1985 witnessed extensive damage to long-period structures in the former lake bed area of Mexico City where the flexible lacustrine deposits caused great amplification of long-period waves (Rosenblueth, 1960; Romo and Seed, 1986). A more typical example of an earthquake where the fundamental period of structures which were most damaged was closely related to depth of alluvium, was that in Caracas in 1967 (Seed *et al.*, 1972). Again, long-period structures were damaged in areas of greater depth of alluvium.

(3) The *slope of the bedding planes* (valleys 2 and 3 in Figure 3.1) of the soils overlying bedrock obviously affects the dynamic response; but it is less easy to deal rigorously with non-horizontal strata.

(4) *Changes of soil types horizontally* across a site (sites F and G in Figure 3.1) affect the response locally within that site, and may profoundly affect the safety of a structure straddling the two soil types.

(5) The *topography of both the bedrock and the deposited soils* has various effects on the incoming seismic waves, such as reflection, refraction, focusing, and scattering. Unfortunately many of these effects always remain suppositional; for instance, while focusing effects in bedrock (valleys 1 and 2 in Figure 3.1) may be amenable to calculation, how are the response modifications at sites G and J to be reliably predicted due to these effects in valley 3?

While there will always be some inherent variability (uncertainty) in the spatial distribution of ground motion, it may well be that geological features such as hidden irregularities in the bedrock topography explain some of the otherwise unexplained differences of response observed at nearby sites. For example, in the 1971 San Fernando earthquake (Housner and Jennings, 1972), at two locations on the campus of the California Institute of Technology, the peak acceleration recorded at one site was 0.21 *g* while only 0.11 *g* was recorded at the other; whereas the local soil profiles at both locations were considered identical.

Local Geology and Soil Conditions

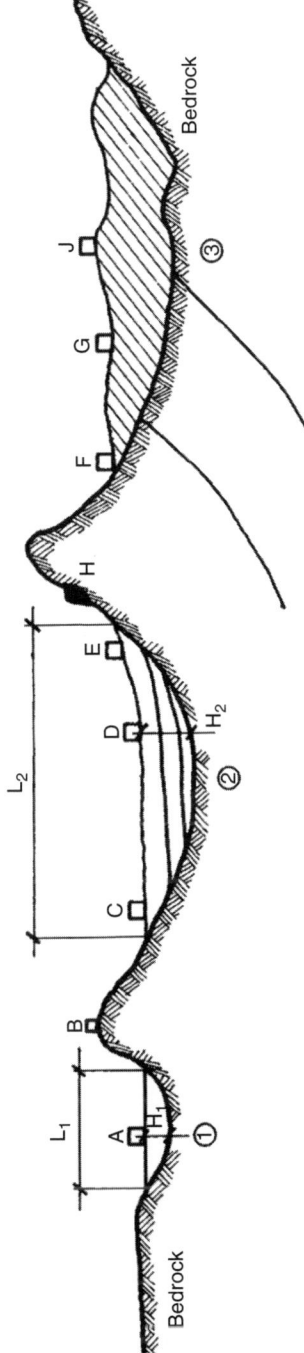

Figure 3.1 Schematic diagram illustrating local geology and soil features

(6) Another topographical feature affecting response is that of *ridges* (site B in Figure 3.1) where magnification of basic motion by factors as high as about 2 may occur (Section 5.2.2).

(7) *Slopes of sedimentary deposits* may, of course, completely fail in earthquakes. In steep terrain (site H in Figure 3.1) failure may be in the form of landslides. This occurred in the Northern Peru earthquakes of 31 May 1970, in which whole towns were buried and about 20,000 people were killed (Cluff, 1971), by one particular avalanche which travelled 18 km at speeds of 200–400 km/h.

(8) Spectacular soil failures can also occur in *gentle slopes*, as seen in the 1964 Alaskan earthquake (Seed, 1968), and again in the 1968 Tokachi-Oki earthquake (Suzuki, 1971). The slope failures in the Alaskan earthquake were mostly related to liquefaction of layers of soil. For instance, landslides occurred in basically clay deposits (Figure 3.2) where liquefaction occurred in thin lenses of sand contained in the clay. In the Tokachi-Oki earthquake, some of the slope failures resulted from upper soil layers sliding on a slippery (wet) supporting layer of clay. This 'greasy back' situation could occur as illustrated in Figure 3.1, site E.

Similar phenomena are known to occur on land in highly sensitive (i.e. quick) clays and on the sea floor, where normally consolidated clays with slopes of less than one degree can fail if subjected to external forces such as earthquakes or waves (Henkel, 1970). During the development of the North Sea oil and gas fields the author was involved in a study (Ove Arup and Partners, 1980) in which it was shown that slopes of less than 1 degree would fail under a ground acceleration of about 0.1 g.

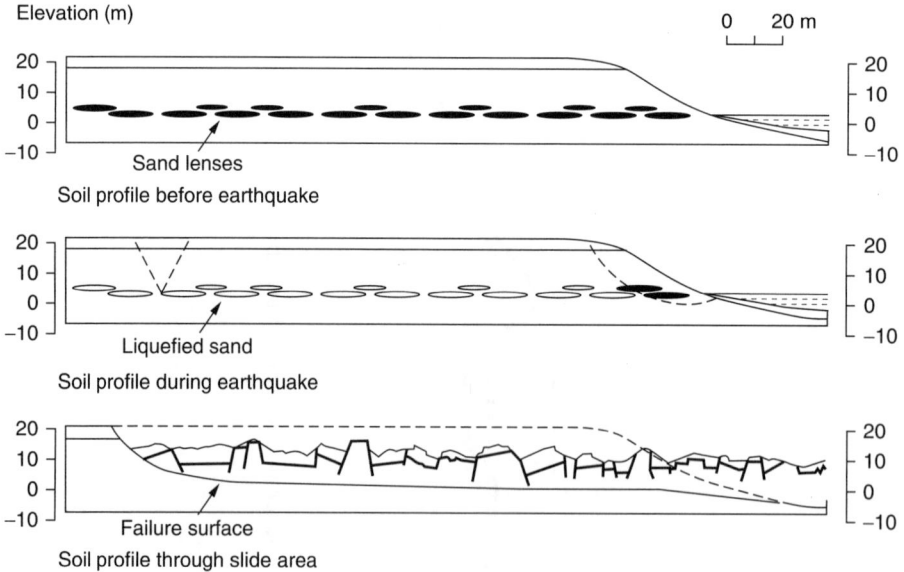

Figure 3.2 Conceptual development of Turnagain Heights landslide, Anchorage, Alaska, due to liquefaction of sand lenses (after Seed, 1968) (reproduced by permission of the American Society of Civil Engineers)

(9) The *water content* of the soil is an important factor in site response. This applies not only to sloping soils as mentioned above, but liquefaction may also occur in flat terrain composed of saturated cohesionless soils (Section 5.2.2). Classical examples of failures of this type occurred in the Alaskan and Tokachi-Oki earthquakes referred to above, and in the much-studied 1964 Niigata earthquake.

(10) *Faults* of varying degrees of potential activity sometimes cross the site of proposed or existing construction and cases of damage have been recorded. The recurrence intervals of given levels of fault displacement, both horizontal and vertical, and the structure's ability to tolerate the design displacement sometimes need to be evaluated (Section 4.8).

(11) *Water waves* are sometimes generated by earthquakes. Those occurring in the sea, called *tsunami*, are caused by vertical displacements of blocks of sea bed. Where the resulting high-velocity, low-amplitude surface wave in the sea reaches the shore, waves of considerable height (10 m) may surge well beyond the normal high-tide limit, hundreds of metres inland in flat terrain. These extreme effects only occur where the topography of the coastline focuses the wave energy, such as the narrow inlets of the southern Alaskan coast, where a disastrous tsunami struck in the 1964 Great Alaska earthquake. Various other coastlines are susceptible to damaging tsunami, particularly the Pacific and Indian Oceans, in which most of the world's tsunami are generated along shallow offshore earthquake belts (Figure 2.1). Tsunami damage can be serious in cases where the causative earthquake has occurred at any distance from local to thousands of kilometres away. This famously and tragically happened in the tsunami which killed more than 283,000 people as a result of the Great Sumatra-Andaman subduction zone interface earthquake of $M_W = 9.3$ on 26 December 2004. A special issue of the *Bulletin of the Seismological Society of America* (vol. 9, no. 1A, 2007) is devoted to this tsunami.

Water waves called *seiches* may also occur in the enclosed waters of lakes and harbours due to resonance effects or landslides, and, while not as large as tsunami, seiches have caused considerable damage.

More information on seismic water waves should be sought in the specialist literature, such as the overview by Wiegel (1976) and Satake *et al.* (2007).

(12) *Changes in groundwater discharge* occur after earthquakes, apparently due to changes in porewater pressure. The discharge may cause local flooding or streams to dry up, extensive sand boils, or erosion, such as observed in the 1983 Borah Peak, Idaho earthquake (Wood *et al.*, 1985).

(13) Finally, the seismic response of a site and structures on it is of course a function of the local *soil types* and their condition (ground classes). This is illustrated by the very different response spectra for different soils shown in Figure 3.3. The dynamic properties of individual soils are described in terms of mechanical properties such as shear modulus, damping, density, and compactability as discussed in Section 5.2.

3.3 Ground Classes and Microzones

As soil types and thicknesses, and to a lesser extent rock, vary widely from site to site in a region and worldwide, many different ways of classifying sites exist. Fortunately, as knowledge of site response to earthquakes has grown in recent years, there has been

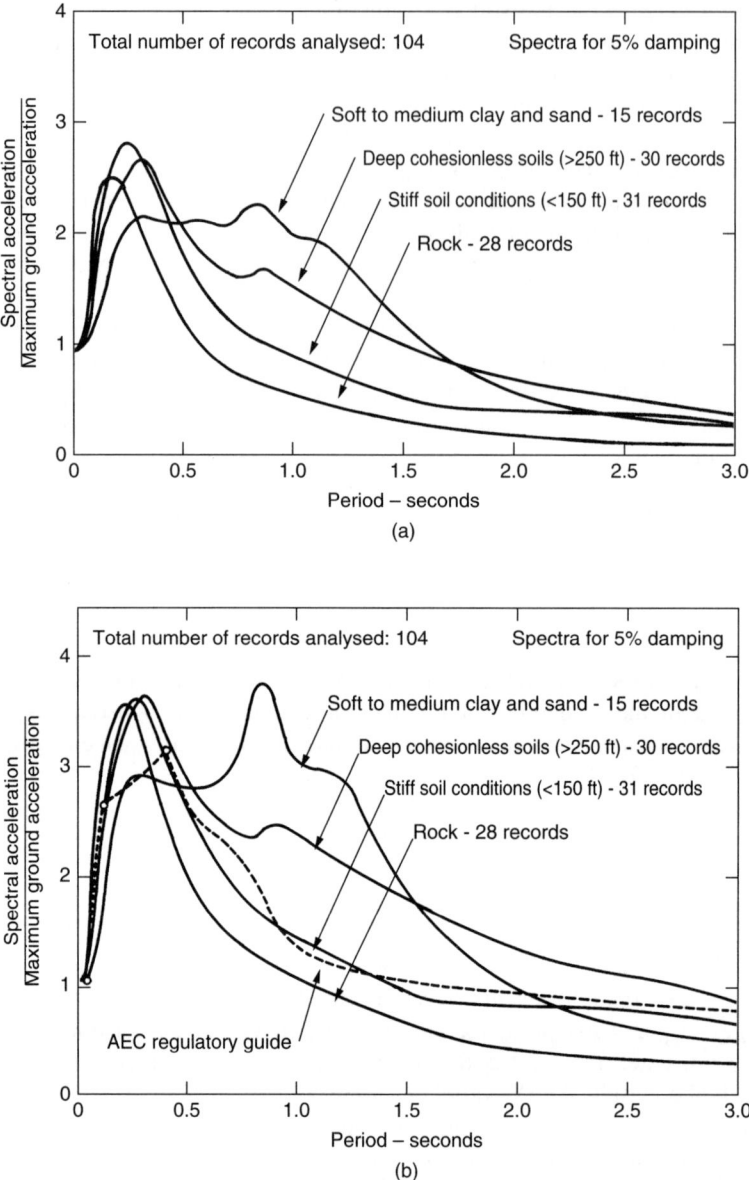

Figure 3.3 (a) Mean acceleration spectra for different site conditions; (b) mean plus one standard deviation (84th percentile) acceleration spectra for different site conditions (after Seed et al., 1974)

a growing consensus on how to classify sites for earthquake hazard purposes. Such a scheme should have several ground classes similar to those given as follows:

Class A: strong rock. Strong to extremely strong rock with unconfined compressive strength greater than 50 MPa and an average shear-wave velocity over the top 30 m greater than 1500 m/s and not underlain by materials having a substantially lower compressive strength less than 18 MPaor a shear-wave velocity less than 600 m/s.

Class B: (weak) rock. Rock (i.e. material with a compressive strength between 1 and 50 MPa) with an average shear-wave velocity over the top 30 m greater than 360 m/s. A surface layer of no more than 3 m depth of highly weathered or completely weathered rock or soil (a material with a compressive strength less than 1 MPa) may be included. The rock should not be underlain by materials having a substantially lower compressive strength less than 0.8 MPa or a shear-wave velocity less than 300 m/s.

Class C: shallow soil sites. Sites that are not class A, B or E and have low-amplitude natural period is less than or equal to 0.6 s, or sites with depths of soil not exceeding those listed in Table 3.1.

Class D: deep or soft soil sites. Sites that are not class A, B or E and have a low-amplitude natural period greater than 0.6 s, or sites with depths of soils exceeding those listed in Table 3.1, or are underlain by less than 10 m of soils with an undrained shear strength less than 12.5 kPa or soils with standard penetrometer test (SPT) N-vales less than 6.

Class E: very soft soil sites. Sites with more than 10 m of very soft soils with undrained shear strength less than 12.5 kPa or SPT N-values less than 6, or with shear-wave velocities of 150 m/s or less, or more than 10 m combined depth of soils with the above properties.

The above classification is that adopted by the New Zealand Loadings Standard, and is very similar to the ground classes used in the USA. The above classes are determined using one or more types of information:

- depth to effective bedrock;
- shear-wave velocity;
- compressive strength;

Table 3.1 Depth limits for site subsoil classes C and D as used in the New Zealand loadings standard NZS 4203 (1992). (Reproduced by permission of Standards New Zealand)

Soil type and description		Maximum depth of soil (m)
Cohesive soil	Representative undrained shear strengths (kPa)	
Soft	12.5–25	20
Firm	25–50	25
Stiff	50–100	40
Very stiff or hard	100–200	60
Cohesionless soil	Representative SPT (N) values	
Loose dry	6–10	40
Medium dense	10–30	45
Dense	30–50	55
Very dense	>50	60
Gravels	>30	100

- shear strength;
- SPT results;
- site period.

Basic information on some of the above ground classes may be available for a given location from geological maps or more specifically from earthquake microzone maps. Various aspects of ground classes and microzones are further discussed in the following section and later chapters of this book.

3.4 Site Investigations and Soil Tests

3.4.1 Introduction

For any construction project it is normal to carry out some investigations of the site, generally using fairly standardized operations in the field and in the laboratory such as drilling boreholes and carrying out triaxial tests. In this section, only those investigating techniques related to the seismic response of soils are discussed.

The scope of the site investigations will depend upon the site and on the budget and importance of the project, but in general it will be desirable to examine *to some degree* the factors relating to local geology and soil conditions discussed above. In Tables 3.1 and 3.2 the main variables in seismic site response have been related to some means for evaluating them. It is not suggested that these tables are exhaustive, but the field and laboratory test methods listed have been chosen because of their availability, reliability, or economy. For some parameters, such as radiation damping and Poisson's ratio, no suitable tests for their evaluation exist.

For a description of the dynamic behaviour of soils see Section 5.2, where the main dynamic design parameters such as shear modulus and damping are defined. The application of the results of the site investigation to soil response and design problems may be found in various parts of the following chapters.

Table 3.2 List of the main soil factors with the most suitable tests used in their evaluation

		Field tests	Laboratory tests
Settlement of dry sands		Penetration resistance	Relative density
Liquefaction		Penetration resistance; Groundwater conditions	Relative density; Particle size
Dynamic response parameters	Shear modulus	Shear-wave velocity; Penetration resistance	Resonant column or cyclic triaxial
	Damping		Resonant column or cyclic triaxial
	Mass Density		Density
	Fundamental soil period	Vibration test	

3.4.2 Field determination and tests of soil characteristics

A brief description of the nature, applications and limitations of those site investigations pertaining to seismic behaviour of soils as listed in Tables 3.1 and 3.2 now follows.

Soil distribution and layer depth

Standard borehole drilling and sampling procedures are satisfactory for determining layer thicknesses for most seismic response analysis purposes as well as for normal foundation design. In the upper 15 m of soil, sampling is usually carried out at about 0.75 or 1.5 m intervals; from 15–30 m depth, a 1.5 m interval may be desirable; while below 30 m depth, 1.5 or 3.0 m may be adequate, depending on the soil complexity. If the site may be prone to liquefaction or slope instabilities, thin layers of weak materials enclosed in more reliable material may need to be identified, requiring more frequent or continuous sampling in some cases.

The depth to which the deepest boreholes are taken will depend, as usual, on the nature of the soils and of the proposed construction. For instance, for the design of a nuclear power plant on deep alluvium, detailed knowledge of the soil is required to a depth of perhaps 200 m, while general knowledge of the nature of subsoil will be necessary down to bedrock or rock-like material.

Depth to bedrock

For use in response calculation, a knowledge of the depth to bedrock or rock-like material is essential. Beyond the ordinary borehole depth of 50–100 m, bedrock may be determined from geophysical refraction surveys, preferably checked by reference to information from geological refraction surveys, preferably checked in turn by reference to information from geological records, artesian water or oil boreholes where available. In areas of deep overburden, for seismic response purposes the depth at which bedrock or equivalent bedrock is reached may have to be defined fairly arbitrarily. For example, on some sites it may be reasonable to say that equivalent bedrock is material for which the shear-wave velocity at low strains ($\leq 0.001\%$) is $v_S \geq 760$ m/s, where such material is not underlain by materials having significantly lower shear-wave velocities (see also Table 5.1). In California on a typical site, effective bedrock would be found within 30 m of the surface, while on virtually all Californian sites it would be found within 150 m depth. The order of accuracy of bedrock depth determination that should be realistic targets for seismic response calculations is as follows:

Bedrock depth (m)	Approximate accuracy (m)
0–30	1.5
30–60	1.5–3.0
60–150	6–15
150–300	15–30
>300	60

The large errors permissible in the measurement of deeper bedrock reflect the great approximations made in soil response analyses at the present time.

Groundwater conditions

Adequate standard borehole installations are available for accurately measuring groundwater conditions at any site. For response calculations, this information is used indirectly through effective confining pressures as they affect both shear modulus and damping of the soil. Those sites which are most susceptible to liquefaction have their water table within 3 m of the surface, while sites with water tables within about 8 m of ground level may also be potentially liquefiable, depending on other soil parameters.

Penetration resistance tests

Penetration resistance tests were originally devised for determining the degree of compaction of granular soils. Because they can be carried out simply and cheaply, their use has been extended widely.

Two basic types of penetrometer are in common use for penetration tests, namely hollow tube samplers and cone penetrometers. Both types may be either driven by a falling weight (dynamic method) or by a static load into the undisturbed soil at the bottom of the borehole as drilling proceeds. In America and some other countries the preferred method is the SPT test, which is a dynamic method having the advantage of sample recovery. The static cone penetrometer tests (CPT), particularly the Dutch cone, have found favour in some countries because of the greater consistency of results deriving from the simple static load application. This advantage is offset, however, because the cone test does not recover samples, so that no visual examination of the material being tested is possible.

The results of penetration tests for assessing the condition of granular soils may be used directly or else indirectly, i.e. after conversion to relative density. As the various penetrometer tests yield different numerical results for the same soils, the exact type of equipment used in each case must be known and appropriate conversions made where necessary for assessing results. For example, Schmertmann (1970) related the static cone penetration resistance, Q_c (kg/cm^2), to standard penetration resistance (blows/0.3 m) for fine sands, but the relationship has been found to vary with grain size. Alternatively CPT results have also been related to relative density, as shown in Figure 3.4.

Other soil properties that have been related to CPT and/or SPT values (e.g. Kulhawy and Mayne, 1990) are shear modulus (Section 5.2.2), undrained shear strength, shear-wave velocity, friction angle and soil classification.

Field determination of shear-wave velocity

Although the shear-wave velocity is often used directly in response analyses (Chapter 5), it may be thought of mainly as a means of determining the shear modulus G of a soil (Section 5.2.1) from the relationship

$$G = \rho v_S^2, \tag{3.1}$$

where ρ is the mass density of the soil. In the geophysical method of determining v_s low-energy waves are propagated through the soil deposit, and the shear-wave velocity is measured directly. Three techniques using boreholes are illustrated in Figure 3.5.

In each case, waves are generated by an explosive charge or a hammer and the time of first arrival of the shear wave travelling from energy source to geophone is recorded.

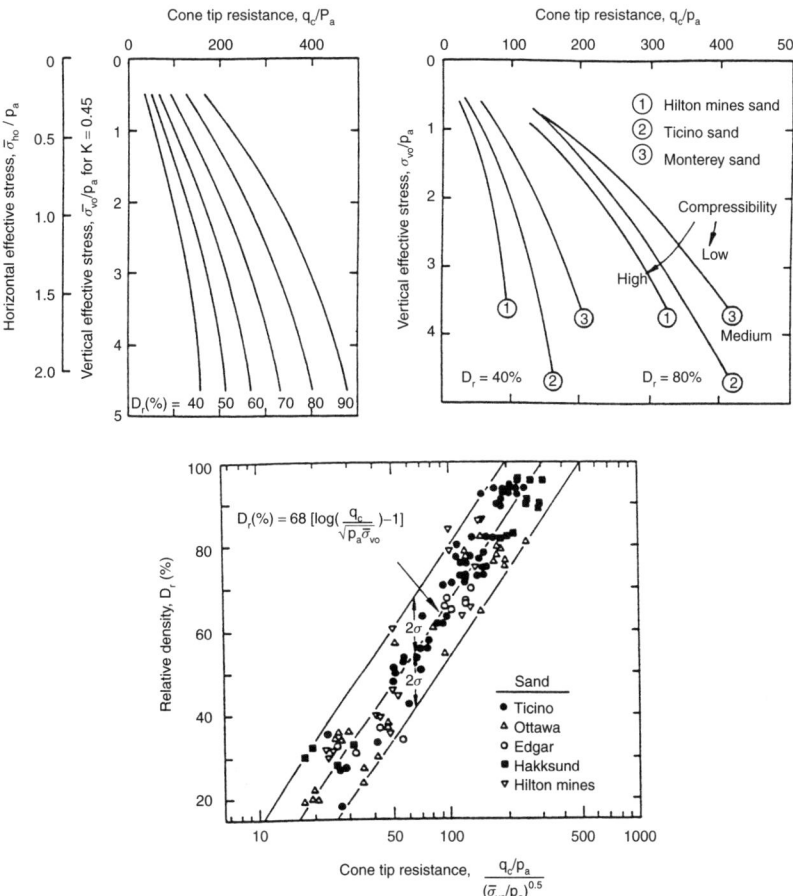

Figure 3.4 Relative density in terms of penetration resistance. Top left: CPT data for uncemented and unaged quartz sands. Top right: effect of compressibility of the sand particle. Bottom: an alternative method of Italian origin for estimating relative density from CPT values. (After Kulhawy and Mayne, 1990)

Difficulties in interpreting results arise from uncertainties in separating the first arrival of shear waves from the faster travelling longitudinal waves. Unfortunately, these latter P-waves are not suitable for shear modulus calculations as they are greatly influenced by the presence of groundwater, whereas shear waves are not.

The cross-hole technique shown in Figure 3.5 measures shear-wave velocities horizontally between adjacent boreholes, and is clearly well suited to response calculations of reasonably homogeneous or thick strata. With thinly bedded deposits, various routes may be taken by waves between source and geophone and the interpretation of arrival times is more problematical and should be viewed with caution. When using the up-hole and down-hole techniques of Figure 3.5 the different wave types can be distinguished more easily, but care must be taken to deal with misleading local borehole effects. For example, where casing has to be used in a borehole, the waves transmitted by the casing

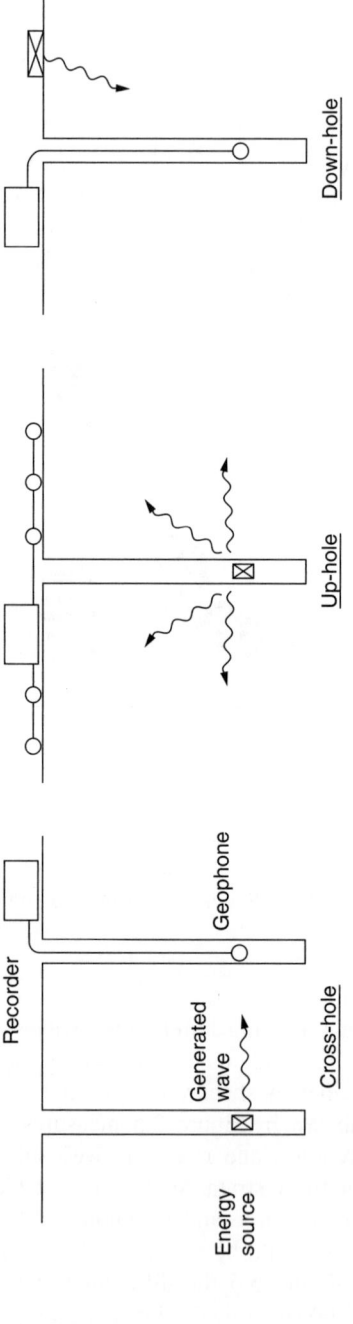

Figure 3.5 Geophysical methods of evaluating shear-wave velocity

may disguise the slower and weaker signals in the soil and experienced resolution of the results is required.

The above geophysical methods of determining v_s are the most applicable field procedures because they involve a large mass of soil, they can be carried out in most soil types, and they permit v_s to be determined as a function of depth. Furthermore, their cost is reasonable and in many countries the necessary equipment is available. Because these tests are only feasible at low levels of soil strain of 10^{-5}–$10^{-3}\%$, compared with design earthquake strains of about 10^{-3}–$10^{-1}\%$, values of shear modulus calculated from these values of v_s will be scaled down for seismic response purposes (Table 5.5). It is also wise to compare values of G computed in this manner with values determined from laboratory tests as discussed in Section 3.3.3.

Field determination of fundamental period of soil

A knowledge of the predominant period of vibration of a given site is helpful in assessing a design earthquake motion (Sections 4.7.4 and 5.2.2) and the vulnerability of the proposed construction to earthquakes (Sections 8.3.6 and 5.3).

Many attempts have been made to measure the natural period of vibration of different sites; the vibrations measured have generally been microtremors, some arising from small earthquakes (Espinosa and Algermissen, 1972) or those induced artificially such as by explosive charges, pile driving, passing trains or nuclear test explosions. The method of Nakamura (1989) seems to give good estimates of site period, but probably only in the presence of Rayleigh waves (W Stephenson, personal communcation, 2000).

For an important or seismically vulnerable project, a vibration test may well be warranted, but problems of interpretation of results arise as such tests involve much lower magnitudes of soil strain than occur in design earthquakes. If a local correlation between soil periods in strong-motion earthquakes and periods recorded during microtremors does not exist, cautious comparisons with strong-motion results on similar soils in different areas will have to be made. In the case of vibration tests carried out for the Parque Central Development in Caracas (Ravarra *et al*., 1971), the measured periods were increased by 50% to convert the microtremor behaviour into strong ground motion. This adjustment factor was derived through comparison of studies of the 1967 Caracas earthquake with the site tests.

It should be noted that the fundamental period of the soil will generally be between about 0.2 and 4.0 s, depending on the stiffness and depth of the soils overlying bedrock (Section 5.2.2).

3.4.3 Laboratory tests relating to dynamic behaviour of soils

A brief description of the nature, applications and limitations of the laboratory tests relating to the dynamic behaviour of soils, as summarized in Tables 3.2 and 3.3, is set out below.

Particle size distribution

This soil property is related to the liquefaction of saturated cohesionless soils as discussed in Section 5.2.2. As the test for its determination is a standard laboratory procedure, it

Table 3.3 List of the best field and laboratory tests related to the evaluation of the seismic response of soils

Field determinations and tests	Related to
Soil distribution and layer depth	Response calculations
Depth to bedrock	Response calculations
Groundwater conditions	Response calculations and liquefaction
Penetration resistance	Settlement and liquefaction
Shear-wave velocity	Shear modulus
Fundamental period of soil	Response calculations
Laboratory tests	
Particle size distribution	Liquefaction
Relative density	Liquefaction and settlement
Cyclic triaxial	Shear modulus and damping
Resonant column	Shear modulus
Unit mass	Response calculations

will not be described here. Although a number of classifications of grain size and standard sieves exist, correlations are straightforward, so that use of any scale of sizes can easily be applied in studies such as those for liquefaction potential.

Relative density test

The *in situ* relative density or degree of compaction is helpful in determining the likely settlement of dry sands and the liquefaction potential of saturated cohesionless soils in earthquakes (Section 5.2.2). As this property has a significant influence on the dynamic modulus, it indirectly relates to response analyses. Relative density for the void ratio must also be assessed to reproduce field conditions in samples which are recompacted in the laboratory for cyclic loading tests. As is well known by geotechnical engineers, larger scatter occurs in the results of relative density tests, the chief reason being the virtual impossibility of retrieving reliable undisturbed samples of granular deposits.

The relative density may be found from

$$D_r = \frac{e_{max} - e}{e_{max} - e_{min}} = \frac{\rho_{max}(\rho - \rho_{min})}{\rho(\rho_{max} - \rho_{min})}, \qquad (3.2)$$

where e_{max} and e_{min} are the maximum and minimum void ratios, e and ρ are the natural (*in situ*) void ratio and unit mass respectively, and ρ_{max} and ρ_{min} are the maximum and minimum unit mass.

In the laboratory e, the void ratio of the undisturbed sample, is first determined by measuring the appropriate quantities in

$$e = \frac{G\rho_w}{\rho_d} - 1, \qquad (3.3)$$

where G is the specific gravity of the solids, ρ_w is the unit mass of water, and ρ_d is the dry unit mass of the sample.

The minimum mass density may be found by pouring oven-dry material gently through a funnel into a mould. For reasonably clean sands, this method is reliable.

More difficulty is experienced in determining the maximum density ρ_{max} with equal consistency, different methods of compaction giving modestly different results. Vibratory compaction techniques seem better for sands with more fines.

If the percentage passing the 200 mesh sieve exceeds approximately 15%, laboratory determination of relative density is of doubtful validity. In this case more reliance will have to be made upon the penetration resistance tests as a measure of relative density as discussed earlier.

Cyclic triaxial test

This test (Figure 3.6) is one of the best laboratory methods at present available for determining the shear modulus and damping of cohesive and cohesionless soils for use in dynamic response analyses (Sections 5.2 and 5.3). In this test, cyclically varying axial compression stress-strain characteristics are measured directly. The compressive modulus E so obtained is converted to the shear modulus G using the relationship

$$G = \frac{E}{2(1+v)},$$

where v is Poisson's ratio. The damping ratio may also be obtained from this test from the resulting hysteresis diagram as illustrated in Figure 5.1. Depending on the range of strains produced in the test, any desired level of strain may be chosen for plotting the hysteresis loops.

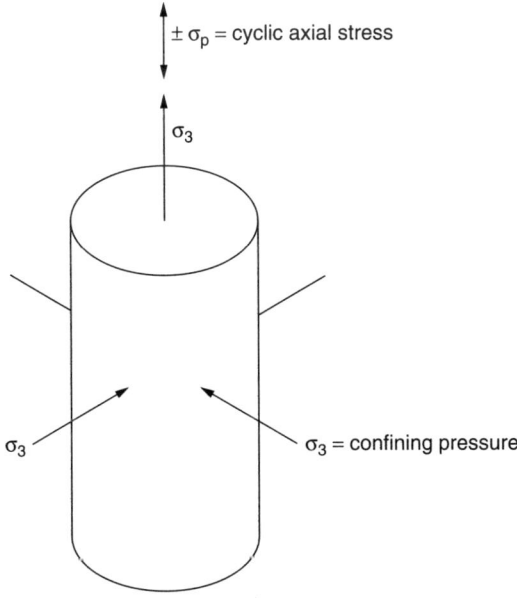

Figure 3.6 Cyclic triaxial test

As well as having the facility for applying a variety of stress conditions, the cyclic triaxial test has the advantages that it can be applied to all types of soils except gravel, that the test equipment is widely available and precise in its control, and that testing is comparatively cheap. The disadvantages of this test are related to its inability to reproduce the stress conditions found in the field, i.e. that the cyclic shear stresses are not applied symmetrically in the test, that zero shear stresses are applied in the laboratory with isotropic rather than anisotropic consolidation, and also that the test involves deformations in the three principal stress directions, whereas in earthquakes the soil in many cases is thought to be deformed mainly unidirectionally in simple shear.

Cyclic shear tests are carried out at high strains (10^{-2}–5%) equal to and larger than the strains occurring in strong earthquakes; since geophysical test involve low strains, values of G at intermediate strains may be determined by interpolating between G values found from these different methods, but as there is no overlap between the strains occurring in these two tests cross-checking between the field and laboratory methods is not possible. It is also to be noted that in the use of this test to determine soil damping characteristics, no field method of evaluating damping is as yet available for comparison, and hence any values of damping coefficient obtained should be treated with appropriate caution.

Resonant column test

This test provides a good alternative to the cyclic triaxial test for the laboratory determination of shear modulus of most soils. A cylindrical column of soil is vibrated at small amplitudes on one end, either torsionally or longitudinally (Figure 3.7), varying the frequency until resonance occurs. Wilson and Dietrich (1960) proposed that the shear or compression modulus for a solid cylinder may be found from

$$G \text{ or } E = 1.59 \times 10^{-8} f^2 h^2 \rho, \qquad (3.4)$$

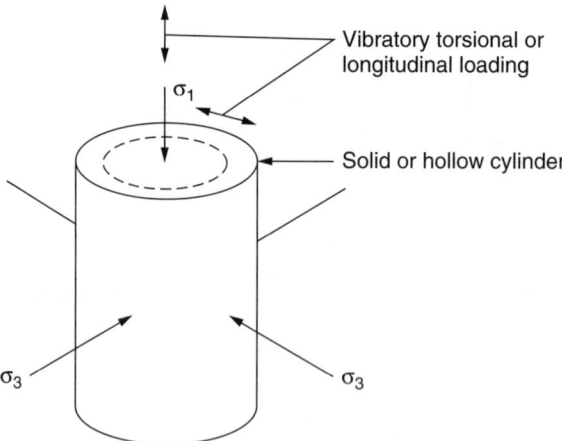

Figure 3.7 Resonant column test

where h is the height of the soil cylinder (mm), ρ is the unit mass of soil (Mg/m^3), and f is the resonant frequency of torsional vibration in cycles per second when determining G, or the resonant frequency of longitudinal vibration in cycles per second when determining E.

It will be seen that by determining E and G separately from these tests, a value of Poisson's ratio v may be determined, but as this test involves low strain and no suitable extrapolation method exists, such values of v are not suitable for most earthquake engineering purposes. Although this test has the disadvantage of being carried out at low strains (10^{-4}–10^{-2}%), it has the advantages of simplicity, cheapness of equipment, and applicability to most soil types.

Details of the equipment used in this test may be found elsewhere (Wilson and Dietrich, 1960; Kramer, 1996).

Other laboratory tests

Various other laboratory tests for soil properties exist, including the cyclic torsional shear test, and the more expensive and elaborate methods of testing soil models, i.e. shake table and centrifuge tests. Such tests are described in the specialist literature (e.g. Kramer, 1996).

References

Cluff LS (1971) Peru earthquake of May 31, 1970; engineering geology observations. *Bull Seism Soc Amer* **61**(3): 5111–5133.

Espinosa AF and Algermissen ST (1972) Soil amplification studies in areas damaged by the Caracas earthquake of July 29, 1967. *Proc. Microzonation Conf*. Seattle, II: 455–464.

Henkel DJ (1970) The role of waves in causing submarine landslides. *Geotechnique* **20**: 75–80.

Housner GW and Jennings P C (1972) The San Fernando California earthquake. *Int J Earthq Eng Struc Dyn* **1**: 5–32.

Kramer SL (1996) *Geotechnical Earthquake Engineering*. Prentice Hall, Upper Saddle River, NJ.

Kulhawy FW and Mayne PW (1990) Manual for estimating soil properties for foundation design. Research Report EL-6800, Geotechnical Eng Group, Civil Eng Dept, Cornell University.

Nakamura Y (1989) A method for dynamic characteristics estimation of subsurface using microtremors on the ground surface. *Quarterly Report of the Railway Technical Research Institute*, Tokyo, **30**: 25–33.

NZS 4203 (1992) *Loadings Standard*. Standards New Zealand, Wellington.

Ove Arup and Partners (1980) Earthquake effects on platforms and pipelines in the UK offshore area. Report to the UK Department of Energy OT-R 7950.

Ravarra A, Pereira J, Oliveira C and Lourtie P (1971) Estudos estruturais dos edifícios de Parque Central – 2° Relatóniò: Análise dinâmica dos edifícios de apartamentos. LNEC Report, Lisbon.

Romo MP and Seed HB (1986) Analytical modelling of dynamic soil response in the Mexico earthquake of September 19, 1985. In MA Cassaro and E Martínez-Romero (eds), *The Mexico Earthquakes – 1985: Factors Involved and Lessons Learned*. ASCE, New York, pp. 148–162.

Rosenblueth E (1960) Earthquake of 28th July, 1957 in Mexico City. *Proc. World Conf Earthq Eng, Japan* **1**: 359–379.

Satake K, Okal EA and Borrero JC (2007) *Tsunami and its Hazards in the Indian and Pacific Oceans*. Birkhäuser, Basel.

Schmertmann JH (1970) Static cone to compute static settlement of sand. *J Soil Mech Found Divn*, ASCE **96**(SM3): 1011–1043.

Seed HB (1968) Landslides during earthquake due to soil liquefaction. *J Soil Mech Found Divn* **94** (SM5): 1053–1122.

Seed HB, Ugas C and Lysmer J (1974) Site dependent spectra for earthquake resistant design. Report No EERC 74-12, Earthquake Engineering Research Center, University of California, Berkeley.

Seed HB, Whitman RV, Dezfulian H, Dobry R and Idriss IM (1972) Soil conditions and building damage in the 1967 Caracas earthquake. *J Soil Mech Found Divn* **90**(SM8): 787–806.

Suzuki Z (ed.) (1971) *General Report on the Tokachi-Oki Earthquake of 1968*. Keigaki Publishing, Tokyo.

Wiegel RL (1976) Tsunamis. In C Lomnitz and E Rosenblueth (eds), *Seismic Risk and Engineering Decisions*. Elsevier, Amsterdam.

Wilson SD and Dietrich RJ (1960) Effect of consolidation pressure on elastic and strength properties of clay. *Proc. ASCE Res Conf on Shear Strength of Cohesive Soils*, University of Colorado: 419–435, discussion: 1086–1092.

Wood SH, Wurts C, Lane T, Ballenger N, Shaleen M and Totorica D (1985) The Borah Peak, Idaho earthquake of October 28, 1983: Hydrologic effects. *Earthq Spectra* **2**(1): 127–150.

4

Seismic Hazard Assessment

4.1 Introduction

Seismic hazard assessment (defined in Section 1.1) involves diverse topics and professional disciplines. Seismic hazard studies are required for the preparation of earthquake loadings regulations, for determining the earthquake loadings for projects requiring special study, for areas where no codes exist, and for various earthquake risk management purposes. Depending on the purpose of a given study, it can involve one or more of many topics which in essence may be summarized as:

- seismic activity,
- attenuation, and
- site response.

When part of a design process, seismic hazard assessment implies all of the studies involved in Boxes 2–4 of Figure 8.1. Thus, as well as the topics covered below, seismic hazard assessment involves also the material in the previous chapter on site characteristics, including the list of geological hazards given in Section 3.1.

Of the above three subject areas, we first turn our attention to seismic activity. Seismic activity is evaluated through studies of three main components: crustal strain (measured by space geodesy), fault activity (geology), and historical seismicity. These three components are complementary, and for the best results should be used together.

4.2 Crustal Strain and Moment Release

Considering any given area of the earth's crust, Kostrov (1974) expressed the relationship between strain rate tensor $\dot{\varepsilon}$ and seismic activity in the form of the sum of the earthquake moment tensors, M_n, as

$$2\mu A H_s \dot{\varepsilon} = \frac{1}{T} \sum_{n=1}^{m} M_n, \qquad (4.1)$$

Earthquake Resistant Design and Risk Reduction D. Dowrick
© 2009, John Wiley & Sons, Ltd

where μ is the rigidity of the crust, A is the area under consideration, H_s is the thickness of the seismogenic crust, and m is the number of earthquakes within the volume over the time T. Thus, the volume strain rate is linearly related to the sum of the seismic moment tensors within that volume. This way of calculating the strain release rate often underestimates the long-term rate, because earthquake catalogues are generally not long enough.

An alternative way of estimating strain release rate is by using an equation similar to (4.1) for faults within the volume, where the moment release rate of each fault passing through the volume is calculated from the style and estimated long-term slip rate of the fault. In this case, the catalogue-length problem is avoided, but there will be errors if the active faulting data are incomplete (due, for example, to missing faults or inaccurate long-term slip rates).

Since about the 1980s it has been possible to directly measure strain accumulation rate over large areas using methods of space geodesy (e.g. GPS). If it is assumed that the accumulated strain is eventually released in earthquakes, space geodesy provides a third method of estimating long-term strain release rates. Consequently there has been a growing contribution of space geodesy to seismic hazard evaluation since the 1980s. In a 1998 paper, Ward described the situation as follows:

> With well-defined faults and sufficiently frequent earthquakes, historical seismicity and palaeoseismology furnish fairly reliable earthquake statistics. More commonly, questionable fault numbers, fault geometries, fault slip rates and scattered seismicity characterize the situation, and earthquake statistics do not reveal themselves readily. For most of the world, historical seismicity and palaeoseismology cannot constrain earthquake statistics to the degree necessary for an acceptable rate assessment. Information from space geodesy patches some of these voids.

For example, in studies of the USA, Ward (1998) found that in California the rate of moment release estimated from faults nearly equals the geodetic estimate, which suggested that the region's geological fault data is nearly complete. In contrast, in the Basin and Range region (northwest and central USA) estimates of moment release rates derived from both geodesy and historical seismicity greatly exceeded that from faults, suggesting a high incidence of understated or unrecognized faults. In addition, in the Basin and Range region, moment release rates estimated from historical seismicity were everywhere less than the geodetic rates, severely so (i.e. 2%) in the slowest straining zones. Although seismic strain may contribute to the situation, it appeared that the existing seismicity catalogues did not reflect the long-term situation.

Somerville (2000a) comments:

> If the seismicity of a source is based solely on historical seismicity, then the seismic moment rate of the source depends on the maximum earthquake magnitude in the earthquake recurrence model. In this case, the larger the maximum magnitude, the higher the calculated hazard. However, if the seismicity of the source is constrained by geologic slip rates or geodetic strain rates, the maximum earthquake in the earthquake recurrence model affects the calculated seismic hazard in a different way. Since there is a fixed seismic moment rate budget, selection of a larger maximum magnitude results in lower rates of occurrence of smaller events, and the calculated seismic hazard may be reduced (Youngs and Coppersmith, 1985).

As stated by Ward, space geodesy's most valuable contributions come from its ability to provide:

- rates of earthquakes on faults that are undocumented or unobservable by traditional methods;
- independent verification of the rates of deformation in regions where geologists have documented faults;
- a means of judging the consistency of the contemporary deformation field and the historical record.

By the start of the twenty-first century, the inclusion of the results of space geodesy studies in seismic hazard models was in its early stages, but appeared certain to become common. Most, if not all, countries with moderate to high seismic hazard had active GPS strain measuring programmes.

4.3 Regional Seismotectonics

In attempting to understand and then quantify seismic activity, whether it be interplate or intraplate in origin, working in the framework of global tectonics (Section 2.2) we try to relate seismicity to quantifiable deformational features such as *faulting, tilting, warping* or *folding*, or to major geological structures such as *basins, grabens* and *platforms*, which are basement rock features. The nature, age, location and movement history of these features need to be known.

The magnitude and frequency of earthquakes in a given area may be estimated in broad terms from the size and strength of the fault blocks (Gubin, 1967). The larger and stronger the block, the larger is the maximum size of earthquake which can be generated along the boundaries of that block. Also, the greater the rate of tectonic movement and the less the competency of the tectonic structures, the more rapid is the build-up of the stress needed for a fault movement, and the more frequent will be the occurrence of the maximum magnitude of earthquake for that structure.

In studies of the intraplate area of the USA, originally because of difficulties in recognizing active faults, the concept of *tectonic provinces* has been used in estimating seismicity. The boundaries of tectonic provinces should be defined by major geological structures relevant to present-day seismogenic (earthquake producing) mechanisms, although in the absence of such knowledge the arbitrary use of old geological structures has traditionally been necessary, and the definition and recognition of appropriate boundaries are problematical. It is clear that the tectonic province concept is inherently weak. As the understanding of any given tectonic province is enriched by data concerning seismogenic features within it, the tectonic province itself becomes, ironically, correspondingly redundant as an analytical tool. However, if tectonic provinces can be defined with clearly defined boundary geometry and boundary conditions, then the behaviour of internal structures may be predicted.

As it is believed that most, if not all, damaging earthquakes occur on faults, faults are the most useful seismogenic feature to identify and evaluate. However, this is often not feasible, either because the rupture plane does not reach the surface or because the fault trace cannot be readily recognized. It is thus necessary to look for other structural features such as grabens, platforms, and folds, and to try to assess their relative seismogenic characteristics.

As an example of the identification of seismogenic features in an intraplate zone, Allen (1976) notes that the faults bounding grabens in China have been, and remain, seismically active features, such as the Shansi graben which gave rise to the 1556 Sian earthquake of about magnitude 8, which caused more than 820,000 deaths. Basin and graben areas in themselves are more unstable than platform areas, and might therefore be expected to exhibit more seismic activity than platforms, at least in the longer term. However, a seismotectonic study (Dowrick, 1981) of the seismically relatively quiet North Sea area (having a maximum magnitude of $M = 6$) found no correlation between these geological structures and known seismic events (Figure 4.1). It may be that for this area there is little or no residual difference between these two structures, or that the time period covered by the known seismic events was too short to show and differentiate their full potential for seismic activity.

In an intraplate zoning study, Klimkiewicz *et al.* (1984) identified the major basement structures of the north central USA using geophysical and geological evidence, and had the benefit of a relatively long historical record of seismic events (200 years for larger

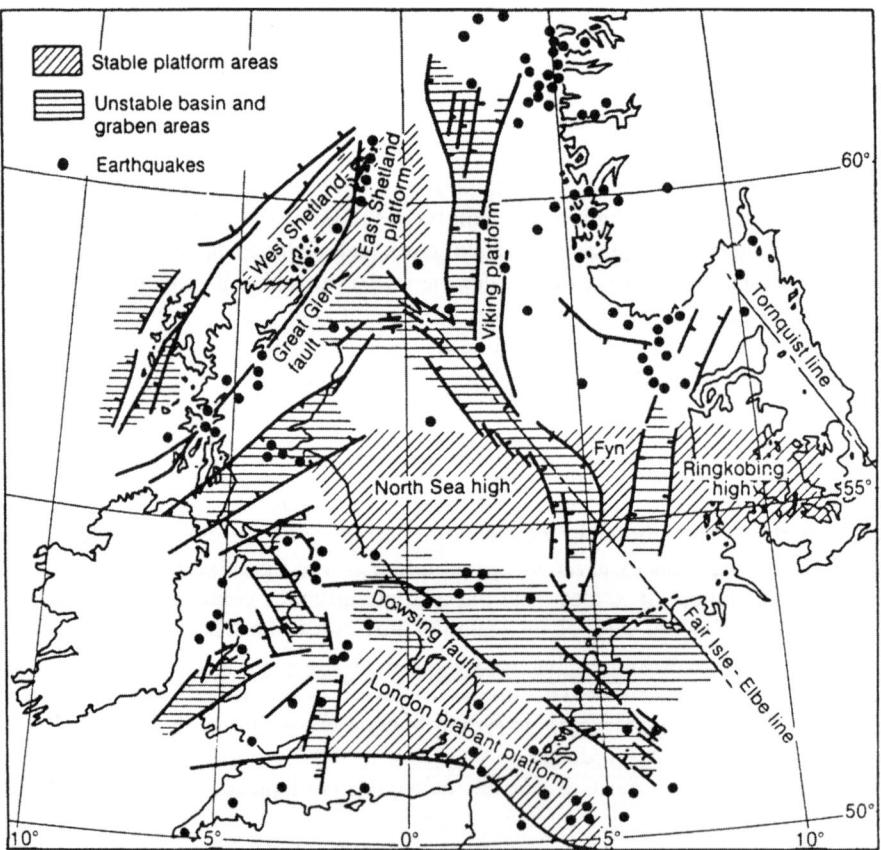

Figure 4.1 Geological structure and seismic events in the North Sea area (from Dowrick, 1981; reproduced by permission of the Institution of Civil Engineers)

magnitudes) which they believed to be sufficient to identify all seismogenic features. However, apart from confirming the seismic potential of the Michigan Basin, they appeared to conclude that further research was necessary to understand the relative seismic potential of other geological features in their large region of study.

In a detailed analysis of the seismotectonic features of western Montana, Waldron *et al.* (1984) divided the area into two major tectonic provinces separated by a major tectonic interplate boundary. Because each of these three areas exhibited unique geological, tectonic, and seismic characteristics, their seismic hazard potentials were seen to be significantly different. Also one seismogenic zone was identified in both of the provinces and in the boundary zone between them. A further outcome of the study was an estimate of the maximum magnitude of earthquake for each of the above three zones, ranging from $M = 6.0$ in the Flathead Lake seismogenic zone, through $M = 6.5$ in the mountain ranges of the interplate boundary zone, to $M = 7.5$ in the Inter-mountain Seismic Belt of the southern tectonic province.

As noted above, *tilting and warping* are further aspects of geological structure which are helpful in seismotectonic studies. In some regions they accompany many large shallow earthquakes. Viewed historically, tilt is helpful in determining the amount and recency of crustal movement in a region, and is measured by the slope of beds which are known to have been originally almost horizontal. The most seismically active regions of the world are in belts of late Tertiary and Quaternary deformation, and by dating sloping beds the age of activity may be estimated. In such a study of New Zealand, Clark *et al.* (1965) plotted the slopes of tilted strata of two periods of geological time. There is good correspondence between the recent seismic activity implied by their map and the evidence from other sources shown in Figure 2.5.

4.4 Faulting

Because faults are the seat of damaging earthquakes, they demand specific attention in addition to the references made to them in the preceding sections of this chapter.

4.4.1 Location of active faults

Large earthquakes are caused by sudden displacements on faults at varying depths. However, in many damaging earthquakes no surface evidence of the fault has been found. In localities where earthquake sources are very shallow and the surface soils are competent, such as California, many fault planes reach the surface. Where the foci are deeper and/or the overburden is not stiff enough to fracture right through, surface manifestations of faults do not occur. These situations occur in New Zealand, where not all earthquake faults reach the surface, but some faults in the basement rocks beneath the sediments have been located by geophysical surveys or by instrumentally detected linear arrays of small earthquakes. In the crust (i.e. of the overlying plates) of New Zealand, of 55 historic earthquakes of M_W 5.0–8.2 (Dowrick and Rhoades, 2005), there were only six onshore events that definitely ruptured the surface, while 10 events of M_W 6.4–7.2 had the tops of their ruptures at depths varying from 1 to 27 km. In highly seismic Chile no fault ruptures have broken through to the land surface in recent geological times, no doubt partly because of the substantial depth of the subduction zone, whereas the shallower events which occur a little offshore will have fault trace expressions on the seabed, where they would obviously not be readily seen.

In some cases, faults may reach the surface but are difficult to recognize, and it may not be possible to identify an active fault from surface traces prior to its next major movement. Apart from the presence of weak superficial deposits, such as in parts of New Zealand or the once glaciated areas of the eastern USA, other factors contribute to the difficulty of identifying faults, such as:

- low degree of fault activity, thus creating less evidence;
- erosion and sediment deposition rates that are higher than the fault slip rates;
- dense vegetation cover disguising faults (although aerial photographs sometimes show up such faults (Allen, 1976));
- some tectonic processes result in dispersed fault zones at the surfaces so that individual features are less pronounced – fault zones vary in width from a few metres to as much as a kilometre or more.

The locations of many active faults in different regions are, of course, shown on the appropriate geology maps, but because of the above-mentioned problems such maps are inevitably incomplete. In New Zealand by the year 2000, geologists believed that they had found over half of that country's active faults which reach the surface, whereas in California perhaps a larger majority of such faults were presumed to be known. Fault maps usually indicate the probable location of those hidden portions of faults, which, although not positively identified, may be inferred by interpolation or extrapolation of outcropping faults plus other evidence. The fact that the known array of active faults in a given area constitutes a minimum array should be allowed for in seismic hazard assessments.

In addition to existing maps, preliminary indications of faulting may be obtained from aerial photographs and Landsat images, prior to investigations on the ground.

4.4.2 Types of fault

The characteristics of strong ground motion are strongly influenced by the type of faulting. There are four main types of fault that should be considered in the study of destructive earthquakes:

(1) Subduction zone interface (underthrust) faults (Figure 4.2(a)). These result from tectonic seabed plates spreading apart and thrusting under the adjacent continental plates, a phenomenon common to much of the circum-Pacific earthquake belt.
(2) Compressive, overthrust faults (Figure 4.2(b)). Compressive forces cause shearing failure forcing the upper portion upwards, as occurred in San Fernando, California, in 1971 (also called reverse or thrust faults).
(3) Extensional faults (Figure 4.2(c)). This is the inverse of the previous type, extensional strains pulling the upper block down the sloping fault plane (also called normal faults).
(4) Strike-slip faults (Figure 4.2(d)). Relative horizontal displacement of the two sides of the fault takes place along an essentially vertical fault plane, such as occurred at San Francisco in 1906 on the San Andreas fault (also called transcurrent faults).

Faulting

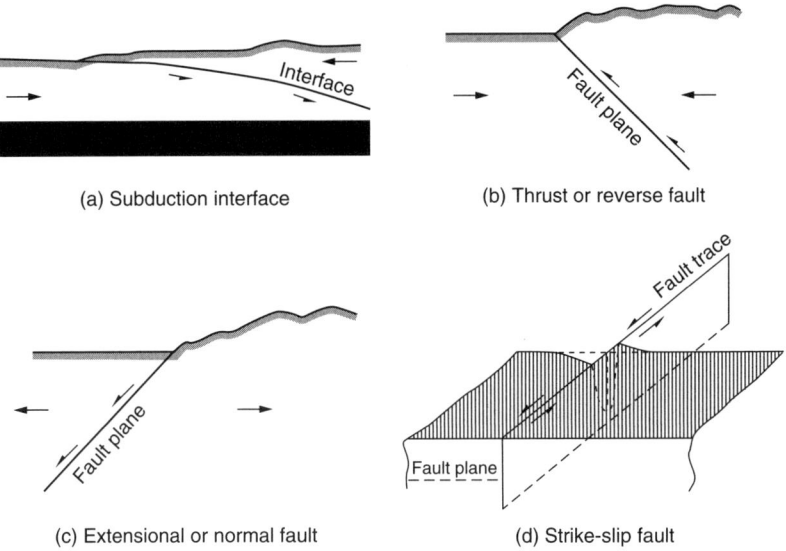

Figure 4.2 The main fault types to be considered in the study of strong ground motion (adapted from Housner, 1973)

Few pure examples of the above occur, many earthquake fault movements being oblique, i.e. having components parallel and normal to the fault trace. Detailed discussions of faulting may be found elsewhere (e.g. Allen, 1976; Bonilla, 1970; Bolt, 1999; Scholz, 1990).

4.4.3 Degree of fault activity

Active faults include any faults which are considered capable of rupturing in the future. Because the amount and frequency of movement can vary enormously, it is important to be able to estimate the degree of activity likely to be exhibited by any fault in the region of interest.

Consider the map of active faults in Figure 4.3, which is part of the computer database developed for New Zealand (Stirling *et al.*, 2000). Each fault segment on the map has an index number linking it to its tabulated parameters, a sample of which is given in Table 4.1. In this sample it is seen that the assigned maximum magnitudes range from 6.9 to 8.1, and the average recurrence intervals between events range from 120 to 34,000 years. Three of the faults shown have ruptured in New Zealand historical periods (i.e. since 1840).

Faults with longer return periods than for fault no. 52 in Table 4.1 are included in the database. The need to consider such relatively inactive faults is illustrated by the fact that this fault (White Creek), with an estimated return period of 34,000 years, ruptured in 1929 in the magnitude M_W 7.7 earthquake.

In zones of crustal convergence or divergence, without underthrusting, where the rates of these movements are known, Anderson (1979) has developed a method of computing slip rates on each fault from the rate at which seismic moment is released. The method

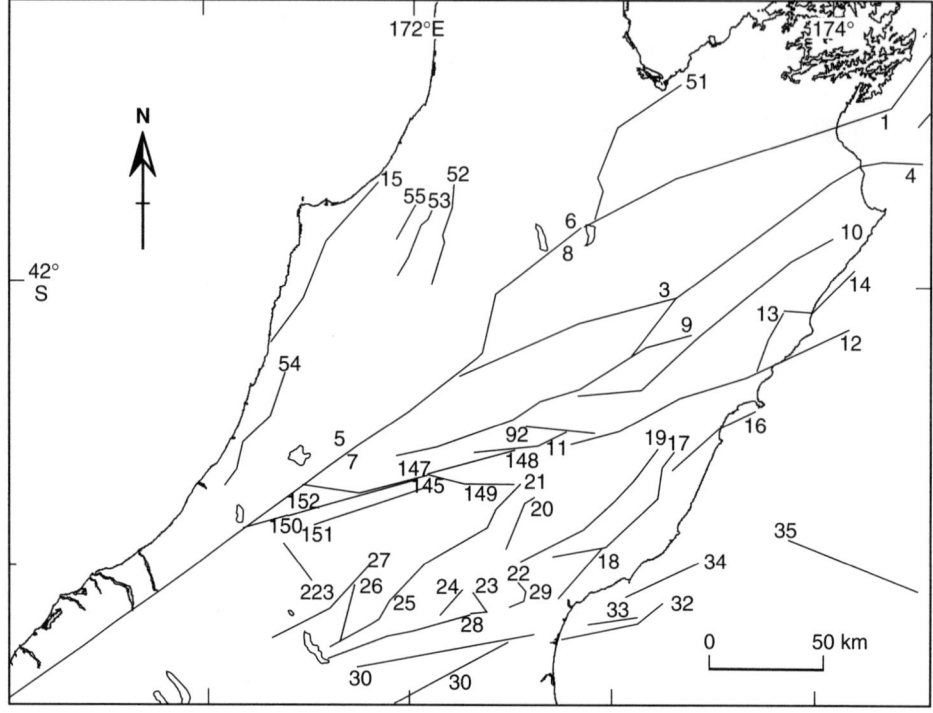

Figure 4.3 Map showing active faults in the northern South Island of New Zealand (after Stirling et al., 2000). Refer also to Table 4.1

needs to be calibrated from data from historical events on some of the faults in the region, and this proved possible for Anderson's study area of southern California.

To estimate the degree of activity of a fault, the mean slip rate and the frequency and size of movements are measured by examining sections through faults, which may occur in natural geomorphological features such as marine or river terraces, cliffs, and slip faces, or in quarries, road cuttings, or other excavations or purpose-dug trenches.

A study of fault activity was carried out by the New Zealand Geological Survey of the Nevis-Cardrona Fault System in relation to hydroelectric development proposals on the Kawarau River in Central Otago, in the eastern ranges of the Southern Alps. Referring to Figure 2.5, the location is about 200 km north-east of the 1938 event at the southern end of the 'seismic gap' in the Southern Alps. A trench dug across a reverse fault at one location illustrates the complex geometrical distortions that sometimes occur in fault zones (Figure 4.4). In an analysis of this fault Beanland and Fellows (1984) positively identified three movements, the sequence of which is illustrated in Figure 4.5. However, there was qualitative evidence for up to six events for the same total displacement of 5.6 m. The age of the displaced surface was estimated to be c. 16,000 years, based on the chronology of the culminating aggradation of the last glacial advance correlated with local terrace surfaces and deposits. As this age embraced three to six fault movements it was concluded that the fault had moved at average intervals of c. 2500 to 7000 years, with mean displacements of c. 1 m to 2 m, respectively.

Table 4.1 Parameters of some South Island faults (Figure 4.3) in the New Zealand database maintained by GNS Science

Index no.	Fault name	Slip type	Dip (°)	Dip. dir. (°)	Depth max (km)	Slip rate (mm/yr)	Displacement (m)	Mmax	Recurrence interval (yrs)
1	Wairau (Onshore)	ss	–	–	15	–	6.0	(7.6)	1650
2	Wairau (Offshore)	ss	–	–	15	–	–	(7.3)	1650
3	Awatere SW	ss	–	–	15	8.0	6.0	(7.5)	2930
4	Awatere NE (1848 rupture)	ss	–	–	15	6.5	6.5	7.5	1000
5	Alpine (Milford-Haupiri)	sr	60	145	12	25.0	8.0	(8.1)	300
6	Alpine (Kaniere-Tophouse)	sr	60	145	12	10.0	6.0	(7.7)	1200
7	Alpine (Kaniere-Haupiri)	sr	60	145	12	10.0	–	(6.9)	1200
8	Alpine (Haupiri-Tophouse)	sr	60	145	12	10.0	6.0	(7.6)	1200
9	Clarence SW	ss	–	–	15	6.0	–	(7.5)	1080
10	Clarence NE	ss	–	–	15	4.7	7.0	(7.7)	1500
11	Hope (1888 rupture)	ss	–	–	15	–	2.0	7.2	120
52	White Creek (1929 rupture)	rv	70	100	15	0.2	6.0	7.8	34000

Notes: Index no.: Cross-reference to the fault sources shown in Figure 4.3.
Slip type: ss = strike-slip; rv = reverse; sr = strike-slip and reverse.
Dip: The preferred or mean value of dip for the fault plane. If no value is given then the dip is either greater than 80° (the case for strike-slip faults) or uncertain.
Dip dir: Azimuth of dip.

In addition to the above-mentioned uncertainties encountered in analysing sections through faults, sometimes there arise even greater difficulties in interpretation due to modifications to the basic fault displacement pattern with the depth of the investigations. Figure 4.6 shows a section about 400 m long by 150 m deep through the Pisa fault in New Zealand close to the Nevis-Cardrona fault discussed above. Here the general fault zone is evident from prominent fault traces (see the backscarp, Figure 4.6) disrupting an alluvial fan c. 70,000 years old (NZ Geological Survey, 1984). Three boreholes at 160 m spacings, with depths ranging from 50 to 110 m, helped to provide a stratigraphic framework, and trenches across two parts of the fault zone were dug for fault activity determination purposed. Seven reverse faults were found (A–G, Figure 4.6) with vertical displacements totalling c. 2.3 m, and were interpreted as antithetic faults. The main 9 m uplift of the alluvial fan was inferred to have occurred on a more important synthetic

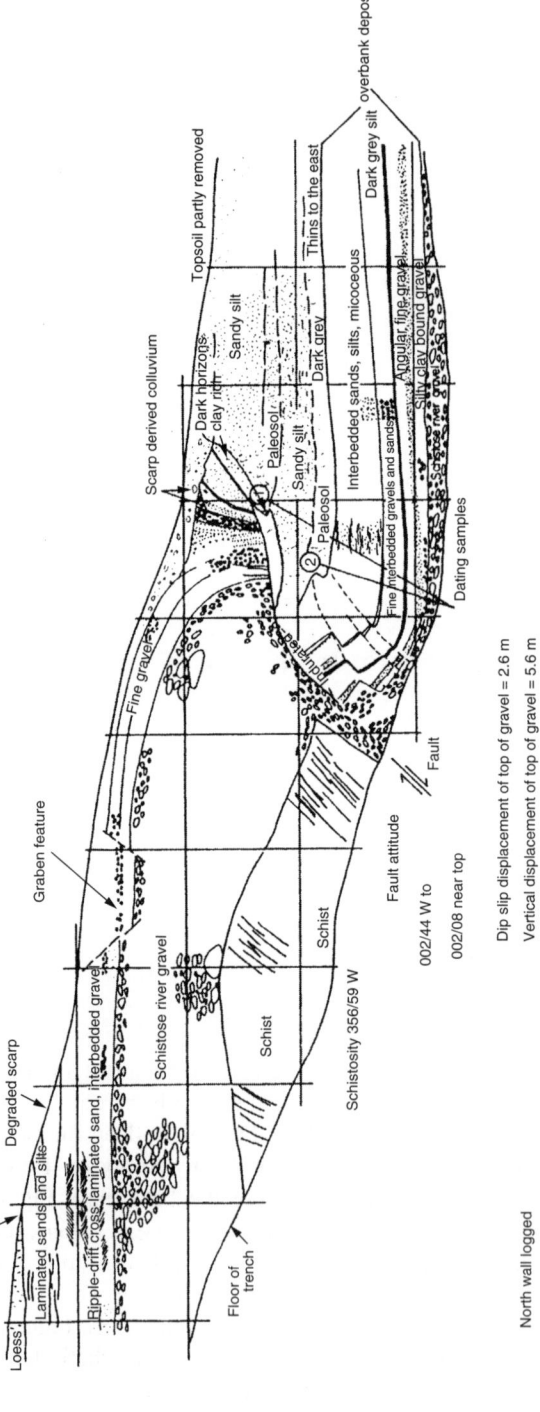

Figure 4.4 Geological log of a trench dug across part of the Nevis-Cardrona fault, Central Otago, New Zealand (after Beanland and Fellows, 1984)

Faulting

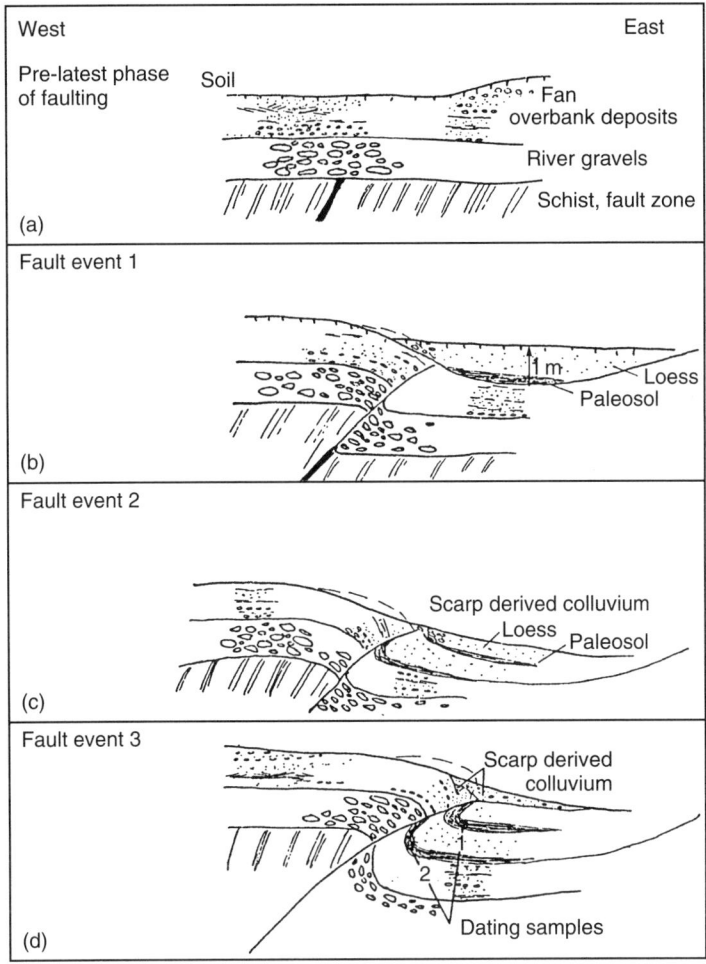

Figure 4.5 Inferred sequence of fault movements on the Nevis-Cardrona fault section shown in Figure 4.4 (after Beanland and Fellows, 1984)

fault which was not exposed at ground surface or in the trenches, possibly because fault movement has been dissipated in the gravels.

The largest vertical displacement for a single event was estimated at 2 m, but there were insufficient data to determine the number of movements which contributed to the 9 m uplift in the 70,000-year period concerned. However, because a related surface (c. 35,000 years) was not tectonically deformed, the faulting activity was thought to be 35,000–70,000 years old.

It is clear from the foregoing discussion that there is likely to be considerable uncertainty in the hazard quantification data obtained from trenches. Unless the sites of trenches are very carefully selected their data yield will be low. The cost-effectiveness of this technique needs to be considered for any given project.

A classification of faults according to their activity level is given in Section 4.8.3.

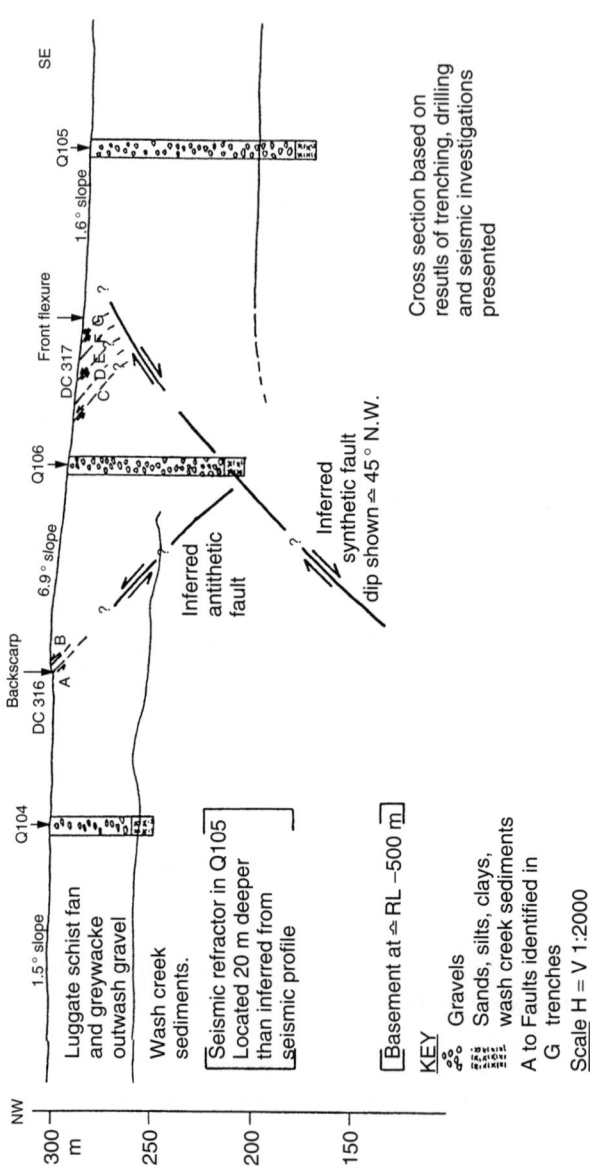

Figure 4.6 Cross-section through Lochar locality of the Pisa fault, Central Otago, New Zealand (from NZ Geological Survey, 1984)

4.4.4 Faults and earthquake magnitudes

In seismic hazard assessment it is important to be able to obtain good estimates with quantified uncertainties of the magnitudes of earthquakes that are likely to be generated on known faults. Models for doing this have been developed where magnitude is estimated from the fault rupture parameters of length, width, displacement and area. Traditionally, empirical models have been based on measurements of lengths and displacements made by geologists on the observed surface traces of the ruptures. The magnitudes of the events associated with the rupture data are obtained either from instrumental data or from estimates of seismic moment (equation (2.2)).

In a study of a multiregion (so-called 'worldwide') database of 244 earthquakes, Wells and Coppersmith (1994) obtained a range of widely used relations between magnitude and fault rupture parameters of length L (km), width W (km), area A (km^2) and displacement D (m). Both surface and subsurface data were considered. (The inherent unreliability of magnitudes estimated from the surface data is discussed by Dowrick and Rhoades (2004).) The majority of the 244 earthquakes occurred in interplate regions. Considering rupture area in km^2 for all their events, Wells and Coppersmith found:

$$M_W = 4.07 + 0.98 \log A, \quad s = 0.24, \qquad (4.2)$$

where s is the residual standard deviation. They also found similar relations for subsurface rupture length (L_{sub}) and width, and for events of different fault mechanisms (strike-slip, reverse and normal), but the latter relations have since been found not to be statistically significant (Hanks and Bakun, 2002; Dowrick and Rhoades, 2004).

Studying rupture data from slip models of 15 earthquakes (11 from California) of different mechanisms, Somerville et al. (1999) derived the expression

$$M_W = 3.95 + \log A, \qquad (4.3)$$

very similar to equation (4.2).

How well do expressions (4.2) and (4.3) explain the relations between magnitude and fault rupture area on a worldwide basis? Wells and Coppersmith found no statistically significant difference between events in compressional and extensional regimes. Other studies (e.g. Nuttli, 1983; Scholz et al., 1986) have concluded that their data show that intraplate events have smaller rupture dimensions than do interplate events. However, contradictions to this arise when examining other data sets. For example, the data of Dowrick and Rhoades (2004) from 29 fault ruptures in New Zealand, which is on a plate boundary, relate better to the above two intraplate models than to the interplate data and/or models of Scholz et al. and Wells and Coppersmith. It may be that the variability in fault rupture data is better explained in terms of some other criterion, such as the slip rate on the fault (Anderson et al., 1996), but more reliable rupture dimensions and more representative region-specific data sets than those used by Wells and Coppersmith are required to resolve this issue.

Considering their New Zealand subsurface rupture dimension,s Dowrick and Rhoades found the following relations:

$$M_W = 4.39(0.03) + 2.0(\text{con}) \log L_{\text{sub}}, \quad s = 0.11, L < 6.0 \text{ km}, \quad (4.4)$$

$$M_W = 4.76(0.13) + 1.53(0.09) \log L_{\text{sub}}, \quad s = 0.15, L \geq 6.0 \text{ km}, \quad (4.5)$$

$$M_W = 4.39(0.03) + 2.0(\text{con}) \log W, \quad s = 0.11, L < 6.0 \text{ km}, \quad (4.6)$$

$$M_W = 3.72(0.32) + 2.86(0.27) \log W, \quad s = 0.23, L \geq 6.0 \text{ km}, \quad (4.7)$$

$$M_W = 4.42(0.03) + 1.0(\text{con}) \log A, \quad s = 0.13, \quad (4.8)$$

$$M_W = 6.09(0.12) + 2.0(\text{con}) \log D, \quad s = 0.23. \quad (4.9)$$

In equations (4.4)–(4.9), the values in parentheses are the standard errors on the estimated coefficients, and con = constrained coefficient.

Four relations for M_W vs. rupture area are plotted in Figure 4.7: New Zealand (Dowrick and Rhoades), Wells and Coppersmith (multiregion), Somerville et al. (predominantly California), and Hanks and Bakun (California). It is obvious that the New Zealand relation is statistically significantly different from the other three relations, and that different relations apply to California and New Zealand.

The possibility that other regions are different again was examined by Dowrick and Rhoades. Using their New Zealand data plus the multiregion database of Hanks and Bakun, the four most highly represented regions of California, New Zealand, China and

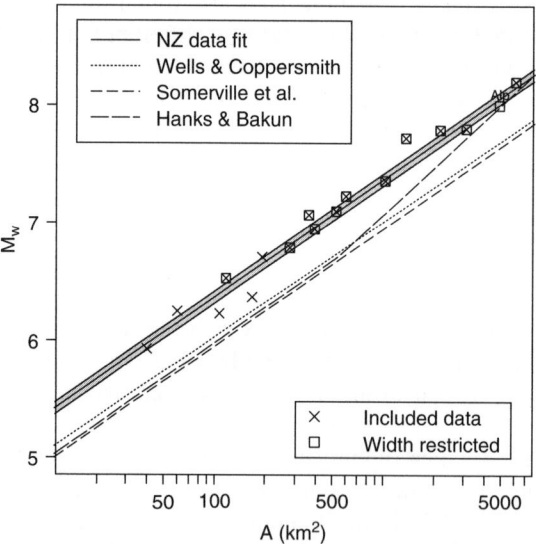

Figure 4.7 Regressions of magnitude versus rupture area of the New Zealand (Dowrick and Rhoades, 2004) and multiregion (Wells and Coppersmith, 1994; Hanks and Bakun, 2002) data sets together with that of Somerville et al. (1999). The 95% confidence limits are plotted for the New Zealand regression (after Dowrick and Rhoades, 2004, reproduced by permission of the Seism Soc Amer)

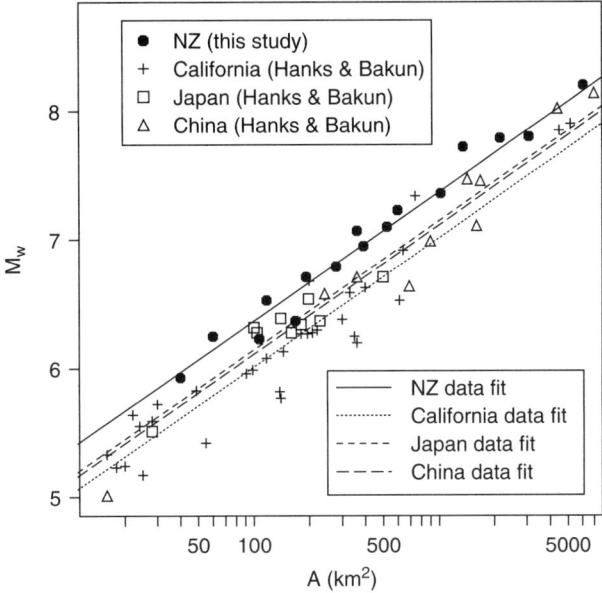

Figure 4.8 Regressions of magnitude versus rupture area for California, China, Japan and New Zealand with their respective data, explaining some of the scatter (after Dowrick and Rhoades, 2004, reproduced by permission of the Seism Soc Amer)

Japan were considered, with the resulting regressions being shown on Figure 4.8. The substantial scatter in the total data set is more satisfactorily explained by the separate regressions than by the use of a single multiregion regression. However, as the China and Japan data sets were small, the relations found for their regions in themselves are not definitive, in contrast to the robust ones for California and New Zealand.

The New Zealand relations (4.4)–(4.7) between M_W and L_{sub} and W shown for $L_{sub} <$ 6 km and $L_{sub} \geq 6$ km are derived from New Zealand earthquakes of $M_W \geq 5.93$. These regressions will not apply at small magnitudes, because it is presumed that ruptures tend on average to become effectively square or circular as magnitude decreases, because there are no physical conditions to constrain either L or W. This effect is modelled in Figure 4.9, where the length and width data and their mean regression line with 95% confidence lines are plotted. It is seen that the regression lines converge at $M_W = 6.0$.

It should be noted that the relationships discussed above are valid for fault ruptures with aspect ratios L/W up to about 6. Thus, very long and narrow ruptures, like that for the 1906 San Francisco earthquake ($L/W = 432/12 = 36$) need to be considered separately (Dowrick and Rhoades, 2004).

4.5 Earthquake Distribution in Space, Size and Time

4.5.1 Introduction

Earthquakes occur at irregular intervals in space, size, and time, and in order to quantify seismic hazard at any given site it is necessary to identify the patterns in the spatial, size,

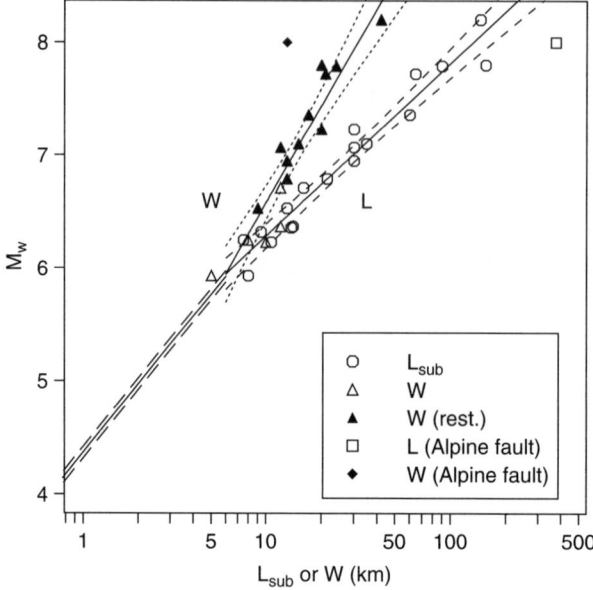

Figure 4.9 Regressions of length L_{sub} and width W for ($L_{sub} = W \geq 6.0$ km) and for $L_{sub} = W = \sqrt{A}$ for ($L_{sub} = W < 6.0$ km). W (rest.) indicates events where W is restricted structurally (after Dowrick and Rhoades, 2004, reproduced by permission of the Seism Soc Amer)

and temporal distributions of seismic activity in the surrounding region. Understanding the tectonic causes of earthquakes and identifying the seismogenic geological features in a region, as discussed in Section 4.3, enable the formulation of distribution patterns of potential sources. These patterns of distributions of occurrence involve the careful plotting or mapping of known historical events, and the correlation of these historical data with the models of crustal structure and deformation. Aspects of the study of the distribution of earthquakes are further discussed in the following sections.

4.5.2 Spatial distribution of earthquakes – maps

Figure 2.1 is a plot of spatial distribution of seismic events with the lowest possible level of information, but has its uses in global tectonic plate recognition. A similar world map, 'Significant earthquakes 1900–1979', produced in the USA, uses different sizes and colours of symbols to indicate the numbers of deaths and cost of damage of the events plotted. Thus a qualitative feel for the relative seismic risk of different localities can be obtained. However, for most analytical purposes the events need to be much more precisely plotted on local maps. Larger-scale seismic event maps of many areas are given in various publications, including the now somewhat dated reference works by Gutenberg and Richter (1965) and Lomnitz (1974). Many countries now have computerized earthquake catalogues, which can be easily revised, and from which data can be readily extracted for making seismicity maps for any chosen study area. Such a map is shown in Figure 4.10, which was prepared for study of the seismic hazard of Christchurch, New Zealand (Dowrick et al., 1998). It shows the epicentres of earthquakes of magnitude at

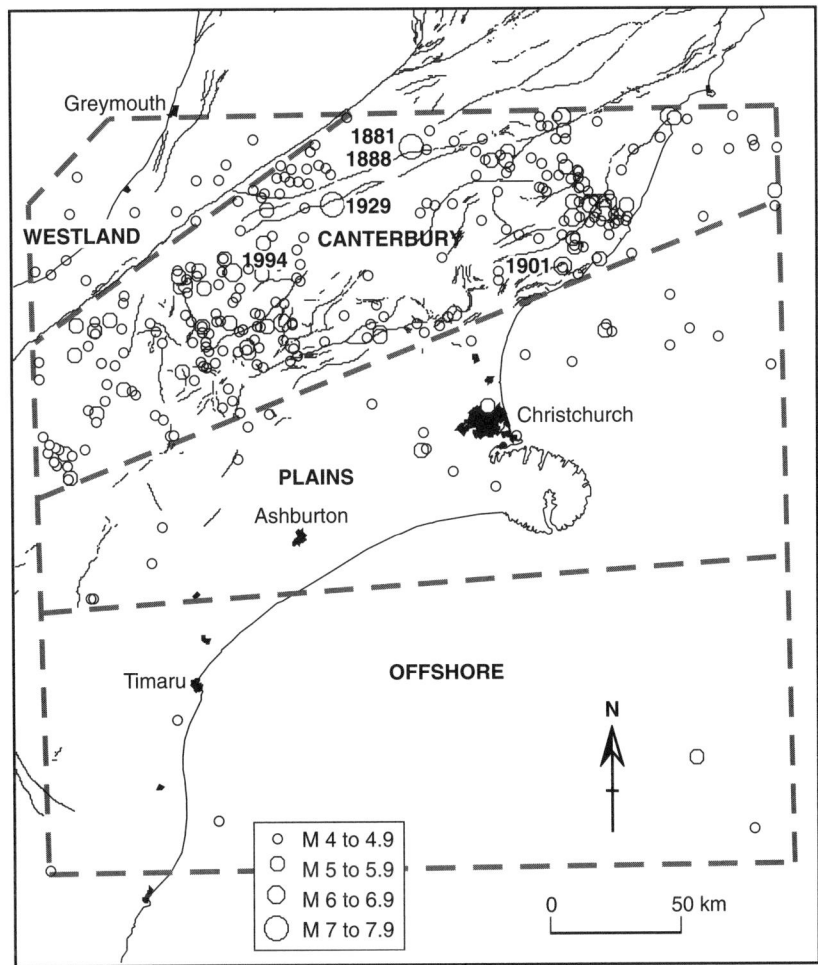

Figure 4.10 Historical shallow (depth ≤ 40 km) seismicity (1840–1994) and active faults near Christchurch. Four zones of differing seismicity have been delineated (from Dowrick *et al.*, 1998)

least 4 in the period 1840–1994. The completeness and accuracy of locations of the events on such maps are greatly affected by the amount and sensitivity of the local seismograph network. For example, in the case of New Zealand it is considered that the catalogue is complete for events of $M_W \geq 6.5$ from 1840, for $M_W \geq 5$ from about 1940 and for $M_W \geq 4$ from 1964.

All of the events plotted in Figure 4.10 are shallow ($h \leq 40$ km), in fact as the seismogenic crust is quite thin over the area shown, the sources for most of the events are less than 15 km deep. Where the depth range is greater than that of Figure 4.10, event maps often need indications of depth, which is either done using different symbols or by having different maps for different depth ranges. For example, this was done by Dowrick and Cousins (2003) for discussing the relative contribution to the seismic hazard of New Zealand of shallow and deep seismicity (down to 100 km). Earthquakes below 70 km in depth contribute only a small percentage of the total hazard, but rare large deep events can be moderately damaging. At perhaps the extreme end of the depth range, the 22 November 1914 Bay of Plenty, New Zealand, earthquake of $M_W = 7.3$ and depth 300 km caused Modified Mercalli intensity MM6 (see Appendix A) over a large area of the North Island (Downes and Dowrick, in preparation). Another deep damaging earthquake was the Bucharest earthquake of 4 March 1977, which had a focal depth of about 90 km and was of magnitude $M = 7.2$.

Another method of identifying spatial patterns of seismic activity is by mapping *strain release*, which can be done by strain measurements as discussed in Sections 2.2 and 4.2 above, or by converting seismic event maps into strain release, as discussed below.

The strain released during an earthquake is taken to be proportional to the square root of its energy release. The relationship between energy E (in ergs) and magnitude M_S for shallow earthquakes has been given by Richter (1958) as

$$\log E = 11.4 + 1.5 M_S. \tag{4.10}$$

The strain release U for a region can be summed and represented by the equivalent number of earthquakes of $M = 4$ in that region, $N(U4)$. The equivalent number of earthquakes $N(U4)$ is divided by the area of the region to give a measurement of the strain release in a given period of time for that region which can be used for comparisons of one region with another, or one period of time with another (Figure 4.11), as discussed by Carmona and Castano (1973).

From equation (4.10), it follows that the energy released by an earthquake of $M_W = 8$ equals that of 1 million $M_W = 4$ events. Thus large shocks constitute the main increments of a cumulative strain energy release plot. This is illustrated by the large step in 1960 on Figure 4.11, which results from one large event, namely the 1960 Chile subduction

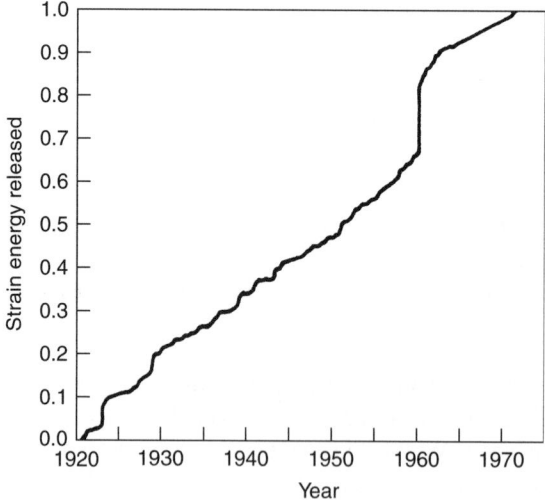

Figure 4.11 Rate of strain energy release in the portion of South America south of Arica, Chile. (after Carmona and Castano, 1973)

zone earthquake of $M_W = 9.5$ with a rupture zone estimated by Cifuentes (1989) as 920 ± 100 km long. From this figure, it is seen that the amount of energy released by this enormous event is approximately equal to one sixth of the total energy released in 50 years (1920–1970) in the 4000 km long southern part of South America south of Arica, Chile.

A plot of strain release against time is a step function to which an upper bound curve can be drawn, giving an indication of the trend in energy release for that region (Figure 4.11). Obviously, if a flattening of the curve tends to be asymptotic to a constant strain value over a significant time, then the faults in the region may have at least temporarily taken up a more stable configuration. On the other hand, a mechanical blockage of strain release may have occurred which only a pending large shock could release. Obviously, as in the case of seismic event maps, information from other sources is required for the proper interpretation of strain release curves and maps.

4.5.3 Earthquake distribution in time and size

On any given fault within any given region, earthquakes occur at irregular intervals in time, and one of the basic activities in seismology has long been the search for meaningful patterns in the time sequences of earthquake occurrence. The longer the historical record, the better is the overall picture that can be obtained. In most places the useful historical record is short, often only a few decades or sometimes one or two centuries, the great exceptions being China and the eastern Mediterranean, which both have useful records going back around 2000 years. Figure 4.12 shows the time distribution of damaging earthquakes in the latter area from the first to the eighteenth century AD, as derived by Ambraseys (1971). The longest quiescent period (c. 250 years) between active periods is worth noting, and similar gaps centuries long have been found in China (Mei, 1960; Allen, 1976).

During any given interval in time, the general underlying pattern or distribution of size of events is that first described by Gutenberg and Richter (1965), who derived an empirical relationship between magnitude and frequency of the form

$$\log N = A - bM, \qquad (4.11)$$

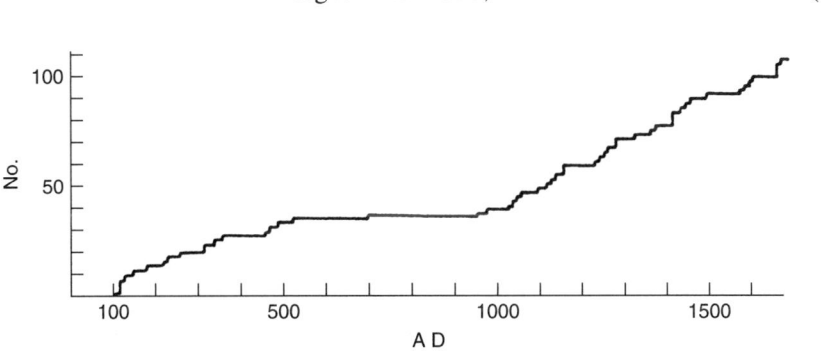

Figure 4.12 Time distribution of damaging earthquakes in the Anatolian fault zone (after Ambraseys, 1971; reprinted with permission from *Nature*, Ambraseys, NN, 232, 375–379, Macmillan Magazines Limited)

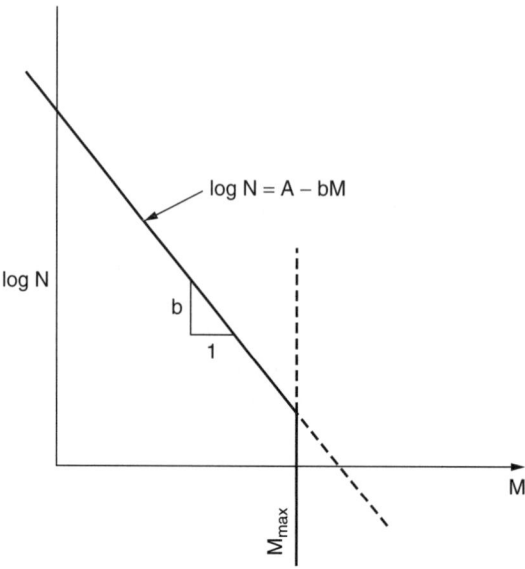

Figure 4.13 Magnitude–frequency relationship for earthquake occurrence

where N is the number of shocks of magnitude at least M per unit time and unit area, and A and b are seismic constants for any given region (Figure 4.13).

In a seismic hazard modelling study of New Zealand (Stirling et al., 2000) the country was divided into 14 shallow and 23 deep seismicity zones, across which b for the distributed seismicity was found to vary from 0.82 to 1.34, with an average of about 1.1. This is typical of the variation found in b around the world.

The slope b of the magnitude–frequency relationship is a key seismicity parameter. A decrease in b over a period of time indicates an increase in the proportion of large shocks. This may be caused by a relative increase in the frequency of large shocks, or by a relative decrease in the frequency of small ones. Some investigators have found that periods of maximum strain release in the earth's crust (see the year 1960 in Figure 4.11) have been preceded and accompanied by a marked decrease in b. From uniaxial compression experiments in the laboratory, Scholz (1968) found that the magnitude–frequency relationship for microfractures in a given rock is characterized by b decreasing when the stress level is raised. Consequently, regional variations in b may indicate variations in the level of compressive stress in the earth's crust.

For any given region, if enough data are available a plot of M against $\log N$ can be made, and the best-fit line of the form of equation (4.11) can be determined by regression analysis. To get the best results, allowance should be made for both incompleteness of the data (Kijko and Sellevoll, 1989) and magnitude uncertainties (Rhoades, 1996).

Although there are arguments for expressions other than equation (4.11) (e.g. quadratic), the empirical log-linear relationship of equation (4.11) fits the data reasonably well in the lower-magnitude range and, because of its simplicity, is the expression in general use. However, it is unsatisfactory at high magnitudes, as demonstrated by Chinnery and North (1975) in Figure 4.14, because there is a maximum achievable magnitude, M_{max}. The latter

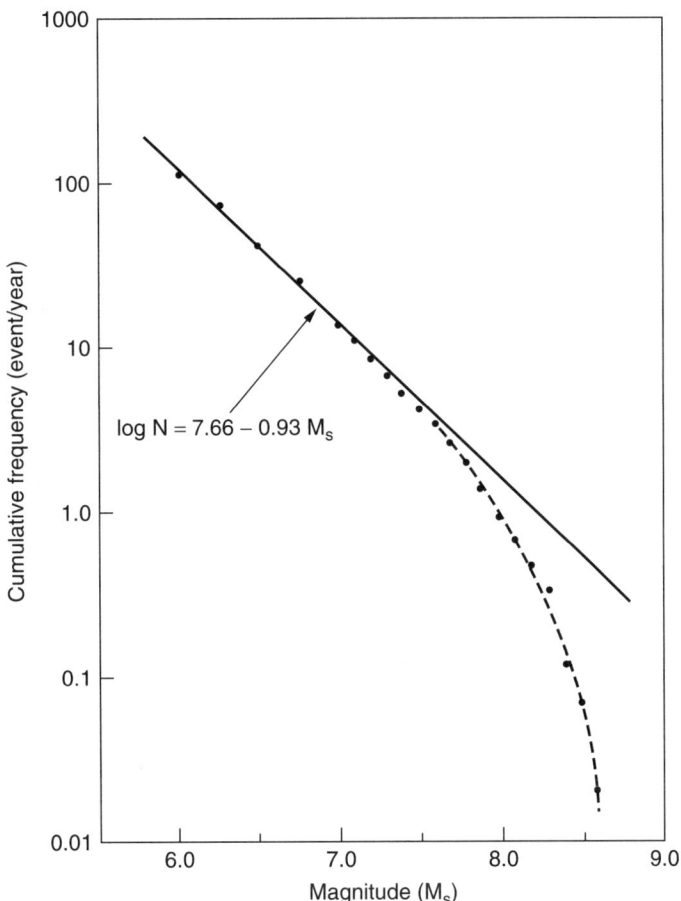

Figure 4.14 Cumulative magnitude–frequency relationship for large earthquakes from worldwide data, showing tendency for M_S to saturate in value (reprinted with permission from Chinnery and North, 1975). *Science* 190, 1197–8, Copyright 1975 American Association for the Advancement of Science

arises because a given fault or tectonic region has physical constraints on the maximum size of event it can generate. Despite the difficulty of reliably estimating values for M_{max} (Section 4.4.4) it is important to have a magnitude cut-off in equation (4.11), as shown in Figure 4.13, when estimating seismic hazard, as such estimates are greatly reduced at lower probability levels. Various expressions have been developed which modify equation (4.11) to incorporate M_{max}, with different transitions between the two lines, some of which are discussed by Anderson and Luco (1983).

Earthquake recurrence relationships taking the above forms are usually derived to fit what is called the 'distributed seismicity' of a region, i.e. the earthquakes that do not occur on known (and modelled) faults. The seismicity associated with faults is best allowed for specifically for each fault using the magnitudes and mean recurrence intervals and their uncertainties as discussed in Section 4.4.

4.5.4 Models of the earthquake process

The following discussion was kindly contributed by David Rhoades.

The earthquakes in a given region are the result of an ongoing physical process, and models may be made of:

(1) the physics of the phenomena producing the earthquakes; and
(2) statistical regularities in the series of times of occurrence, magnitudes, and locations of the resulting events.

These two separate approaches are complementary; both are needed to improve our understanding of the earthquake generation process. Wyss et al. (1999) provide a useful overview of this field.

Many physical models are built to describe the accumulation of stress over time in parts of the crust, as a result of slow but inexorable plate motion and the associated crustal deformation, and its release and redistribution by earthquakes. One perceived result of redistribution is the short-term triggering of aftershock activity in the immediate source area of a main shock. Another is the increase of stress at the ends of the associated fault rupture, which is held to increase the likelihood of rupture of neighbouring faults in the short to medium term. Simple constructs such as spring and block models (Burridge and Knopoff, 1967) are commonly used for physical modelling of the stress-transfer processes.

A very different idea, that earthquakes occurrence can be modelled as a self-organized critical phenomenon (Bak and Tang, 1989), i.e. a kind of deterministic chaos analogous to avalanches in a steadily growing sand-pile, has been used to explain important aspects of the earthquake process. For example, it is consistent with evidence that the drop in stress resulting from the occurrence of an earthquake is only a small fraction of the ambient stress in the crust. The power-law decay of aftershocks in time according to Omori's law (Utsu, 1961), and the Gutenberg–Richter frequency–magnitude relation (Section 4.5.3) are also consistent with self-organized criticality. Models of earthquakes as a critical phenomenon have been used to suggest that an accelerating stress release should occur in the approach of criticality, i.e. an accelerating occurrence of minor earthquakes leading up to a major earthquake. There is some empirical support for this theory. For example, Bowman et al. (1998) found precursory accelerating patterns in a wide area surrounding the sources of the eight most recent large ($M > 6.5$) earthquakes on the San Andreas fault system. The duration of the acceleration was found to vary from about a year to about a century, but to bear no obvious relation to earthquake magnitude. There is stronger empirical support for a more general long-term precursory increase in seismicity not usually accelerating, taking place in a region not much larger than the source area of major shallow earthquakes. This has been called the precursory scale increase, and is explained by a process in which the three stages of faulting – crack formation, fracture and healing – are separated in time (Evison and Rhoades, 2001), and in which self-organized criticality plays a background role. A study of 47 major earthquakes in four well-catalogued regions has shown that there are simple scaling relations among the magnitude of the major earthquake and the magnitude, duration and area of the precursory scale increase (Evison and Rhoades, 2004).

Real sequences of earthquakes in a given region are typified by those shown in Figures 4.12 and 4.15, with the magnitudes, the locations and the time intervals all varying, and various stochastic models have been applied by different researchers in attempts

Earthquake Distribution in Space, Size and Time 67

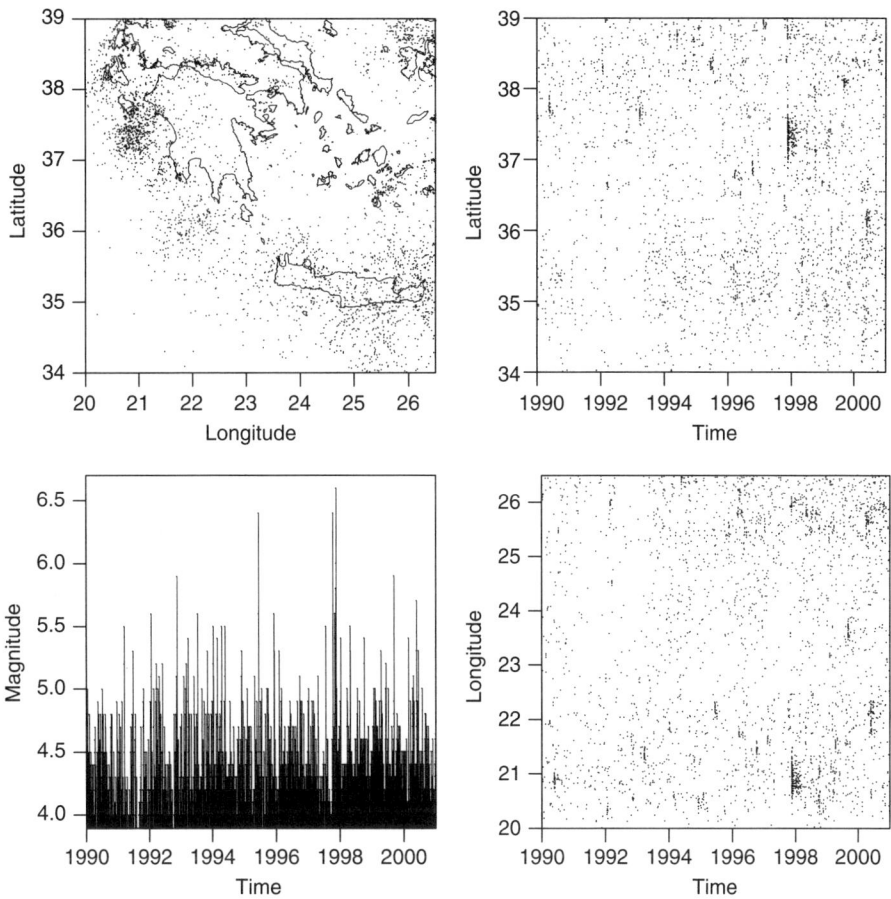

Figure 4.15 Epicentral locations, magnitudes and times of occurrence of earthquakes in a region of Greece (Aristotle University of Thessaloniki catalogue; plotted by D.A. Rhoades)

to find analytical descriptions of the earthquake process. They originate in some cases from physical ideas, and others from empirical observations.

Some statistical models describe only the time intervals between successive events. For example, in considering the successive ruptures of a particular fault segment in earthquakes of supposedly characteristic magnitude, it is common to adopt a renewal process framework. The hazard rate at elapsed time t since the last rupture is quantified by the *hazard function* $h(t)$, defined so that $h(t)dt$ is the conditional probability that an event will take place in the time interval $(t, t+dt)$. If $F(t)$ is the cumulative probability distribution of the time between successive events, then

$$h(t) = f(t)/[1 - F(t)], \qquad (4.12)$$

where $f(t) = dF(t)/dt$.

For the Poisson process, $h(t)$ is a constant equal to the mean rate of the process and the distribution of the time between successive events is exponential. Other commonly used distributions are the Weibull and lognormal, and recently the inverse Gaussian (i.e. Brownian passage-time) distribution (Ellsworth et al., 1999). All of these distributions have different shaped hazard functions (see Figure 4.16). The Weibull hazard function is monotone increasing with the parameter values usually adopted, the lognormal hazard function rises to a peak and then tends asymptotically to zero, and the inverse Gaussian rises to a peak and tends asymptotically to a positive value. Physical arguments have been advanced in support of each of these models, and each implies some degree of regularity in the time intervals between earthquakes in contrast to the pure randomness of the Poisson process model. The data available at present on earthquake recurrence do not support a clear preference for a particular distribution. Moreover, detailed studies have shown that the choice of distribution may have only a small impact on the estimated hazard when all the data and parameter uncertainties are accounted for (Rhoades et al., 1994). However, the Poisson process (exponential) model can be rejected for some data sets.

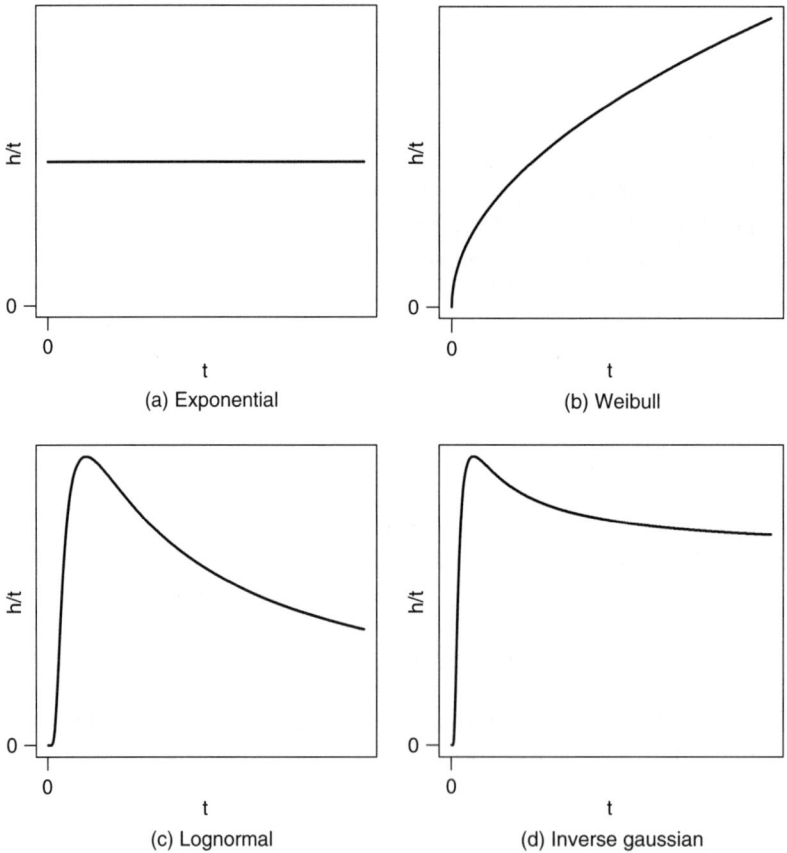

Figure 4.16 Typical shapes of the hazard functions of four different processes modelling the distribution of time between successive earthquakes (prepared by D. A. Rhoades)

The renewal process framework is quite limited as a model for earthquake occurrence: it does not allow for interactions between neighbouring faults, and ignores the obvious tendency for earthquakes to occur in clusters. Early attempts at creating better models include trigger models, and branching renewal processes (Vere-Jones, 1970). In trigger models, the earthquake process is treated as the superposition of a number of sub-processes, each having a different origin. In branching renewal processes, the intervals between cluster centres, as well as those between cluster members, constitute renewal processes.

Models involving time and magnitude include the slip-predictable and time-predictable models (Kiremidjian and Anagnos, 1984). Like the renewal process model, these models are based on the stress-release concept, but the characteristic magnitude assumption is relaxed. In the time-predictable model the time to the next earthquake is allowed to depend on the magnitude of the most recent earthquake, and in the slip predictable model the magnitude of the next earthquake depends on the elapsed time. Another stress-release model is that of Zheng and Vere-Jones (1991) in which the hazard rate steadily increases over time, but drops suddenly upon the occurrence of an earthquake by an amount that depends on its magnitude. This model was applied to a long sequence of historical earthquakes in north China. A coupled version of this model, allowing for stress transfer between discrete regions, has also been developed.

An important model of earthquake clustering is the epidemic-type aftershock (ETAS) model of Ogata (1989). In this model, each earthquake has its own aftershock sequence which decays over time according to the modified Omori relation

$$\lambda(t) = K/(t+c)^p, \qquad (4.13)$$

where time t is measured from the occurrence of the earthquake, λ is the rate of occurrence of aftershocks above some magnitude threshold, p is a constant with a value typically between 1 and 1.5, K is a function of the earthquake magnitude, and c is a small time constant included to avoid the function going to infinity at $t = 0$. Some earthquakes occur independently according to a stationary Poisson process, and the magnitudes of all earthquakes follow the Gutenberg–Richter relation. The ETAS model fits earthquake data much better than a stationary Poisson process, and better than a model in which only the larger events have aftershocks.

Space-time-magnitude point process models are now receiving more attention. Extensions of the ETAS model to include the spatial variable (e.g. Console and Murru, 2001) can be used to describe the short-term variation of hazard, by means of the rate density $\lambda(t, m, x, y)$ of earthquake occurrence, which can be defined for any time t, magnitude m, and location (x, y) within a large region of surveillance. For a given magnitude and location, the rate density can fluctuate over a short time by several orders of magnitude. Progress is also being made with long-term variation of the rate density. Rhoades and Evison (2004) have proposed a model (EEPAS) in which every earthquake is regarded as a long-term precursor according to scale. The contribution that an individual earthquake makes to the future distribution of hazard in time, magnitude and location is on a scale determined by its magnitude, through the scaling relations associated with the precursory scale increase phenomenon. The EEPAS model has been fitted to the New Zealand earthquake catalogue, and verified by application, with unchanged parameters, to the Californian region. Under this model the rate density can fluctuate slowly over time by about 1.5 orders of magnitude.

Despite these developments, time-varying hazard models have not yet been widely adopted for practical purposes. The stationary Poisson model and the Gutenberg–Richter magnitude distribution are still the basis for most seismic hazard models in practical use. However, this may soon change. The California-based Collaboratory for the Study of Earthquake Predictability (CSEP) is now spearheading an international collaborative effort to advance the science of earthquake forecasting. Regional earthquake forecast testing centres have been established in California, New Zealand, Europe and Japan. Their purpose is to provide transparent, verifiable prospective tests of time-varying earthquake occurrence models, using defined testing regions and rules of the game. To be included in the tests, a model must provide estimates of the expected number of earthquakes within grid cells of earthquake times, locations and magnitudes. Several different time steps are used for short-range, medium-range and long-range forecasts, e.g. 24 hours, 3 months, and 5 years. The tests will be continued for a minimum of 5 years, but the intention is that they will be ongoing. A recent issue of *Seismological Research Letters* (vol. 78, no. 1, 2007) describes the models initially submitted and the testing procedures that will be applied.

4.6 The Nature and Attenuation of Ground Motions

To obtain a complete predictive model for the ground motion at a given site, it is necessary (1) to describe fully the ground motion at the source, and (2) to describe the modifications to the ground motion as it propagates from source to site, i.e. the attenuation. The sources and the attenuation are not the same for all regions, and hence the appropriate regional descriptions need to be determined from assessing the seismic hazard at a given site.

4.6.1 Earthquake source models

The subject of source models is an area of study for seismologists, the results of which are fundamental to our understanding of the nature of ground motion. From amidst the complexities of this major study area a number of key parameters are evident as being of interest to earthquake engineers, some of which have already been introduced, such as fault length, fault width, fault displacement (or slip), stress drop on a fault, and, of course, earthquake magnitude. Some regional differences in fault length have been noted in Section 4.4. A few further features of source models are briefly described below, and for further reading specialist textbooks should be consulted, such as Kasahara (1981).

An earthquake is the product of a displacement discontinuity sweeping across a fault surface. The shape of the rupture surface and the resistance across it are variable, the areas where the resistance and subsequent slip are high being known as asperities. The slip distribution on the fault in the 1994 Northridge, California, earthquake is given by the source model of Wald and Heaton (1994), illustrated here in Figure 4.17. The regions of higher slip are seen to be grouped into three areas which could be described as separate asperities.

In recent years a large amount of work has been done to estimate the distribution of slip on the fault surface during past earthquakes. In a study of 15 earthquakes of M_W between 6.0 and 7.2, Somerville *et al.* (1997) found that on average asperities comprised 22% of the total rupture area. Their definition of an asperity was a rectangular area where

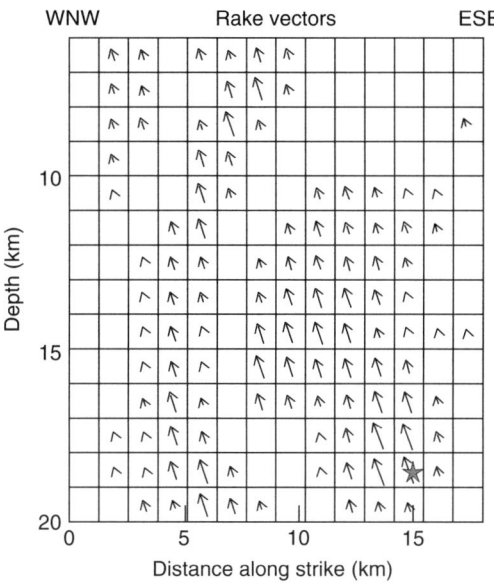

Figure 4.17 Rupture model of the Northridge earthquake. The vertical scale shows the downdip distance, which ranges from 0 to 21 km and is measured from the top of the fault at 7.5 km depth (after Wald and Heaton, 1994)

the average slip exceeded 1.5 times the average slip over the whole rupture area. It was also found that there were 2.6 asperities on average in each earthquake.

Brune (1970) gave a simple expression for estimating the stress drop,

$$\Delta \sigma = \frac{7M_o}{16r^3}, \qquad (4.14)$$

where the rupture surface is assumed to be circular with radius r. This equation and the previous methods give estimates of stress drop which differ for the same event by as much as an order of magnitude. Thus the maximum value of stress drop that is likely to occur in any earthquake is uncertain, but values higher than several hundred bars do not seem likely.

Rupture velocity, the velocity at which fault rupture propagates, is a basic parameter of source modelling, with estimates typically varying from about half to about equal to the shear-wave velocity of the ruptured material, yielding rupture velocities v_r of 2–3 km/s.

Rise time is the time required for the slip or stress change on a fault to take place, and is most simply expressed as a ramp function so that the displacement u at time t at a point x on the fault is

$$u(t) = u_\infty G\left(t - \frac{x}{v_r}\right), \qquad (4.15)$$

where G is a ramp function which increases linearly from zero at $t = 0$ to unity at $t = T$ (where T is the rise time), and u_∞ is the final displacement. The rise time may

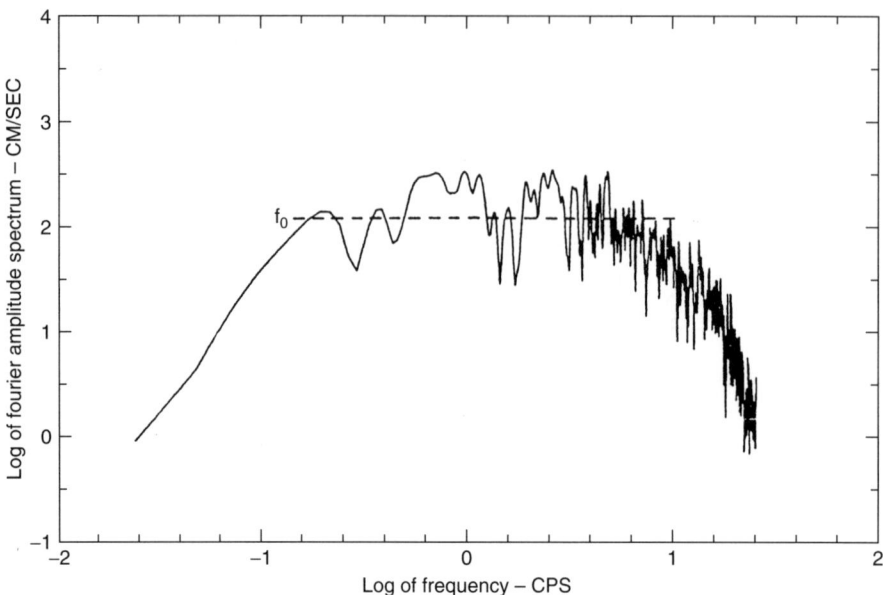

Figure 4.18 f_0 estimated from the spectrum of the San Fernando earthquake of 9 February 1971, $M_L = 6.4$ at the Pacoima Dam site (adapted from Hanks, 1982, reproduced by permission of the Seism Soc Amer)

be determined directly if the ground motion near the fault is recorded completely. Rise times computed from two theoretical models including equation (4.15) for 41 events in different parts of the world (Kasahara, 1981) range in value from 0.7 to 36 s, while Brune (1976) postulates a value as low as $T = 0.1$ s in estimating an upper bound for peak ground acceleration. Clearly, peak ground velocity and acceleration are dependent on the rise time.

The *frequency parameter*, f_0, that arises in source modelling is most readily described by reference to Figure 4.18. Far-field shear-wave acceleration spectra are characteristically flat at frequencies greater than the *corner frequency* f_0, which has been defined by Brune (1970) as the frequency at the intersection of the low- and high-frequency asymptotes of the spectrum. Corner frequencies are calculated for both P- and S-waves, and, despite controversy (Aki, 1984) about the relative magnitudes of $f_0(P)$ and $f_0(S)$, the corner frequency is an important feature of source models.

4.6.2 The characteristics of strong ground motion

Introduction

In earthquakes the motion of any particle of the ground follows in general a complex three-dimensional path having rapidly changing accelerations, velocities, and displacements and a broad band of frequency content. Strong ground motion is measured by a large-amplitude type of seismograph called an *accelerograph*, in the form of an accelerogram, which is an acceleration history typically of the form shown in Figure 5.19, while

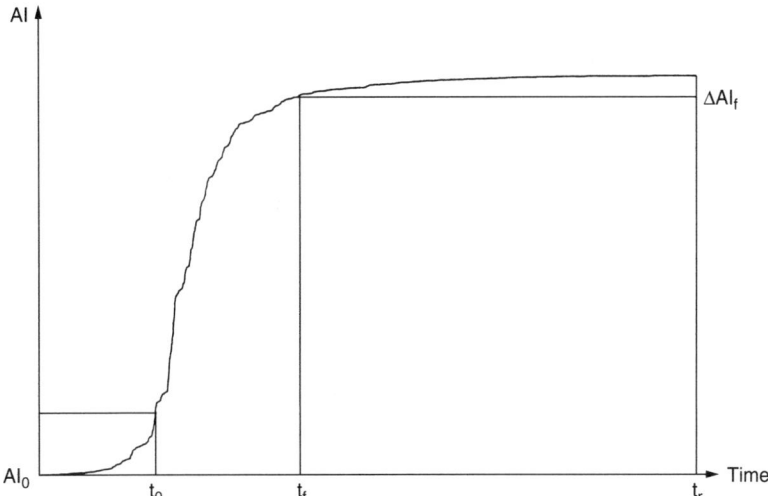

Figure 4.19 Definition of effective duration (after Bommer and Martínez-Pereira, 1999. Reproduced by permission of Imperial College Press)

the velocity and displacement histories (also in Figure 5.19) are obtained after the earthquake by integration of the acceleration record. Not only is earthquake ground motion complex and irregular, but no two earthquakes are the same. In an attempt to obtain an adequate understanding of earthquake ground motion a number of parameters have been used to characterize it, particularly:

- peak ground acceleration, PGA;
- peak ground velocity, PGV;
- peak ground displacement, PGD;
- root mean square acceleration, a_{rms};
- response spectrum parameters – spectral acceleration, SA(T, ξ), spectral velocity, SV(T, ξ) and spectral displacement, SD(T, ξ);
- Fourier spectra;
- duration of strong motion.

PGA, PGV, PGD and a_{rms} are measured from the respective ground motion records, often referred to as time-histories. As seen in Figure 5.19, the PGA for the north–south component of the 1940 El Centro record is 0.33g. SA, SV and SD are the maximum responses of a single-degree-of-freedom structure of chosen period of vibration (T, second) and damping (ξ, % of critical), subjected to the recorded ground motion (e.g. Figure 5.20).

Further discussion of the above factors is given below.

Duration of strong motion

This variable is important because the amount of cumulative damage incurred by structures increases with number of cycles of loading, and also because the duration of strong motion is used in evaluating one of the measures of strength of shaking, namely the root mean

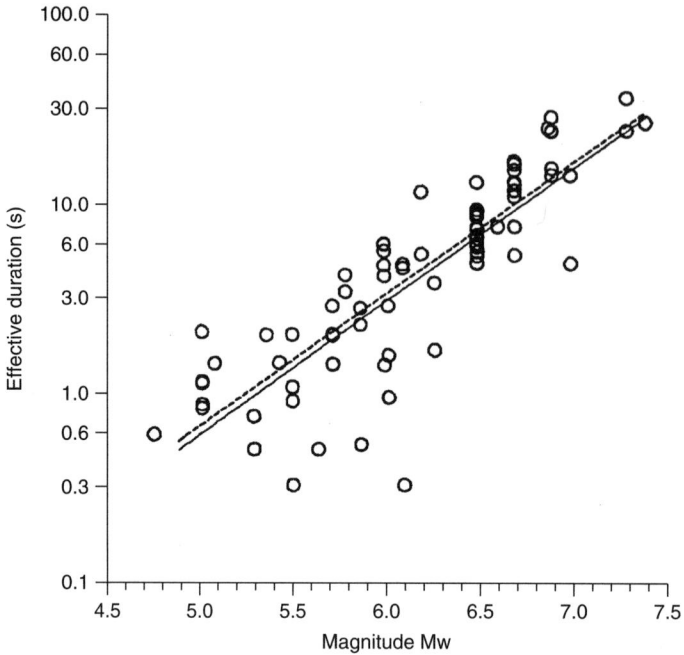

Figure 4.20 Effective durations (mean of horizontal components) as a function of moment magnitude for records from soil sites at distances of less than 10 km. The solid line is the best-fit found by regression analysis (equation (4.18)). The dashed line is the rupture duration (after Bommer and Martínez-Pereira, 1999. Reproduced by permission of Imperial College Press)

square acceleration (discussed below). Duration of strong motion is usually defined in relation to ground accelerations, and many difference definitions exist.

A comprehensive definition of a predictive model for duration of strong motion should incorporate (at least) the influence of earthquake magnitude, travel path (attenuation) and site ground class. In a review of 30 different definitions of duration, Bommer and Martínez-Pereira (1999) found that they all had shortcomings and yield very different estimates of duration. They offered their own 'preliminary' definition called 'effective duration', limited to near-source situations.

Their definition is based on the Husid plot (Husid, 1969) as shown in Figure 4.19, where the effective duration, D_E, is given by

$$D_E = t_f - t_o. \tag{4.16}$$

The start of the strong-motion phase is the time, t_o, when the Husid plot reaches an Arias intensity (Arias, 1970) value of $AI_o = 0.01$ m/s. The end of the strong-motion phase, t_f, is the time when the remaining energy in the record $\Delta AI_f = 0.125$ m/s.

Considering 90 accelerograms from soil sites and 32 records from rock sites, all located at distances of less than 10 km from the source, Bommer and Martínez-Pereira found that

D_E varied with magnitude as follows:

$$\log(D_E) = 0.69 M_W - 3.70 \qquad (4.17)$$

for rock sites, and

$$\log(D_E) = 0.70 M_W - 3.70 \qquad (4.18)$$

for soil sites. The standard deviation for soil sites $\log(D_E)$ is 0.30.

Thus, durations for soil sites were found to be fractionally greater (20% for $M_W = 7.5$) than those for rocks sites. Figure 4.20 shows plots of the soil data and the corresponding model, plus a plot of their model for duration of rupture. Interestingly, the models of the mean effective duration and mean rupture duration are seen to be very similar.

Estimates of duration of strong motion as a function of source distance are heavily dependent on the definition used for duration. Thus the generalization of expressions (4.17) and (4.18) to include a distance term would be a valuable extension of Bommer and Martínez-Pereira's work.

The fundamental issue relating to duration of strong motion is of course its influence on structural damage. In their state-of-knowledge review of this topic, Hancock and Bommer (2006) noted that widely differing conclusions have been made in different studies, with those employing damage measures related to cumulative damage usually finding a positive correlation between strong-motion duration and structural damage, while those following the current-state-of-practice approach which uses maximum response generally do not do so.

Root mean square acceleration

A measure of the strength of ground shaking which in the past has been used in strong-motion seismology is the root mean square acceleration, which is defined as

$$a_{\text{rms}} = \left\{ \frac{1}{T_2 - T_1} \int_{T_1}^{T_2} a^2(t) \, dt \right\}^{1/2}. \qquad (4.19)$$

For a stationary process (Section 4.7.5) the location and size of the duration $(T_2 - T_1)$ over which the squared accelerations are averaged is relatively unimportant, but for a transient signal like an earthquake record, a_{rms} is obviously strongly dependent upon which portion of the record is included. Commonly, T_1 and T_2 are chosen to exclude the (arbitrarily defined) insignificant shaking and use has been made of one or other of the definitions of duration of strong motion such as noted in the preceding section. Two further approaches to duration that have been tried both take T_1 as the time of the S-wave arrival, McCann and Boore (1983) taking $T_2 = T_1 + 10$ s and Trifunac and Brady (1975) taking $T_2 = T_1 + T_d$, where T_d is the duration of faulting. In the latter study it was assumed that $T_d = f_0^{-1}$, which is an approximation that was observed to be accurate to $\pm 50\%$, and T_d varied from about 2 to 20 s for earthquakes in the magnitude range $M = 5$ to 7.7. The use of $T_1 + T_d$ has the merit that it was found that within this window T_d the earthquake records were essentially stationary processes, thus improving the stability of the a_{rms} values calculated. Obviously, with duration being variably defined, care is necessary in comparing a_{rms} values to ensure that they have been calculated on the same basis.

The root mean square acceleration has long been of interest to earthquake engineers as a measure of strength of ground motion, partly because the averaging involved could be expected to lead to a more stable parameter than peak ground acceleration. However, in some studies (McCann and Boore, 1983) it has been found, contrary to expectation, that a_{rms} is not less variable than PGA, and it has not found a regular place among design parameters.

Upper bounds to peak ground motions

Expressions of the type given in equation (4.19) do not always give appropriate estimates of peak ground motions that may be expected at a given site, because there may be limitations on the amplitudes achievable at a given site.

From considerations of stress, frequency content and rupture velocity, Brune (1976) gives two arguments for an upper bound of about $2g$ for horizontal acceleration in solid rock near to the source. An elaboration of these results comes from considering the ground motion at the surface due to an S-wave radiated vertically during the failure of the most heavily loaded asperity on a fault, whence McGarr (1982) found the expression for peak ground horizontal acceleration

$$a < 1.58 \Delta \tau_i / \rho z, \qquad (4.20)$$

where $\Delta \tau_i$ is the stress drop on the asperity (bars), ρ is the density of the rock (gm/cm^3), and z is the depth (km).

McGarr also found upper bounds for this stress drop of

Compressional stress state	$\Delta \tau_i = 334z,$	(4.21)
Perfect strike-slip state	$\Delta \tau_i = 112z,$	(4.22)
Extensional stress state	$\Delta \tau_i = 67z.$	(4.23)

Substituting equations (4.21)–(4.23) into equation (4.20) yields upper bounds on peak horizontal accelerations as follows:

Compressional state	$a = 2.0g,$
Perfect strike-slip state	$a = 0.7g,$
Extensional state	$a = 0.4g.$

If the strike-slip state of stress is other than 'perfect', the upper bound for a can lie anywhere between $0.4g$ and $2.0g$, depending on the ratio of horizontal to vertical principal stress (McGarr, 1982).

A method of estimating peak horizontal accelerations which is independent of a source mechanism and location comes from considering the maximum acceleration that can be transmitted according to the strength of the soil. Consider a seismic shear wave being

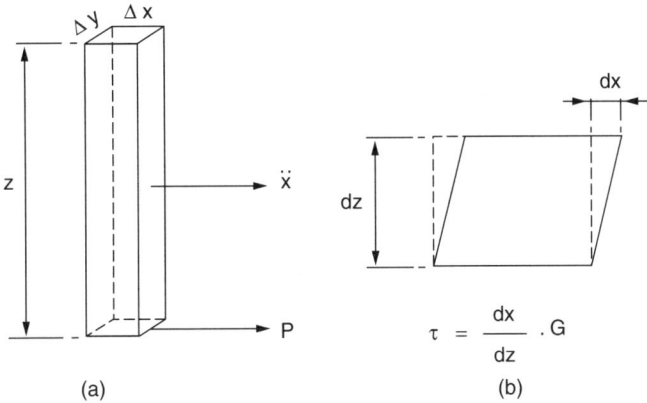

Figure 4.21 Shear behaviour of soil column assumed for calculating bounds on peak ground motion: (a) free body; (b) shear distortion

transmitted upwards through an elementary column of soil with forces and motions as shown on Figure 4.21(a). From dynamic equilibrium,

$$\tau \, \Delta x \, \Delta y = \rho z \, \Delta x \, \Delta y \, \ddot{x},$$

whence

$$a = \ddot{x}_{max} = \frac{\tau}{\rho z}, \qquad (4.24)$$

where τ is the shear stress at failure and ρ is the density.

Assuming the mean specific gravity $\bar{\gamma} = \gamma/2$, we find the following approximate relationship for the maximum horizontal acceleration that can be transmitted by a saturated, normally consolidated (NC) soil:

$$\frac{a}{g} \approx \frac{c_u}{2p'}, \qquad (4.25)$$

where c_u is the undrained shear strength and p' is the effective overburden pressure.

From dynamic soil response analyses which incorporated similar limiting criteria for a variety of soils, Ambraseys (1973) found the following upper bounds on horizontal accelerations:

low plasticity NC clays, $0.1g \le a \le 0.15g$;

high-plasticity NC clays, $0.25g \le a \le 0.35g$;

saturated sandy clays and dense medium sands, $a \le 0.6g$.

Overconsolidated clays can transmit very high accelerations. In a study of the seabed in part of the Gulf of Alaska in which the author was involved, using a method based in part on that outlined above for normally consolidated clay, it was estimated that 6 m below the seabed the maximum acceleration would be $1.85g$. The soil shear strength was 215 kPa at the surface and increased in strength at a rate of 4.7 kPa per metre of depth. It is of interest that this value is nearly as high as the upper bound of $2g$ predicted above for rock considering source characteristics.

Recently Rhoades *et al.* (2008) have developed a simple statistical test for inhibition of very strong shaking in ground motion models. Studying a very large Japanese data set of PGAs, they found that the ratio of actual to expected number of times given accelerations are exceeded declines in a statistically significant and regular fashion from about 1 at $0.3g$ to about 0.15 at $1.0g$. If these results are indicative of ground-motion models in general, the implications for probabilistic seismic hazard analyses may be far reaching, particularly in projects where very long return periods are of interest.

A simple means of *obtaining* upper bounds on *peak ground velocity*, v, comes from considering the shear distortion of an element of soil as shown in Figure 4.21(b), from which

$$\tau = \frac{dx}{dy}G = \frac{dx}{v_s dt}G$$
$$= \frac{v}{v_s}G,$$

i.e.

$$v = \frac{v_s \tau}{G}, \qquad (4.26)$$

where v_s is the velocity of the shear wave and G is the shear modulus for the soil. A limiting value for v reached when the shear-wave stress τ equals the shear strength of the soil c, and using the relationship $v_s = \sqrt{G/\rho}$, equation (4.26) becomes

$$v_{max} = \frac{c}{\sqrt{G\rho}}. \qquad (4.27)$$

Thus the maximum particle velocity is a simple function of the mechanical properties of the local soil or rock. For a strong rock such as limestone or basalt (Table 5.3), using $E = G/3$, $E = 10^5$ MN/m^2, $E/c = 600$, $\rho = 2700$ kg/m^3, it follows that $v_{max} \approx 1800$ cm/s. However, it has been shown (Ambraseys and Hendron, 1968) that the initial velocity may be expressed as

$$v = \frac{cv_s(1 - c_r/c)}{2G}, \qquad (4.28)$$

where c is the average strength of the material on the fault surface at rupture and c_r is the residual strength after rupture. According to Ambraseys (1973), velocities near the source may reach as high as 480 cm/s, but because c_r is unlikely to be zero, and because there has been neither observable melting on the fault planes nor heat-flow anomalies along the reactivated faults, the upper bound for velocities in the rock near a fault rupture should be about 150 cm/s. However, Sorensen *et al.* (2007) estimated that bedrock velocity in bedrock near the highest slip regions of the great 2004 Sumatra-Andaman earthquake of M_w between 9 and 9.3 was c. 200 cm/s.

Returning to equation (4.27), soft clay could have $E = 10$ MN/m^2 and $\rho = 1600$ kg/m^3, which gives $20 \leq v_{max} \leq 40$ cm/s, depending on the ratio E/c. Thus equations (4.27) and (4.28) have shown that for ground conditions covering the practical range of building sites, the peak ground velocities that are physically sustainable at the site lie approximately between 20 and 150–200 cm/s.

The reader may wish to relate the above maximum ground motions to the upper bound on Modified Mercalli intensity, i.e. damage, as discussed in Section 6.2.

Frequency content

The frequency of vibration associated with the amplitudes of motion is, of course, an intrinsic characteristic of earthquake ground motion, and frequency content is commonly studied in spectral form. As an example, Figure 4.18 shows a Fourier amplitude spectrum for the Pacoima Dam recording of the San Fernando 1971 earthquake, where the Fourier amplitude spectrum is given by

$$|F(\omega)| = \left\{ \left[\int_0^t \ddot{u}_g(\tau) \cos \omega \tau \, d\tau \right]^2 + \left[\int_0^t \ddot{u}_g(\tau) \sin \omega \tau \, d\tau \right]^2 \right\}^{1/2} \quad (4.29)$$

where ω is the circular frequency, $\ddot{u}_g$ is the acceleration of the ground, and τ is the time.

Earthquake Fourier spectra are mainly used in seismology, while another spectral method of examining frequency content, namely the response spectrum, is generally used by engineers. Response spectra are referred to throughout this book.

The frequency content of ground motions is a function of a number of phenomena, notably source mechanism, depth, distance from the source nature of the travel path and site soil, and the magnitude of the event. Considering source mechanisms, basic physics tells us that a rapid rupture in strong rock will produce more high-frequency than low-frequency vibration.

Intraplate earthquakes have smaller rupture areas (Section 4.4.4) and more high-frequency content than interplate earthquakes (Hermann and Nuttli, 1984). This difference in frequency content is illustrated in Figure 4.22 which shows the normalized uniform hazard spectra for four cities in different parts of the USA, derived as part of the development of the national US earthquake code (Leyendecker et al., 2000). It is evident that for the intraplate city of Charleston on the east coast of the USA there is much more high-frequency and less low-frequency content than there is for the interplate cities of San Francisco and Los Angeles. It is interesting to note that Salt Lake City has similar frequency content to that of these two interplate cities, although it is about 1000 km from San Francisco and Los Angeles on the western side of the seismicity associated with that part of the circum-Pacific plate boundary. This implies that the interplate region is very wide and diffuse in the western USA, as reflected in the seismicity distribution shown in Figure 2.5.

Because higher frequencies attenuate more rapidly than low ones, there is a tendency for the predominant period to increase with distance from the source. This effect is illustrated in Figure 4.23, where the response spectra at distances of 10 and 100 km are compared

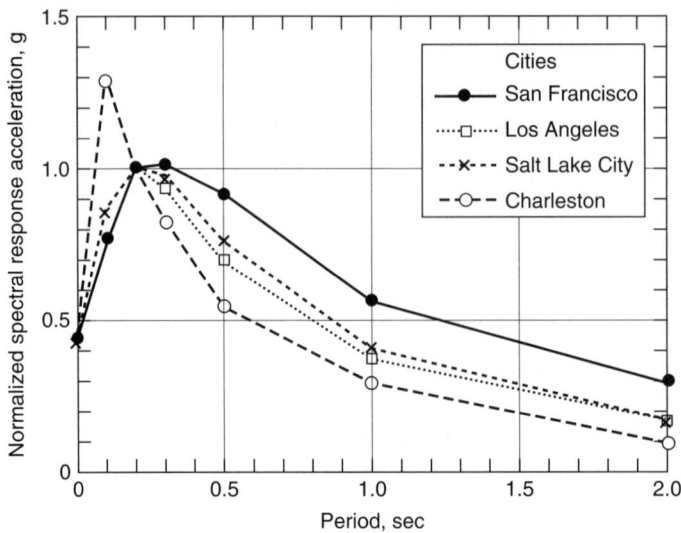

Figure 4.22 Normalized uniform hazard response spectra for four US cities for a mean return period of 2500 years showing difference in frequency content of interplate and intraplate events (adapted from Leyendecker *et al.*, 2000). Reproduced with permission of the Earthquake Engineering Research Institute

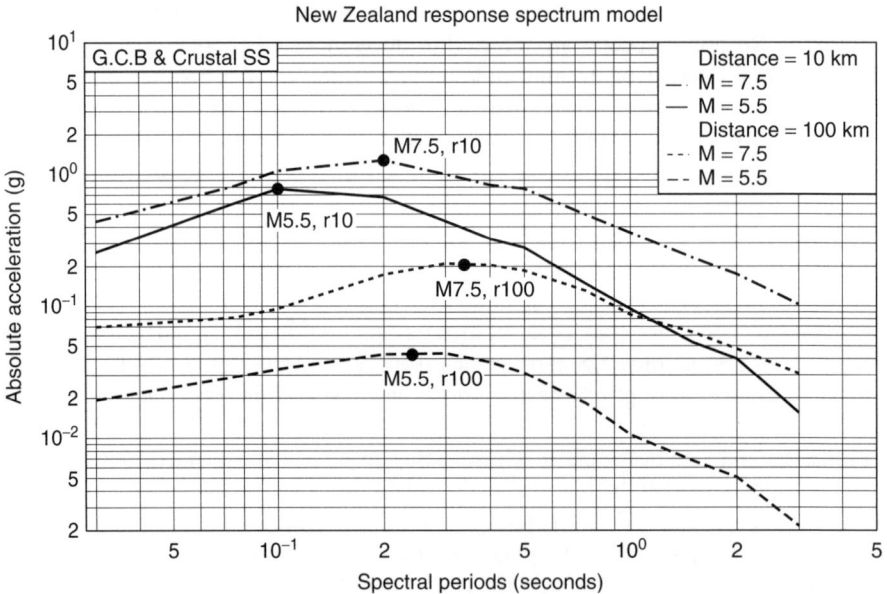

Figure 4.23 New Zealand acceleration response spectra for M_W 5.5 and 7.5 earthquakes at source-site distances of 10 and 100 km, showing how the predominant period increases with both magnitude and distance (after GH McVerry, personal communication, 2001)

for small and large interplate earthquakes. This figure also illustrates how the predominant period increases with magnitude. This effect occurs because the longer duration of shaking in large earthquakes allows more time for the generation of longer-period vibration.

The effect of local soil conditions on frequency content is considerable, as shown in the response spectra of Figure 3.3, and as discussed in Section 5.4.

Strong ground motion versus intensity

Many researchers have developed relationships between strong ground motion parameters (especially PGA) and subjective intensity scales, such as the Modified Mercalli or European Macroscopic scales, which are quite similar (Dowrick, 1996). This is done because of the valuable qualitative and quantitative descriptions that intensities give of various things, e.g.:

- the effects of strong motion on the built and natural environments (Appendix A and Table 6.1);
- spatial distributions of ground motions (Section 4.6.3);
- attenuation in regions with few or no strong-motion records;
- regional and tectonic variations in attenuation (Section 4.6.4);
- estimate of magnitudes, macroseismic centre and depths of earthquakes having inadequate or no instrumental data;
- estimate of the strike of earthquakes, i.e. the alignment of the fault rupture where no surface trace is found;
- comparison of modelled and historical seismic hazard (e.g. Dowrick and Cousins, 2003).

In addition the vulnerabilities of different items of the built environment may be measured as functions of intensity where few or no strong-motion recordings have been made (Chapter 6).

Considering relations between PGA and MM intensity, wide differences are found between the relations obtained by different researchers (e.g. Murphy and O' Brien, 1977). This may be attributed largely to (a) the natural variability of intensities which are normally distributed and the lognormal distribution of PGAs, and (b) the difference in interpretation of the intensity scale in difference countries or by different people. This is shown by Figure 4.24(a), the maximum ground acceleration at a given intensity from data from 40 years of earthquakes.

A more robust relationship between PGA and MMI can be obtained by relating the observations from one country only, of PGA to isoseismals using decimalized intensity values. This reduces the scatter from both between-earthquake and within-earthquake sources. This is illustrated by considering the PGA versus MM isoseismal data for New Zealand plotted in Figure 4.24(b), where it is seen that the scatter of PGA values for a given intensity has reduced to about $1\frac{1}{2}$ orders of magnitude, rather than the 2+ orders of magnitude of Figure 4.24(a).

In developing empirical relationships between ground motions and intensities the most common form used has been log-linear expression, such as that found based on a worldwide data set by Murphy and O' Brien (1977), i.e.

$$\log PGA = -2.74 + 0.25 I_{mm}, \qquad (4.30a)$$

where PGA is in units of g, and I_{mm} is MM intensity.

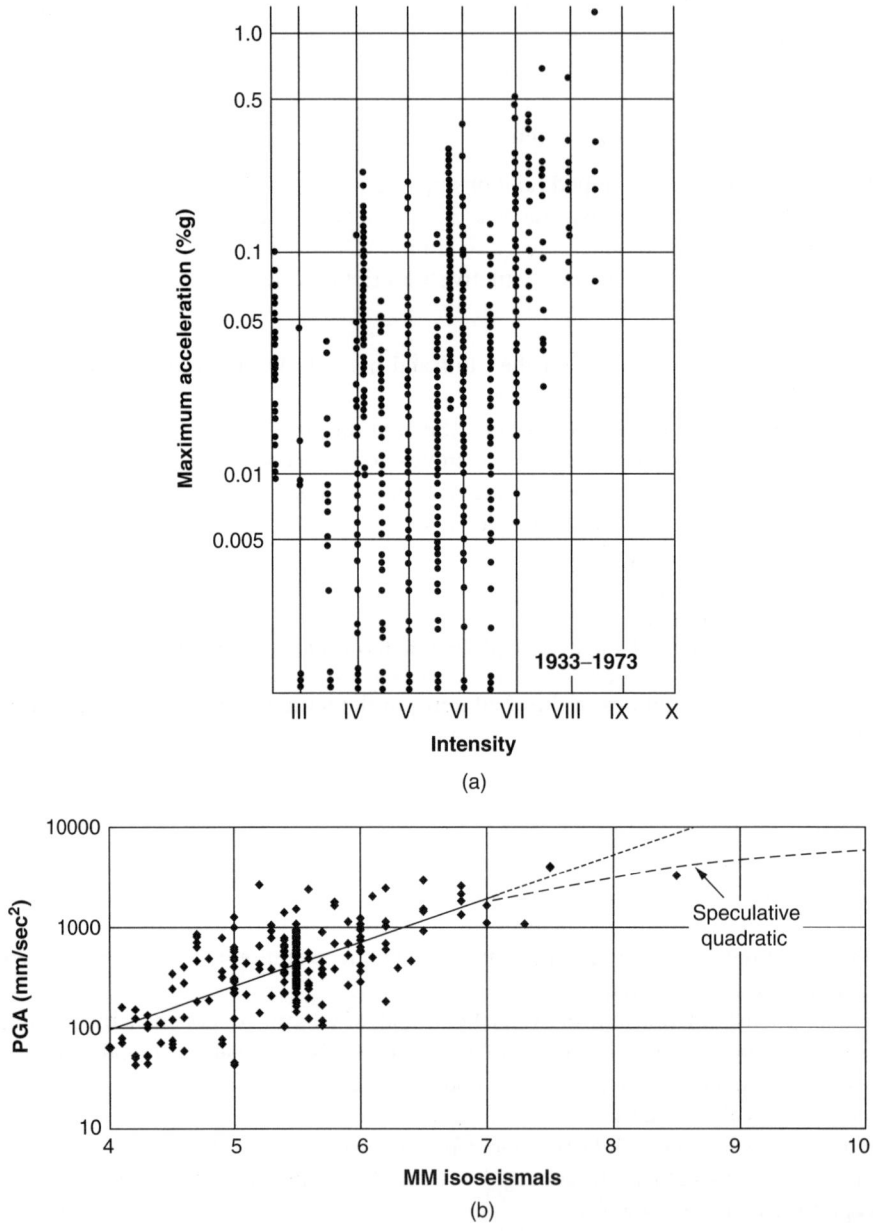

Figure 4.24 (a) Maximum ground acceleration vs. local intensity data from various countries (Ambraseys, 1973). (b) PGA vs. interpolated isoseismal value from New Zealand earthquakes (adapted from Davenport, 2003)

Such a relationship is valid for intensities up to about MM8, but for stronger ground shaking allowance should be made for saturation of PGA. This may be done by using a quadratic expression, such as was used in a study by Menu (1991) whose model is approximately

$$\log \text{PGA} = -0.63 + 0.55 I_{mm} - 0.018 I_{mm}^2. \qquad (4.30b)$$

In a study by Davenport (2003), a linear regression analysis of the New Zealand isoseismal intensity, I_{isos}, data in Figure 4.24(b) found the relationship:

$$\log \text{PGA} = -3.69 + 0.426 I_{isos}, \qquad s = 0.29, \qquad (4.30c)$$

where PGA is in units of g.

The data and regression line of equation (4.30c) plotted in Figure 4.24(b) support the concept of a quadratic expression, as the log-linear regression expression predicts the excessively high PGA of $1.86g$ at intensity MM10. As near-source PGAs saturate at high intensities, a mean PGA of about $0.7g$ is to be expected at MM10, as indicated by the speculative quadratic curve also shown on the figure.

It should be noted that as equation (4.30a) is based on local intensity values and equation (4.30c) is based on isoseismals, I_{mm} is approximately equal to $I_{isos} + 0.5$. Thus (for example) we should compare the PGA of $0.10g$ predicted for MM7 by equation (4.30a) is equivalent to the PGA of $0.32g$. The large difference between these two predictions is consistent with the large differences found in the predictions of models produced by various researchers, as shown, for example, by Murphy and O' Brien (1977). This demonstrates the strong regional variations in relationships between PGA and felt intensity.

Uses of Modified Mercalli intensities in seismic hazard studies

Over many decades felt intensities have been put to a variety of scientific purposes relating to seismic hazard, such as:

- estimating the magnitude, epicentre and depth of earthquakes, in the absence of any, or suitable, instrumental records (Bakun, 2005; Downes and Dowrick, in preparation);
- estimating the strike of the earthquake, i.e. the alignment of fault rupture (see Figures 4.26 and 4.33(b));
- estimating attenuation rates and their variations, regionally and directionally (see Section 4.6.3 and Dowrick and Rhoades, 2005);
- comparing modelled and historical seismic hazard (see Section 4.10);
- spatial distribution of strength of shaking near the source of large shallow earthquakes (see Section 4.6.3 and Dowrick and Rhoades, in preparation);
- determining the upper bound on damage levels (and, by implication, the strength of shaking) in earthquakes (see Section 4.5.3);
- MM intensities describe in words what has happened, or will happen, in earthquakes, something that instrumental measures of ground shaking do not do.

4.6.3 Spatial patterns of ground motions

The spatial pattern of ground motions in the $M_W = 6.7$ Northridge, California, earthquake of 1994 is shown in Figure 4.25, as expressed by Dewey *et al.* (1995) in terms of Modified Mercalli intensity. The variability of the intensities at any given distance from the centre of the source is seen to be considerable. This irregularity is made particularly apparent as the isoseismals have been drawn without much smoothing.

This inherent variability in the pattern of ground motions arises because the ground motions at a given site shaken by an earthquake depend on a number of usually complex factors, principally:

(i) the nature depth and geometry of the source (fault rupture);
(ii) location of the site in relation to the source;
(iii) the attenuation characteristics of the wave travel path;
(iv) the local geology (geometry and ground conditions) at the site;
(v) the heterogeneity of the asperities on the rupture surface.

Some of these effects have been well demonstrated by Aagaard *et al.* (2001).

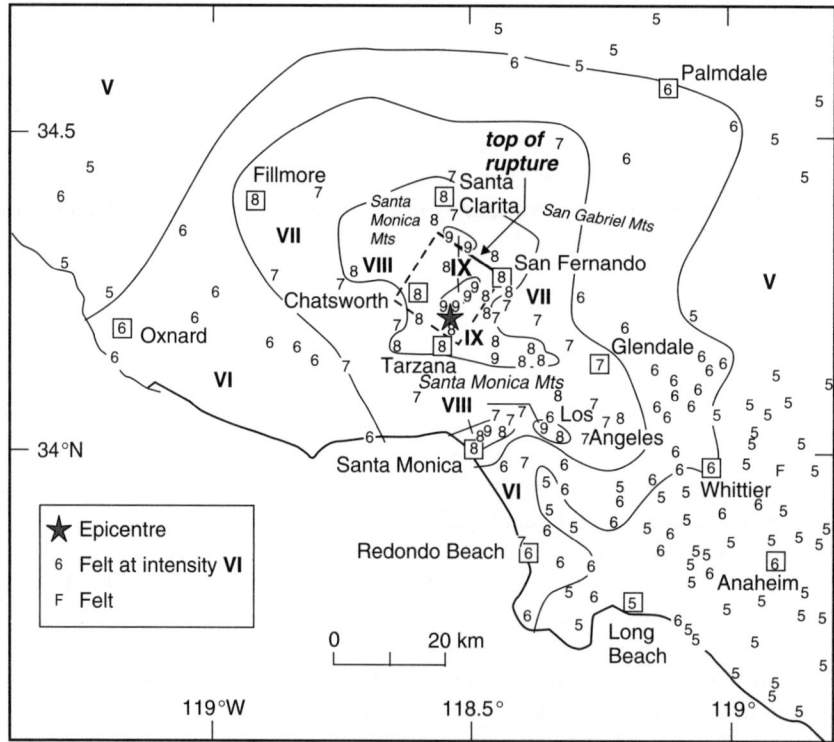

Figure 4.25 Distribution of Modified Mercalli intensities in the epicentral region of the 1994 Northridge earthquake. Roman numerals give average intensities within isoseismals. Arabic numerals represent intensities at specific locations. Squares denote towns labelled in the figure (adapted from Dewey *et al.*, 1995)

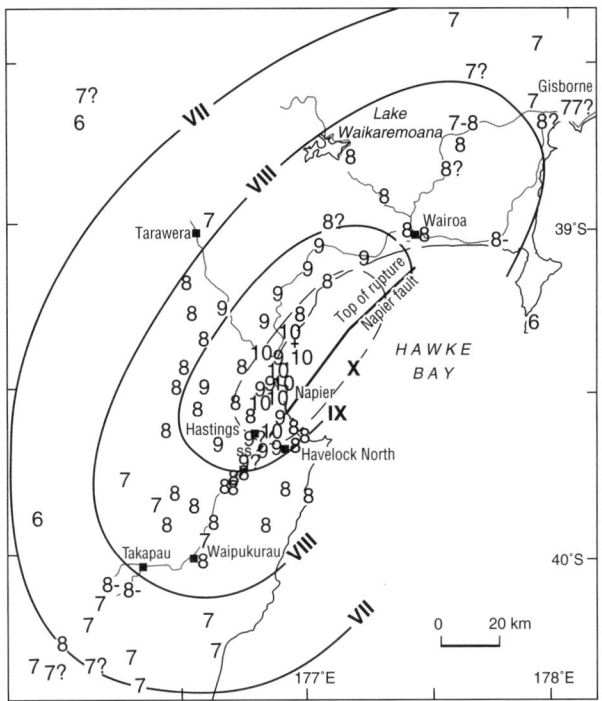

Figure 4.26 Isoseismal map of the $M_w = 7.8$ Hawke's Bay, New Zealand, earthquake of 1931 (modified from Dowrick, 1998)

The spatial patterns of intensities are presented very differently from Figure 4.25 in New Zealand, where the practice is to draw smoothed isoseismals which are mostly fairly elliptical in shape, ignoring possible microzoning effects. This is shown by the isoseismal map of the inner isoseismals of the $M_W = 7.8$ Hawke's Bay earthquake, shown in Figure 4.26 as prepared by Dowrick. The modelling of such isoseismals is discussed later in this section (under 'Elliptical models').

Point-source models

When modelling ground-motion patterns, the simplest modelling assumes that the source is concentrated at a point, and source-to-site distances are measured from the point-source. In the past, this point has mostly been modelled as either the point of initiation of the rupture (the focus or hypocentre) or the point on the surface vertically above the focus (i.e. the epicentre). Thus the distance measures would either be r or d in Figure 4.27. The point-source assumption is satisfactory for small events, or moderate sized distant ones, and implies circular patterns of any given level of shaking.

Line or planar source models

In the near field of larger events the point-source assumption is clearly inadequate, so that it has become normal practice for the distance to be taken as the shortest distance

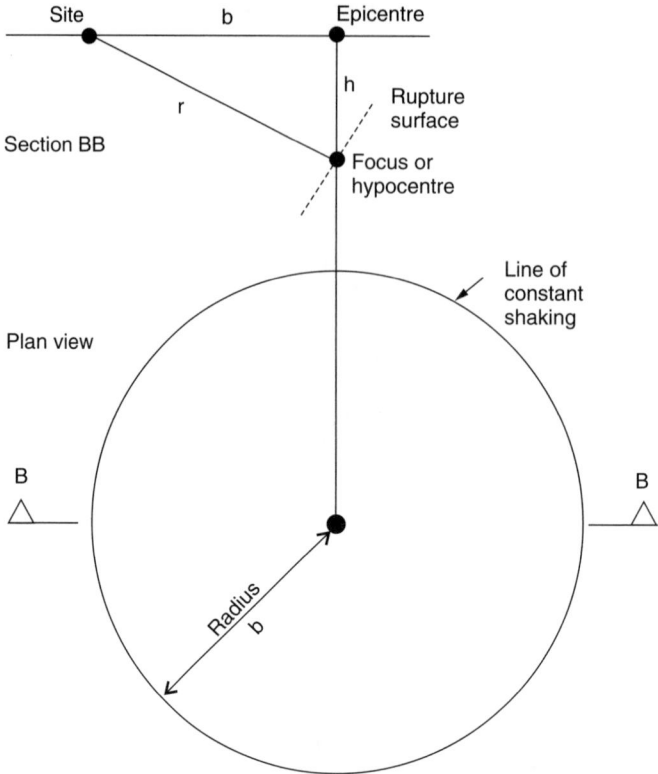

Figure 4.27 Point-source earthquake model which assumes that all of the seismic waves emanate uniformly in all directions from a point, resulting in a circular spatial distribution of ground shaking

between the site and the fault rupture in empirical models of attenuation. For vertical faults (strike-slip) this assumption results in spatial patterns of ground motion shaped like racetracks (Figure 4.28(a)), with the line of constant strength of shaking always surrounding the rupture over its whole length. If the fault dips at an angle $\beta < 90°$, the pattern becomes asymmetrical about the strike of the top of the fault, except in the near-source region, i.e. at horizontal distances $f \leq h_t \tan \beta$ from the vertical projection of the top of the rupture (Figure 4.28(b)). The two spatial patterns of Figure 4.28 are nearest to correct in the special case that occurs when two conditions are fulfilled: (i) the whole rupture surface consists of one uniform asperity, and (ii) the fault ruptures bilaterally from the centre outwards.

In some studies the spatial distribution of ground motions is found by estimating the ground motions at any point as the sum of the vibrations arriving from all parts of the rupture surface. An example of the resulting spatial distribution pattern is shown in Figure 4.29. This two-dimensional source model has been developed by Dowrick and Rhoades (in preparation). The source is represented by a rectangular fault rupture plane of chosen dip, discretized into small rectangles each with its own share of the total seismic moment, and modelling chosen distributions of asperities. In this case the asperities occupy

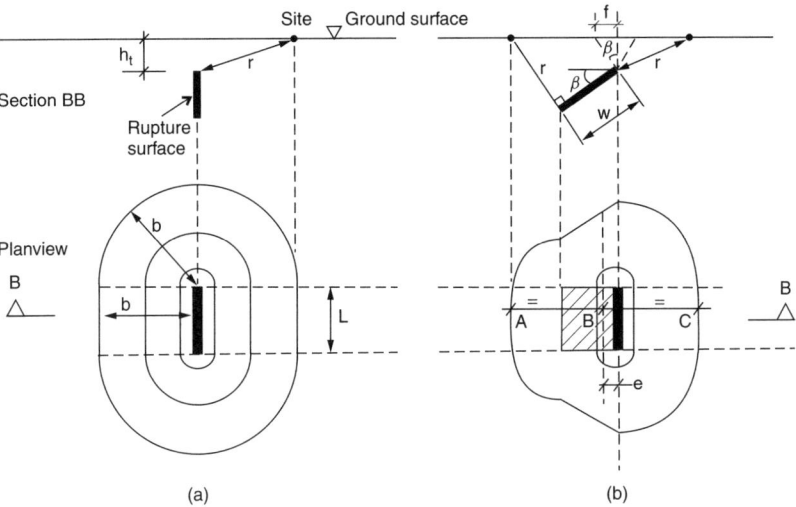

Figure 4.28 Spatial patterns of ground shaking constrained by measuring distance as shortest path between source and site for (a) a vertical surface rupture and (b) a surface rupture dipping at angle β

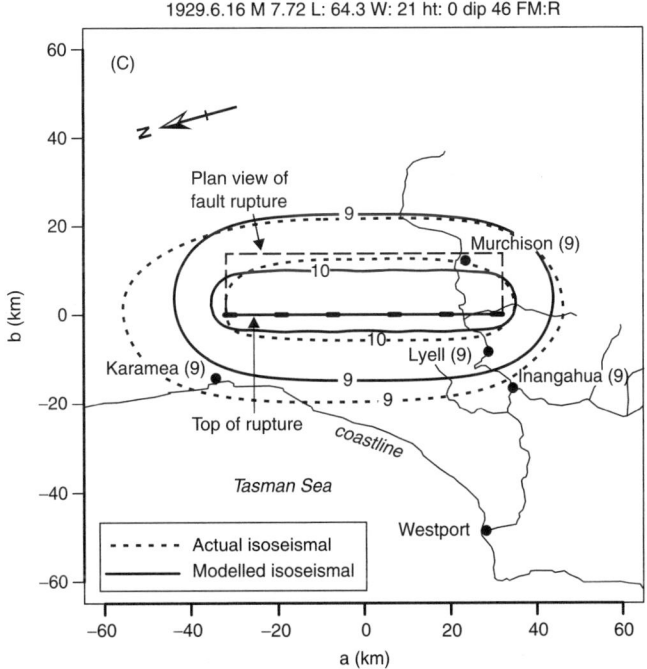

Figure 4.29 Map of the actual inner isoseismals for the $M_W = 7.72$ Buller, New Zealand, earthquake of 1929 compared with those predicted by the distributed (two-dimensional) source model of Dowrick and Rhoades (in preparation)

21% of the total rupture surface and are uniformly distributed. On the figure it is seen that the isoseismals are asymmetrical about the top of the fault, being offset in the direction of the dip of the fault plane.

This model is discussed further in relation to Figure 4.33 later in this section and in Section 4.6.6.

Elliptical models

A third way of modelling spatial patterns of ground motions is to treat the patterns as elliptical. A revealing way of doing this is to model decay in strength of shaking (i.e. the attenuation) in two orthogonal directions a and b with respect to the strike of the fault rupture, as shown in Figure 4.30. Here a and b are the horizontal distances from the centre of the pattern to a line of any given strength of shaking, or isoseismal, measured parallel and normal respectively to the strike of the fault. As real isoseismals are generally not symmetrical along or normal to the strike (e.g. Figure 4.26), a and b are taken as half the overall lengths or widths of the isoseismals.

Resulting from studies of intensities in New Zealand earthquakes, Dowrick and Rhoades (1999, 2005) found that the shapes of the isoseismals, measured by the ratio b/a, become more circular with increasing source distance, increasing depth, and decreasing earthquake magnitude (see Figure 4.31). These trends are to be expected, as in each case the source size becomes smaller in relation to the distance from it, so that the model approaches that of a point source, which creates circular spatial patterns of ground motions (Figure 4.27).

Comparison of models of spatial distribution

First compare three models assembled in Figure 4.32. These comprise (a) the elliptical model for intensities with attenuation different in two orthogonal directions (from Dowrick and Rhoades, (1999, 2005)), (b) the shortest-distance-from-source strong-motion model (from Figure 4.28(b)), and (c) the two-dimensional source model (intensities) (from Figure 4.29). It is interesting that not only are the three models fairly different from each other, but also different from the plot of actual ground shaking intensity. This is not so surprising considering the differences in the assumptions made in each of the models. The elliptical model (a) predicts the mean shapes of the isoseismals for the magnitude, depth and focal mechanism of the chosen event, but unavoidably underestimates the length of the innermost isoseismals of surface (or near-surface) rupturing events of M_W less than about 7.0 because of the functional form of the regression model. Model (b) artificially constrains the shapes of its iso-strong-motion lines with its attenuation assumption that the ground motion is determined by the shortest distance from the source. Model (c) does not have such limitations, but is not a closed-form solution and hence more effort is required for finding the spatial distribution for each event scenario.

Direct comparisons of the elliptical model and the two-dimensional source model are shown in Figures 4.33(a) and (c) for two New Zealand earthquakes, the $M_W = 6.95$ strike-slip Arthur's Pass earthquake of 1929 and the $M_W = 7.72$ reverse faulting Buller earthquake of the same year. As noted above in relation to Figure 4.32, it is seen that the elliptical model and the two-dimensional source model are very similar for the $M_W = 6.95$ event, but the MM9 and MM10 isoseismals of the elliptical model are too short in the along-strike direction for the $M_W = 7.72$ event. As shown in Figures 4.33 (b) and (d),

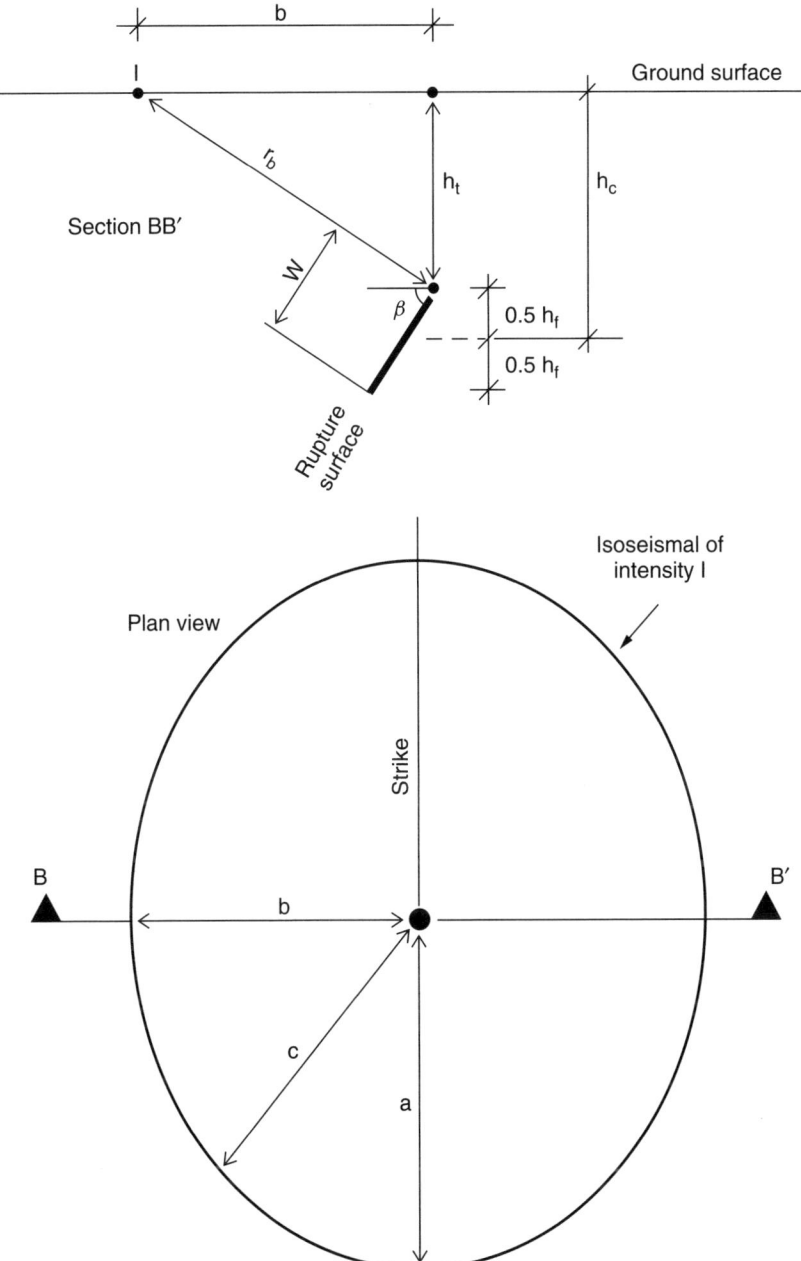

Figure 4.30 Elliptical pattern of isoseismals resulting from deriving separate attenuation models for the along-strike direction (dimension a) and strike-normal direction (dimension b) (after Dowrick and Rhoades, 2005)

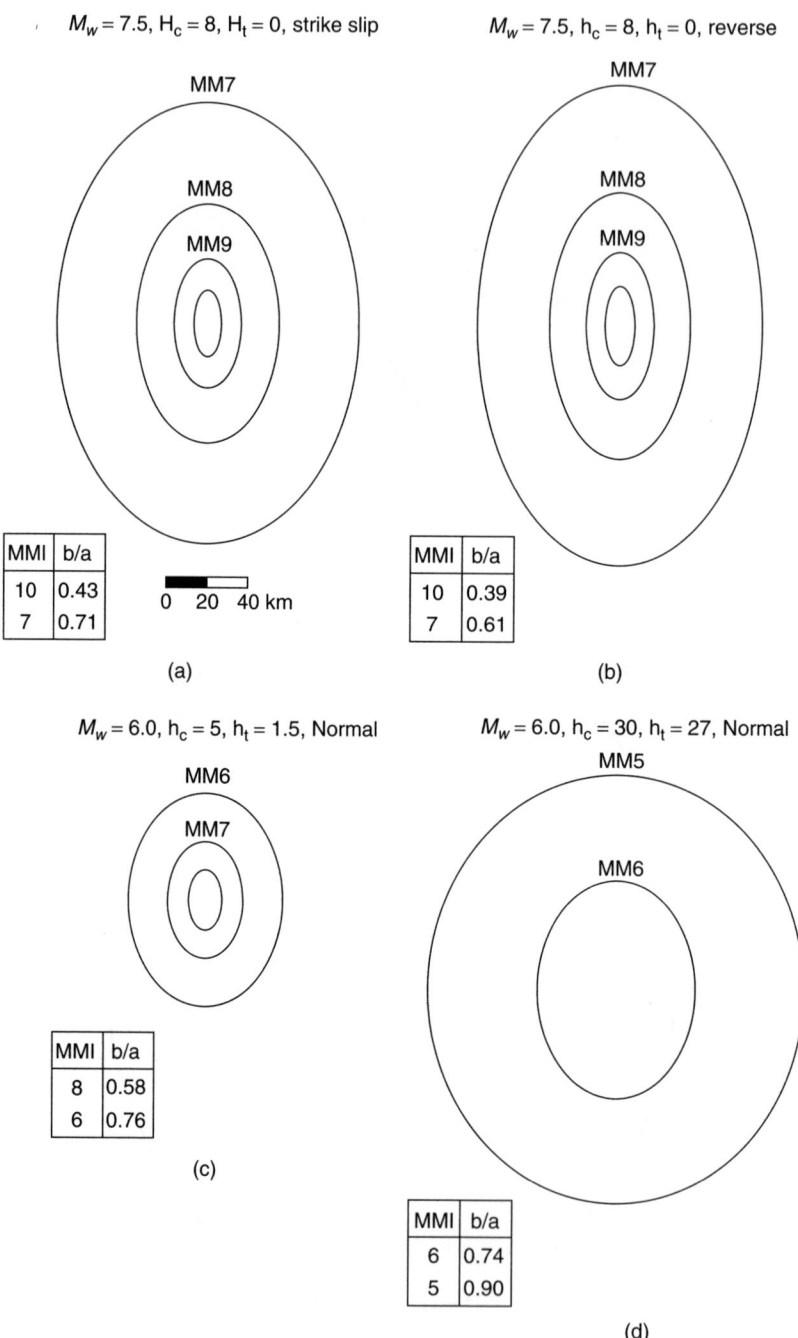

Figure 4.31 Isoseismal dimensions and shapes predicted by the attenuation model of Dowrick and Rhoades (2005), for earthquakes of various source parameters. The ratio b/a is seen to increase with (i) source distance, (ii) depth, and (iii) decreasing magnitude

Nature and Attenuation of Ground Motions

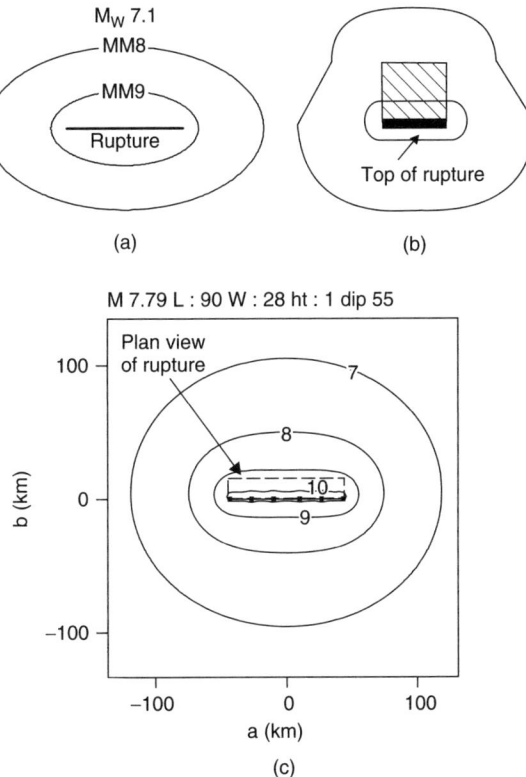

Figure 4.32 Comparison of models of spatial patterns of ground motions: (a) elliptical model, (b) model using shortest distance to planar source, and (c) Dowrick and Rhoades two-dimensional source model

the two-dimensional source model closely fits the actual inner MM9 and MM10 isoseismals of both earthquakes. The fit of both the elliptical model and the two-dimensional source model to the actual isoseismal dimensions is further illustrated below in relation to Figure 4.39.

4.6.4 Attenuation of ground motions, spectral response and intensity

Introduction

Attenuation of ground motions is the term applied to the decrease in strength of shaking as the distance from the earthquake source increases, as illustrated in Figure 4.34. Attenuation is mostly modelled empirically from recorded data of one or other description, Y, of strong ground motion as listed in Section 4.6.2 (e.g. PGA), or from intensity data. Various functional forms have been used for modelling the attenuation of Y in terms of distance, x, the simplest general form being

$$\log Y = b_1 + b_2 M_w - b_3 \log(x + b_4)^n - b_5 x. \qquad (4.31)$$

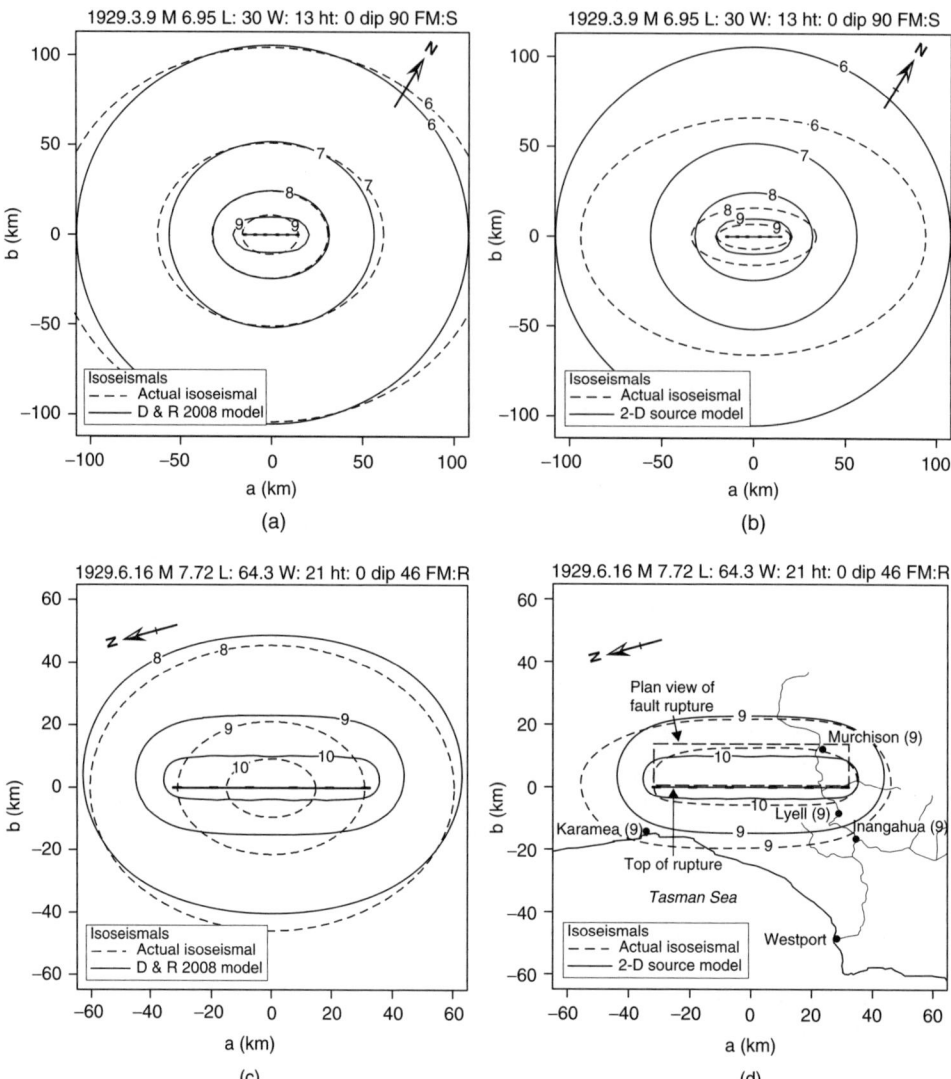

Figure 4.33 Plot of isoseismals and fault ruptures for two surface-rupturing New Zealand earthquakes: (a) $M_W = 6.95$ Arthur's Pass earthquake of 1929, isoseismals from Dowrick and Rhoades (2005) model and from the Dowrick and Rhoades two-dimensional source model; (b) Arthur's Pass earthquake, actual isoseismals vs. those from Dowrick and Rhoades two-dimensional source model; (c) $M_W = 7.72$ Buller earthquake of 1929, isoseismals from Dowrick and Rhoades (2005) model and from the Dowrick and Rhoades two-dimensional source model; and (d) Buller earthquake, actual isoseismals vs. those from Dowrick and Rhoades two-dimensional source model

Nature and Attenuation of Ground Motions

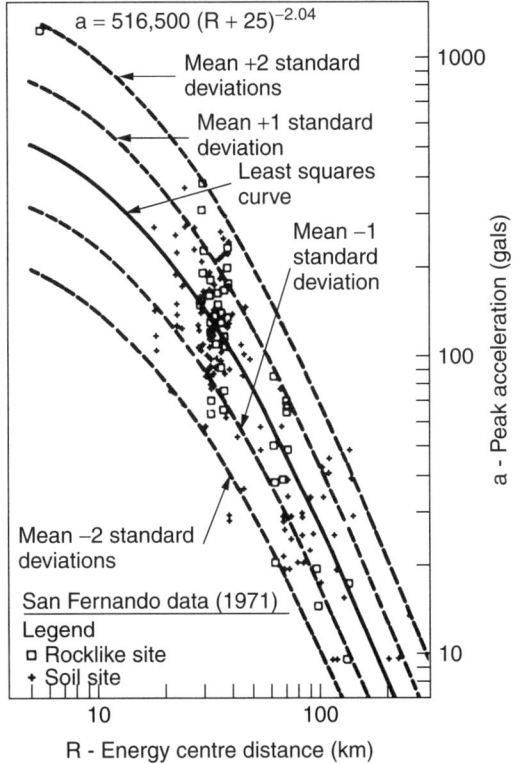

Figure 4.34 Attenuation of peak ground acceleration. Least squares and standard deviation curves for the 1971 San Fernando, California, earthquake (after Donovan, 1973)

The regression coefficients $b_1, \ldots, b_5$ vary depending on the data being fitted. The values of the coefficients are found to be dependent on various phenomena, particularly:

(1) interplate or intraplate region;
(2) tectonic type (crust, slab or interface events);
(3) volcanic or non-volcanic rock region;
(4) local mechanism (reverse, strike-slip or normal);
(5) depth of source;
(6) ground class;
(7) horizontal or vertical motion.

In subduction zone areas, all but the first of the above seven phenomena are found to be influential (Dowrick and Rhoades, 2005). The geometry of this situation is illustrated in Figure 4.35 by the subduction zone in relation to the North Island, New Zealand. Here attenuation modelling needs to take account of several factors, and becomes geometrically complicated for deep events in the subducting Pacific plate, as shown by Dowrick and Rhoades (2005) in terms of MM intensity and by Zhao et al. (1997) in terms of PGA.

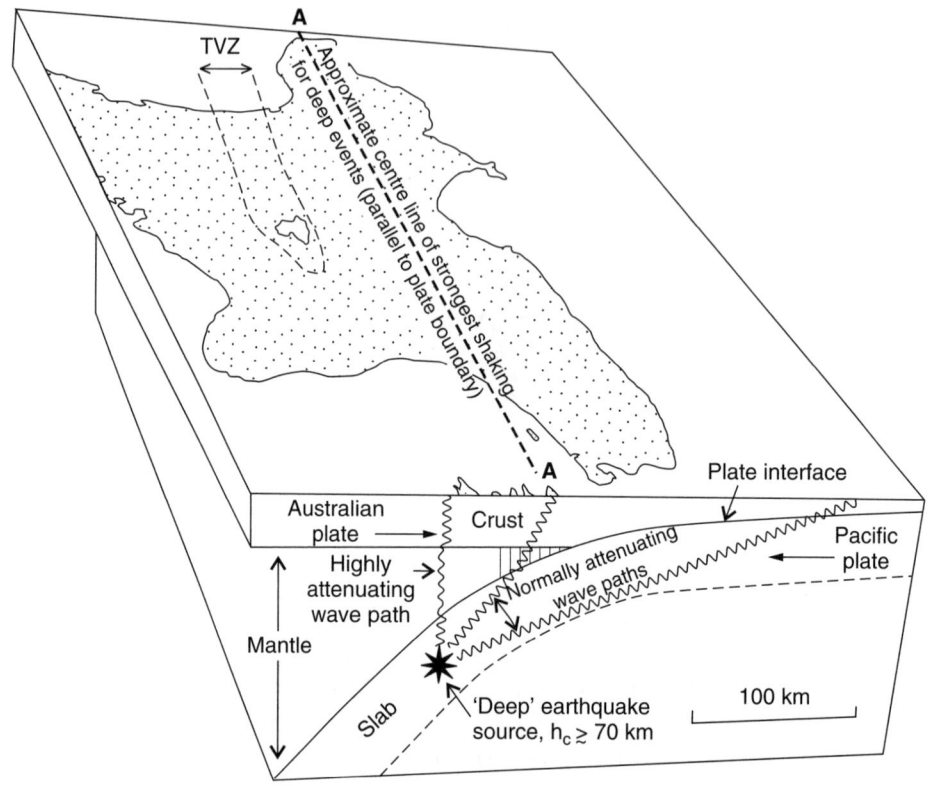

Figure 4.35 Perspective view of the Australian plate and the subducting Pacific plate under the North Island of New Zealand, with the highly attenuating wedge of mantle protecting surface locations vertically above deep earthquakes, and constraining the effective 'strike' of the isoseismals (adapted from Dowrick and Rhoades, 1999)

In developing attenuation relations, the most favoured method for the regression analysis is to use the random effects models as described by Abrahamson and Silva (1997):

$$Y_{ij} = f(M_i, r_{ij}) + \varepsilon_{ij} + \eta_i, \qquad (4.32)$$

where Y_{ij} is the ground motion for the jth recordings from the ith earthquake, M_i is magnitude of the ith earthquake, and r_{ij} is the distance for the jth recordings from the ith earthquake. There are two stochastic terms in the model: both ε_{ij} and η_i are assumed to be normally distributed with mean zero. The random effects models uses the maximum likelihood method to partition the residual for each recording into the ε_{ij} and η_i terms. There are two parts to the standard error for the model: an inter-event term, τ, which is the standard error η_i; and the intra-event term ε_{ij} and its standard error σ. The total standard error of the model is

$$s = (r^2 + \tau^2)^{1/2}. \qquad (4.33)$$

Attenuation of strong motion in interplate regions

In modelling the attenuation of response spectrum ordinates (S_a), we first consider the model of Abrahamson and Silva (1997). They used 655 recordings from 58 shallow crustal earthquakes of magnitude 4.9–7.4 from active tectonic regions (i.e. interplate regions) of the world, excluding subduction zone interface events. Of their events, 49 were from the western USA and nine were from Canada, Iran, Italy, Taiwan and the USSR. The functional form that they used is

$$\ln Sa(g) = f_1(M, r_{\text{rup}}) + F f_3(M) + HW f_4(M, r_{\text{rup}}) + S f_5(\overline{pga}_{\text{rock}}) \qquad (4.34)$$

where $Sa(g)$ is the median spectral acceleration in units of g, M is moment magnitude, r_{rup} is the closest distance to the rupture plane in kilometres (resulting in the spatial pattern of Figure 4.28(b)), modified by the hanging wall effect, F is the fault type (1 for reverse, 0.5 for reverse/oblique, and 0 otherwise), HW is the dummy variable for hanging wall sites (1 for sites over the hanging wall, 0 otherwise), and S is a dummy variable for the site class (0 for rock or shallow soil, 1 for deep soil). For the horizontal component, the geometric mean of the two horizontal components is used. The hanging wall is discussed in Section 4.6.6.

The function $f_1(M, r_{\text{rup}})$ is the basic functional form of the attenuation for strike-slip events recorded at rock sites. For $f_1(M, r_{\text{rup}})$, the following is used: for $M \leq c_1$,

$$f_1(M, r_{\text{rup}}) = a_1 + a_2(M - c_1) + a_{12}(8.5 - M)^n + [a_3 + a_{13}(M - c_1)] \ln R; \qquad (4.35)$$

for $M > c_1$,

$$f_1(M, r_{\text{rup}}) = a_1 + a_4(M - c_1) + a_{12}(8.5 - M)^n + [a_3 + a_{13}(M - c_1)] \ln R, \qquad (4.36)$$

where

$$R = \sqrt{r_{\text{rup}}^2 + c_4^2}.$$

The above model results in smoothed (averaged) response spectra, as illustrated in Figure 4.36, which shows spectra estimated for rock and deep soil sites at source distances of 1, 10, 30 and 100 km for a magnitude 7.0 earthquake. It is of interest that whereas soil motions are mostly stronger then rock motions, the reverse is predicted for short periods ($T < 0.2$ s) at short distances (<25 km) from the source rupture. This effect is attributed to non-linear material behaviour occurring in soils when the amplitudes of shaking are strong, i.e. in the near field of large earthquakes.

The Abrahamson and Silva model involves a number of functions and many coefficients which vary with the period of vibration, T. Their model is therefore presented in tabular form, for which reference should be made to the source work (Abrahamson and Silva, 1997). This includes tabulation for the quantification of the standard errors, which are smaller for larger events, and are modelled to vary linearly between values which are constant (at 0.70) for $M \leq 5.0$ and (at 0.43) for $M \geq 7.0$. The variability of the data for any given event, i.e. the intra-event uncertainty, is illustrated by the large scatter of the PGAs recorded in the 1971 San Fernando earthquake (Figure 4.34). Even though some of this scatter is explained in terms of the effects of ground classes, the inter-event uncertainty

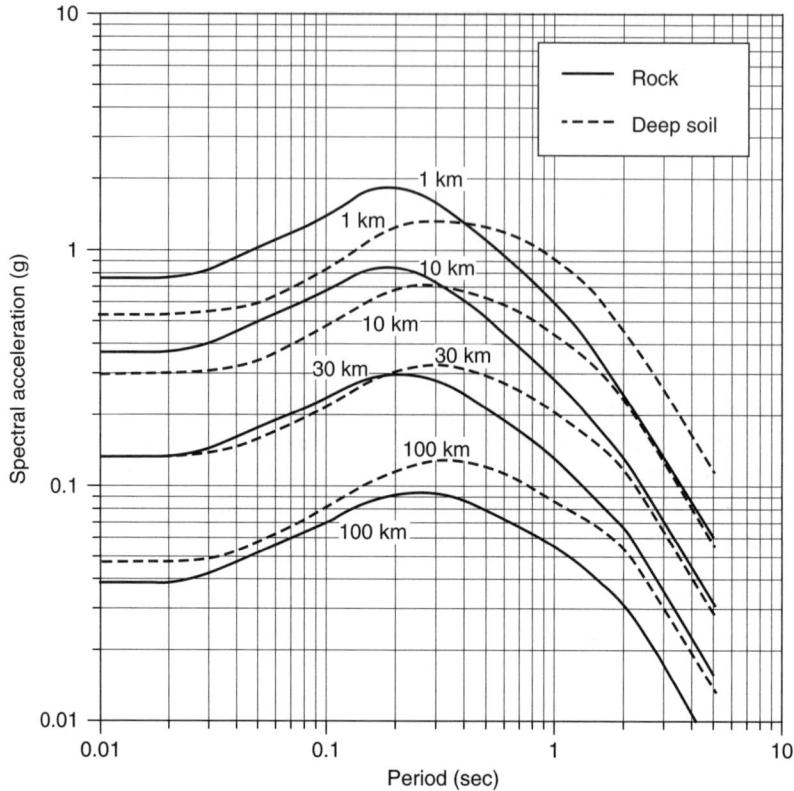

Figure 4.36 Median response spectra for $M_W = 7.0$ earthquakes showing how the relationship between the spectral shapes and amplitudes of rock and deep soil sites varies with source-site distance (after Abrahamson and Silva, 1997)

is substantial. For example, for PGAs the inter-event variance τ^2 is about 35% of the total variance $\tau^2 + \sigma^2$, for $M \leq 5$ and about 15% for $M \geq 7$, (Abrahamson, personal communication). The mean of these two values (24%) is consistent with the modelling of attenuation of intensities by Dowrick and Rhoades (2005) where it was found that τ^2 is about 23% of the total variance (no variation with magnitude being considered).

The next generation of ground-motion attenuation models project

The objective of the Next Generation of Ground-Motion Attenuation Models (NGA) project is to develop new ground-motion prediction relations applicable to the western USA and similar tectonic regimes, through a comprehensive research programme. Five sets of models were developed by teams working independently but interacting with each other through the development process. These models and supporting projects are reported in the special issue of *Earthquake Spectra* (vol. 24, no. 1, 2008) dedicated to this project. For the convenience of comparison with the above 1997 model of Abrahamson and Silva, here we consider the NGA model of Abrahamson and Silva (2008). Drawing on the large NGA ground-motion database, they used 2754 records from 135 earthquakes with magnitude range 4.27–7.68.

The model is applicable to magnitudes 5–8.5, distances 0–200 km, and spectral period 0–10 s. In place of generic site categories (soil and rock), the site is parameterized by average shear-wave velocity in the top 30 m (V_{S30}) and the depth to engineering rock (depth to $V_S = 1000$ m/s). In addition to magnitude and style of faulting, the source term is also dependent on the depth to top of rupture (for the same magnitude and distance, buried ruptures lead to larger short-period ground motions than surface ruptures). The hanging wall effect is included with an improved model that varies smoothly as a function of the source properties (magnitude M, dip, depth), and site location. The standard deviation is magnitude-dependent, with smaller magnitudes leading to larger standard deviations. The short-period standard deviation model for soil sites is also distance-dependent due to non-linear site response, with smaller standard deviations at short distances.

A sample of spectral plots from the (more complex) NGA model of Abrahamson and Silva may be compared with those from their 1997 model (Figure 4.37). Here it is seen that the peak responses (at T around 0.3–0.4 s) for $M = 8$ events predicted by the NGA model are c. 40% less than those of the 1997 model for both rock and soil sites, while for $M = 5$ events the peak spectral responses predicted by the NGA model are 2.0–2.7 times greater than those of the 1997 model.

Comparisons of the ground-motion relations found by the five teams in the NGA project (Abrahamson et al., 2008) show challengingly wide differences in their predictions. For example, the median spectral values for strike-slip earthquakes are within a factor of 1.5 for magnitudes between 6.0 and 7.0 for distances less than 100 km, and differences increase to a factor of 2 for earthquakes with magnitudes 5 and 8, for buried ruptures, and for distances more that 100 km. For soil sites, the differences in modelling of soil/sediment depth effects increase the range in median long-period spectral values for magnitude 7 strike-slip earthquakes to a factor of 3.

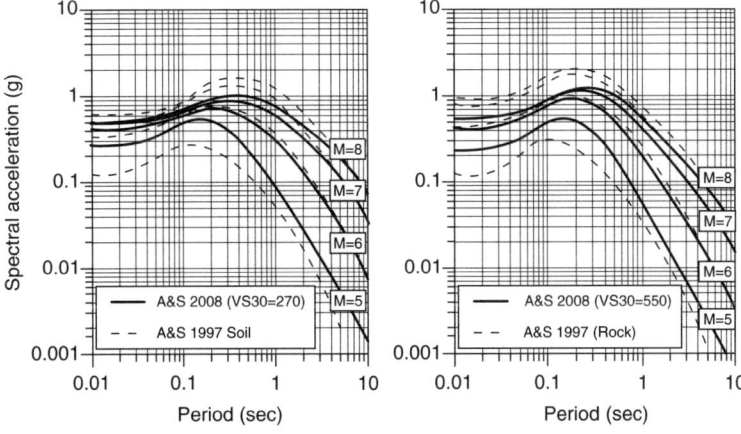

Figure 4.37 Comparison of the median spectral acceleration from the Abraham and Silva (1997, 2008) models for vertical strike-slip earthquakes for $R_{JB} = 1$ km. (For this case, R_{JB} is the closest distance to the coseismic rupture plane). Reproduced by permission of the Earthquake Engineering Reseach Institute

Attenuation of MM intensity in interplate regions

The attenuation of MM intensity in interplate regions is represented here by the modelling of Dowrick and Rhoades (2005) which produces elliptical isoseismals as illustrated by Figure 4.31. The attenuation plots of such models are shown in Figure 4.38, which shows that at a given distance from the source the shaking is strongest for reverse faulting and weakest for normal faulting. This effect is also found by strong-motion studies such as those discussed above.

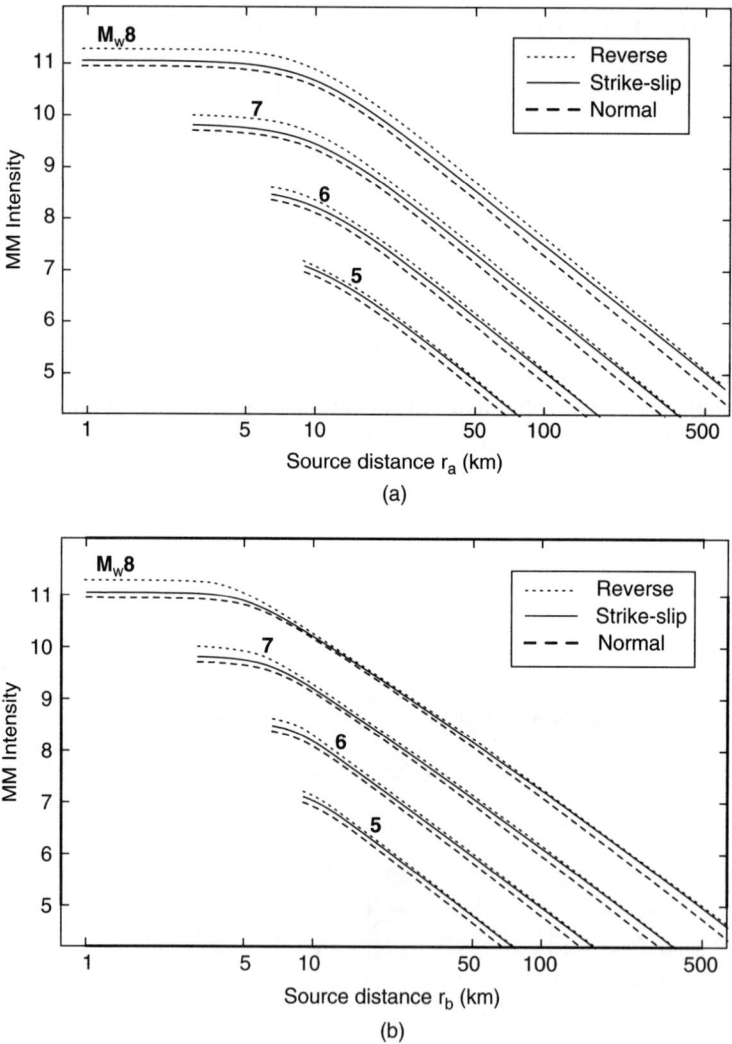

Figure 4.38 Effect of focal mechanism. Intensity plotted against source distance (a) along-strike and (b) normal to strike, as predicted for strike-slip, reverse- and normal-faulting earthquakes of M_W from 5 to 8, all of centroid depth $h_c = 10$ km (after Dowrick and Rhoades, 2005)

Nature and Attenuation of Ground Motions

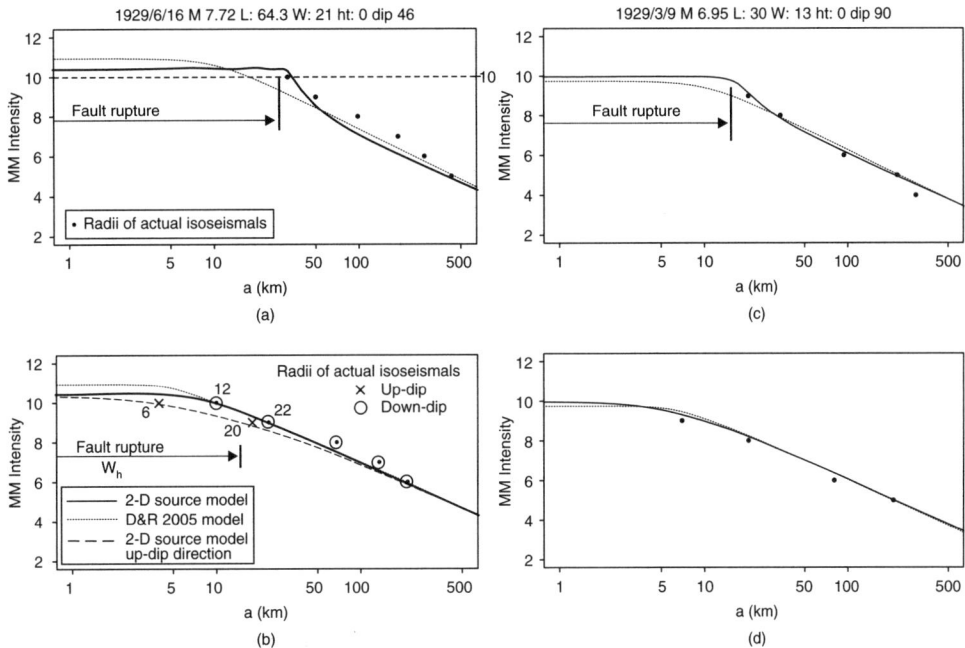

Figure 4.39 Attenuation plots of the distributed (two-dimensional) source model and the elliptical Dowrick and Rhoades (2005) models against distance a and b (along-strike and strike-normal) data for the Buller earthquake ((a) and (b)), and the Arthur's Pass earthquake ((c) and (d)). W_h is the horizontal component of the rupture width (after Dowrick and Rhoades, in preparation)

The effect of focal mechanism, through differences in fault dip, is illustrated by the attenuation plots for two historical New Zealand earthquakes in Figure 4.39. These are the same two events for which the isoseismal patterns were discussed above in relation to Figure 4.33. In Figure 4.39(b) we see how the Dowrick and Rhoades distributed (two-dimensional) source model fits well the actual attenuation in the up-dip and down-dip directions of the Buller earthquake. In Figures 4.39(a) and (c) the better fit of the MM10 isoseismal achieved by the two-dimensional source model than by the elliptical model in the along-strike direction (a) can be seen.

Attenuation in intraplate regions

The attenuation of ground motions in intraplate regions appears to be very different from that of interplate regions, and differs from one intraplate area to another. However, because of the natural scarcity of seismic activity in intraplate regions, empirical verification of their attenuation models is greatly hampered by shortage of strong-motion data, particularly from earthquakes with magnitudes of engineering interest.

The problem is illustrated by even the current best of such models, such as that of Atkinson and Boore (1995) for eastern North America. Their model, for SA, PGA and PGV, over distances of 10–500 km and magnitudes 4.0–7.25, is derived from an empirically based stochastic ground-motion model. The verification of the model in terms of their data was demonstrated for M_W between 4 and 5, where they had ample data. However, there were insufficient data to adequately judge the relations at larger magnitudes, although they were consistent with data from the Saguenay ($M_W = 5.8$) and Nahanni ($M_W = 6.8$) earthquakes. The above model is not expressed in terms of a single mathematical function, but in an extensive tabular form, for which reference should be made to the source paper of Atkinson and Boore (1995). A simpler (non-tabular) alternative approach is that developed by Somerville et al. (1998) for Australia.

The attenuation rates in New Zealand (interplate) and south China (intraplate) have been compared in terms of MM intensity by Dowrick and Rhoades (2005), where it was found that that the attenuation rate in south China was lower than that of New Zealand, consistent with the above observations.

Attenuation in volcanic regions

In some actively volcanic regions the crustal rock is relatively soft and hence has higher attenuation rates than either the interplate or intraplate regions discussed above. Such a region is the Taupo Volcanic Zone of the North Island, New Zealand (Figure 4.35). Here, in a zone about 70 km wide, the brittle crust is only about 8 km thick and is evidently made of more highly attenuating (softer?) rock than the surrounding crust. This zone has been shown to be highly attenuating in independent studies of PGAs by Cousins et al. (1999) and of MM intensity by Dowrick and Rhoades (1999, 2005). The difference in the attenuation rates for the volcanic and non-volcanic rock is illustrated in Figure 4.40, where it is seen that over a distance of about 80 km the intensity attenuates by an extra unit in the volcanic zone compared with the attenuation in the non-volcanic Main Seismic region. A typical example of the effect of these different attenuation rates is seen in the shapes of isoseismals of the $M_W = 6.3$ Bay of Plenty earthquake of 1956 from Dowrick (2007a) shown here in Figure 4.41. This issue has been more fully discussed by Dowrick (2007b).

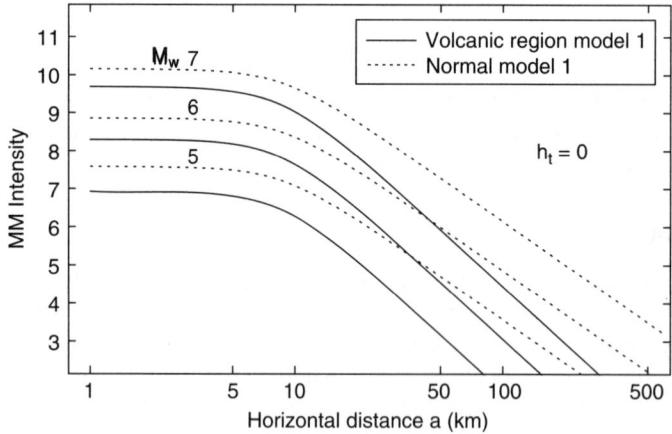

Figure 4.40 Comparison of New Zealand attenuation models showing the much higher attenuation in the volcanic region (TVZ in Figure 4.35) than for the rest of the country (after Dowrick and Rhoades, 2005)

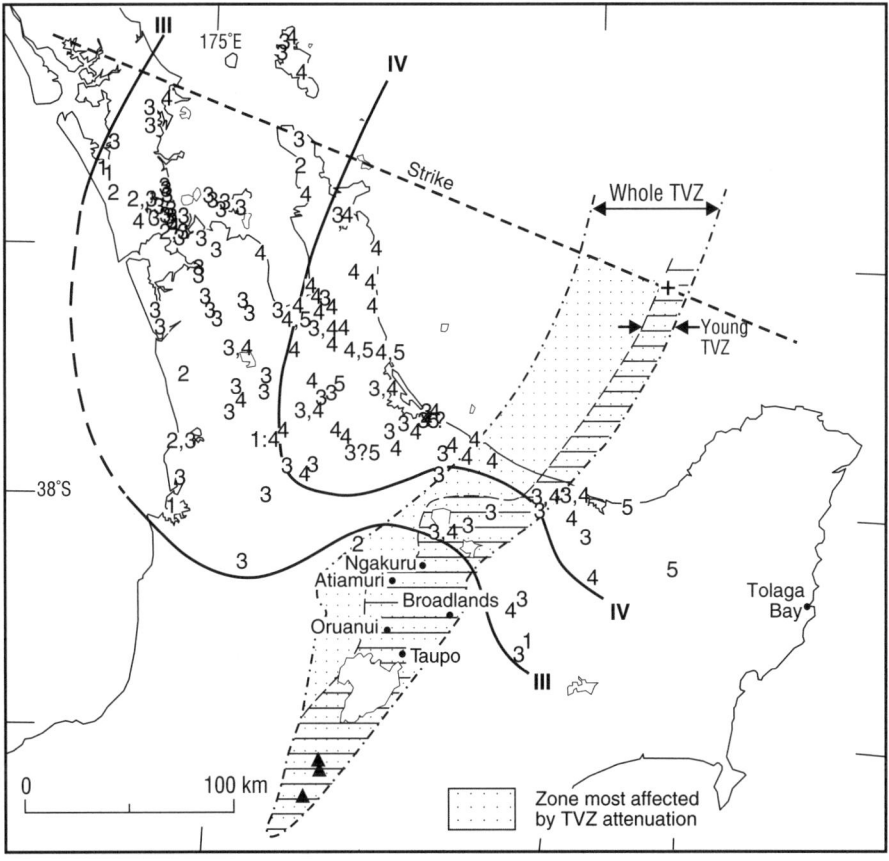

Figure 4.41 Isoseismal map of the $M_W = 6.3$ Bay of Plenty, New Zealand, earthquake of 1956, showing the effect of the high attnuation within the Taupo Volcanic Zone (after Dowrick, 2007a)

Allowance for the high attenuation characteristics of the New Zealand volcanic zone obviously reduces the estimated hazard within the zone compared with simpler estimates which ignore this phenomenon. The high attenuation is also a welcome buffer for sites north-west of the volcanic zone when earthquakes occur in the much more seismic region to the south-east.

Attenuation in other regions

In the previous three sections we have discussed the attenuation in three very different types of region: interplate, intraplate and volcanic. It is likely that all parts of the world have attenuation rates not very different from one or other of those regions. But care must be taken to use expressions and ground classes that are consistent with what a given region's attenuation model has been developed from. For Europe the model of Ambrascys et al. (1996) differs from western US models, but some of the difference may be due to different definitions of ground class used in the different regions and by different authors.

4.6.5 Attenuation of displacement

With the growing interest in displacement-based design (Section 10.1.1) there is a need for predictive models of the attenuation of displacement spectra. One of the first such models is that of Bommer *et al.* (1998) which is based on ground-motion data from the European area.

4.6.6 Other conditions that influence ground motions

In addition to the factors discussed above (Sections 4.6.1–4.6.4), the following further conditions can strongly influence ground motions.

Near-source directivity effects

The following three paragraphs come from Somerville (2000a):

> The propagation of fault rupture toward a site at a velocity that is almost as large as the shear-wave velocity causes most of the seismic energy from the rupture to arrive coherently in a single large long-period pulse of motion which occurs at the beginning of the record. This pulse of motion represents the cumulative effect of most of the seismic radiation from the fault. The radiation pattern of the shear dislocation on the fault causes this large pulse of motion to be oriented in the direction perpendicular to the fault, causing the strike-normal peak velocity to be larger than the strike-parallel peak velocity. The enormous destructive potential of near-fault ground motions was manifested in the 1994 Northridge and 1995 Kobe earthquakes. In each of these earthquakes, peak ground velocities as high as 175 cm/sec were recorded. The period of the near-fault pulses recorded in both of these earthquakes lie in the range of 1 to 2 seconds, comparable to the natural periods of structures such as bridges and mid-rise buildings, many of which were severely damaged.
>
> Forward rupture directivity effects occur when two conditions are met: the rupture front propagates toward the site, and the direction of slip on the fault is aligned with the site. The conditions for generating forward rupture directivity effects are readily met in strike-slip faulting, where the rupture propagates horizontally along strike either unilaterally or bilaterally, and the fault slip direction is oriented horizontally in the direction along the strike of the fault. However, not all near-fault locations experience forward rupture directivity effects in a given event. Backward directivity effects, which occur when the rupture propagates away from the site, give rise to the opposite effect: long duration motions having low amplitudes at long periods.
>
> Although the response spectrum provides the basis for the specification of design ground motions in all current design guidelines and code provisions, there is a recognition that the response spectrum is not capable of adequately describing the seismic demands presented by brief impulsive near-fault ground motions. This indicates the need to use time histories to represent near-fault ground motions. Since the strike of the controlling fault is usually known, the differences between ground motions in the directions normal to and parallel to the fault strike can be readily taken into account.

As an example of near-fault directivity, when modelling a reverse fault with a depth to top of rupture of 8 km and a dip of 23°, Aagaard *et al.* (2001) found that the strongest displacements and velocities were about 5 km up-dip of the surface projection of the fault (Figure 4.42), i.e. on the foot wall not the hanging wall. Clearly, more work is needed to

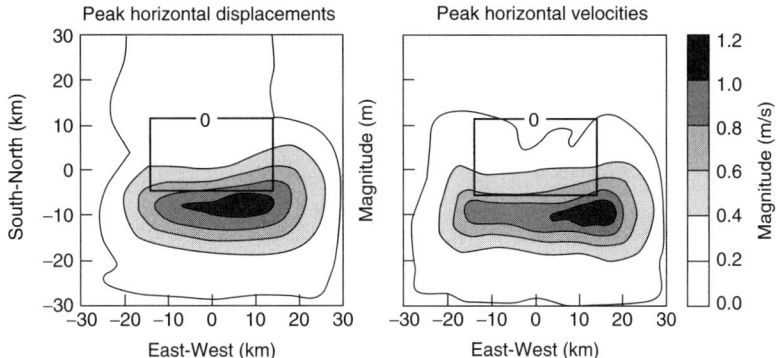

Figure 4.42 Models showing spatial distribution of longer-period motion resulting from near-fault directivity. Maximum magnitudes of the horizontal displacement and velocity vectors at each point on the ground surface. The solid line indicates the vertical projection of the fault plane on to the ground surface, and the hollow circle identifies the epicentre. The south-south-east slip direction (rake angle of 105°) creates the asymmetry in the east–west direction. Fault dip = 23°, top of fault 8 km deep (from Aagaard et al., 2001). Reproduced by permission of the Earthquake Engineering Research Institute

provide a robust generalization for the modelling of this problem in a reasonably simple way.

A method for allowing for near-source directivity effects using standard strong-motion attenuation modelling techniques has been given by Somerville et al. (1997). This model shows enhanced ground motions for periods of vibration greater than 0.6 seconds at designated locations in relation to strike-slip and dip-slip fault ruptures. However, near-source directivity effects are complex and not well quantified, and thus are not incorporated into most attenuation models.

Hanging wall effects

As shown in Figure 4.43, sites on the hanging wall of a non-vertical fault have closer proximity to the fault as a whole than do sites at the same closest distance to the foot wall. Depending on the geometry of the rupture surface, this sometimes causes short-period motions on the hanging wall that are 30% or more larger than on the foot wall at the same closest distance from the rupture (Abrahamson and Somerville, 1996). They propose that the effect is greatest at a factor of 1.45 in the distance range 8–18 km and the period range 0–0.6 seconds, and decreases to unity at 5 seconds. This simplified model of hanging wall effects is incorporated into the ground-motion model of Abrahamson and Silva (1997), as discussed in Section 4.6.4. However, as there are few strong-motion data available in the relevant distance range, the hanging wall effect has so far not been included in most attenuation models.

Crustal waveguide effects

Explaining anomalously strong shaking in the San Francisco Bay area 80–90 km from the source of the 1989 Loma Prieta earthquake, Somerville (2000a) writes that:

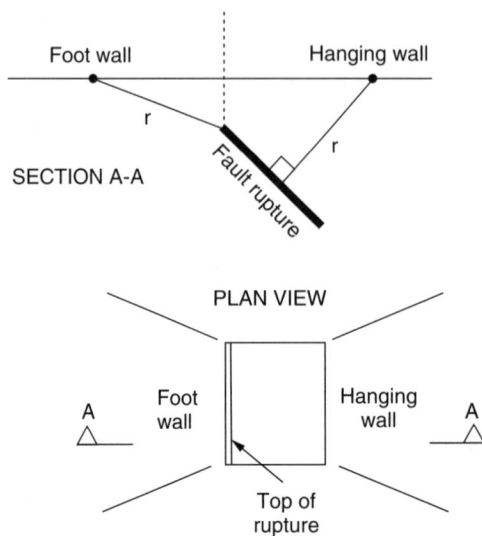

Figure 4.43 Illustration of the hanging wall and foot wall of a non-vertical fault (from Abrahamson and Somerville, 1996). Reproduced by permission of the Seismological Society of America

as distance from the source increases, the direct wave becomes weaker, and the reflections of downgoing waves from interfaces below the source reach the critical angle and undergo total internal reflection. The strong contrast in elastic moduli at these interfaces, especially the Moho, causes these critical reflections to have large amplitudes. The arrival of these critical reflections, beginning at a distance of about 50 km, causes a reduction in the rate of attenuation of ground motion out to distances of about 100 km or more (Burger et al., 1987). While the elevated ground motion amplitudes in this distance range are usually not large enough by themselves to cause damage, they may produce damage if combined with the amplifying effects of soft soils.

The above phenomenon has not been well documented in other earthquakes (e.g. it has not happened in New Zealand) but appears to be a real effect, which helps to explain the apparently lower attenuation rate observed in intraplate as compared to interplate earthquakes. Whereas earlier seismological models used simple half-space approximations for the attenuation of ground motion ($1/R$ for body waves at close distances and $1/R^{1/2}$ for surface waves at larger distances, where R is distance), all of the current models now take account of the effect of the crustal wave guide; see Lam et al. (2000) (refer to Section 4.7.5), Atkinson and Boore (1995) and Somerville et al. (1994).

Basin edge effects

In both the 1994 Northridge and 1995 Hyogo-ken Nanbu (Kobe) earthquakes larger than average ground motions were generated by the geological structure of fault-controlled basin edges (Graves et al., 1998; Pitarka et al., 1998). In both cases concentrations of damage in relatively small areas as a result of amplifications caused by the constructive interference of direct waves with the basin edge generated diffraction at the basin edge.

Clearly, hazard modelling of other regions should seek to identify similar geological subsurface geometry.

Effects of sedimentary basins on ground motions

Many urban areas around the world are situated on sediment-filled basins, with sediment thickness ranging from tens of metres up to 10 or more kilometres. Depending on the angle of incidence, seismic waves entering the basin can become trapped in it, causing resonances resulting in very strong motions that are very damaging. This effect explains much of the damage to buildings and the collapse of the elevated I-10 freeway structure in the Los Angeles basin in the 1994 Northridge earthquake (Graves *et al.*, 1998).

The most dramatic example of this effect occurred in the 1985 Mexico earthquake. In Mexico City, 400 km from the source, while the earthquake was generally not felt at sites away from the basin, in the Lake Zone of the basin resonances were extremely large, causing catastrophic damage to longer-period structures that were resonant with the predominant period of the resonating ground motion (Sections 8.3.6 and 5.2.2).

4.7 Design Earthquakes

4.7.1 *Introduction*

As defined by the EERI Committee on Seismic Risk (1984), a *design earthquake* is *a specification of the seismic ground motion at a site, used for the earthquake resistant design of a structure*. In general, this definition implies specifying the probability of occurrence of the ground motions. The ground motions may be specified in a number of ways, i.e. by peak accelerations, velocities, and displacements, by accelerograms, and by response spectra. In some cases, the differential surface displacement of the ground due to fault displacement may be important, such as the design uplift or downthrow of the area, or in occasional instances where an active fault crosses the construction site, the design relative displacement across the fault will need to be specified.

As set out in the design flowchart (Figure 8.1), specifying design earthquakes requires information on seismic activity and on the site. It is then necessary to establish the acceptable risk so that the appropriate rarity of event may be chosen. Again, quoting the EERI Committee of Seismic Risk, *acceptable risk* is *a probability of social or economic consequences due to earthquakes that is low enough to be judged by appropriate authorities to represent a realistic basis for determining design requirements for engineered structures*.

In general, establishing design earthquakes involves both deterministic and probabilistic considerations of various aspects contributing to the hazard assessment. The ratio between the two types of argument varies widely, but obviously there should always be some probability content, implicit or explicit, in a rational assessment of seismic hazard. An introduction to methods of determining the probabilities of ground motion is therefore given below.

Several widely differing ways exist for specifying the design earthquake, the principal ones being:

(1) equivalent-static loadings in codes;
(2) response spectra

- from the design event (various methods),
- uniform risk spectra;
(3) accelerograms
- from records of real earthquakes,
- from theoretical simulation.

Equivalent-static loadings are discussed in Section 5.4.7 and are of no particular interest here, so only response spectra and accelerograms will be considered below. In establishing design earthquakes, two major factors which need early consideration are:

(1) the nature of the site;
(2) the type(s) of seismic response analysis to be carried out.

As discussed in Chapter 3, the site investigation should have determined the nature of the soil and the topography at the site. Regarding soil conditions, the main issue is the location of bedrock; does it occur at the surface, or is it overlain by sedimentary soil? As shown on the flowchart in Figure 4.44, sites having surface bedrock will not require a site response analysis, while sites having subsurface bedrock may or may not. The need to carry out a site response analysis will increase with increasing thickness and softness of the overburden, and with the size and sensitivity of the structures under consideration.

The types of response analysis to be used for both the soil and the structure dictate whether design earthquakes are specified as accelerograms or as response spectra, and whether they are to be located at or below the ground surface (Figure 4.44). While response spectra are satisfactory for the majority of projects, more sophisticated analyses requiring the use of accelerograms, such as studies of non-linear material behaviour, are sometimes required, depending on the nature of the site and the size and sensitivity of the structure.

Dynamic response analysis of soils or any type of structure may be carried out using accelerograms or response spectra as input (Chapter 5). Whichever form of dynamic input is used, a number of earthquakes should be used or implied. Because of the random nature of earthquake shaking, it seems unlikely that any single seismic event can be shown to be safely representative of the design risk, without choosing an uneconomically powerful ground motion.

The use of accelerograms in a time-dependent analysis is analytically more powerful than response spectrum analysis and may be significantly more informative about the dynamic response of the structure. Individual accelerograms may induce local response peaks (Figure 5.20) in elastic analyses which may be difficult to interpret or justify. In non-elastic analyses this difficulty is partly overcome, but a number of accelerograms should be taken in all cases. It is common practice to take three or four accelerograms in a given study, these sometimes being a mixture of real events (sometimes scaled) and simulated records.

When using response spectra as input, either several response spectra from individual events (e.g. Figure 5.20) or a single spectrum which is the average of several events (Figure 4.45) should be taken. This will help to allow for the randomness of earthquakes, and smoothed average spectra will eliminate undue influence of local peaks in response. Figure 4.45 indicates the scatter of response from five simulated earthquakes; it is worth

Design Earthquakes

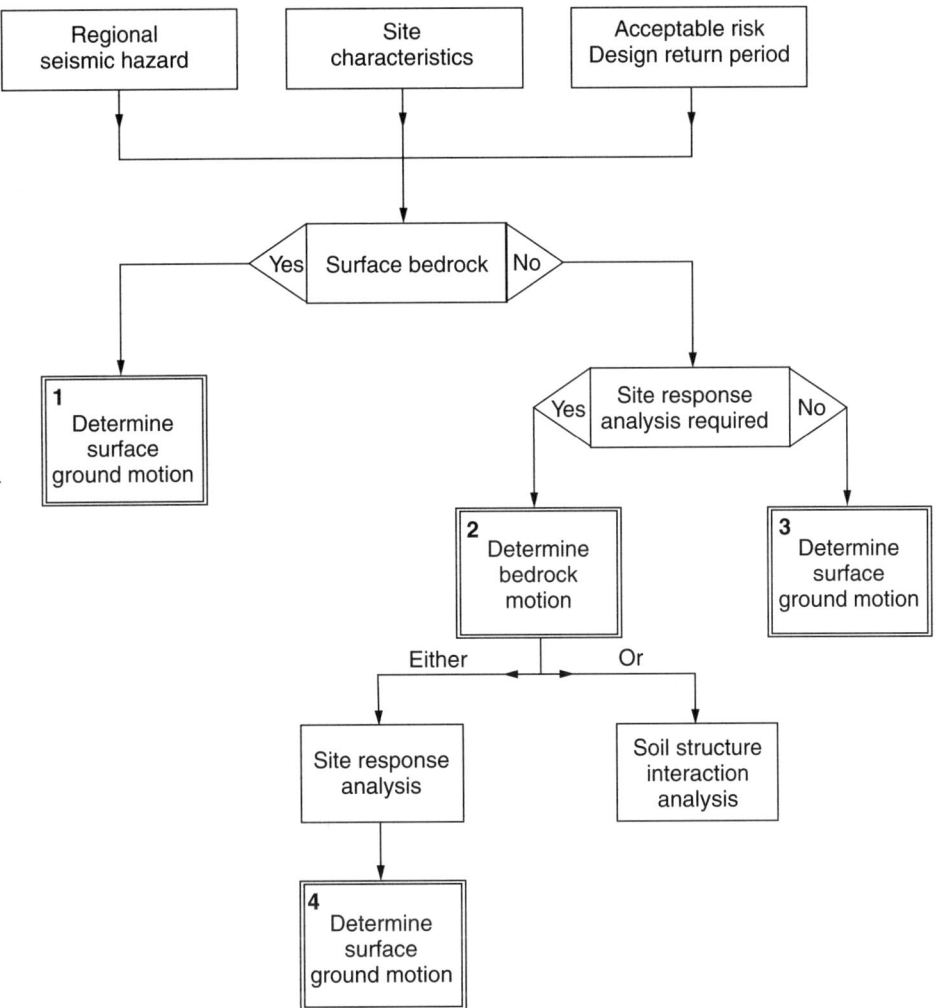

Figure 4.44 Flowchart showing four different procedures for determining design earthquakes (boxes 1–4), depending on the situation

noting that the standard error in the response spectra is likely to have been much greater if real rather than simulated accelerograms had been used.

It is argued that where surface motions have been computed from bedrock motions as in Section 5.2.2, these surface motions should be used for structural analysis in the form of averaged response spectra rather than accelerograms. It is considered that so many simplifying assumptions are made in site response analyses, that very sophisticated use of the computed surface accelerograms can scarcely be justified for practical design purposes.

Further discussions on different methods of seismic response analysis are given in Chapter 5, including a comparative review in Section 5.4.7.

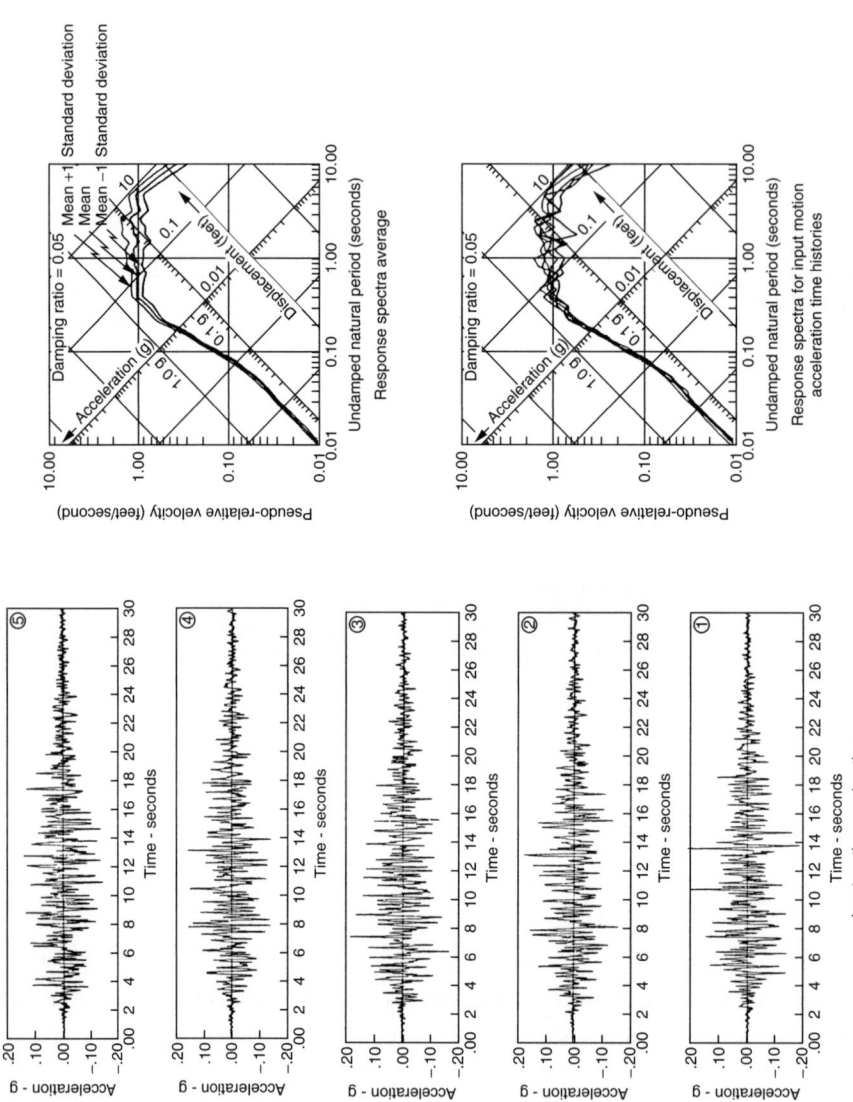

Figure 4.45 Set of simulated input accelerograms and response spectra for a magnitude 7.0 earthquake with a closest distance to the causative fault of 50 km (after Valera and Donovan, 1973)

4.7.2 Defining design events

Referring to the EERI Committee on Seismic Risk (1984), the *design event* is defined as a *specification of one or more earthquake source parameters, and of the location of energy release with respect to the site of interest*, which is *used for the earthquake resistant design of a structure*. Thus a design event specification may consist simply of a magnitude M and a source distance R.

Design events are required when normal code loadings are inappropriate or unavailable, as often arises in the case of very large, or critical, or novel structures such as tall buildings, large dams, liquefied gas storage depots, cooling towers, nuclear power facilities, or offshore oil platforms. The size and location of design events will depend upon the establishment of the design levels of seismic hazard, which may be related to some code or regulatory requirement or may need to be agreed with the owner.

In some cases more than one level of hazard may be required, say one for operational and one for survival performance, the equivalents in nuclear facility terminology being the operating basis earthquake and the safe-shutdown earthquake. Also, two design events may need to be considered for a particular hazard level. For example, tall buildings with long natural periods will be more sensitive to larger, more distant events because of the greater long-period content of such events, while short-period structures designed to the same hazard level may be more sensitive to smaller, closer events.

To further illustrate the point, Kuala Lumpur in Malaysia is located at the rather remote distance of about 400 km from the Alpide earthquake belt, and historically has therefore had little concern for earthquakes. However, from the 1970s, with the advent of taller buildings some instances of alarming swaying and cracking occurred, but *only* in tall buildings. The author has been involved in investigations which causally linked these phenomena with large-magnitude events occurring 400 km away. Many other sites around the world located at similar distances from large earthquake sources, while safe for most traditional construction, may merit seismic design checks for longer-period construction.

Studies of the seismic activity of a region as discussed in the previous sections of this chapter supply the material for defining the design earthquake for a given project. The characteristics of the design earthquake may be used in conjunction with the dynamic characteristics of the site to determine the dynamic design criteria for the project as discussed below.

An adequate definition of a design earthquake is very elusive, even prior to consideration of site conditions, because of difficulties in defining past earthquake behaviour and difficulties in predicting future seismic events. The main variables derived or implied in this chapter for use in defining the design event are magnitude, return period, source distance, source depth, fault positions, fault types, and rupture length, while the associated dependent variables, such as peak ground acceleration, peak ground velocity, peak ground displacement, duration of strong shaking, dominant period of shaking and attenuation relations, are used for ground-motion specification (see below).

Data on the above aspects of earthquakes are variable, often inaccurate, and scarce. This means that the interpretation of the data sometimes must be highly subjective, and the use of mean values or some other value such as the 90% confidence level may be open to argument. A considerable amount of idealization is necessary.

To illustrate the definition of design earthquakes for a given site, reference will be made to Figure 4.46. Assume that studies of the region have suggested the use of two design

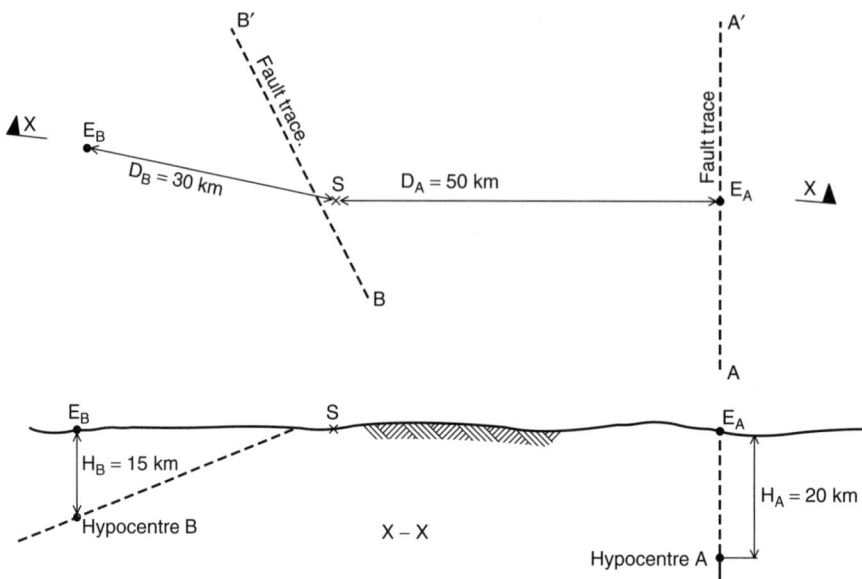

Figure 4.46 Hypothetical example of site relationship to two design earthquakes A and B with epicentres E_A and E_B, respectively

earthquakes, A and B, with source geometry as shown in Figure 4.46. It is quite common practice to consider two different design earthquakes with magnitudes and return periods as suggested above; normally the larger, less frequent, earthquake would be considered the worst design condition for use as ultimate loading, while the smaller, more frequent, earthquake might be used as a criterion for control of non-structural damage or the serviceability limit state. However, in the situation illustrated in Figure 4.46, the associated fault types might render this use of the design earthquakes inappropriate, depending on the magnitudes of events A and B. If the fault trace BB' had been undetected or not allowed for at the time of the design, the intensity of ground motion at the site would be underestimated assuming normal attenuation from a source 30 km away, rather than the shortest distance from the fault rupture.

In regions where there are no known faults posing a substantive design threat to a given site, it is common practice to select one or more magnitudes for the design event and establish the corresponding design source distance(s) from the required design probability of occurrence and the magnitude–frequency relationship for the local distributed seismicity.

A common case is for one design event to be associated with the nearest active fault (say $M_W = 7$ at a source distance of 20 or more kilometres), with a second design event being a smaller local earthquake established from the distributed seismicity as noted above (of, say, $M_W = 5.5$ at a distance of 5 km).

In conclusion, it is recommended that the proposed design events are carefully reviewed prior to use. All of the assumptions used in deriving them should be listed and their degrees of conservatism or non-conservatism should be noted so that a rational attempt at a balanced assessment can be made.

4.7.3 Sources of accelerograms and response spectra

Earthquake engineers experienced at working outside basic code requirements have developed sources of information of their own, through government and university organizations specializing in seismology and earthquake engineering. As the problem of availability of information varies so widely from place to place and as the situation is changing so rapidly, this section will simply discuss a few of the chief sources of data presently existing.

(i) *Accelerograms of real earthquakes.* The major source of accelerograms is a worldwide collection of strong-motion records for dissemination (in various forms) to the scientific and engineering community, which is available from the World Data Center for Solid Earth Geophysics. A list of their available data is obtainable from NOAA National Data Center, 325 Broadway, E/GC4, Dept A05, Boulder, CO 80303-3328, USA. Many countries contribute to the strong-motion data base.

(ii) *Accelerograms of simulated earthquakes.* Many earthquake engineering research organizations throughout the world have computer programs for generating artificial earthquakes. Software for the generation of simulated earthquakes such as PSEQGN is available at a small cost from the National Information Service for Earthquake Engineering (NISEE) at the University of California, Berkeley, PEER Building 451 RFS, 1301 South 46th Street, Richmond, CA 94804-4698, USA (email: info@nisee.berkeley.edu).

(iii) *Response spectra of real earthquakes.* Response spectra are more readily available than accelerograms as they are easily described in diagram form in the literature. Response spectra can be computed from earthquake accelerograms by computer programs such as SPECEQ, which is available through NISEE at the address given in (ii) above.

(iv) *Response spectra of simulated earthquakes.* Response spectra may readily be computed from simulated accelerograms from computer programs such as SPECEQ (see (ii)).

4.7.4 Response spectra as design earthquakes

As noted in Section 4.7.1, response spectra used as design earthquakes may be derived in a number of ways, all of which have considerable uncertainties and need subjective input. Some of these methods are described below.

Elastic response spectra derived from design events

Design events (Section 4.7.2) provide a base from which response spectra may be readily determined in a number of ways.

Elastic response spectra from selected records of real earthquakes

Having determined the magnitude and focal distance of the design event, ideally it may be possible to select a number of records of earthquakes with similar M and R values, and with appropriate source mechanisms and similar site soil conditions. These spectra may be applied to the analytical model individually, or the average or an envelope may be determined by creating a smoothed design spectrum as originally proposed by Housner (1959). As well as trying to match the M and R values, it is often considered appropriate to scale the individual events to have either the same peak ground acceleration or the same peak ground velocity. While acceleration is probably more commonly used as the scaling criterion, velocity may often be the better arbiter of damage to the structure.

Scaling of earthquake accelerograms (or spectra) should be done with caution if the change in amplitudes is large (changes of more than 50%, say). Large changes imply either that the scaled event is much larger or smaller than the original event or that the focal distance is different. These conditions imply a different source-controlled or attenuation-controlled frequency content (Section 4.6), and also imply possible non-linearities regarding soil behaviour. Leading earthquake loadings codes give detailed recommendations on the scaling of accelerograms.

While peak ground motions are commonly used for scaling response spectra, they are not very satisfactory for this purpose. Hall *et al.* (1984) report that a three-parameter system, using response spectrum intensities, may offer a better means of scaling response spectra.

Uniform hazard response spectra

Using the techniques noted in Section 4.7.3, response spectra may be generated for a given site such that the spectral ordinates for all of the periods of vibration have the same probability of occurrence. Such a response does not represent just one design event (M, R), as given by any of the curves on Figure 4.23, but represents all of the (M, R) pairs contributing to the distribution of spectral values at each period and damping value for which they are calculated. Such design spectra are therefore usually referred to as *uniform hazard spectra*, and may be used for specific sites or for codes.

An example of a set of uniform hazard spectra for deep soil in Christchurch, New Zealand, and using four different attenuation models is given in Figure 4.47. These spectra were calculated for a 10% probability of exceedance in a 50-year period, and were based on the distributed seismicity and fault activity in the hinterland, illustrated in Figure 4.10. Also shown is the spectrum for Christchurch for the intermediate soil class derived on the same basis as the New Zealand loadings standard NZS 4203 (1992), which was under revision when the other spectra were produced. It is cautionary to observe the large differences between the spectra obtained using the different attenuation models, i.e. of Abrahamson and Silva (1997), Sadigh *et al.* (1997) and an early version of McVerry *et al.* (2000) labelled NZ in Figure 4.47. This problem is diminishing as the differences between alternative attenuation models for a given tectonic environment decrease (as strong-motion databases grow).

Special features of design earthquake response spectra

In establishing design earthquake response spectra by the two methods outlined above, it will be important to ensure that the resulting spectra relate to the site soil conditions. As shown by Figure 3.3, the soil conditions at the site have a profound influence on the shape of the response spectrum; the longer the predominant period of vibration of the site, the greater will be the period at which the peak in the response spectrum occurs,

Design Earthquakes

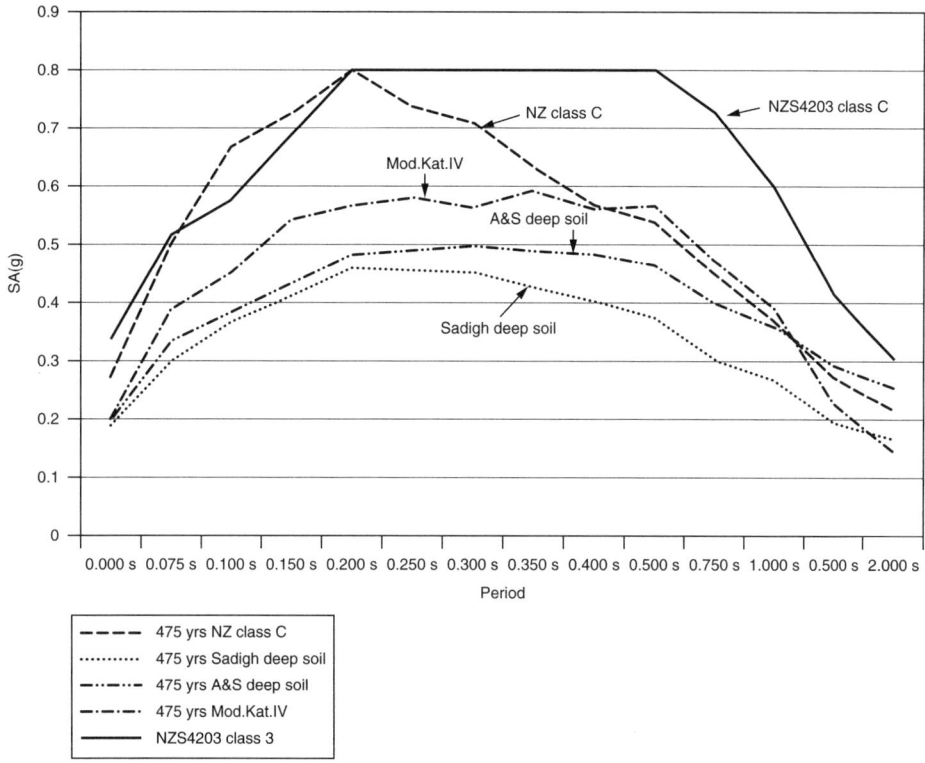

Figure 4.47 Uniform hazard acceleration response spectra for deep soil for Christchurch, New Zealand, derived from four attenuation models and one from the 1992 New Zealand loadings standard (G McVerry, personal communication, 2000)

although the values of these two parameters are likely to be the same only in cases of site resonance. This effect should be allowed for by using a data set derived from earthquakes recorded on sites with similar soil conditions to that of the sites in question.

Inelastic response spectra as design earthquakes

In the foregoing discussion the response spectra presented have been those derived assuming linear elastic structural behaviour. In design practice, in some cases the structure will be required to remain elastic in the design earthquake, but more commonly some degree of inelastic behaviour will be assumed. As discussed in Chapter 5, inelastic behaviour is often expressed in terms of the ductility factor μ, where $\mu = 1$ represents elastic behaviour and $\mu = 6$ is about the greatest degree of inelastic deformation that can readily be achieved in most structures (Table 10.1).

To arrive at a response spectrum corresponding to the desired degree of inelasticity, i.e. the design ductility level μ, the usual (least effort) technique is simply to divide the elastic response spectrum for the design earthquake by factors which allow for μ. This is considered to be inaccurate at short periods, and other schemes exist for allowing for ductility. A representative one comes from Berrill *et al.* (1981), where the ordinates of the elastic spectrum are multiplied by a factor R for any given level of ductility μ, as a function of the period of vibration T, as follows:

$$\text{for } T = 0, \qquad R = 1.0; \tag{4.37}$$

$$\text{for } 0 < T < 0.7 \text{ s}, \qquad R = \frac{0.7}{(\mu - 1)(T + 0.7)}; \tag{4.38}$$

$$\text{for } T \geq 0.7 \text{ s}, \qquad R = \frac{1}{\mu}. \tag{4.39}$$

A discussion of the reasons for using the above values of R is given in Section 5.4.7.

The well-known Newmark–Hall procedure gives alternatives to the reductions given by equations (4.38) and (4.39) as described in various publications (e.g. Newmark and Hall, 1982; Chopra and Goel, 2001).

4.7.5 Accelerograms as design earthquakes

Accelerograms used as design earthquakes may be derived using the parameter values (M, R) or (a, v, d) representing the design event (Section 4.7.2), or may be made to match a target design response spectrum. Using such criteria, the accelerograms are obtained either from records of real events or by simulation techniques as discussed below.

Accelerograms of selected real earthquakes

In choosing from amongst real earthquake records (Section 4.7.3) it will be desirable to match as nearly as possible the design conditions of magnitude, source distance, source depth, source mechanism, tectonic regime (i.e. interplate or intraplate), and soil profile with those of the real earthquakes. Close matching of magnitude and distance is desirable for minimizing scaling errors. Unfortunately, not all of the above factors may be known or be readily available for the real events. However, even when soil conditions are reasonably matched, an individual earthquake record has strong features characteristic only of that particular earthquake and site.

Major difficulties may arise over the choice of a suitable suite of real earthquakes, with suitable peak accelerations and frequency content. Advice on appropriate scaling methods is given in leading earthquake loadings codes, but care should be taken to include enough earthquakes in the suite of records used. It is helpful to use records that have spectral shapes as similar as possible to the shape of the target response spectrum, so that the required number of records does not become excessive. Further background information is available in recent papers such as that by Naeim *et al.* (2004).

The above remarks should be read in conjunction with the discussion on response spectra from real earthquakes given in Section 4.7.4.

For sites with surface bedrock there is an added difficulty, because relatively few strong ground motions have as yet been recorded on rock sites, and it may be appropriate to derive synthetic accelerograms to supplement the real ones.

Synthetic accelerograms

For most design purposes it can be assumed that ground motion is a random vibratory process, and that accelerograms can be mathematically simulated with random vibration theory. This will be most true at distances from the causative fault sufficient to ensure that the details of the fault displacement are not significant in the ground shaking. Because of the scarcity of actual bedrock recordings, at present the modelling of simulated earthquakes is necessarily based largely on the more numerous accelerograms recorded on softer soils. The main difference between bedrock and soft-soil motions is one of frequency content; this difference can be dealt with in the simulation process.

By far the most common pattern of ground motion is one of an abrupt transition from zero to maximum shaking, followed by a portion of more or less uniformly intense vibration, and finally a rather gradual attenuation (Figure 4.45). In the terminology of random vibration theory, the middle portion may be considered as a stationary random process, whereas the initial and final phases, being transitional, are non-stationary.

As described in a review by Lam *et al.* (2000), synthetic accelerograms can be generated by three methods, namely (i) deterministic, (ii) stochastic, or (iii) seismological model. Among the *deterministic* simulation methods, the two most popular are the empirical Green's function method and the ray-theory method which uses theoretical Green's functions (Atkinson and Somerville, 1994). Both methods involve superimposing small impulses (of small events) to simulate large events.

Stochastic methods consist typically of defining the frequency content by a deterministic target Fourier amplitude spectrum and defining the phase arrivals with a set of random phase angles. Controls may be created using magnitudes, source distance and site classifications (e.g. Trifunac, 1989). Another approach is to derive the Fourier spectrum iteratively from a target response spectrum, as in the computer program SIMQKE-1 (1990).

A *seismological model* originally developed by Brune (1970) and developed by others (e.g. Boore and Atkinson, 1987) derives a Fourier amplitude spectrum $A_x(f)$ for any given situation as the product of five main factors:

$$A_x(f) = S(f)G An(f) P(f) V(f), \qquad (4.40)$$

where $S(f)$ is the source factor, G is the geometric attenuation factor, $An(f)$ is the inelastic whole path attenuation factor, $P(f)$ is the upper crust attenuation factor, and $V(f)$ is the upper crust amplification factor. The definitions and quantification of these terms are conveniently described by Lam *et al.* (2000). They consider the above model to be generic, simple to use, and suitable for seismic hazard modelling of low-seismicity regions where details of potential earthquake sources are mostly unknown.

Having defined the frequency content of the earthquake ground motion by the Fourier amplitude spectrum above, synthetic accelerograms can be generated by combining it with random phase angles in a stochastic process. The required procedure (Boore, 1983; Lam *et al.*, 2000) consists of four steps:

(i) generation of Gaussian band-limited white noise $nt(t)$;
(ii) windowing the white noise to obtain $st(t)$;
(iii) derivation of the frequency filter $\Lambda_{at}(f)$;
(iv) generation of the synthetic accelerograms $at(t)$.

Lam (1999) has developed a computer program called GENQKE which generates accelerograms by implementing the above four steps. In step (ii) a window function is introduced to shape the accelerogram to resemble a real accelerogram. Considering real accelerograms of earthquakes of magnitude around 6, Lam (1993) empirically derived the window function expressed as

$$\omega t(t) = e^{-0.4t(6/t_d)} - e^{-1.2t(6/t_d)}, \qquad (4.41)$$

where t_d is the effective duration within which about 90% of the total energy is released.

4.8 Faults – Hazard and Design Considerations

4.8.1 Introduction

Intuitively the thought of building across an active fault is alarming, and obviously in general it is best avoided. However, in some circumstances structures can safely ride a fault rupture. For example, in the 1972 Managua earthquake the Banco Central de Nicaragua was astride a fault which moved 17 cm (horizontally only), and its foundation was strong enough to deflect the rupture around itself and survive intact (Wyllie et al., 1977). Indeed, the situation not infrequently arises when it is highly desirable to build across, or immediately beside, an active fault. Typically this happens when the location of a structure is conditioned by factors such as:

(1) topography, e.g. with dams;
(2) the structure's function, e.g. tunnels or pipelines; or
(3) where land is particularly valuable, e.g. in city centres.

In such circumstances seismic risk evaluation of alternative designs may be necessary.

In evaluating the seismic hazard of the displacement of a particular fault it will be important not only to know the probability of fault rupture during the lifetime of the structure, but also to differentiate between the likely amounts of vertical and horizontal displacement. Obviously some structures may be much more severely affected by vertical than by horizontal displacements. In such a case the possibility of building across a fault may have to be abandoned, unless the implied risks to both people and property are acceptable.

4.8.2 Probability of occurrence of fault displacements

In order to carry out a hazard analysis of fault displacements an investigation of the degree of activity of the fault in question, along the lines discussed in Section 4.4, may be required. The more significant the structure, the greater will be the effort that will be appropriate in the hazard study. For major structures, the study will hopefully result in recurrence intervals being associated with fault displacements of given magnitudes with an appropriate level of confidence. An example of an extensive investigation of fault displacement hazard is that carried out for the site of a very large water-filtration plant at Sylmar near Los Angeles, California (Spellman et al., 1984). Site exploration consisted of logging 550 m of trench, and four faults were observed and dated by age of soil profile development. One of these faults was judged to be an active reverse fault. Its displacement increased with depth, indicating that more than one fault had occurred within the age span of the soil profile, the maximum single displacement being $c.$ 0.2 m of a total observed cumulative vertical displacement of 0.8 m. This maximum displacement event of 0.2 m occurred about 5000 years ago, and the probability of a recurrence of this size of displacement during a period of 100 years was estimated as being less than 2%, but the method of obtaining this estimate was not stated.

4.8.3 Designing for fault movements

Surface fault displacements in large earthquakes, i.e. of magnitude 7.5 or more, can be very large, horizontal and vertical displacements of 10 m and more having been measured in various parts of the world. For example, horizontal displacements of up to $c.$ 13 m have been found by measuring the offsets of streams crossing the surface trace of the 1855 Wairarapa, New Zealand, earthquake (Grapes and Downes, 1997). While designing against such extreme displacements would seldom be contemplated, the feasibility of designing against more modest movements has been demonstrated by the Banco Central (Section 4.8.1), which survived 17 cm of horizontal rupture. As well as sudden rupture, allowance for slow creep movements is also sometimes desirable.

As occurred for the Banco Central in Managua, the possibility exists that a horizontal fault rupture may be diverted by the structure, and that this situation may be predicted at the design stage. For rupture diversion to occur the structure must be sufficiently strong and heavy in relation to the underlying soil, and in general diversion is likely to be realized only for soil (and not rock) sites and for horizontal (strike-slip) fault displacements.

As noted previously, large dams are sometimes constructed on sites traversed by active faults. This was the case with the Clyde Dam in New Zealand. The minor fault in the river channel under the dam site was estimated to be capable of 0.2 m horizontal displacement with a probability of occurrence of 1 in 1000 to 1 in 100 in a period of 100 years. This local fault displacement would occur as a consequence of an event of $M_{max} = 7.4$ occurring on a major fault located within about 3 km of the site. The dam, of concrete gravity construction, was provided with a vertical construction joint along the line of the river channel fault, which is designed to slip sympathetically to the design displacement of the fault, for which the conservative values of 2 m horizontal and 1 m vertical were adopted (Hatton and Foster, 1987).

Long *pipelines* for water supply, sewerage, or oil and gas supplies quite often have to cross active fault zones, e.g. in Alaska, California, and the Himalayas. Fortunately, because of their configuration, pipes are relatively amenable to design solutions for fault displacements. This may be achieved by providing loops and/or flexible connections, such as is done for pipework in buildings as illustrated in Section 11.3.6, or flexible joints and pipework on offshore oil and gas platforms and single-buoy moorings.

In the majority of cases, however, it is best to avoid building on or very close to an active fault, a issue likely to arise for buildings in urban areas. Risk-based guidelines (Kerr *et al.*, 2003) have been developed in New Zealand for addressing this problem. A fault avoidance zone 20 m wide either side of a fault is recommended, in which different classes of buildings are allowable depending on the importance category of the building and the average recurrence interval of the fault (Table 4.2). The building importance categories

Table 4.2 Relationship between fault recurrence interval and building importance category (after Kerr et al., 2003)

Recurrence interval class	Fault recurrence interval	Building importance category (BIC) limitations (allowable buildings)	
		Previously subdivided or developed sites	'Greenfield' sites
I	≤ 2000 years	BIC 1	BIC 1
II	> 2000 years to ≤ 3500 years	BIC 1 and 2a	
III	> 3500 years to ≤ 5000 years	BIC 1, 2a and 2b	BIC 1 and 2a
IV	> 5000 years to ≤ 10,000 years		BIC 1, 2a, and 2b
V	> 10,000 years to ≤ 20,000 years	BIC 1, 2a, 2b and 3	BIC 1, 2a, 2b and 3
VI	> 20,000 years to ≤ 125,000 years	BI Category 1, 2a, 2b, 3 and 4	

Note: Faults with average recurrence intervals in excess of 125,000 years are not considered active.

comprise a modified version of the New Zealand Loadings Standard classifications (NZS 1170: 2004). Here buildings range in importance from Class 1, i.e. structures presenting a low degree of hazard to life and other property (e.g. farm buildings); through Class 2a, i.e. residential timber framed buildings; Class 2b i.e. normal-use structures; Class 3, i.e. structures that may pose risks to people in crowds or contain contents of high value; to Class 4, i.e. structures with special post-disaster functions.

4.9 Probabilistic Seismic Hazard Assessment

The determination of design earthquake ground-motion criteria from seismic hazard analyses on a probabilistic basis was formulated by Cornell (1968). The method involves two separate models: a seismicity model describing the geographical distribution of event sources and the distribution of magnitudes; and an attenuation model describing the effect at any given site as a function of magnitude, source-to-site distance, and ground class.

The *seismicity model* may comprise a number of source regions the seismicity of which may be expressed (Section 4.5.3) in the form

$$\log N = A - bM, \tag{4.42}$$

where N is the annual number of earthquakes of magnitude exceeding M. The source regions may be described as areas of diffuse (i.e. distributed) seismicity, so that N relates to a unit area, and faults are represented as lines with their own activity rates treated according to what is known of them (Section 4.4.3). The value of N will also generally be found assuming that M has upper and lower bounds M_1 and M_0.

Attenuation models relate the effect Y at a site to magnitude and distance, so that in general

$$Y = Y(M, r), \qquad (4.43)$$

from which we have the inverted expression for magnitude

$$M = M(Y, r). \qquad (4.44)$$

More specifically, the attenuation of peak ground motion amplitudes (a, v, d) and also response spectrum ordinates (S_a, S_v, S_d), are commonly expressed in the form of equation (4.33), discussed in Section 4.6.4, such that

$$\log Y = b_1 + b_2 M - b_3 \log(r + b_4) - b_5 r, \qquad (4.45)$$

where $b_1, \ldots, b_5$ are empirically derived constants, such as those of Abrahamson and Silva (1997) (Section 4.6.2).

Rewriting equation (4.45) in the form of equation (4.45), we have

$$M = \frac{1}{b_2} \{\log[Y(r + b_4)^{b_3}] - b_1\}. \qquad (4.46)$$

Combining the above two models leads to the probability p_Y that any earthquake occurring at random in the source region will produce an effect with strength exceeding Y at the site:

$$p_Y = P[Y' > Y] = \int_{\text{source}} 10^{-bM(Y,r)} f_R(r) \, dr, \qquad (4.47)$$

where $f_R(r)$ is the probability density function of distance r.

As there are on average N earthquakes per year in the source region, the average annual probability of Y being exceeded at the site is

$$p_D = p_Y N \qquad (4.48)$$

and hence the average return period of the effect exceeding Y is

$$T_R = \frac{1}{p_D} = \frac{1}{p_Y N}. \qquad (4.49)$$

As discussed in Section 4.5.3, the rate of occurrence of earthquakes varies considerably over different time periods (Figure 4.12), and the great uncertainty involved in extrapolating the curve in this figure into the future is very apparent. Is a quieter or more active phase approaching?

In a study of the c. 3000-year long Chinese earthquake catalogue, McGuire (1979) found that 50- and 100-year data intervals provided better estimates of probabilities of felt shaking in the 50-year period following each time segment than 200-year data intervals.

McGuire argued that 'at a specific time, the most recent seismic activity is therefore the best data base to use for calculation of probabilities of shaking in the near future'. However, these results concern the global reliability for an ensemble of data sets for 62 cities in China, and the reliability of such projections for individual cities would presumably be less. Based purely on the statistics, it would not be wise to argue that the 50-year long quieter phase in New Zealand, for c. 1950–2000 (Dowrick and Cousins, 2003), is likely to be a more reliable basis for estimating seismic activity in the coming 50 years than taking into account the known higher seismic activity of the century prior to 1950.

In the preceding paragraphs, we have discussed three sources of uncertainty in the probabilistic evaluation of design ground-motion criteria:

(1) the earthquake recurrence relationship;
(2) the attenuation expression;
(3) site response.

Uncertainties are associated with every parameter that we use in probabilistic seismic hazard assessment (PSHA), the notable additions to these given above are uncertainties in the magnitude estimates (Rhoades and Dowrick, 2000) and the spatial distribution of future earthquakes. An overview of how uncertainties in PSHA may be reduced has been given by Somerville (2000b).

PSHA in essence comprises only a few components, while in practice there are many sub-components as is evident from the foregoing parts of this chapter and from state-of-the-art papers such as Somerville (2000a). The products of PSHA studies are estimates of ground motion parameters for chosen probabilities of occurrence, for a particular site (e.g. Christchurch, New Zealand (Section 4.5.2)) or a region. An example of the latter is the probabilistic seismic hazard map of New Zealand developed by Stirling *et al.* (2000) and presented in Figure 4.48. This model includes both uniformly distributed seismicity and the activity of the faults. The strong influence on regional variations in seismic hazard that can be exerted by faults is illustrated by the narrow band of very high hazard running down the South Island where a number of long, very active faults are located (see Figure 4.3).

Figure 4.48 is taken a stage further in Figure 4.49(b), where it is in the form of a seismic hazard zoning map for use in a loadings code. Here it is compared with the 1992 code zoning map (Figure 4.48(a)) which made only a small allowance for fault hazard. The large difference between the two maps is remarkable, especially for the South Island which is dominated by the very active Alpine fault system part of which is shown in Figure 4.3.

A question to be asked about fully probabilistic hazard assessment is whether it overestimates the time hazard. Using data of the historical incidence of Modified Mercalli intensity in New Zealand from Dowrick and Cousins (2003), Stirling and Petersen (2006) compared the historical record of earthquake hazard from 78 sites distributed across New Zealand and continental USA with the hazard estimated from the national probabilistic seismic hazard models for the two countries. A tendency was found for the probabilistic seismic hazard model to slightly exceed the historical hazard in New Zealand and westernmost continental US interplate regions, but show lower hazard than that of the historical

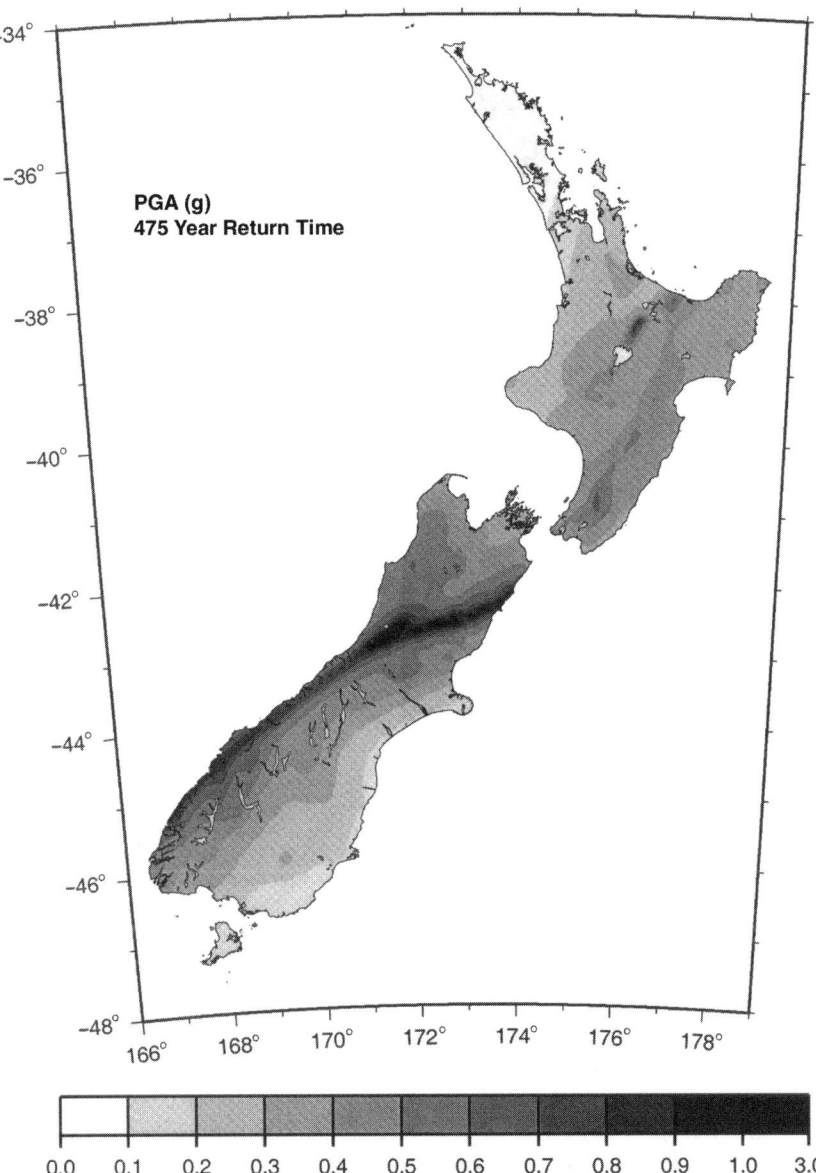

Figure 4.48 Probabilistic seismic hazard map of New Zealand, showing peak ground accelerations with a 10% probability of exceedance in 50 years, derived using Stirling *et al.* (2000) (M Stirling, personal communication, 2001)

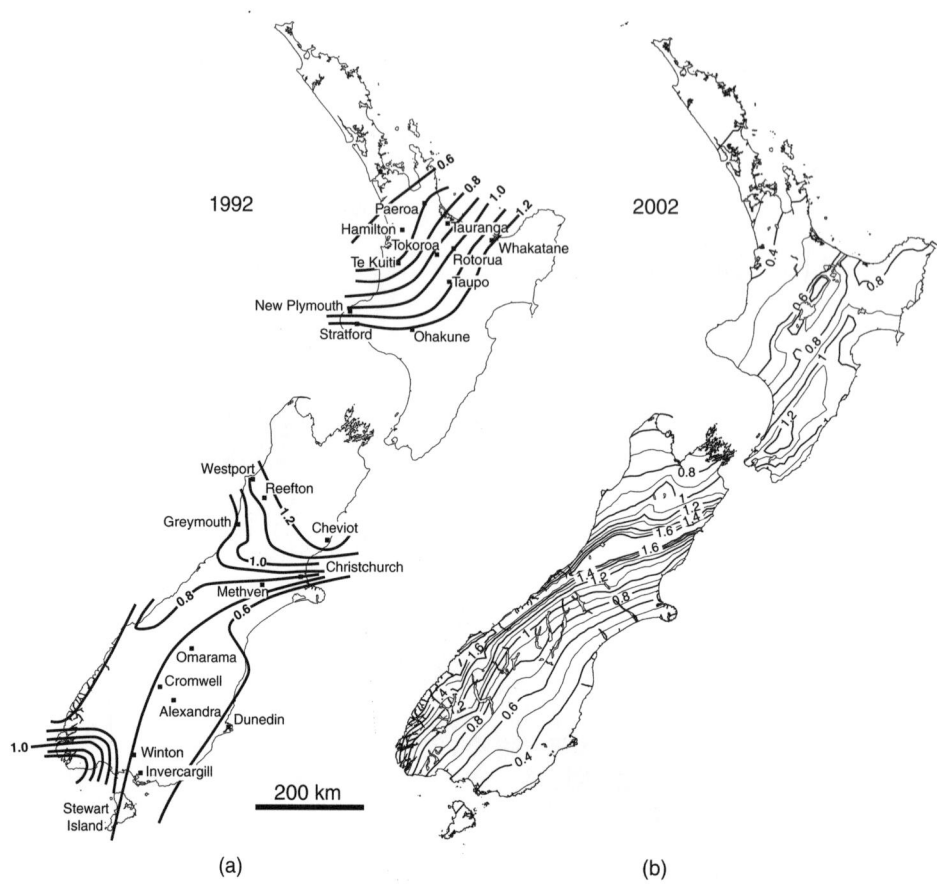

Figure 4.49 Probabilistic seismic hazard zoning maps for the New Zealand loadings standard: (a) 1992 version, and (b) 2002 version fully accounting for fault hazard (from GH McVerry personal communication, 2002)

record in the continental US intraplate region. The higher than historical levels of hazard at the interplate sites may have arisen from factors such as non-Poisson behaviour and parameterization of active fault data ion the probabilistic seismic hazard calculations. In contrast, the less than historical hazard for the intraplate USA may be largely due to site conditions that not have been considered at these sites and uncertainties in correlating ground-motion levels to historical felt intensities.

Geoscience Australia has developed a program (EQRM) for estimating seismic hazard and risk, suitable for any part of the world, which is available for free download at https://source-forge.net/projects/eqrm.

4.10 Probabilistic vs. Deterministic Seismic Hazard Assessment

In the assessment of seismic hazard the terms *deterministic* and *probabilistic* do not have exact or unequivocal meanings in the way they are commonly used. This is recognized by the frequent use of the terms *semi-deterministic* or *semi-probabilistic*, where the word *semi* means (vaguely) *part* rather than (precisely) *half*. As discussed earlier in this chapter, the probabilistic approach to seismic hazard assessment accounts quantitatively for the uncertainties surrounding the values of the variables. Thus, in what would be called a *fully probabilistic* assessment the uncertainties in *all* of the variables used explicitly and implicitly would be formally taken into account. This implies, for example, modelling the uncertainties in the location, size and strength of the asperities that would rupture on any given fault.

In practice, of course, not enough is known about some of the parameters to define their uncertainties, and the complexities of a fully probabilistic analysis would generally be technically excessive and financially unrealistic. Thus some components of hazard assessments are necessarily deterministic, e.g. the choice of the magnitude of the design earthquake is often to some extent a matter of judgement. In any given study the approach should be chosen according to the nature of the project and also should be tailored to the seismicity of the region, including the quantity and quality of the seismicity data available. For example, in the New Zealand loadings standard NZS 1170.5 (2004) the minimum earthquake shaking required to be considered (for survival against collapse) in the region of lowest seismicity (i.e. from Auckland northwards) was determined to be equivalent to that caused by the 84th percentile motions of a magnitude 6.5 normal-faulting earthquake at a distance of 20 km from the site in question. In Modified Mercalli intensity terms this scenario would cause intensity MM8.0, whereas the highest historical intensity to have occurred (once) in this region since 1840 is MM7.

For further reading on the issues involved in how probabilistic a seismic hazard assessment should be, see the wide-ranging paper by Bommer (2002) and McGuire (2004).

References

Aagaard BT, Hall JF and Heaton TH (2001) Characterization of near-source ground motions with earthquake simulations. *Earthq Spectra* **17**(2): 177–207.

Abrahamson N and Silva W (1997) Empirical response attenuation relations for shallow crustal earthquakes. *Seism Res Letters* **68**(1): 94–117.

Abrahamson N and Silva W (2008) Summary of the Abrahamson & Silva NGA ground-motion relations. *Earthq Spectra* **24**(1): 67–97.

Abrahamson NA and Somerville PG (1996) Effects of hanging wall and foot wall on ground motions recorded during the Northridge earthquake. *Bull Seism Soc Amer* **86**: S93–99.

Abrahamson N, Atkinson G, Boore D, Bozorgnia Y, Campbell K, Chiou B, Idriss IM, Silva W and Youngs R (2008) Comparisons of the NGA ground-motion relations. *Earthq Spectra* **24**(1): 45–66.

Aki K (1984) Prediction of strong motion using physical models of earthquake faulting. *Proc. 8th World Conf Earthq Eng*, San Francisco **II**: 433–440.

Allen CR (1976) Geological criteria for evaluating seismicity. In *Seismic Risk and Engineering Decisions* (eds C Lomnitz and E Rosenblueth). Elsevier, Amsterdam.

Ambraseys NN (1971) Value of historical records of earthquakes. *Nature* **232**(5310): 375–379.

Ambraseys NN (1973) Dynamics and response of foundation materials in epicentral regions of strong earthquakes. *Proc. 5th World Conf Earthq Eng, Rome* **1**: CXXXVI–CXLVIII.

Ambraseys NN and Hendron AJ (1968) Dynamic behaviour of rock masses. In *Rock Mechanics in Engineering Practice* (eds KG Stagg and OC Zienciewicz). John Wiley & Sons, Ltd, London.

Ambraseys NN, Simpson KA and Bommer JJ (1996) Prediction of horizontal response spectra in Europe. *Earthq Eng Struc Dyn* **25**: 371–400.

Anderson JG (1979) Estimating the seismicity from geological structure for seismic risk studies. *Bull Seism Soc Amer* **69**(1): 135–158.

Anderson JG and Luco JE (1983) Consequences of slip rate constraints on earthquake occurrence relations. *Bull Seism Soc Amer* **73**(2): 471–496.

Anderson JG, Wesnousky SG and Stirling MW (1996) Earthquake size as a function of fault slip rate. *Bull Seism Soc Amer* **86**: 683–690.

Arias A (1970) A measure of earthquake intensity. In *Seismic Design for Nuclear Power Plants* (ed. R Hanson). MIT Press, Cambridge, MA, pp. 438–483.

Atkinson GM and Boore DM (1995) Ground-motion relations for eastern North America. *Bull Seism Soc Amer* **85**: 17–30.

Atkinson G and Somerville P (1994) Calibration of time history simulation methods. *Bull Seism Soc Amer* **84**: 400–414.

Bak P and Tang C (1989) Earthquakes as a self-organized critical phenomenon. *J Geophys Res* **94**: 15635–15637.

Bakun WH (2005) Magnitude and location of historical earthquakes in Japan and implications for the 1855 Ansei Edo earthquake. *J Geophys Res* **110**(B2): 1–12.

Beanland S and Fellows DL (1984) Late quaternary tectonic deformation in the Kawarau River area, Central Otago. NZ Geol Survey/EDS Immediate Report 84/019, Dept Scientific and Industrial Research, New Zealand.

Berrill JB, Priestley NJM and Peek R (1981) Further comments on seismic design loads for bridges. *Bull NZ Nat Soc Earthq Eng* **14**: 3–11.

Bolt B (1999) *Earthquakes*, 4th edn. WH Freeman and Co, New York.

Bommer J (2002) Deterministic vs. probabilistic hazard assessment: an exaggerated and obstructive dichotomy. *J Earthq Eng* **6** (Special Issue 1): 43–73.

Bommer JJ and Martínez-Pereira A (1999) The effective duration of earthquake strong motion. *J Earthq Eng* **3**: 127–172.

Bommer JJ, Elnashai AS, Chliminatzas GO and Lee D (1998) Review and development of response spectra for displacement-based design. ESEE Research Report No. 98-3, Imperial College, London.

Bonilla MG (1970) Surface faulting and related effects. In *Earthquake Engineering* (ed. RL Wiegel). Prentice Hall, Englewood Cliffs, NJ.

Boore DM (1983) Stochastic simulation of high-frequency ground motions based on seismological model of radiated spectra. *Bull Seism Soc Amer* **73**(6): 1865–1894.

Boore DM and Atkinson G (1987) Stochastic prediction of ground motion and spectral response parameters at hard-rock sites in eastern North America. *Bull Seism Soc Amer* **77**(2): 440–467.

Bowman DD, Ouillon G, Sammis CG, Sornette A and Sornette D (1998) An observational test of the critical earthquake concept. *J Geophys Res* **103**: 24359–24372.

Brune JN (1970) Tectonic stress and spectra of shear waves from earthquakes. *J Geophys Res* **75**: 4997–5009.

Brune JN (1976) The physics of earthquake strong motion. In *Seismic Risk and Engineering Decisions* (eds C Lomnitz and E Rosenblueth). Elsevier, Amsterdam.

Burger RW, Somerville PG, Barker JS, Hermann RB and Helmberger DV (1987) The effect of crustal structure on strong ground motion attenuation relations in eastern North America. *Bull Seism Soc Amer* **77**: 420–439.

Burridge R and Knopoff L (1967) Model and theoretical seismicity. *Bull Seism Soc Amer* **57**: 341–371.

References

Carmona JS and Castano JC (1973) Seismic risk in South America to the south of 20 degrees. *Proc. 5th World Conf Earthq Eng*, Rome **2**: 1644–1653.

Chinnery MA and North RJ (1975) The frequency of very large earthquakes. *Science* **190**: 1197–8.

Chopra AK and Goel RK (2001) Direct displacement-based design: Use of inelastic vs. elastic design spectra. *Earthq Spectra* **17**(1): 47–64.

Cifuentes IL (1989) The 1960 Chilean earthquakes. *J Geoph Res* **94**(B1): 665–680.

Clark RH, Dibble RR, Fyle HE, Lensen GJ and Suggate RP (1965) Tectonic and earthquake risk zoning in New Zealand. *Proc. 3rd World Conf Earthq Eng* New Zealand 1: I-107–I-124.

Console R and Murru M (2001) A simple and testable model for earthquake clustering. *J Geophys Res* **106**(B5): 8699–8711.

Cornell CA (1968) Engineering seismic risk analysis. *Bull Seism Soc Amer* **58**(5): 1583–1606.

Cousins WJ, Zhao JX and Perrin ND (1999) A model for the attenuation of peak ground accelerations in New Zealand earthquakes based on seismograph and accelerographic data. *Bull NZ Soc Earthq Eng* **32**(4): 193–220.

Davenport PN (2003) Instrumental measures of earthquake intensity in New Zealand. *Proc. Pacific Conf Earthq Eng*, Christchurch.

Dewey JW, Reagor BG, Dengler L and Moley K (1995) Intensity distribution and isoseismal maps for the Northridge, California, earthquake of July 17, 1994. US Geological Survey Open File Report #95-2.

Donovan NC (1973) A statistical evaluation of strong motion data including the February 9, 1971 San Fernando earthquake. *Proc. 5th World Conf Earthq Eng*, Rome **1**: 1252–1261.

Downes GL and Dowrick DJ (in preparation) *Atlas of Isoseismal Maps of New Zealand Earthquakes*, 2nd edn. GNS Science, Lower Hutt, New Zealand.

Dowrick DJ (1981) Earthquake risk and design ground motions in the UK offshore area. *Proc. Inst Civ Engrs* **71**: 305–321.

Dowrick DJ (1996) The Modified Mercalli earthquake intensity scale – Revisions arising from recent studies of New Zealand earthquakes. *Bull NZ Nat Soc Earthq Eng* **29**(2): 92–106.

Dowrick DJ (1998) Damage and intensities in the magnitude 7.8 1931 Hawke's Bay, New Zealand, earthquake. *Bull NZ Nat Soc Earthq Eng* **31**(3): 139–163.

Dowrick DJ (2007a) Revised isoseismal maps for the 1956 Bay of Plenty and 1987 Edgecumbe, New Zealand, earthquakes – Implications for seismic hazard and risk. *Bull NZ Soc Earthq Eng* **40**(4): 200–206.

Dowrick DJ (2007b) Effects of attenuation in the Taupo Volcanic Zone on patterns of spatial distribution of ground shaking in New Zealand earthquakes. *NZ J Geol Geoph* **50**: 315–325.

Dowrick DJ and Cousins WJ (2003) Historical incidence of Modified Mercalli intensity in New Zealand and comparisons with hazard models. *Bull NZ Nat Soc Earthq Eng* **36**(1): 1–24.

Dowrick DJ and Rhoades DA (1999) Attenuation of Modified Mercalli intensity in New Zealand earthquakes. *Bull NZ Soc Earthq Eng* **32**: 55–89.

Dowrick DJ and Rhoades DA (2004) Relations between earthquake magnitude and fault rupture dimensions: How regionally variable are they? *Bull Seism Soc Amer* **94**(3): 776–788.

Dowrick DJ and Rhoades DA (2005) Attenuation of Modified Mercalli intensity in New Zealand earthquakes. *Bull NZ Soc Earthq Eng* **38**(4): 185–214.

Dowrick DJ and Rhoades DA (in preparation) A distributed source approach to modelling the spatial distribution of MM intensities resulting from large crustal New Zealand earthquakes. *Bull NZ Soc Earthq Eng*.

Dowrick DJ, Berryman KR, McVerry GH and Zhao JX (1998) Earthquake hazard in Christchurch. *Bull NZ Nat Soc Earthq Eng* **31**(1): 1–23.

EERI Committee on Scientific Risk (1984) Glossary of terms for probabilistic seismic-risk and hazard analysis. *Earthq Spectra* **1**(1): 33–40.

Ellsworth WL, Matthews MV, Nadeau RM, Nishenko SP, Reasenberg PA and Simpson RW (1999) A physically based earthquake recurrence model for estimation of long-term probabilities. US Geological Survey Open File Report #99-522.

Evison F and Rhoades D (2001) Model of long-term seismogenesis. *Annali di Geofisica* **44**: 81–93.

Evison FF and Rhoades DA (2004) Demarcation and scaling of long-term seismogenesis. *Pure Appl Geophys* **161**(1): 21–45.

Grapes R and Downes G (1997) The 1855 Wairarapa, New Zealand, earthquake – Analysis of historical data. *Bull NZ Natl Soc Earthq Eng* **30**: 271–368.

Graves RW, Pitarka A and Somerville PG (1998) Ground motion amplification in the Santa Monica area: effects of shallow basin edge structure. *Bull Seism Soc Amer* **88**: 1224–1242.

Gubin IE (1967) Earthquakes and seismic zoning. *Bull Int Inst Seism Earthq Eng* **4**: 107–126.

Gutenberg B and Richter CF (1965) *Seismicity of the Earth*. Hafner, New York.

Hall WJ, Nau JM and Zahrah FT (1984) Sealing of response spectra and energy dissipation in SDOF systems. *Proc. 8th World Conf Earthq Eng*, San Francisco **IV**: 7–14.

Hancock J and Bommer JJ (2006) A state-of-knowledge review of the influence of strong-motion duration on structural damage. *Earthq Spectra* **22**(3): 827–845.

Hanks TC (1982) f_{max}. *Bull Seism Soc Amer* **72**(6): Part A, 1867–1879.

Hanks TC and Bakun WH (2002) A bilinear source-scaling model for M-log A observations of continental earthquakes. *Bull Seism Soc Amer* **92**: 1841–1846.

Hatton JW and Foster PF (1987) Seismic considerations for the Clyde Dam. *IPENZ Trans* **14**(3/CE): 129–140.

Hermann RB and Nuttli OW (1984) Scaling and attenuation relations for strong ground motion in eastern North America. *Proc. 8th World Conf Earthq Eng*, San Francisco **II**: 305–309.

Housner GW (1959) Behaviour of structures during earthquakes. *J Eng Mech Divn* ASCE **85**(EM4): 109–129.

Housner GW (1973) Important features of earthquake ground motions. *Proc. 5th World Conf Earthq Eng*, Rome **1**: CLIX–CLXVIII.

Husid LR (1969) Características de terremotos. Análysis general. *Revista del IDIEM* **8**, Santiago de Chile, 21–42.

Kasahara K (1981) *Earthquake Mechanics*. Cambridge University Press, Cambridge.

Kerr J, Nathan S, Van Dissen R, Webb P, Brunsdon D and King A (2003) *Planning for Development of Land Close to Active Faults*. Ministry for the Environment, Wellington, New Zealand, ME No 483. http:/mfe.resultspage.com/search?p=Q&ts=c2&w=fault.

Kijko A and Sellevoll MA (1989) Estimation of earthquake hazard parameters from incomplete data files. Part I: Utilization of extreme and incomplete catalogues with different threshold magnitudes. *Bull Seism Soc Amer* **79**: 645–654.

Kiremidjian A and Anagnos T (1984) Stochastic slip-predictable model for earthquake occurrences. *Bull Seism Soc Amer* **74**: 739–755.

Klimkiewicz GC, Leblanc G, Holt RJ and Thiruvengodam TR (1984) Relative seismic hazard assessment for the north central United States. *Proc. 8th World Conf Earthq Eng* **I**: 149–156.

Kostrov BV (1974) Seismic moment and energy of earthquakes, and seismic flow of rock. *Izv Acad Sci USSR Phys Solid Earth* **1**: 23–40.

Lam NTK (1993) Assessment of torsion in regular multi-storey buildings subjected to earthquake ground motion. PhD thesis, University of Melbourne.

Lam NTK (1999) Program 'GENQKE' users guide: Program for generating accelerograms based on stochastic simulation of seismological models. Dept Civil and Environmental Eng, University of Melbourne.

Lam N, Wilson J and Hutchinson G (2000) Generation of synthetic earthquake accelerograms using seismological modelling: a review. *J Earthq Eng* **4**: 321–354.

Leyendecker EV, Hunt RJ, Frankel AD and Ruckstales KS (2000) Development of maximum considered earthquake ground motion maps. *Earthq Spectra* **16**: 21–40.

Lomnitz C (1974) *Global Tectonics and Earthquake Risk*. Elsevier, Amsterdam.

McCann MW and Boore DM (1983) Variability in ground motions: root mean square acceleration and peak acceleration for the 1971 San Francisco earthquake. *Bull Seism Soc Amer* **73**(2): 615–632.

McGarr A (1982) Upper bounds on near-source peak ground motion based on a model of inhomogeneous faulting. *Bull Seism Soc Amer* **72**(6) Part A: 1825–1841.

McGuire RK (1979) Adequacy of simple probabilistic models for calculating felt-shaking hazard, using the Chinese earthquakes catalog. *Bull Seism Soc Amer* **69**(3): 877–892.

McGuire RK (2004) *Seismic Hazard and Risk Analysis*, MNO-10. EERI, Oakland, CA.

McVerry GH, Zhao JX, Abrahamson NA and Somerville PG (2000) Crustal and subduction zone attenuation relations for New Zealand earthquakes. *Proc. 12th World Conf Earthq Eng*, Auckland.

Mei S-Y (1960) Characteristics of earthquake activity in China. *Acta Geophys Sinica* **9**: 1–9 (in Chinese).

Menu J-HM (1991) Influence de la prise en compte des mouvements forts des seismes. Rapport CEA/SM009/F.

Murphy JR and O'Brien JL (1977) The correlation of peak ground acceleration amplitude with seismic intensity and other physical parameters. *Bull Seism Soc Amer* **67**(3): 877–915.

Naeim F, Alimoradi A and Pezeshk S (2004) Selection and scaling of ground motion time histories for structural design using genetic algorithm. *Earthq Spectra* **20**(2): 413–426.

Newmark NM and Hall WJ (1982) *Earthquake Spectra and Design*. Earthquake Engineering Research Institute, Berkeley, CA.

Nuttli OW (1983) Average seismic source-parameter relations for mid-plate earthquakes. *Bull Seism Soc Amer* **73**(2): 519–535.

NZ Geological Survey (1984) Seismotectonic hazard evaluation for Upper Clutha lower development. NZGS Rept EG 377, Dept Scientific and Industrial Research, New Zealand.

NZS 1170.5 (2004) *Structural Design Actions Part 5: Earthquake Actions*. Standards New Zealand, Wellington.

NZS 4203 (1992) *Loadings Standard*. Standards New Zealand, Wellington.

Ogata Y (1989) Statistical models for standard seismicity and detection of anomalies by residual analysis. *Tectonophysics* **169**: 159–174.

Pitarka A, Irikura K, Iwata T and Sekiguchi H (1998) Three-dimensional simulation of the near-fault ground motion for the 1995 Hyogo-ken Nanbu (Kobe), Japan, earthquake. *Bull Seism Soc Amer* **88**: 428–440.

Rhoades DA (1996) Estimation of the Gutenberg-Richter relation allowing for individual earthquake uncertainties. *Tectonophysics* **258**: 71–83.

Rhoades DA and Dowrick DJ (2000) Effects of magnitude uncertainties on seismic hazard estimates. *Proc. 12th World Conf Earthq Eng*. Auckland.

Rhoades DA and Evison FF (2004) Long-range earthquake forecasting with every earthquake a precursor according to scale. *Pure Appl. Geophys* **161**(1): 47–72.

Rhoades DA, Van Dissen RJ and Dowrick DJ (1994) On the handling of uncertainties in estimating the hazard of rupture on a fault segment. *J Geophys Res* **99**: 13701–13712.

Rhoades DA, Zhao JX and McVerry GH (2008) A simple test for inhibition of very strong shaking in ground-motion models. *Bull Seism Soc Amer* **98**: 448–453.

Richter CF (1958) *Elementary Seismology*. Freeman, San Francisco.

Sadigh K, Chang C-K, Egan JA, Makidisi F and Youngs RR (1997) Attenuation relationships for shallow crustal earthquakes based on Californian strong motion data. *Seism Res Letters* **68**(1): 180–189.

Scholz CH (1968) The frequency-magnitude relationships of microfacturing in rock and its relationship to earthquakes. *Bull Seism Soc Amer* **58**(2): 399–415.

Scholz CH (1990) *The Mechanics of Earthquakes and Faulting*. Cambridge University Press, Cambridge.

Scholz CH, Aviles C and Wesnousky S (1986) Scaling differences between large intraplate and interplate earthquakes. *Bull Seism Soc Amer* **76**: 65–70.

SIMQKE-1 (1990) A program for artificial motion generation: User's Manual and Documentation (revised). Dept Civil Eng, Massachusetts Institute of Technology.

Somerville P (2000a) Seismic hazard evaluation. *Bull NZ Soc Earthq Eng* **33**: 371–386 (and figures in **33**: 484–491).

Somerville (2000b) Reducing uncertainty in strong motion prediction. *Proc. 12th World Conf Earthq Eng*, Auckland.

Somerville PG, Smith NF and Graves RW (1994) The effect of critical Moho reflections on the attenuation of strong motion from the 1989 Loma Prieta earthquake. In *The Loma Prieta, California, Earthquake of October 17, 1989 – Strong Ground Motion*, US Geol Survey Prof Paper 1555-A, A67–A75.

Somerville PG, Smith NF, Graves RW and Abrahamson NA (1997) Modification of empirical strong motion attenuation relations to include the amplitude and duration effects of rupture directivity. *Seism Res Letters* **68**(1): 199–222.

Somerville M, McCue K and Sinadinovski C (1998) Response spectra recommended for Australia. *Proc. Australasian Struc Eng Conf*, Auckland, 439–444.

Somerville PG, Irikura K, Graves R, Sawada S, Wald D, Abrahamson N, Iwasaki Y, Kagawa T, Smith N and Kowada A (1999) Characterizing earthquake slip models for the prediction of strong ground motion. *Seism Res Letters* **70**: 59–80.

Sorensen MBK, Atakan K and Pulido N (2007) Simulated ground motions for the great M 9.3 Sumatra-Andaman earthquake of 26 December 2004. *Bull Seism Soc Amer* **97**(1A): S139–151.

Spellman HA, Stellar JR and Shlemon RJ (1984) Risk evaluation for construction over an active fault, Sylmar, California. *Proc. 8th World Conf Earth Eng*, San Francisco **I**: 15–22.

Stirling M and Petersen M (2006) Comparison of the historical record of earthquake hazard with the seismic-hazard models for New Zealand and the continental Unites States. *Bull Seism Soc Amer* **96**(6): 1978–1994.

Stirling M, McVerry G, Berryman K, McGinty P, Villamor P, Van Dissen R, Dowrick D, Cousins J and Sutherland R (2000) Probabilistic seismic hazard assessment of New Zealand. Report prepared for Earthquake Commission Research Foundation, Institute of Geological and Nuclear Sciences Client Report 2000/53.

Trifunac MD (1989) Dependence of Fourier amplitude spectrum amplitudes of record earthquake accelerations on magnitude, local soil conditions and on depth of sediments. *Earthq Engrg Struct Dyn* **18**: 999–1016.

Trifunac MD and Brady AG (1975) A study on the duration of strong earthquake ground motion. *Bull Seism Soc Amer* **65**(3): 581–626.

Utsu T (1961) A statistical study on the occurrence of aftershocks. *Geophys Mag* **30**: 521–605.

Valera JE and Donovan NC (1973) Incorporation of uncertainties in the seismic response of soils. *Proc. 5th World Conf Earthq Eng*, Rome **1**: 370–379.

Vere-Jones D (1970) Stochastic models for earthquake occurrence (with discussion). *J Roy Statist Soc B* **32**: 1–62.

Wald DJ and Heaton TH (1994) A dislocation model of the 1994 Northridge, California, earthquake determined from strong ground motions. US Geological Survey Open File Report #94-278.

Waldron HH and Galster RW (1984) Comparative seismic hazards study of western Montana. *Proc. 8th World Conf Earthq Eng*, San Francisco **I**: 31–38.

Ward SN (1998) On the consistency of earthquake moment rates, geological fault data, and space geodetic strain: the United States. *Geophys J Int* **134**: 172–186.

Wells D and Coppersmith K (1994) New empirical relationships among magnitude, rupture length, rupture width, rupture area and surface displacement. *Bull Seism Soc Amer* **84**: 974–1002.

Wyllie LA, Chamorro F, Cluff LS and Niccum MR (1977) Performance of Banco Central relating to faulting. *Proc. 6th World Conf Earthq Eng*, New Delhi: 2417–2422.

Wyss M, Shimazaki K and Ito A (eds) (1999) Special issue: Seismicity patterns: their statistical significance and physical meaning. *Pure Appl Geophys* **155**: 203–726.

Youngs RR and Coppersmith KJ (1985) Implications of fault slip rates and earthquake recurrence models to probabilistic seismic hazard estimates. *Bull Seism Soc Amer* **75**: 939–964.

Zhao JX, Dowrick DJ and McVerry GH (1997) Attenuation of peak ground accelerations in New Zealand earthquakes. *Bull NZ Nat Soc Earthq Eng* **30**(2): 133–158.

Zheng X and Vere-Jones D (1991) Application of stress-release models to historical earthquakes from North China. *Pure Appl Geophys* **135**: 559–576.

5

Seismic Response of Soils and Structures

5.1 Introduction

This chapter is principally concerned with the determination of seismic motions, stresses, and deformations necessary for detailed design. The design earthquake (Chapter 4) is applied to the soil and/or the proposed form and materials of the structure (Chapter 8).

In earthquake conditions the subsoil–substructure–superstructure–non-structure relationship should ideally be analysed as a structural continuum. Although in practice this is seldom feasible, each of the parts should be seen as part of the whole when considering boundary conditions.

The problems involved in adequately representing seismic behaviour in theoretical analysis are numerous, and many compromises have to be made. To obtain the maximum benefit from any method of seismic analysis, an understanding of the dynamic response characteristics of materials is essential. For the adequate earthquake resistance of most structures, satisfactory post-elastic performance as well as elastic performance must occur.

5.2 Seismic Response of Soils

5.2.1 Dynamic properties of soils

Soil behaviour under dynamic loading depends upon many factors, including:

- the nature of the soil;
- the environment of the soil (static stress state and water content); and
- the nature of the dynamic loading (strain magnitude, strain rate, and number of cycles of loading).

Some soils increase in strength under rapid cyclic loading, while others such as saturated sands or sensitive clays may lose strength with vibration.

This section provides background information on soil and rock properties required for dynamic response analysis of soil or soil–structure systems. Ways of estimating the basic parameters of shear modulus, damping, and shear-wave velocity are suggested, and typical values of these and other parameters are given. To obtain appropriate design values of these parameters for a given site, suitable field and laboratory tests as discussed in Section 3.3 may be necessary.

Shear modulus

For soils the stress–strain behaviour of most interest in earthquakes is that involving shear, and, except for competent rock, engineering soils behave in a markedly non-linear fashion in the stress range of interest.

For small strains the shear modulus of a soil can be taken as the mean slope of the stress-strain curve. At large strains the stress-strain curve becomes markedly non-linear so that the shear modulus is far from constant but is dependent on the magnitude of the shear strain (Figure 5.1).

There are various field and laboratory methods available for finding the shear modulus G of soils. Field tests may be used for finding the shear-wave velocity, v_s, and calculating the maximum shear modulus from the relationship

$$G_{max} = \rho v_s^2, \tag{5.1}$$

where ρ is the mass density of the soil. Typical values of v_s and ρ are given in Tables 5.1 and 5.2, respectively.

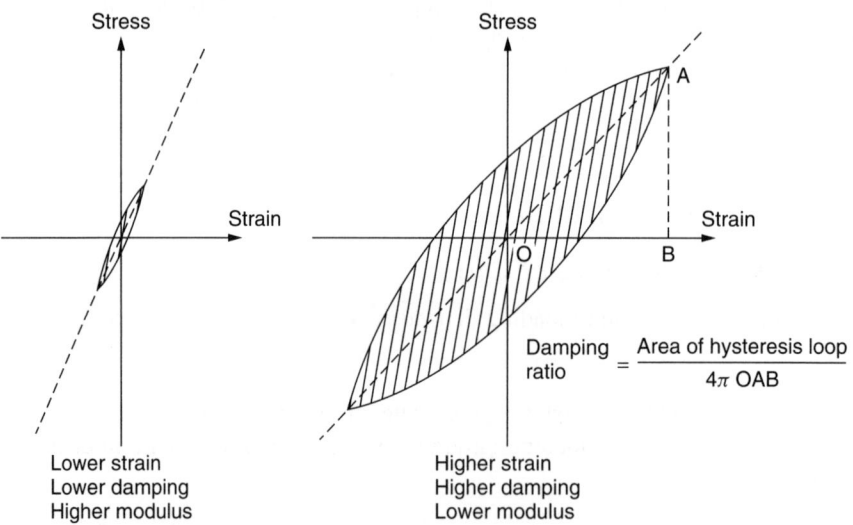

Figure 5.1 Illustration defining the effect of shear strain on damping and shear modulus of soils. (Reprinted from Seed and Idriss (1969), Influence of soil conditions on ground motions during earthquakes. *J Soil Mech and Found Divn* **95**(SM1): 99–137, by permission of the American Society of Civil Engineers)

Table 5.1 Mean shear-wave velocities (m/s) for the top 30 m of ground (mainly from Borcherdt, 1994)

General description	Mean shear-wave velocity		
	Minimum	Average	Maximum
Firm and Hard Rocks			
Hard Rocks	1400	1620	
(e.g. metamorphic rocks with very widely spaced fractures)			
Firm to Hard Rocks*	700	1050	1400
(e.g. granites, igneous rocks, conglomerates, sandstones, and shales with close to widely spaced fractures)			
Gravelly Soils and Soft to Firm Rocks	375	540	700
(e.g. soft igneous sedimentary rocks, sandstones, and shales, gravels, and soils with >20% gravel)			
Stiff Clays and Sandy Soils	200	290	375
(e.g. loose to v. dense sands, silt loams and sandy clays, and medium stiff to hard clays and silty clays ($N > 5$ blows/ft))			
Soft Soils			
Non-special Study Soft Soils	100	150	200
(e.g. loose submerged fills and very soft ($N < 5$ blows/ft) clays and silty clays <37 m (120 ft) thick)			
Very Soft Soils	50?	75?	100
(e.g. loose saturated sand, marshland, recent reclamation)			

*The NGA project (Section 4.6.4) has adopted $v_s = 1000$ m/s as the threshold for engineering rock.

Table 5.2 Typical mass densities of basic soil types

Soil type	Mass density ρ(Mg/m^3)*			
	Poorly graded soil		Well-graded soil	
	Range	Typical value	Range	Typical value
Loose sand	1.70–1.90	1.75	1.75–2.00	1.85
Dense sand	1.90–2.10	2.00	2.00–2.20	2.10
Soft clay	1.60–1.90	1.75	1.60–1.90	1.75
Stiff clay	1.90–2.25	2.07	1.90–2.25	2.07
Silty soils	1.60–2.00	1.75	1.60–2.00	1.75
Gravelly soils	1.90–2.25	2.07	2.00–2.30	2.15

*Values are representative of moist sands and gravels and saturated silts and clays.

The idea that the stress–strain behaviour of a soil can be modelled as a linear elastic material is a very considerable idealization. First, the stiffness of a soil is dependent on the effective stresses. It is generally agreed that the small-strain stiffness is proportional to the square root of the mean principal stress. For example, Seed *et al.* (1986) proposed the following relation for the small-strain shear modulus of normally consolidated

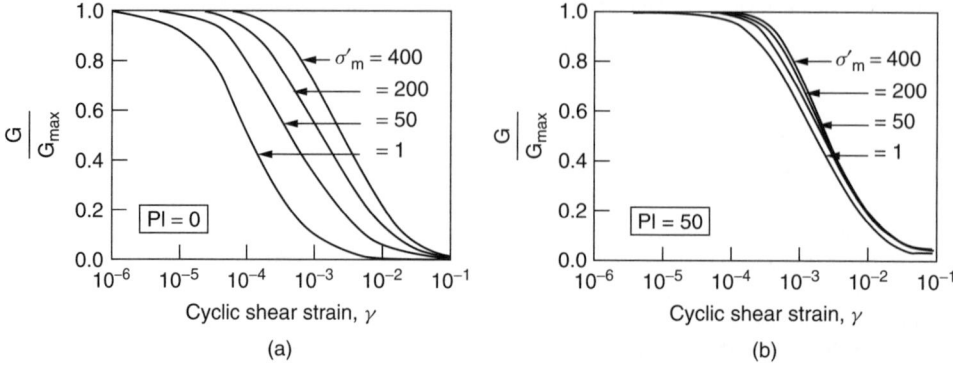

Figure 5.2 Influence of mean effective confining pressure (kPa) on modulus reduction curves for (a) non-plastic (PI = 0) soil, and (b) plastic (PI = 50) soil. (Reprinted from Ishibashi (1992). Discussion to: Effect of soil plasticity on cyclic response, by M Vucetic and R Dobry, *J Geotech Eng* **118**: 830–832, by permission of the American Society of Civil Engineers)

sands:

$$G_{\max} = 3.6\sqrt{\sigma'_m[(N_1)_{60}]^{1/3}} \quad \text{(MPa)} \tag{5.2}$$

where $G_{\max}$ denotes the small-strain shear modulus (the maximum value that it may take for a given material and effective stress), σ'_m is the mean principal effective stress (kPa) and $(N_1)_{60}$ is a corrected N value.

Laboratory methods generally measure G more directly from stress–strain tests. It is clear from Figure 5.1 that the level of strain at which G is measured must be known. This is further illustrated in Figure 5.2 which shows how G also varies with confining pressure and plasticity index (PI). In a study of normally consolidated and moderately overconsolidated soils, Dobry and Vucetic (1987) found that $G/G_{\max}$ depends also upon other factors, i.e. void ratio, number of cycles of loading, and sometimes geologic age and cementation.

The difficulties involved in finding a reliable shear modulus model for any given project are compounded by the fact that there is no simple linear relationship between laboratory and field tests (Tani, 1995; Yasuda *et al.*, 1994). The latter found that the ratio of G from laboratory tests to G from field tests decreases markedly with increasing shear stiffness.

Shear strains developed during earthquakes may increase from about $10^{-3}\%$ in small earthquakes to $10^{-1}\%$ for large motions, and the maximum strain in each cycle will be different. For earthquake design purposes a value of two-thirds G measured at the maximum strain developed may be used. Alternatively, an appropriate value of G can be calculated from the relationship

$$G = \frac{E}{2(1+v)} \tag{5.3}$$

where E is Young's modulus and v is Poisson's ratio. In the absence of any more specific data, low strain values of E may be taken from Table 5.3. Values of Poisson's ratio from Table 5.4 may be used in the above formula.

Table 5.3 Typical modulus of elasticity values for soils and rocks

Soil type	$E(\text{MPa/m}^2)$	E/c_u
Soft clay	up to 15	300
Firm, stiff clay	10–50	300
Very stiff, hard clay	25–200	300
Silty sand	7–70	
Loose sand	15–50	
Dense sand	50–120	
Dense sand and gravel	90–200	
Sandstone	up to 50,000	400
Chalk	5,000–20,000	2000
Limestone	25,000–100,000	600
Basalt	15,000–100,000	600

Note that the values of E vary greatly for each soil type depending on the chemical and physical condition of the soil in question. Hence the above wide ranges of E value provide only vague guidance prior to test results being available. The ratio E/c_u may be helpful, if the undrained shear strength c_u is known, although the value of this ratio also varies for a given soil type.

Table 5.4 Typical values of Poisson's ratio for soils

Soil type	Poisson's ratio, v
Clean sands and gravels	0.33
Stiff clay	0.40
Soft clay	0.45

A value of 0.4 will be adequate for most practical purposes.

Damping

The second key dynamic parameter for soils is damping. Two fundamentally different damping phenomena are associated with soils, namely material damping and radiation damping.

Material damping (or internal damping) in a soil occurs when any vibration wave passes through the soil. It can be thought of as a measure of the loss of vibration energy resulting primarily from *hysteresis* in the soil. Damping is conveniently expressed as a fraction of critical damping, in which form it is referred to as the damping ratio.

Considering the hysteresis loop on the right-hand side of Figure 5.1, it can be shown that the equivalent viscous damping ratio may be expressed as

$$\xi = \frac{W}{4\pi \Delta W} \qquad (5.4)$$

where W is the energy loss per cycle (area of hysteresis loop), and ΔW the strain energy stored in equivalent perfectly elastic material (area OAB).

Published data on damping ratios are sparse, and consist only of values deduced from tests on small samples, or theoretical estimates. It should be appreciated that to date no *in situ* determinations of material damping have been made, and that damping ratios may only be used in analyses in a comparative sense. As dynamic soils analyses are required for some projects, at least for its qualitative information, a means of choosing values of material damping is required. Some material damping values are therefore given in Figure 5.3. These represent average values of laboratory test results on sands and saturated clays. In the absence of any other information it may be reasonable to take the damping of gravels as for sand.

The variability of damping hidden by the 'average' curves in Figure 5.3 is illustrated by the large number of factors found by Dobry and Vucetic (1987) to affect soil damping. As well as increasing with strain, damping decreases with confining pressure, void ratio, geologic age, plasticity index and sometimes with cementation.

In considering the vibration of foundations radiation damping is present as well as material damping. Radiation damping is a measure of the energy loss from the structure through radiation of waves away from the footing, i.e. it is a purely geometrical effect. Like material damping, it is very difficult to measure in the field. The theory for the elastic half-space has been used to provide estimates for the magnitude of radiation damping. Whitman and Richart (1967) have calculated approximate values of radiation damping for circular footings for machines by this method and their results are reproduced in Figure 5.4.

As with the values for material damping, the limitations of the values in Figure 5.4 must be emphasized. First, they involve the approximation that radiation damping is frequency independent, a reasonable assumption in some cases; second, because they are only theoretical values for a particular type of footing, they should be applied with circumspection. In the analysis of foundations of buildings the usefulness of Figure 5.4 may be for qualitative rather than quantitative assessments, but the following generalizations may be helpful. For horizontal and vertical translations, radiation damping may be quite

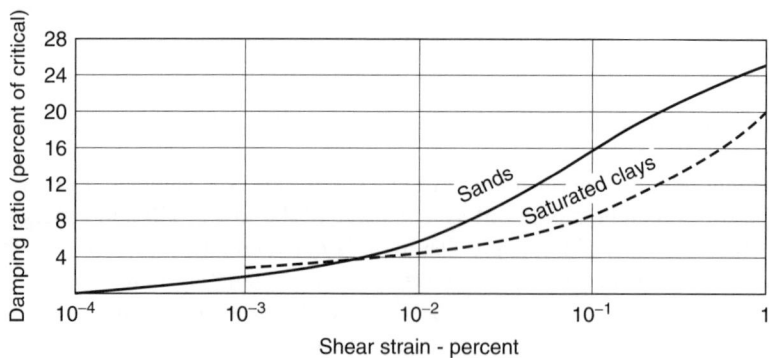

Figure 5.3 Average relationship of internal damping to shear strain for sands and saturated clays (after Seed and Idriss, 1970; Seed *et al.*, 1984)

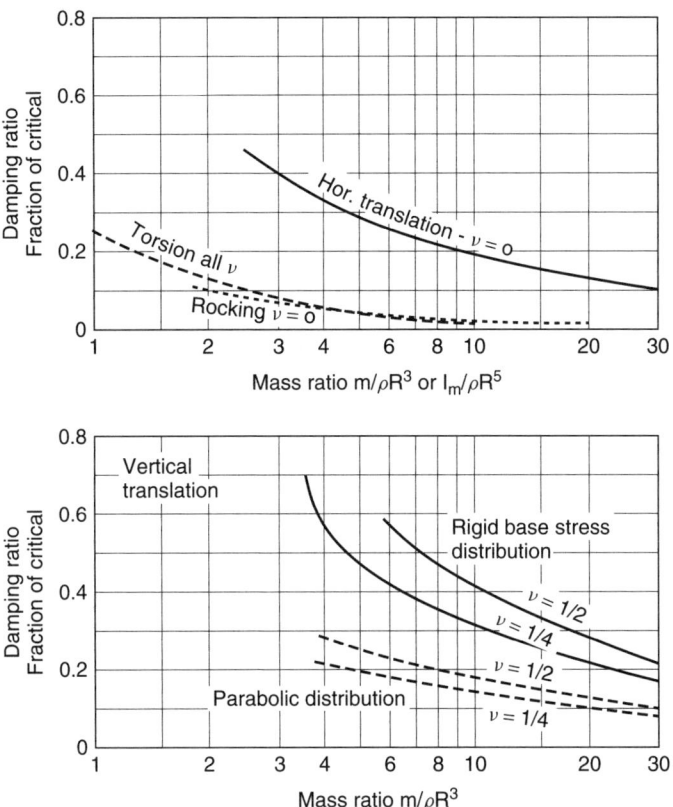

Figure 5.4 Values of equivalent damping ratio for radiation damping for machines, derived from the theory of circular footings on elastic half-space. (Reprinted from Whitman and Richart (1967), Design procedures for dynamically loaded foundations. *J Soil Mech and Found Divn*, **93**(SM6): 169–91, by permission of the American Society of Civil Engineers)

large (>10% of critical), while for rocking or twisting it is quite small (about 2% of critical) and may be ignored in most practical design problems.

A further limitation of the half-space theory is that it takes no account of the reflective boundaries provided by harder soil layers or by bedrock at some distance vertically or horizontally from the structure. Any such reflection of radiating waves will naturally reduce the beneficial radiation damping effect. Various aspects of radiation damping are discussed in Section 5.3.

In Figure 5.4, m is the mass of foundation block plus machinery, R is the radius (or equivalent radius) of the soil contact area at the foundation base, ρ is the mass density of the soil and v is Poisson's ratio for the soil. For rectangular bases of plan size $B \times L$ the equivalent radius is given by

$$R = \left(\frac{BL}{\pi}\right)^{1/2} \quad (5.5)$$

for translation,

$$R = \left(\frac{BL^3}{3\pi}\right)^{1/4} \tag{5.6}$$

for rocking, and

$$R = \left\{\frac{BL(B^2 + L^2)}{6\pi}\right\}^{1/4} \tag{5.7}$$

for twisting. The above method is comparable in ease of application to that given in under 'Effective damping' in Section 5.3.4. However, their limitations may make them inappropriate for use in various circumstances. If dashpot damping coefficients, c, are required, frequency-independent approximations to the half-space values for circular foundations are given by the simple formulae in Table 5.8. Alternatively, more widely applicable means of allowing for radiation damping, and a method of combining material and radiation damping, are given in Section 5.3.3.

5.2.2 Site response to earthquakes

Introduction

As outlined in Chapter 3, there is a great variety of possible geological and soil conditions at construction sites, which give rise to a variety of responses in earthquakes. The basic response phenomena which will be considered below are:

- modification of bedrock excitation during transmission through the overlying soils (amplification or attenuation);
- topographical effects;
- settlement of dry sands;
- liquefaction of saturated cohesionless soils.

The methods of analysing these responses vary in complexity, from simple empirical criteria to highly sophisticated analytical techniques. Regardless of the resources available, it should be borne in mind that knowledge of the real dynamical characteristics of the underlying soils is always incomplete, and the sophistication of the analyses used should not exceed the quality of the available data.

In the following discussion, emphasis will be placed more on practical design procedures than on research methods. As has been stressed by Ambraseys (1973), there is a great need for simple methods correlated to field experience in the subject of soil dynamics.

Effect of soil on bedrock excitation

The presence of soil overlying bedrock modifies the excitation in a complex manner, with conflicting effects dependent on dynamic characteristics of the soil layers and the strength of the excitation. In many earthquakes, the degree of damage to structures situated on soils has been reported as worse than that occurring on adjacent bedrock sites. Measured on the subjective intensity scales, the intensity may increase by 1 or 2 units (or occasionally more) compared with bedrock, depending on the soil type. Such measures of soil effects are very crude, but give a broad indication of the effect of soil layers when amplification, or sometimes attenuation, occurs.

The modifications to the incoming bedrock motions are dependent on several factors, notably:

- amplitude of shaking;
- frequency of vibration;
- properties of soil (modulus and damping);
- geometry, depth and stratification of the soils;
- water level (liquefaction).

It follows from the above that it is essential to understand the dynamical properties of soils as structures in order to predict their response.

Period of vibration of soil sites

The natural frequencies or periods of vibration of any dynamical system comprise a fundamental indicator of the dynamic response characteristics of the system. In the case of soil systems, we have seen in Section 3.3 that the distinction between two of the ground classes (C and D) is whether the site periods are less than or greater than 0.6 s, respectively.

If we consider a stratum of uniform thickness H, we find that the period of vibration T is a simple function of stratum stiffness and density parameters,

$$T_n = \frac{4H}{(2n-1)v_s}, \qquad (5.8)$$

where N is an integer, 1, 2, 3, ..., and v_s is the mean shear-wave velocity in the layer and a function of stiffness and density (equation (5.1)). The fundamental period, corresponding to $N = 1$, occurs when a shear wave of wavelength $4H$ passes through and is reflected in the stratum, while the larger integers $N = 2, 3, \ldots$, correspond to the higher harmonics.

Where a site is composed of more than one layer of soil the period of the soil may be estimated by using a weighted average value for the shear-wave velocity in equation (5.8) such that

$$v_s = \frac{\sum_{i=1}^{n} v_{si} H_i}{H}. \qquad (5.9)$$

In practice, in attempting to assess the fundamental period T_1 of a given site it is difficult to obtain a value from equation (5.8), unless reliable periods of similar sites are available for tuning purposes. The chief difficulty arises in deriving a suitable value for the shear-wave velocity, which should be that related to the level of shear strain in the soil, G, during the design earthquake. The value of T for soil increases with increasing strength of shaking (just as it does for structures when stressed beyond the elastic state), because G decreases (Figures 5.1 and 5.2). The shear-wave velocity is measured at low strains (0.0001%), and

Table 5.5 Factors for reducing shear-wave velocity measured at low shear strain ($\leq 0.001\%$), from Applied Technology Council (1978, p. 66). These values are representative of the San Francisco Bay area

Effective peak ground acceleration	$\dfrac{v_s\text{(high strain)}}{v_s\text{(low strain)}}$
$a_{max} \leq 0.1g$	0.9
$a_{max} = 0.15g$	0.8
$a_{max} = 0.2g$	0.7
$a_{max} \geq 0.3g$	0.65

to convert such values to those appropriate to strong shaking they may be multiplied by the factors given in Table 5.5.

The values of T for soil calculated from equation (5.8) are likely to be higher than reality, unless due allowance is made for stiffening effects of geometrical features such as the restraint imposed by sides of valleys on alluvial deposits, and by properly judging the appropriate depth H to bedrock. For example, for using equation (5.8) bedrock may be defined as a low-strain shear-wave velocity of about 700 m/s (see also Table 5.1).

In the Lake Zone of Mexico City, the depth to the stiff soil layer which constitutes effective bedrock ranges up to about 60 m, and at this location the superficial clays are so flexible that the site periods reach as high as $T = 5$ s. This may be taken as a worldwide upper bound for buildable sites. Values of fundamental periods representative of some common soil deposits are given in Table 5.6.

In addition to the above methods of determining T, field measurements are sometimes also made (Section 3.4.2), or site period may be estimated from soil properties determined from borehole records (Lam and Wilson, 1999).

One-dimensional site response analysis

One-dimensional analysis is suitable for near horizontal sites and strata. As such conditions are common at building sites, one-dimensional analyses are widely used. Consider the three analytical situations indicated in Figure 5.5 for sites with regular geometry

Table 5.6 Typical values of fundamental period for soil deposits (for rock motions with $a_{max} = 0.4$ g), from Structural Engineers Association of California (1980)

Soil depth (m)	Dense sand (s)	5 m of fill over normally consolidated clay* (s)
10	0.3–0.5	0.5–1.0
30	0.6–1.2	1.5–2.3
60	1.0–1.8	1.8–2.8
90	1.5–2.3	2.0–3.0
150	2.0–3.5	

*Representative of San Francisco Bay area.

Soils

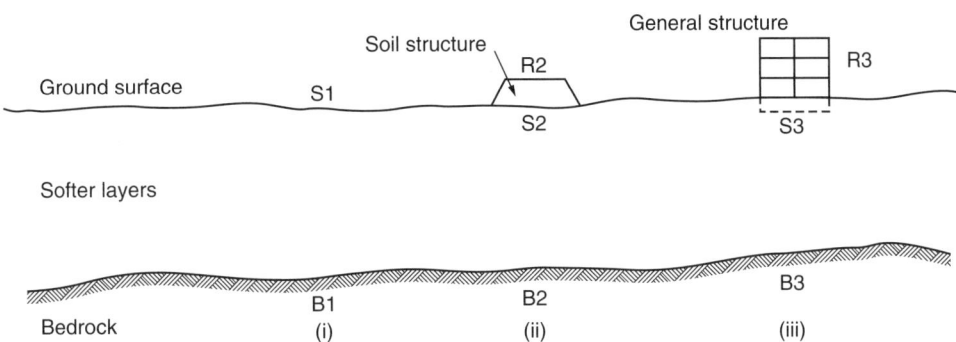

Figure 5.5 Schematic diagram of sites with subsurface rock and near horizontal strata

(i.e. near horizontal layers and flat topography). Site (i) represents the general site evaluation problem; here the stability of the overburden in earthquakes is to be determined with regard to phenomena such as settlement, liquefaction, or landslides, in relation to the feasibility of future construction on this site or the safety of adjacent sites. For the dynamic response analysis of this site, an accelerogram or response spectrum must be applied at B1 to the soil system between B1 and S1. This necessitates the choice of a suitable bedrock motion. Attempts have also been made to compute bedrock motion from surface motion from another site, as discussed later in this section.

Sites (ii) and (iii) in Figure 5.5 represent any site with any structure. The dynamic analysis of a structure on such a site may be carried out in either of two ways. First, the total soil and structure system from bedrock to the top of the structure may be analysed together with applying bedrock motion at B2, B3 and determining the responses of the whole system, including that of the structure R2, R3. This is the ideal means of analysis, as full allowance for interaction between *in situ* soils and constructed soils is included. The dynamic input at bedrock is chosen as for site (i). Or second, the structure may be analysed by applying a dynamic input at its base (S2, S3) or at some arbitrary distance below ground surface. The dynamic input appropriate for application at S2, S3 may be either: (1) surface motion accelerograms or spectra derived specially for the site by computing the modifications caused by the overlying soils on the bedrock motions input at B2, B3; or (2) surface motion accelerograms or spectra derived without specific dynamic analysis of the soil layers, as described in Section 4.7.

At present, the most common and probably most practical technique for modelling the dynamic behaviour of the soil above bedrock is that of the *vertical shear beam model*, which is so called because of the use of shear-wave theory. Several types of errors or limitations apply to the shear beam model as discussed below.

(1) Errors arise in representing a three-dimensional problem by a one-dimensional model.
(2) Nearly all shear beam models assume linear material behaviour as a crude approximation to the real non-elastic behaviour.
(3) Errors arise from the use of viscous rather than hysteretic damping.
(4) Errors from the use of approximate mathematical solutions.

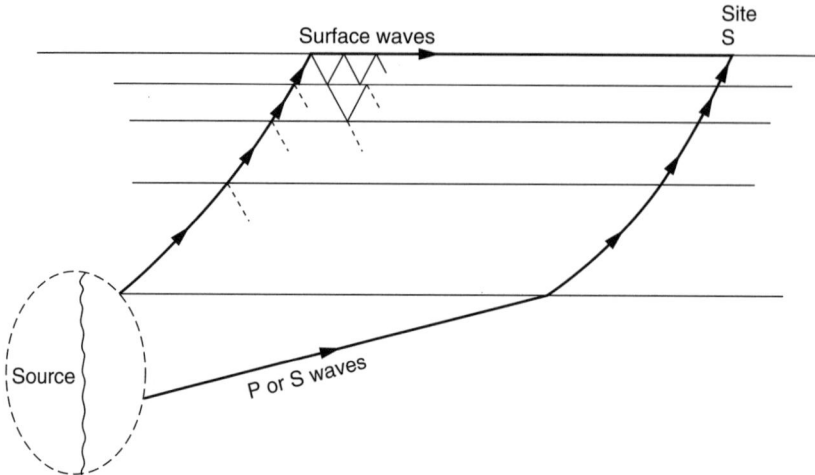

Figure 5.6 Schematic relationship of source, travel paths, and site as assumed in one-dimensional shear wave studies

(5) The shear beam model is valid only for sites where ground motion is dominated by shear waves propagating vertically through the soil. This is reasonably true at many sites related to the earthquake focus as illustrated in Figure 5.6. The shear waves will be approximately vertical for deep-focus earthquakes, but this will be less true at sites near the source of shallow earthquakes.
(6) For the shear beam model to be applicable the boundaries of the site must be essentially horizontal, allowing the soil profile to be treated as a series of semi-infinite layers.
(7) Finally, the effect of the presence of the proposed structure (or other structures) is not readily included in the computation of surface motion. For further discussion of the soil–structure interaction problem, see Section 5.3.

Two main types of vertical shear beam model are in use: first, the lumped mass methods; and second, continuous solutions in the frequency domain. The chief characteristics of each of these methods are now discussed briefly.

In the *lumped mass model* the soil profile is idealized with discrete mass concentrations interconnected by stiffness elements which represent the structural properties of the soil. A modal analysis is commonly used because of the familiarity of modal superposition to earthquake engineers. In modal analysis, it is necessary to assume linear material behaviour and viscous damping. As the damping of soils is more nearly hysteretic it is common to use an *equivalent* viscous damping which assumes constant damping in all modes, rather than true viscous damping in which the critical damping ratio would increase in proportion to the natural frequency of each mode.

Further allowances for the non-linearity of soil behaviour and hysteretic damping have been made by Seed and Idriss (1969), who used an iterative procedure to adjust the soil properties according to the level of strain. Even so they make the considerable simplification of averaging the properties of all layers in the soil profile. Further discussion on the problems involved in determining suitable soil properties for use in modal solutions may be found in Section 5.2.1 and in papers by Whitman *et al.* (1972) and Ambraseys (1973).

A widely used computer program using this iterative linear lumped mass shear beam model is SHAKE, which has been made more user-friendly since its original release in 1972.

A *continuous solution in the frequency domain* provides an 'exact' alternative to the lumped mass treatment of the vertical shear beam model. In this method the transfer of the bedrock motion to the surface is derived by consideration of the equation of motion of one-dimensional wave propagation in a continuous medium,

$$\rho \frac{\partial^2 u}{\partial t^2} = G \frac{\partial^2 u}{\partial x^2} + \eta \frac{\partial^3 u}{\partial t \partial x^2}, \qquad (5.10)$$

where ρ is the density of a semi-infinite soil layer, G the shear modulus, η the viscosity constant, and $u(x, t)$ the displacement of a point in the soil layer.

Transfer functions may be derived which modify input bedrock harmonic motions into corresponding surface motions in terms of the elastic properties of the intervening layers and of the bedrock layer itself. By multiplying the Fourier spectrum corresponding to the time-dependent bedrock motion by the transfer function, the surface Fourier spectrum is found. This Fourier spectrum may then be converted into the surface accelerogram. Fuller discussion of the continuous solution to the shear beam problem has been given by Roesset (1972) and Schnabel *et al.* (1971). Computer programs involving Fourier analysis and transfer functions are simple and may be more economical than those using the lumped mass solution if output at only a few points is desired. The continuous solution has the advantages that it readily handles many soil layers with different properties, including the bedrock layer, and any linear damping may be used, but it has the disadvantage of handling linear properties only.

An interesting feature of the transfer function technique is the facility with which bedrock motions can be estimated from surface motion recorded at a given site. This is a useful source of input bedrock data at another site. The main problem with transferring surface motion at one site to surface motion at another site lies in the incorporation of *two* sets of errors implicit in modelling ground motion transfer downwards through one soil profile as well as upwards through another.

Measured responses of soil sites

The most common effect of soils, amplification, was observed, for example, in the 1980 magnitude 6.1 Chiba-ken Chubu earthquake in Japan (Tazoh *et al.*, 1984); at one site the peak ground acceleration increased from 31 cm/s^2 at a depth of 60 m to 104 cm/s^2 at the surface vertically above, while at another site a_{max} increased from 64 cm/s^2 at 42 m depth to 194 cm/s^2 at the surface. As a further example, amplification was particularly strong in part of Mexico City in the September 1985 earthquake. At the one instrumented site in the zone of interest, the 30 m deep soft clay remained essentially elastic (and hence had low damping) throughout the long excitation, such that the peak bedrock acceleration was amplified about five times. The spectral accelerations at the site period of 2 s was amplified even more (Seed and Romo, 1986).

The opposite effect, *attenuation*, appears to have occurred in the 1957 magnitude 5.3 San Francisco earthquake, as shown in Figure 5.7. Here there were several sites all about equidistant from the earthquake focus, and the peak ground acceleration at two of the soil sites was only half those at the adjacent rock sites. The increase in the response ordinates at longer periods on the soil sites is also evident in Figure 5.7 and has been further illustrated by Seed (1975). It is interesting and surprising that attenuation of peak ground accelerations occurred in soils at the low amplitudes of shaking of this small-magnitude event.

Two- and three-dimensional dynamic site response analysis

For a range of sites of interest the geometry of soil sites cannot be adequately modelled in one dimension, and two-, or even three-dimensional models are appropriate. These include irregular or sloping sites, earth dams, retaining walls, or sites with buried structures. Computer modelling using finite elements (linear and non-linear) and shear beam approaches are used for such problems. For an introduction to these topics, see Kramer (1996) or other specialist literature.

Topographical effects

When the surface topography is not flat, the hills or valleys constitute structures which obviously will have dynamical characteristics different from a flat plain. Such effects have been demonstrated by the evidence of amplification on ridges or hilltops when compared to adjacent flat ground. For example, the high ground motions recorded at the Pacoima Dam site in the 1971 San Fernando, California, earthquake were believed to be due in part to the ridge location (Boore, 1973). In subsequent studies, Brune (1984) found that both amplification or attenuation was possible, depending on the angle of incident waves. However, the high accelerations appear to be at least in part due to near-fault rupture directivity effects. This illustrates the complexity of the subject.

An apparently unequivocal example of amplification is that given by several groups of similar houses on topographically different sites in the 1985 San Antonio, Chile, earthquake. The houses on hilltops or ridges were heavily damaged while those on nearly flat or valley sites were only slightly damaged (Dowrick, 1985). A further example is given by accelerograph recordings of two earthquakes at three adjacent rock sites across a valley in New Zealand. It was found (McVerry *et al.*, 1984) that amplitudes on the two hilltop sites on opposite sides of the valley were much the same as each other, but they were about twice the size of those recorded for the third site, which was near the bottom of the valley. In another study of ground-motion records, from three hilltop sites in Israel, Zaslavsky and Shapira (2000) found amplifications by factors of up to 4 due to topographic effects. In New Zealand an instrumental study by Beuch *et al.* (2007) was made of a mountain ridge 180 m high, 500 m wide and 800 m long. Time- and frequency-domain studies found that amplification was concentrated along the elongated crest, where amplifications were found of up to 11 times relative to the motion at the base, and the main response frequency was around 5 Hz. It is important to be aware that these very large amplifications occurred in low-amplitude shaking, and are not expected to be maintained in strong shaking, which would cause non-linear effects such as cracking (and landsliding) of the rock.

Soils

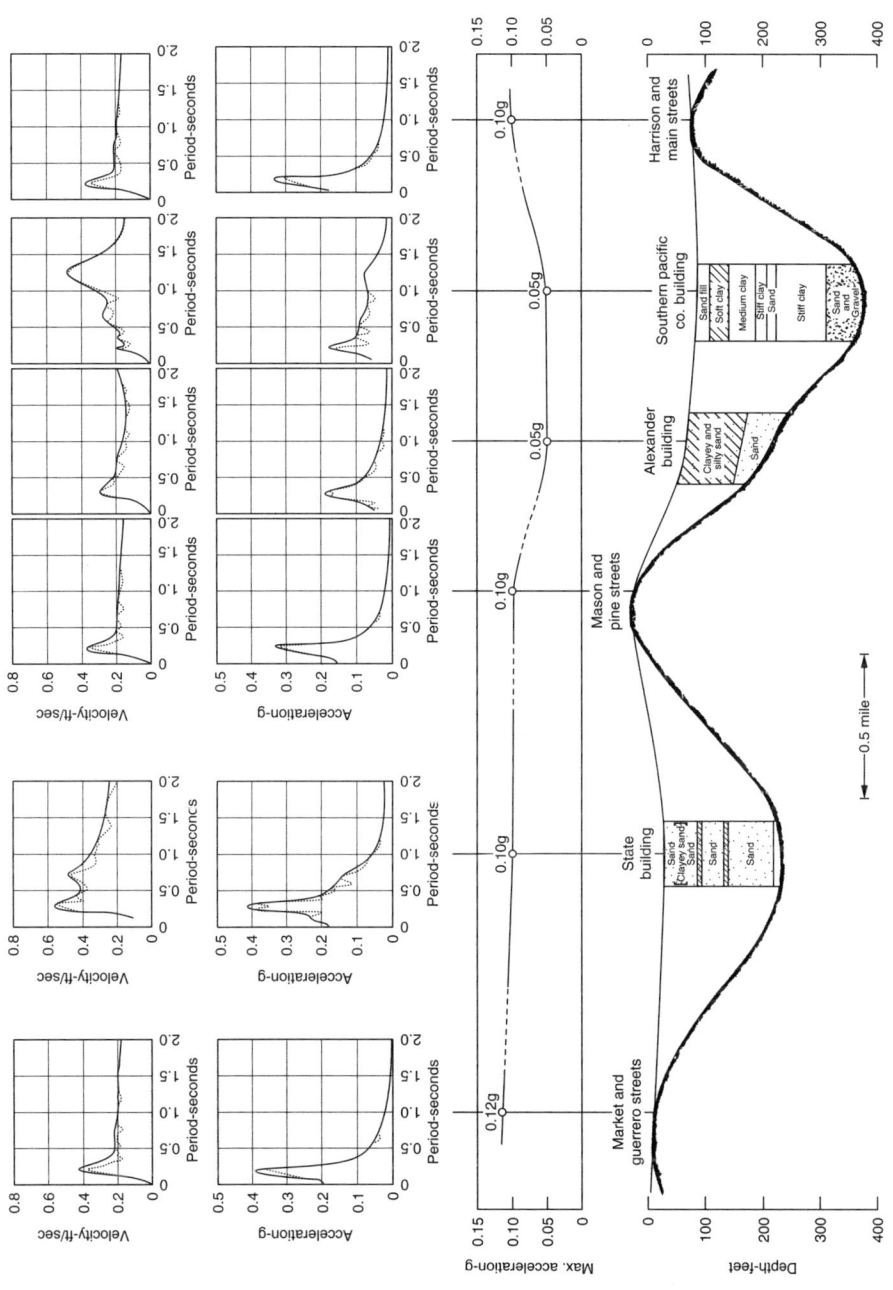

Figure 5.7 Soil conditions and response spectral characteristics of ground motions of six sites, San Francisco earthquake, 1957 (after Seed, 1975). (Reproduced by permission of John Wiley & Sons Originally Van Nostrand Reinhold)

Many analytical studies have been made of topographical effects, different shapes of hills being modelled. Amplification factors of up to 10 have been estimated. Such studies have been reviewed by Geli *et al.* (1988).

Settlement of dry sands

It is well known that loose sands can be compacted by vibration. In earthquakes, such compaction causes settlements which may have serious effects on all types of construction. It is therefore important to be able to assess the degree of vulnerability to compaction of a given sand deposit. Unfortunately this is difficult to do with accuracy, but it appears that sands with relative density less than 60%, or with standard penetration resistance less than 15, are susceptible to significant settlement. The amount of compaction achieved by any given earthquake will obviously depend on the magnitude and duration of shaking as well as on relative density, as demonstrated by the laboratory test results plotted in Figure 5.8.

Attempts have been made to predict the settlement of sands during earthquakes and a simple method from Newmark and Rosenblueth (1971) is presented below. It should be noted that this ignores the effect of important factors such as confining pressure and number of cycles, but no fully satisfactory method of settlement prediction as yet exists.

There is a critical void ratio e_{cr} above which a granular deposit will compact when vibrated. If the void ratio of the stratum is $e > e_{cr}$ the maximum amount of settlement possible can be shown to be

$$\Delta H = \frac{e_{cr} - e}{1 - e} \tag{5.11}$$

where H is the depth of the stratum.

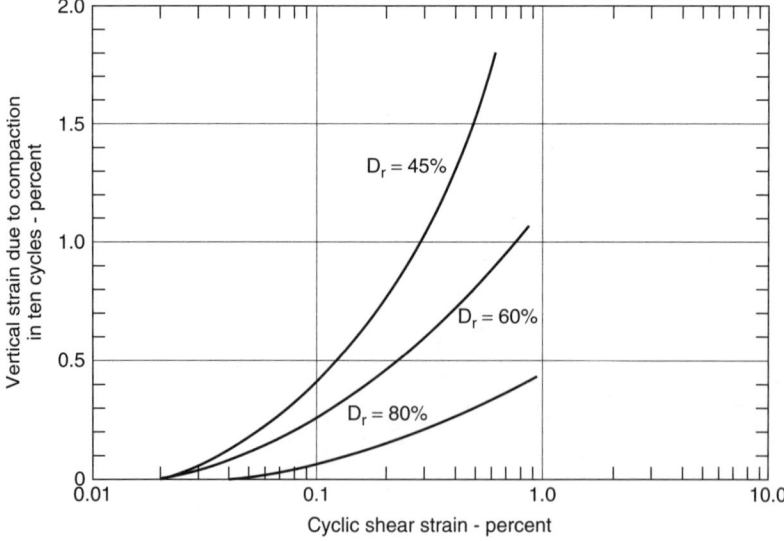

Figure 5.8 Effect of relative density on settlement in ten cycles (after Silver and Seed, 1969)

The critical void ratio can be obtained from

$$e_{cr} = e_{min} = (e_{max} - e_{min})\exp[-0.75\,a/g] \tag{5.12}$$

where e_{min} is the minimum possible void ratio as determined by testing, e_{max} the maximum possible ratio, a the amplitude of applied acceleration, and g the acceleration due to gravity.

Liquefaction of saturated cohesionless soils

During earthquake shaking, some saturated granular soils may compact, increasing the porewater pressure, thereby decreasing the effective stress which results in a loss of shear strength. This phenomenon is generally referred to as *liquefaction*. It is usually confined to sands and cohesionless coarse-grained silts, and is more severe in looser, uniformly graded soils, and those with more rounded particles. Also, in truly undrained conditions, gravelly soils can be susceptible to liquefaction.

Engineering interest in liquefaction has been high since 1964, when the Great Alaska earthquake and the Niigata earthquake both caused extensive and spectacular liquefaction damage to ground and structures alike. According to Kramer (1996), liquefaction is best understood if its phenomena are divided into two groups: *flow liquefaction* and *cyclic mobility*. Flow liquefaction produces dramatic flow failures which are driven by static shear stresses, such as are seen in the failure of earth dams. Cyclic mobility produces deformations which develop incrementally during an earthquake, as a result of cyclic shear stresses with or without a static shear stress regime. Although in cyclic mobility situations, the static shear stresses remain less than the shear strength of the liquefied soil, their presence contributes to *lateral spreading* on gently sloping ground or essentially flat land adjacent to bodies of water.

Level-ground liquefaction is the name given to the special case of cyclic mobility which arises when the ground is flat and static horizontal shear stresses are (consequently) zero. Large ground movements known as *ground oscillations* occur during the shaking, and little permanent lateral displacement occurs. However, level-ground liquefaction failure may occur when hydraulic equilibrium is reached after shaking has stopped, and is experienced as large vertical settlement.

Given that susceptible soils exist at a given site, an indication of whether liquefaction need be considered may be obtained by examination of Figure 5.9, prepared by Ambraseys (1988), where it is seen that liquefaction occurs within about 2 km of a shallow earthquake of $M_W = 5$, and up to about 300 km from event of $M_W = 8$. Implicit in Figure 5.9 is the importance of duration of shaking as well as amplitude criteria for susceptibility of a soil deposit to liquefaction, including the composition of the deposit (particle size, shape and gradation), the initial state of the soil (its stress and density at the time of the earthquake), and its geologic environment. Regarding the latter, soils susceptible to liquefaction come from a narrow range of geologic environments (Youd, 1991). In particular, recent Holocene soils laid down in low-energy environments are susceptible to liquefaction.

The crucial issue is whether liquefaction will be initiated in a given susceptible soil when subjected to the earthquake shaking of the chosen or specified hazard level. Various methods of assessing the likelihood of initiation of liquefaction have been proposed, using

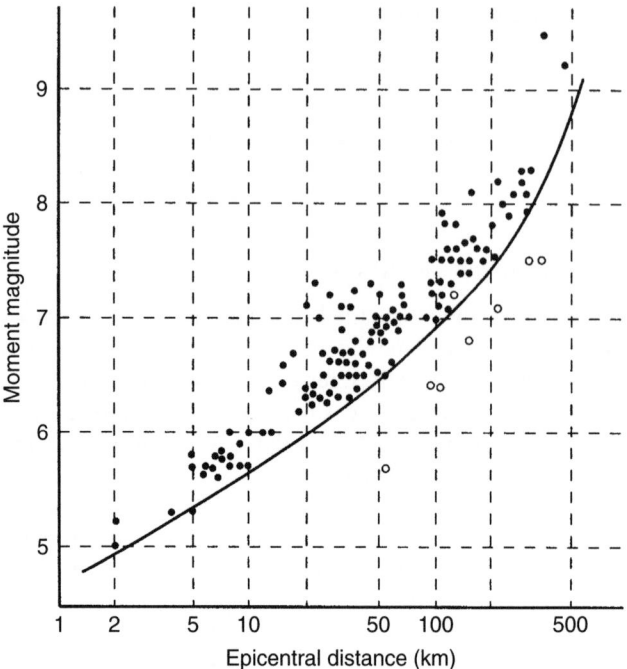

Figure 5.9 Relationship between limiting epicentral distance of sites at which liquefaction has been observed and moment magnitude for shallow earthquakes. Deep earthquakes (focal depths >50 km) have produced liquefaction at greater distances (after Ambraseys, 1988). (Reproduced by permission of John Wiley & Sons, Ltd)

a range of descriptors of soil properties (e.g. SPT values) and strength of shaking. The results of such methods do not necessarily give the same answer for a given situation. Only one method will be discussed here, namely that of Liao *et al*. (1988). This method is attractive because (unlike other methods) it gives a measure of the uncertainties involved in liquefaction assessments, being a probabilistic approach. Liao *et al*. analysed 278 case studies and developed an expression for the probability of liquefaction, P_L, where the strength of shaking is expressed in terms of the cyclic stress ratio, CSR, and the soil properties in terms of the SPT value, $(N_1)_{60}$, such that

$$P_L = \frac{1}{1 + \exp[-(\beta_0 + \beta_1 \ln(\text{CSR}) + \beta_2 (N_1)_{60})]} \tag{5.13}$$

where the parameters β_0, β_1, β_2 are shown in Table 5.7. Liquefaction probability curves for clean sand (<12% fines) and silty sands (>12% fines) are shown in Figure 5.10.

For further information on liquefaction see the specialist literature, including the lucid treatment given by Kramer (1996), the report by Kramer and Elgamal (2001) and the EERI monograph by Idriss and Boulanger (2008).

Table 5.7 Regression parameters for calculating probability of liquefaction, derived by Kramer (1996) from Liao *et al.* (1988). (Reprinted from *Geotechnical Earthquake Engineering* by Kramer, © by permission of Pearson Education, Inc., Upper Saddle River, NJ)

Data*	Number of cases	β_0	β_1	β_2
All cases	278	10.2	4.19	−0.244
Clean sand cases only	182	16.4	6.46	−0.398
Silty sand cases only	96	6.48	2.68	−0.182

*A fines content of 12% is used as the boundary between clean and silty sands.

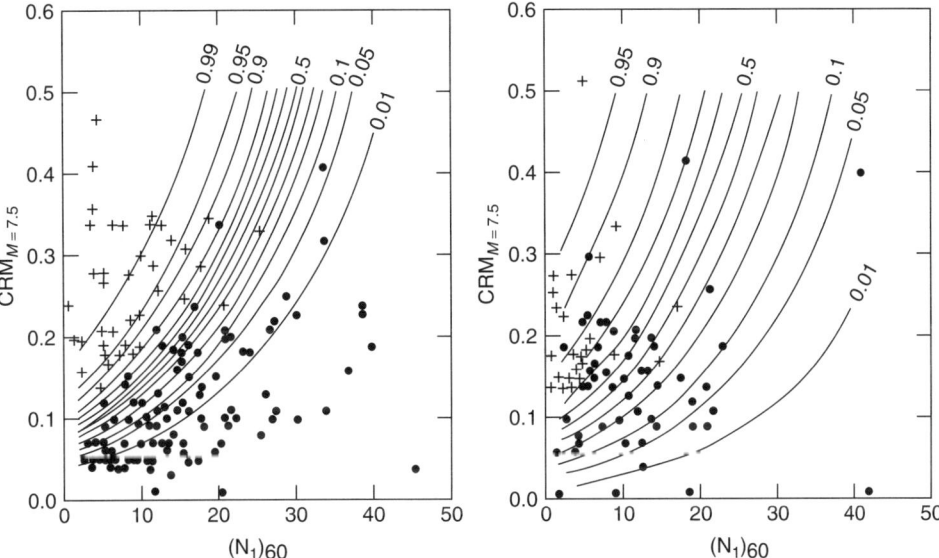

Figure 5.10 Contours of equal probability of liquefaction for (a) clean sand (less than 12% fines), and (b) silty sands (greater than 12% fines). (Reprinted from Liao *et al.* (1988), Regression models for evaluating liquefaction probability. *J Geotech Eng*, **114**: 389–411, by permission of the American Society of Civil Engineers)

5.3 Seismic Response of Soil–Structure Systems

5.3.1 Introduction

The importance of the nature of the subsoil for the seismic response of structures has been demonstrated in many earthquakes. For example, it is clear from studies of earthquakes that the relationship between the periods of vibration of structures and the period of the supporting soil is of profound importance with regard to the seismic response of the structure. An example from Mexico City is given in Section 3.2, item (2). In the case of the 1970 earthquake at Gediz, Turkey, part of a factory was demolished in a town 135 km from the epicentre while no other buildings in the town were damaged. Subsequent investigations revealed that the fundamental period of vibration of the factory was approximately equal to that of the underlying soil. Further evidence of the importance of periods of vibration was derived from the medium-sized earthquake of Caracas in 1967, which completely destroyed four buildings and caused extensive damage to many others. The pattern of structural damage has been directly related to the depth of soft alluvium overlying the bedrock (Seed *et al.*, 1972). Extensive damage to medium-rise buildings (5–9 storeys) was reported in areas where depth to bedrock was less than 100 m, while in areas where the alluvium thickness exceeded 150 m the damage was greater in taller buildings (over 14 storeys). The depth of alluvium is, of course, directly related to the periods of vibration of the soil (equation (5.8)).

To evaluate the seismic response of a structure at a given site, the dynamic properties of the combined soil–structure system must be understood. The nature of the subsoil may influence the response of the structure in five ways:

(1) The seismic excitation at bedrock is modified during transmission through the overlying soils to the foundation. This may cause *attenuation* or *amplification* effects (Figure 3.3, page 144 and Figure 5.7).
(2) The fixed base dynamic properties of the structure may be significantly modified by the presence of soils overlying bedrock. This will include changes in the mode shapes and periods of vibration.
(3) A significant part of the vibrational energy of the flexibly supported structure may be dissipated by material damping and radiation damping in the supporting medium.
(4) The increase in the fundamental period of moderately flexible structures due to soil–structure interaction may have detrimental effects on the imposed seismic demand.
(5) Structures sited on soft alluvium may be damaged by differential vertical displacements occurring before and/or during earthquakes. Although this phenomenon is not properly understood it seems logical that structures with relatively low horizontal strength will suffer worst from this phenomenon, i.e. low-rise structures will be most vulnerable. This effect is in contrast to resonance which, in the case of soft ground, will, of course, occur for longer-period (taller) structures.

Items (2)–(4) above are investigated under the general title of *soil–structure interaction*, which may be defined as the interdependent response relationship between a structure and its supporting soil. The behaviour of the structure is dependent in part upon the nature of the supporting soil, and similarly, the behaviour of the stratum is modified by the presence of the structure.

It follows that *soil amplification* and *attenuation* (item (1) above) will also be influenced by the presence of the structure, as the effect of soil–structure interaction is to produce a difference between the motion at the base of the structure and the free-field motion which would have occurred at the same point in the absence of the structure. In practice, however, this refinement in determining the soil amplification is seldom taken into account, the free-field motion generally being that which is applied to the soil–structure model. Because of the difficulties involved in making dynamic analytical models of soil systems, it has been common practice to ignore soil–structure interaction effects, simply treating structures as if rigidly based regardless of the soil conditions. However, intensive study in recent years has produced considerable advances in our knowledge of soil–structure interaction effects and also in the analytical techniques available, as discussed below.

5.3.2 Dynamic analysis of soil–structure systems

Comprehensive dynamic analysis of soil–structure systems is the most demanding analytical task in earthquake engineering. The cost, complexity, and validity of such exercises are major considerations. There are two main problems to be overcome. First, the large computational effort which is generally required for the foundation analysis makes the choice of foundation model very important; five main methods of modelling the foundation are discussed in the next section. Secondly, there are great uncertainties in defining a design ground motion which not only represents the nature of earthquake shaking appropriate for the site, but also represents a suitable level of risk.

Ideally, the earthquake motion should be applied at bedrock to the complete soil–structure system. This is not a very realistic method at present, because much less is known about bedrock motion than surface motion, and there is a great scatter in possible results for the soil amplification effect defined above. At present, the most realistic methods of analysis seem to be those which apply the *free-field* motion to the base of the structure, the free-field motion being that which would occur at the surface in the absence of the structure. This may be done most simply using simple springs at the base of the structure (Figure 5.11), as described in Section 5.3.3, or using a substructuring technique in which the foundation dynamic characteristics are predetermined and superposition of soil and structure response is carried out. The latter technique has been described by Penzien and Tseng (1976), using half-space modelling of the soil, and by Vaish and Chopra (1974), who illustrate their presentation with finite-element modelling. These two types of soil model are discussed in the next section.

It should be noted that where the dynamic behaviour is expressed in frequency-dependent terms, the problem must be analysed in the *frequency domain*, not the *time domain*. For this purpose, acceleration-time records must be transformed into acceleration-frequency terms using Fourier transform methods before application to the system. An inverse transformation is required to obtain the response time record. These techniques are described in the above two papers.

For projects in which soil–structure interaction effects are likely to be important, the choice of analytical method requires careful consideration. The reader will find useful extra guidance in Wolf and Song (1999) and the books of Wolf (1985, 1988).

5.3.3 Soil models for dynamic analysis

A dynamic model of the soil which attempts to fully model reality requires the representation of soil stiffness, material damping and radiation damping, allowing for strain dependence (non-linearity) and variation of soil properties in three dimensions. While various analytical techniques exist for handling different aspects of the above soil behaviour, they all suffer from varying combinations of expensiveness or inaccuracy. Therefore, there is some difficulty for any given project in choosing an analytical model for the soil which will permit an appropriate level of understanding of the soil–structure system.

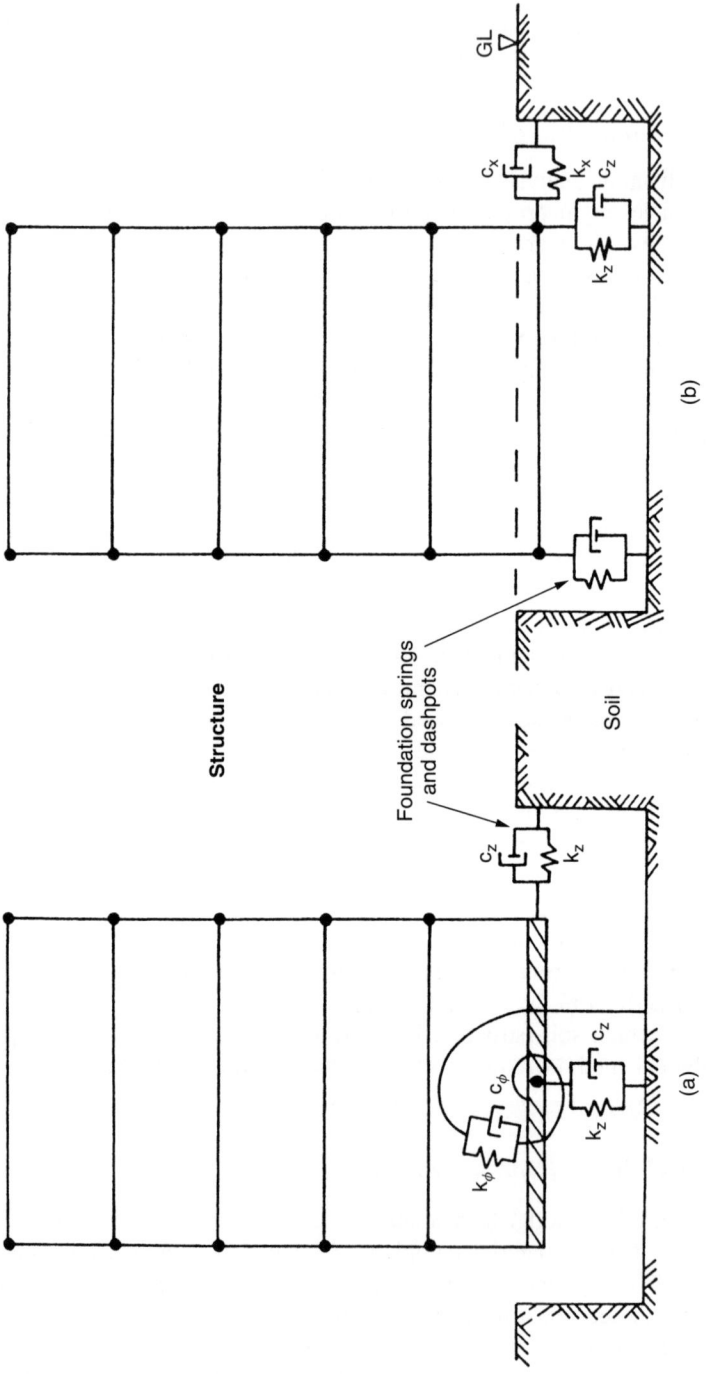

Figure 5.11 Rudimentary soil–structure analytical models representing soil properties by springs and dashpots

Soil–Structure Systems

The methods of modelling the soil may be divided into five categories of varying complexity:

(1) equivalent-static springs and viscous damping located at the base of the structure only;
(2) shear beam analogy using continua or lumped masses and springs distributed vertically through the soil profile;
(3) elastic or viscoelastic half-space;
(4) finite elements;
(5) hybrid model of (3) and (4).

A brief discussion of each of the above modelling methods follows below.

Springs and dashpots at the base of the structure

The most rudimentary method of modelling the soil is to use only springs, located at the base of the structure, to represent the appropriate selection of horizontal, rocking, vertical, and torsional stiffnesses of the soil (Figure 5.11). An increase in the rigorousness of the model may be effected by adding dashpots at the same location. In the system shown in Figure 5.11(b), the stiffness of the individual vertical springs must be chosen to sum to either the global rocking stiffness or the global vertical stiffness, as used in Figure 5.11(a), as it is unusual to achieve both conditions simultaneously. The same is true for damping. This discrepancy may not matter in analyses in which horizontal and vertical excitations are not applied simultaneously, but generally a conflict arises.

As a simple illustration, consider modelling a circular disc foundation by 32 vertical springs located around its perimeter (Figure 5.12). The total vertical spring stiffness is

$$k_{z0} = \frac{4GR}{(1-v)}. \tag{5.14}$$

This stiffness would therefore be provided by the sum of 32 vertical springs of stiffness

$$k_{zi} = \frac{GR}{8(1-v)}. \tag{5.15}$$

These springs give a rotational stiffness

$$k_\phi = 16R^2 k_{vi} = \frac{2GR^3}{(1-v)}. \tag{5.16}$$

However, this value is only three-quarters of that given by the half-space rocking spring formula

$$k_\phi = \frac{8GR^3}{3(1-v)}. \tag{5.17}$$

In these circumstances, the stiffness value for the vertical springs will need to be chosen to give a conservative result depending on the nature of the loading. This will usually

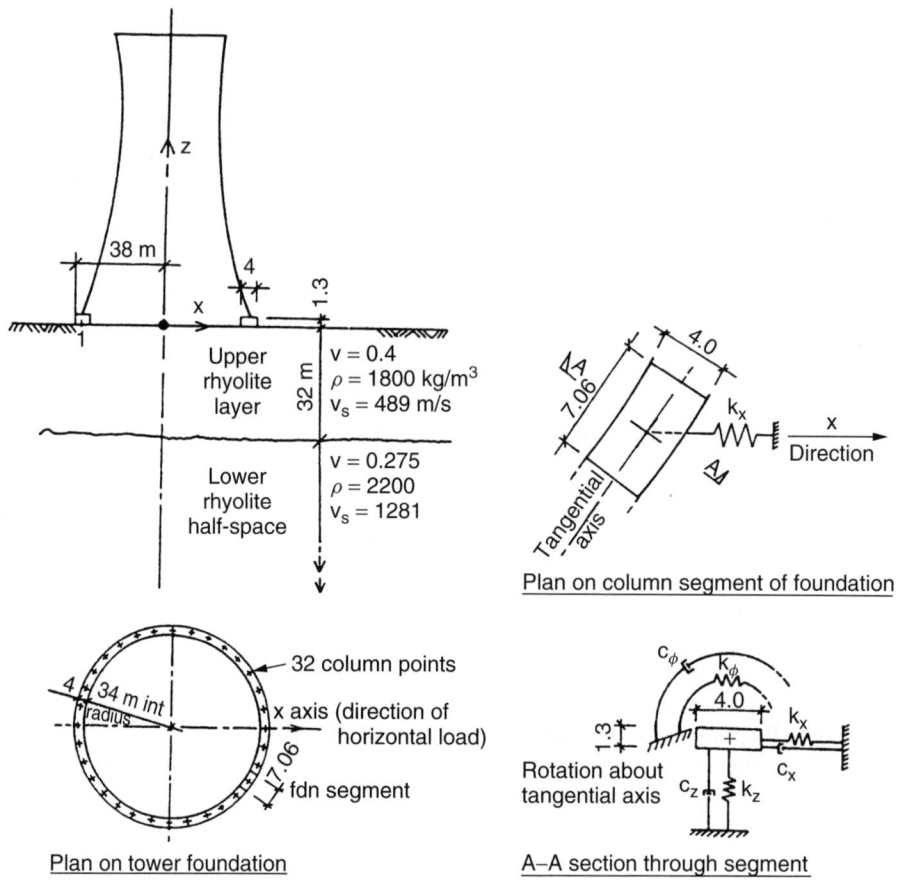

Figure 5.12 Cooling tower with soil properties approximating a layered half-space, represented by a two-dimensional model of springs and dashpots

be done by increasing k_{zi} so that the value of k_ϕ equates to the half-space solution. In some cases, it may be possible to equate the vertical and rotational stiffness criteria by locating the vertical springs on an increased radius, but this necessitates introducing very stiff dummy members into the foundation model, which may lead to numerical or local modelling problems.

A convenient method for determining the overall foundation spring stiffnesses is to use the zero-frequency (static) stiffnesses derived from elastic half-space theory as given in Table 5.8 and Figure 5.13. It should be noted that the values in Table 5.8 are for a homogeneous elastic half-space, but need to be factored to give some equivalence to layered soils or to allow for a given degree of non-linearity in the soil behaviour. Solutions for the stiffness of various shapes of footings may be found elsewhere, conveniently collected by Poulos and Davis (1974).

As an example of layered soils, consider a circular disc footing of 70 m radius on soils consisting of a layer of depth $H = 32$ m overlying a half-space. The soil properties of

Soil-Structure Systems

Table 5.8 Discrete foundation properties for rigid plate on elastic half-space

Motion	Circular footings			Rectangular footings
	Spring stiffness k	Viscous damper*	Added mass*	Spring stiffness k
Vertical	$\dfrac{4GR}{1-v}$	$1.79\sqrt{k\rho R^3}$	$1.5\rho R^3$	$\dfrac{G}{1-v}\beta_z\sqrt{BL}$
Horizontal	$\dfrac{8GR}{2-v}$	$1.08\sqrt{k\rho R^3}$	$0.28\rho R^3$	$2G(1+v)\beta_x\sqrt{BL}$
Rocking	$\dfrac{8GR^3}{3(1-v)}$	$0.47\sqrt{k\rho R^5}$	$0.49\rho R^5$	$\dfrac{G\beta_\phi BL^2}{1-v}$
Torsion	$\dfrac{16GR^3}{3}$	$1.11\sqrt{k\rho R^5}$	$0.7\rho R^5$	†

G is the shear modulus for the soil, where $G = E/\{2(1+v)\}$, v is Poisson's ratio for soil, ρ is mass density for soil, R is radius of footing, B, L, are the plan dimensions of rectangular pads, and β_x, β_z, β_ϕ are coefficients given in Figure 5.13.
*The properties come from Clough and Penzien (1993).
†For torsional spring stiffnesses of rectangular footings, see Newmark and Rosenblueth (1971, p. 98).

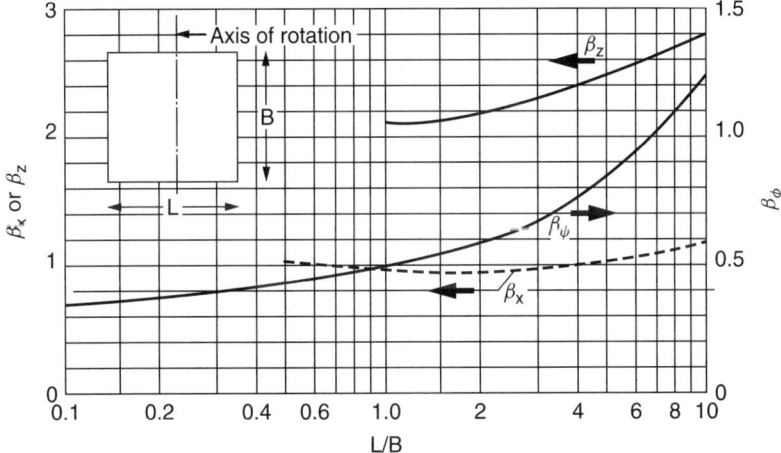

Figure 5.13 Coefficients β_x, β_z and β_ϕ for estimating spring stiffness of rectangular footings as in Table 5.8. (Reprinted from Whitman and Richart (1967), Design procedures for dynamically loaded foundations *J Soil Mech and Found Divn*, **93**(SM6): 169–91, by permission of the American Society of Civil Engineers)

these two elements are as follows:

Soil element	v	G(MPa)
(1) Layer	0.4	430
(2) Half-space	0.275	3610

To find the approximate equivalent half-space spring stiffness of the foundation system by simple hand calculation, first determine the spring stiffness of the two elements. Considering vertical stiffness for a layer on a rigid base, Bycroft (1956) gives

$$k_{z1} = \frac{4G_1 R_1}{1 - v_1}\left(1 + 1.4\frac{R_1}{H_1}\right) = 315 \times 10^9 \text{ N/m}. \tag{5.18}$$

For the lower half-space, make the approximate assumption that the stress from the disc footing spreads out through the layer at an angle of $45°$, so that the stiffness of element (2) relates to an effective disc radius of

$$R_2 = 40 + 32 = 72 \text{ m}.$$

Thus

$$k_{z2} = \frac{4G_2 R_2}{1 - v_2} = 1434 \times 10^9 \text{ N/m}. \tag{5.19}$$

The vertical spring stiffness for the combined soil system is obtained by adding the flexibilities of elements (1) and (2), so that

$$k_z(\text{system}) = \frac{k_{v1} k_{v2}}{k_{v1} + k_{v2}} = 258 \times 10^9 \text{ N/m}.$$

It can be seen that the presence below the layer of a half-space of moderate stiffness makes the foundation more flexible than if the layer had been underlain by effectively rigid rock.

The spring stiffnesses are dependent on the shear modulus, which in turn varies with the level of shear strain. Hence for linear elastic calculations, spring stiffnesses should be calculated corresponding to a value of shear strain which is less than the maximum expected shear strain. For instance, if the spring stiffness at low strain is k_0, then a value of k equal to $0.67k_0$ may be used in the analysis. Alternatively a series of comparative analyses may be done using a range of values of k, particularly if *in situ* tests have not been made; in this case it may be appropriate to select values of k from the range

$$0.5k_0 \leq k \leq k_0 \tag{5.20}$$

for translation, or

$$0.3k_0 \leq k \leq k_0 \tag{5.21}$$

for rocking. Table 5.8 gives viscous damper values equivalent to radiation damping in a half-space foundation, where the degrees of freedom are represented by single discrete dashpots, as in Figure 5.11(a). These values will generally be reduced (often substantially) if layering exists in the upper regions of the soil, due to wave reflections at the interfaces (see Figure 5.15 and related text).

When the chosen method of analysis does not allow the use of foundation dashpots, difficulties arise in accurately representing the effects of material damping and radiation damping in the foundation, as the total amount of equivalent viscous damping for the foundation in some cases exceeds considerably that for the superstructure. A conservative compromise between the structural and soil damping values will generally be necessary.

Also the damping in the soil in different modes of vibration varies considerably. When using dynamic analysis computer programs written for equal damping in all modes, some intermediate value of damping has to be chosen which hopefully will lead to the most realistic result. The value of damping used should not vary too greatly from that of the mode in which most of the vibrational work is done. Hence, a trial mode shape analysis may have to be done to determine which modes predominate. Use of too high or too low a value of damping will lead to unconservative or conservative results, respectively.

Shear beam

The shear beam approach may be used to model the soil layers overlying bedrock (Figure 5.14), although difficulties arise in choosing appropriate stiffness and damping values for the soil. Non-linearity may be allowed for by using iterative linear analyses such as those used in soil amplification studies (Section 5.2.2), or by non-linear foundation springs.

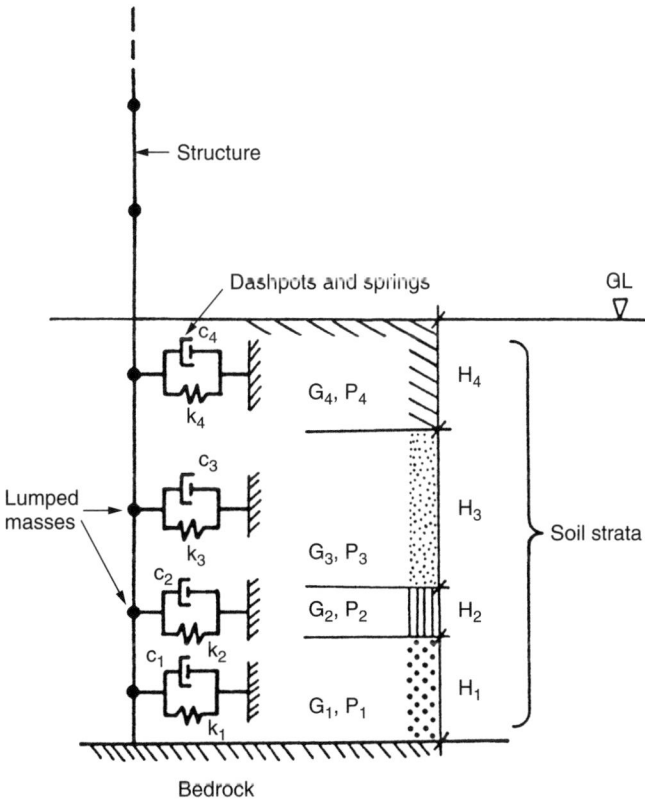

Figure 5.14 Soil–structure analytical model representing the soil vertical profile by a lumped parameter system of masses, springs and dashpots

Elastic or viscoelastic half-space

Modelling the foundation as a homogeneous linear elastic or viscoelastic half-space in which the stiffness and damping are treated as frequency-dependent provides a very useful means of allowing for the radiation damping effect. Various numerical and partly closed-form formulations of the theory have been made, such as those of Luco and Westmann (1971) and Veletsos and Wei (1971), and others as noted in the following discussion.

Consider a rigid circular plate of radius R on the surface of an *elastic homogeneous half-space* of density, ρ, Poisson's ratio v, and shear-wave velocity v_s. Let u_x, u_ϕ and u_z be the amplitudes of the horizontal, rotational and vertical displacements of the plate. Neglecting the small coupling between the horizontal and rocking motions, the relationship between forces and displacements may be stated as

$$F_j = K_j u_j \qquad (5.22)$$

where the subscript j denotes x, ϕ or z, and K_j are complex-valued stiffness (impedance) functions of the form

$$K_j = k_j(\beta_{kj} + ia_0 \beta_{cj}). \qquad (5.23)$$

The symbol k_j in equation (5.23) is the zero-frequency stiffness of the foundation, as given by the expressions for spring stiffness k in Table 5.8 and a_0 is a dimensionless frequency parameter $a_0 = \omega R/v_s$, where ω is the forcing frequency. Veletsos and Verbic (1973) have found analytical expressions approximating to the 'exact' numerical solutions such that β_{kj} and β_{cj} in equation (5.23) are given by

$$\beta_{kj} = 1 - \left[\frac{b_1 b_2^2}{1 + b_2^2 a_0^2} + b_3\right] a_0^2, \qquad (5.24)$$

$$\beta_{cj} = \frac{b_1 b_2^3 a_0^2}{1 + b_2^2 a_0^2}, \qquad (5.25)$$

where the parameters $b_1, \ldots, b_4$ are dimensionless functions of Poisson's ratio, and vary for horizontal, vertical and rocking motions, as given in Table 5.9. In this equivalent spring–dashpot representation of the supporting medium, i.e. the half-space, β_k is a measure of the dynamic stiffness of the spring and β_c is a measure of the damping coefficient of the dashpot. In this case, the damping is solely due to radiation damping.

Although the general expressions for β_{kj} and β_{cj} are both functions of frequency (a_0 in equations (5.24) and (5.25), it should be noted that in horizontal motion the terms involving a_0 are zero, so that $\beta_{kx} = 1$ and $\beta_{cx} = b_4$.

The expression for β_{cj} given in equation (5.25) may be used for evaluating the foundation dashpots for a structure supported on an elastic half-space (e.g. Figure 5.11) using the dashpot coefficient c_j, obtained from

$$c_j = \frac{R k_j \beta_{cj}}{v_s}. \qquad (5.26)$$

Table 5.9 Coefficients $b_1, \ldots, b_4$ for use in equations (5.24) and (5.25), from Veletsos and Verbic (1973). (Reprinted by permission of John Wiley & Sons, Ltd)

Motion	Poisson's ratio	b_1	b_2	b_3	b_4
Horizontal	0	0	0	0	0.775
	1/3	0	0	0	0.65
	1/2	0	0	0	0.60
Vertical	0	0.25	1.0	0	0.85
	1/3	0.35	0.8	0	0.75
	1/2	0	0	0.17	0.85
Rocking	0	0.8	0.525	0	0
	1/3	0.8	0.5	0	0
	1/2	0.8	0.4	0.027	0

This dashpot coefficient is used for obtaining the dynamic damping forces F_{Dj} from

$$F_{Dj} = c_j \dot{u}_j$$

where $\dot{u}_j$ is the velocity experienced by the dashpot.

A more widely applicable alternative to the above method of estimating radiation damping has been given by Gazetas and Dobry (1984). They found closed-form expressions for determining the frequency-dependent radiation damping coefficients for footings of various shapes and also for piles, and extended their method to deal with inhomogeneous soil conditions as well as the idealized half-space. For example, they found for a strip footing of width $2B$, on a uniform elastic half-space, that the radiation damping coefficient is given by

$$c_j = \rho V A \Re \left[-i \frac{H_1^{(2)}(a)}{H_0^{(2)}(a)} \right] \tag{5.27}$$

with

$$a = \frac{\omega B}{V}, \tag{5.28}$$

where ρ is the density of the soil; $A = 2B$ is the area (per unit length) of the footing; ω is frequency (rads/s); $H_1^{(2)}$ is the first-order Hankel function of second kind; $H_0^{(2)}$ is the second-order Hankel function of second kind; $\Re$ denotes the real part of the complex quantity implied by $i = \sqrt{(-1)}$; and V is wave velocity, where for obtaining c_z for vertical motion, *Lysmer's analogue* V_{La} is appropriate, i.e.

$$V = V_{La} = \frac{3.4 v_s}{\pi (1 - v)}, \tag{5.29}$$

while for finding c_x for horizontal motion, $V = v_s$.

For *inhomogeneous soil* – as well as the above uniform half-space solution, Gazetas and Dobry (1984) considered semi-infinite elastic soils having stiffness varying with depth z, from a value of G_0 at the surface, in the form

$$G = G_0 \left(\frac{z}{B}\right)^m, \quad 0 \le m \le 1. \tag{5.30}$$

The resulting expressions for c_j should be compared with those derived by Werkle and Waas (1986), who used a semi-analytical method to find stiffness and damping coefficients for the four modes of vibration for a half-space of linearly increasing stiffness with depth.

Non-linear soil behaviour cannot be explicitly modelled in the frequency-domain solutions used for the above formulations, but the *viscoelastic* hysteretic model may be thought of as representing a limited degree of non-linearity.

In the above formulations for an elastic half-space the *material damping* is neglected. The viscoelastic formulation of foundation impedance improves on this by allowing for material damping through the parameter $\tan \delta$ defined by

$$\tan \delta = \frac{\Delta W}{2\pi W}. \tag{5.31}$$

ΔW and W are defined in Section 5.2.1, where it will be seen that $\tan \delta$ is equal to twice the equivalent viscous damping for soil as defined by equation (5.31).

Veletsos and Verbic (1973) modified the elastic parameters β_{kj} and β_{cj} into the viscoelastic terms β^v_{kj} and β^v_{cj}, which are given here in a rearranged form due to Danay (1977), such that

$$\beta^v_{kj} = \beta_{kj} - a_0 \beta_{cj} \left[\frac{(1+\tan^2 \beta)^{1/2} - 1}{2}\right]^{1/2}, \tag{5.32}$$

$$\beta^v_{cj} = \beta_{cj} - a_0 \beta_{cj} \left[\frac{(1+\tan^2 \beta)^{1/2} + 1}{2}\right]^{1/2} + \beta_{kj} \frac{\tan \delta}{a_0}. \tag{5.33}$$

As may be expected, the inclusion of hysteretic damping increases the overall damping of the system and reduces the deformations (Veletsos and Nair, 1975). However, inspection of equation (5.33) shows that in cases where radiation damping is large, the effect of including the material damping will often be negligible.

The values obtained for c_x and c_z for circular footings on homogeneous half-spaces by the various methods outlined above are virtually *frequency-independent*, and an appropriate value of frequency has to be chosen when using equation (5.32) or (5.33). In some cases it may be deemed sufficient simply to take the frequency of the dominant mode of vibration, or perhaps a mean of the main modes weighted according to their participation factors. In a sophisticated analysis where foundation dashpots were required for non-linear analysis of an offshore oil platform, Watt et al. (1978) carried out a series of constant dashpot analyses until a constant rocking dashpot value was found with which the peak response of the system was similar to that obtained from an analysis using the 'exact' frequency-dependent impedance.

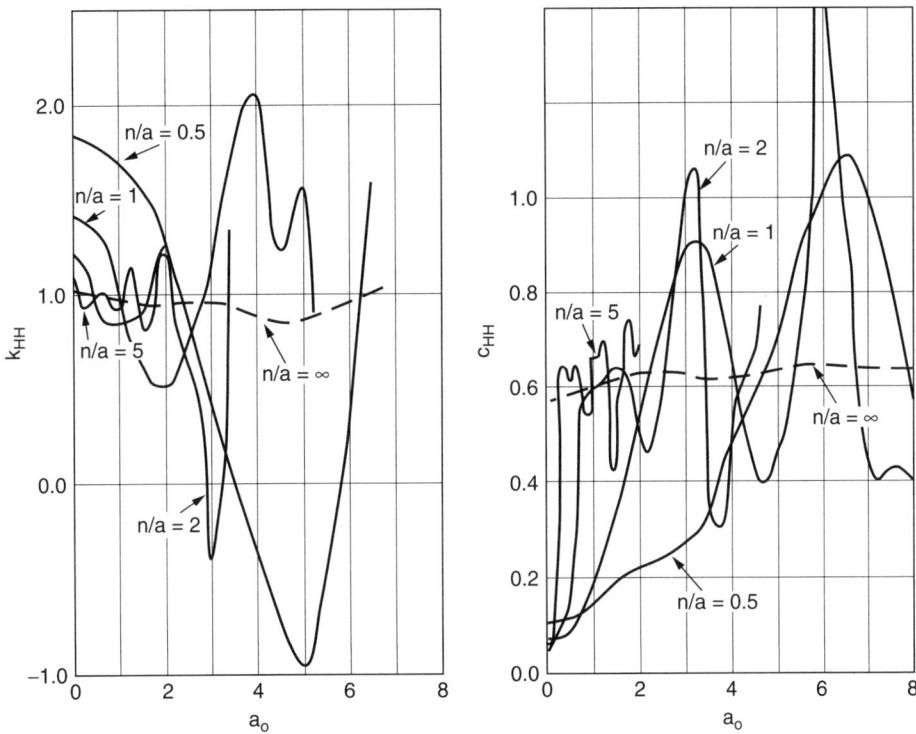

Figure 5.15 Frequency-dependent stiffness and damping coefficients for a layer on a half-space. H/a equates to H/R in the text. (Reprinted from *Nuclear Eng and Design*, **31**(2), Luco, JE (1974), Impedance functions for a rigid foundation on a layered medium, with permission from Elsevier Science)

The effects of *soil layers* may be studied using the half-space model. The work of Luco (1974) has demonstrated the importance of shallow reflective layer interfaces, i.e. low values of H/R, where H is the layer depth and R is the radius of the footing. This causes wave energy radiating away from the footing to be reflected back, thereby reducing radiation damping. In some cases, the damping oscillates very rapidly with frequency, so that caution with selection of frequencies is required. As an example, consider a cooling tower structure with an annular footing supported on a layered half-space (Figure 5.12). For horizontal vibrations we may assume that the annular footing behaves like a solid circular plate, so that we use $H/R = 32/38 = 0.85$ for entering Luco's graph (Figure 5.15). From a modal response analysis it was found that the first three horizontal modes of vibration had periods of vibration $T_1 = 0.44$ s, $T_2 = 0.22$ s, $T_3 = 0.11$ s. From these values the dimensionless frequency parameter $a_0 = \omega R/v_s$ has respective values 1.1, 2.2, and 3.3.

Entering Figure 5.15 using these values of H/R and a_0, we can determine that the damping in the first three modes is reduced compared to the unlayered case ($H/R = \infty$), by multiplying by the factors 0.3, 0.7 and 1.0. Finally, the effective reduction factor for horizontal motion is estimated by weighting the above reduction factors according to the contributions of each mode to the earthquake horizontal forces, which were 0.71, 0.28

and 0.01 for the first, second and higher modes, respectively. Thus the effect of the layer is to reduce the radiation damping by multiplying it by the reduction factor

$$(0.3 \times 0.71) + (0.7 \times 0.28) + (1.0 \times 0.01) = 0.42.$$

Using the same argument for the vertical mode, a reduction factor for the vertical radiation damping dashpot because of the layer was estimated at 0.51. For rocking, however, the situation was somewhat different, because the foundation rocking dashpots relate to rotations about the centreline of the annular strip footing as shown in Figure 5.12, so that the appropriate radius to use is the half width of the annulus. Hence $H/R = 32/2 = 16$. For these physical proportions negligible reflection of radiating energy occurs, and hence no reduction is required in damping values in the rocking mode due to layering.

Embedment of footings into a half-space is another situation which may deserve attention. For practical purposes, modifications to the results of the above methods for surface footings may be relatively simply made if the increased foundation stiffness caused by embedment is used. An example of this approach is given in Section 5.3.4.

Finally, in modelling soil–structure systems for dynamic analysis the *added mass of soil* which participates with the vibration of the footing may sometimes be significant. Estimates of this quantity have been derived in various studies of footings supported on a half-space, such as those of Veletsos and Verbic (1973) or those due to Clough and Penzien (1993) given above in Table 5.8.

Finite elements

The use of finite elements for modelling the foundation of a soil–structure system is the most comprehensive (if most time-consuming) method available. Like the half-space model, it permits radiation damping and three-dimensionality, but has the major advantage of easily allowing changes of soil stiffness both vertically and horizontally to be explicitly formulated. Embedment of footings is also readily dealt with. Although a full three-dimensional model may be too expensive, three dimensions should be simulated. This can be achieved either by an equivalent two-dimensional model, or for structures with cylindrical symmetry an analysis in cylindrical coordinates can be used.

To simulate radiation of energy through the boundaries of the element model three main methods are available:

(1) *Elementary boundaries* that do not absorb energy and rely on the distance to the boundary to minimize the effect of reflection waves.
(2) *Viscous boundaries* which attempt to absorb the radiating waves, modelling the far field by a series of dashpots and springs, as used by Lysmer and Kuhlemeyer (1969). The accuracy of this method is not very good for thin surface layers or for horizontal excitation, although an improved version has been developed by Ang and Newmark (1971).
(3) *Consistent boundaries* are the best absorptive boundaries at present available, reproducing the far field in a way consistent with the finite-element expansion used to model the core region. This method was generalized by Kausel (1974). The whole problem of finite-element modelling of unbounded media has been described by Wolf and Song (1996).

Non-linearity of soil behaviour can be modelled with non-linear finite elements, and *time-domain* analysis, but is very time-consuming. In *frequency-domain* solutions (for example, when using consistent boundaries), non-linearity can be approximately simulated using an iterative approach. In a study of a nuclear containment structure, Kausel (1976) showed that the iterative linear approach was adequate for structural response calculations, the full non-linear analysis only being warranted for detailed investigation of soil behaviour at or near failure.

5.3.4 Useful results from soil–structure interaction studies

There have been intensive theoretical investigations of the dynamics of soil–structure systems using soil modelling techniques as described above. Although many of the conclusions of these studies are still tentative, requiring experimental or field verification, some of the results are physically or intuitively sound. A brief summary of the more important conclusions is therefore included here.

When to include soil–structure interaction

Perhaps the leading question to be answered about soil–structure interaction is: 'For what soil conditions will the rigid base assumption lead to significant errors in the response calculations?' Veletsos and Meek (1974) have suggested that consideration of soil–structure interaction is only warranted for values of the ratio

$$\frac{v_s}{fh} < 20, \qquad (5.34)$$

where v_s is the shear-wave velocity in the soil half-space, f is the fixed-base frequency of the single-degree-of-freedom structure, and h is its height. Substituting $f \approx 30/h$ for framed buildings and $f \approx 45/h$ for shear wall buildings in the above equation implies that soil–structure interaction effects may be important for framed buildings when $v_s \leq 600$ m/s, or for shear wall buildings when $v_s \leq 900$ m/s. As seen from Table 5.1, these shear-wave velocities cover the full range of ground conditions softer than bedrock. Obviously, this is too general to be of much use in predicting when soil–structure interaction effects are likely to be substantial.

It is of interest that equation (5.34) correctly predicts that soil–structure interaction is important for the concrete gravity oil platforms studied by Watt *et al.* (1976). Radiation damping effects were found to reduce the base shear of a platform on 'very hard' ground ($v_s = 480$ m/s) by about 50% (the relevant value of v_s/fh was 6.6), despite the fact that the foundation was effectively rigid regarding its effect on the mode shapes and periods. It is relevant that these offshore structures have a high mass density factor, $m/\rho \pi R^2 h$, where ρ is the density of the soil and m is the participating mass of the structure.

Research by Zhao (1990) has shown that if the fundamental period of the structure is less than the fundamental period of the site, ignoring the soil–structure interaction may sometimes be dangerous, while if the fundamental period of the structure is longer than the fundamental period of the site, the soil–structure effects would reduce the response of the structure even without the effects of radiation damping. If the fundamental periods of the structure and the site are similar, the displacement response of the structure relative to the free-field responses are generally very large. It has been found that, in most cases, the period shift is the more important factor affecting the structural response than is the energy dissipation by plastic deformations in the structure and radiation of energy into the flexible soil. Radiation damping is only significant when the natural frequency of the soil–structure system is greater than the natural frequency of the site itself.

In a study by Mylonakis and Gazetas (2000), it was found that soil–structure interaction has detrimental effects in certain seismic and soil conditions, as follows:

- By comparing conventional code design spectra to actual response spectra, it was shown that an increase in fundamental natural period of a structure due to soil–structure interaction does not necessarily lead to smaller response, and that the prevailing view in structural engineering of the always beneficial role of soil–structure interaction is an oversimplification which may lead to unsafe design.
- Ductility demand in fixed-base structures is not necessarily a decreasing function of structural period, as suggested by traditional design procedures. Analysis of motions recorded on soft soils has shown increasing trends in ductility demand at periods higher than the predominant period of the motions.
- Soil–structure interaction in inelastic bridge piers supported on deformable soil may cause significant increases in ductility demand in piers, depending on the characteristics of the motion and the structure. However, inappropriate generalization of ductility concepts and geometric considerations may lead to the wrong conclusion when assessing the seismic performance of such structures.

Periods of vibration

The *periods of vibration* of a given structure increase with decreasing stiffness of the subsoil. This logical phenomenon has been widely noted, for example, by Veletsos and Meek (1974) and Watt *et al*. (1976). The latter found this effect to be very marked for a large offshore oil platform, where the fundamental period was 2.95 s for the rigid foundation condition and 5.9 s when allowance was made for a substratum of 'firm' overconsolidated clay.

Most of the rest of this section was derived from ATC (1978), whose recommendations have been adopted by others, notably by the National Earthquake Hazards Reduction Program (Building Seismic Safety Council, 2001).

In general form, the effective fundamental period (horizontal translation) of a structure as modified by the soil has been given (Veletsos and Meek, 1974) as

$$\tilde{T} = T\sqrt{1 + \frac{\bar{k}}{k_x}\left(1 + \frac{k_x \bar{h}^2}{k_\phi}\right)} \qquad (5.35)$$

where T is the fundamental period of the fixed base structure; $\bar{k}$ is the stiffness of the structure when fixed at the base, i.e. $\bar{k} = 4\pi^2 \bar{W}/gT^2$; k_x and k_ϕ are the horizontal and rocking stiffnesses of the foundation in the direction being considered, such as given in Table 5.8; $\bar{h}$ is the effective height of the structure, which for buildings may be taken (ATC, 1978) as 0.7 times the total height h, except that where the gravity load is concentrated at a single level it should be taken as the height to that level; and $\overline{W}$ is the effective or generalized weight of the structure vibrating in its fundamental natural mode, which for buildings may be taken as 0.7 times the total gravity load used in the earthquake analysis, except that where the gravity load is concentrated at a single level, the total gravity load should be used.

For simple consideration of buildings which are square in plan, equation (5.35) may be restated as

$$\frac{\tilde{T}}{T} = \sqrt{1 + \frac{1.47Jb^2}{v_s^2 T^2}(1 + 1.65J^2)} \quad (5.36)$$

where b is the width of the building, and J is the aspect ratio h/b.

Considering a building of height 80 m and width 20 m, then $J = 4$. The fundamental period of the building is $T = 1.8$ s. If the building is sited on soils for which the shear-wave velocity $v_s = 100$ m/s, then from equation (5.36) the effective period of the structure as modified by the soil is found to be $\tilde{T}/T = 1.73$, i.e. $T = 3.11$ s. Clearly, this soft soil (Table 5.1) has a substantial effect on the vibrational characteristics of the building.

The same building is significantly affected even when sited on fairly stiff soils of $v_s = 200$ m/s as the effective period is $\tilde{T} = 1.22\ T$.

Effective damping

The *effective damping* of a soil–structure system incorporates the combined material and radiation damping in the soil, the radiation damping in some cases leading to substantial reductions in response. For very stiff massive structures such as large offshore concrete gravity platforms, this reduction may be as much as 50%.

The effective damping factor for structure-foundation systems has been proposed by Veletsos and Nair (1975) as

$$\tilde{\beta} = \beta_0 + \frac{\beta}{(\tilde{T}/T)^3} \quad (5.37)$$

where β is the damping ratio for the fixed base structure, and β_0 is the foundation damping factor given in Figure 5.16.

The quantity r in Figure 5.16 is a characteristic foundation length given by

$$r = \begin{cases} (A_0/\pi)^{1/2}, & \text{for } \overline{h}/b_0 \leq 0.5, \\ (4I_0/\pi)^{1/4}, & \text{for } \overline{h}/b_0 \geq 1.0, \end{cases}$$

where b_0 is the length of the foundation in the direction being analysed, A_0 is the area of the foundation, and I_0 is the static moment of inertia of the foundation about a horizontal axis normal to the direction being analysed.

The ATC (1978) adopted equation (5.37), and makes the structural damping constant by letting $\beta = 0.05$ and ruling that the effective damping is never less than this value, i.e. $\tilde{\beta} \geq 0.05$. For the example building discussed above in this section, the effective damping is $\tilde{\beta} \geq 0.065$ when $v_s = 100$ m/s, and $\tilde{\beta} = 0.052$ when $v_s = 200$ m/s.

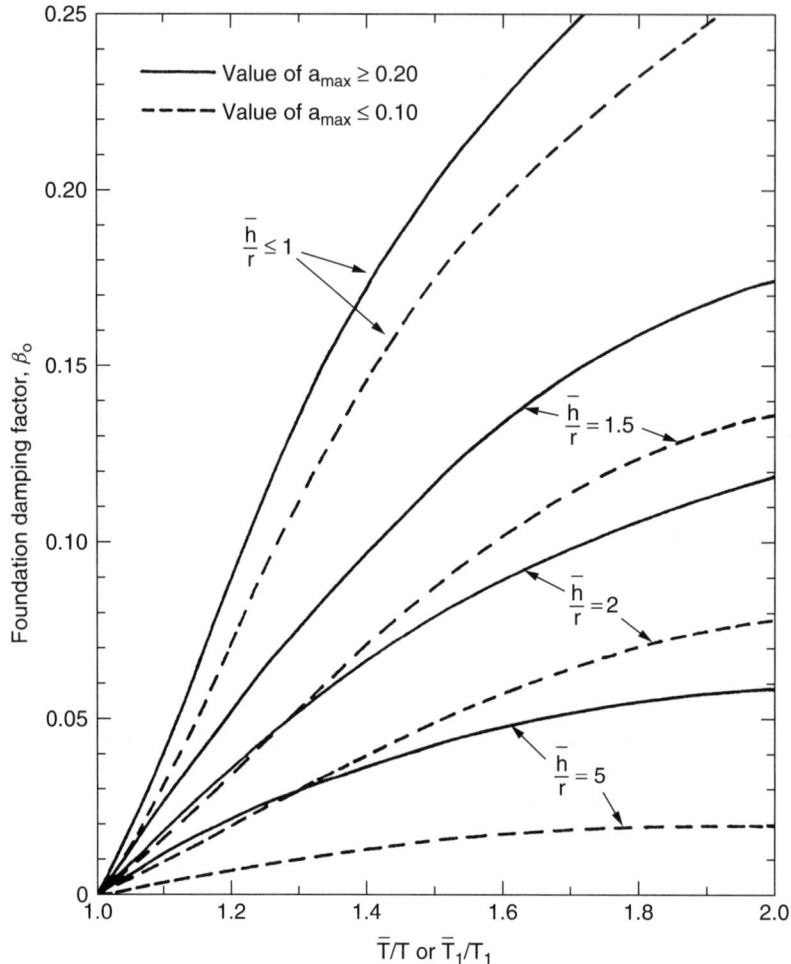

Figure 5.16 Foundation damping factor $\beta_0 \cdot a_{max}$ is the effective ground acceleration

The Building Seismic Safety Council (2000) recommends that for structures supported on point bearing piles and in all other cases where the foundation soil consists of a soft stratum of reasonably uniform properties underlain by a much stiffer, rock-like deposit with an abrupt increase in stiffness, the factor β_0 in equation (5.37) should be replaced by

$$\beta_0 = \left(\frac{4D_s}{V_s \tilde{T}}\right)^2 \beta_0 \tag{5.38}$$

if $4D_s/v_s\tilde{T} < 1$, where D_s is the total depth of the stratum.

The value of $\tilde{\beta}$ computed from equation (5.37), both with and without the adjustment represented by equation (5.38), should in no case be taken as less that 0.05 or greater than 0.20.

Equivalent viscously damped response

Because of the complexity and expense of rigorously computing the effects of radiation damping in the foundation, an *equivalent viscously damped* response spectrum technique would be desirable. For estimating an equivalent viscous damping for a soil–structure system, the foundation damping (radiation plus hysteretic) is not directly additive to the structural damping, as described above. For structures with more uniform stiffness and mass distributions the equivalent viscous damping concept may be reliable. Therefore, a simple method of modifying the results of analyses of fixed-base buildings has been proposed by the ATC (1978) such that the base shear may be reduced to the value $\tilde{V}$:

$$\frac{\tilde{V}}{V} = 1 - \frac{\Delta V}{V}, \qquad (5.39)$$

where V is the base shear for a fixed base structure and ΔV is the reduction in base shear given by

$$\Delta V = \left[C_s - \tilde{C}_s \left(\frac{0.05}{\tilde{\beta}} \right)^{0.04} \right] \overline{W}, \qquad (5.40)$$

in which $C_s = V/W$ is the seismic design coefficient for the fixed base structure of period T; and $\tilde{C}_s$ is the seismic design coefficient for the flexibly supported structure of period $\tilde{T}$.

These expressions may be used to modify the equivalent-static lateral forces derived from the code, or the moment and shears derived from a fixed-base modal analysis (the ATC restricts the reduction to the first mode forces). The above expressions relate to conditions where the soil may be regarded as a homogeneous half-space, and will hence be unconservative when radiation damping is reduced by shallow reflective soil layer interfaces (see Figure 5.12 and the related text).

The above expressions may be used to arrive at

$$\frac{\tilde{V}}{V} = \frac{W - \overline{W}}{W} + \frac{\overline{W}S}{W} \left(\frac{1}{\tilde{T}/T} \right)^{2/3} \left(\frac{0.05}{\tilde{\beta}} \right)^{0.4}, \qquad (5.41)$$

where $\tilde{T}/T$ and $\tilde{\beta}$ are given in equations (5.35) or (5.36) and (5.37), and S is the appropriate value of the soil profile coefficient as given in Table 5.10.

For a building of uniformly distributed mass and stiffness $\overline{W} \approx 0.7W$ (Section 5.3.4), so that equation (5.41) reduces to

$$\frac{\tilde{V}}{V} = 0.3 + 0.75 \left(\frac{1}{\tilde{T}/T} \right)^{2/3} \left(\frac{0.05}{\tilde{\beta}} \right)^{0.4}. \qquad (5.42)$$

In the above equations, $\tilde{\beta}$ has a minimum value of 0.05, and the reduction in base shear due to soil–structure interaction is limited to 30%, giving the range of values $0.7 \leq \tilde{V}/V \leq 1.0$.

Table 5.10 Soil profile coefficients, S, proposed by the ATC (1978)

S	Soil profile type*		
	S_1	S_2	S_3
	1.0	1.2	1.5

*S_1 denotes rock of any characteristic, and having a (low strain) $v_s \geq 760$ m/s, or where sands, gravels, or stiff clay deposits less than 60 m thick overlie rock. S_2 denotes profiles where sands, gravels and stiff clays exceeding 60 m thick overlie rock. S_3 denotes soft to medium stiff clays and sands exceeding 9 m in thickness.

Referring again to the example building discussed above, and values of $\tilde{T}/T$ and $\tilde{\beta}$ derived, these may be inserted into equation (5.39), and $\overline{W}$ may be taken as equal to 0.75. Taking $S = 1$ for soils with $v_s = 200$ m/s, and $S = 1.2$ for $v_s = 100$ m/s, the 80 m high by 20 m square building would have $\tilde{V}/V = 0.90$ and $\tilde{V}/V = 0.81$, respectively, for the two soil conditions, both representing significant favourable soil–structure interaction. It is interesting to observe that for a building of quite high flexibility (a fixed-base period of $T_1 = 1.8$ s) sited on reasonably stiff soils with $v_s = 200$ m/s, there is a predicted reduction of 10% in base shear compared with the fixed-base condition.

Effects of embedment

The *effects of embedment* have been studied by various workers, such as Bielak (1975) and Luco et al. (1975). There is general agreement that increasing embedment increases the static stiffness of the system, decreases the periods of vibration, and decreases the displacement responses. These effects are evident in all four modes of vibration, i.e. vertical, horizontal, rocking and torsion. Where backfill is softer than the undisturbed soil, the effects of embedment are obviously reduced. In cases where theoretical results have been compared with experimental, agreement is qualitative rather than quantitative.

Approximate factors for estimating the increase in horizontal and rocking foundation stiffness have been proposed by the ATC (1978) as follows:

$$K_x(\text{embedded}) = k_x \left(1 + \frac{2}{3}\frac{d}{r}\right), \tag{5.43}$$

$$K_\phi(\text{embedded}) = k_\phi \left(1 + 2\frac{d}{r}\right), \tag{5.44}$$

where k_x and k_ϕ are the horizontal and rocking stiffnesses for surface footings such as those given in Table 5.8, R is the equivalent radius for the footing, and d is the effective depth of embedment for the conditions that would prevail in the design earthquake. Because of the above-noted lack of good quantitative correlation between experiment and theory, the selection of values for d will be somewhat subjective.

5.4 Seismic Response of Structures

5.4.1 Elastic seismic response of structures

Dynamic loading comprises any loading which varies with time, and seismic loading is a complex variant of this. The way in which a structure responds to a given dynamic excitation depends on the nature of the excitation and the dynamic characteristics of the structure, i.e. on the manner in which it stores and dissipates vibrational energy. Seismic excitation may be described in terms of displacement, velocity, or acceleration varying with time. When this excitation is applied to the base of a structure it produces a time-dependent response in each element of the structure, which may be described in terms of motions or forces.

The simplest dynamical system which we can consider is a single-degree-of-freedom system (Figure 5.17) consisting of a mass on a spring which remains in the linear elastic range when vibrated. The dynamic characteristics of such a system are simply described by its natural period of vibration T (or frequency ω) and its damping ξ. When subjected to a harmonic base motion described by $u_g = a \sin \omega t$, the response of the mass at top of the spring is fully described in Figure 5.18. The ratios of response amplitude to input amplitude are shown for displacement R_d, velocity R_v and acceleration R_a, in terms of the ratio between the frequency of the forcing function and the natural frequency of the system ω_n.

The significance of the natural period or frequency of the structure is demonstrated by the large amplifications of the input motion at or near the resonance conditions, i.e. when $\omega/\omega_n = 1$. Figure 5.18 also shows the importance of damping particularly near resonance. When the damping $\xi = 0.01$, the resonant amplification of the input motion is 50-fold for this system, but if the damping is increased to $\xi = 0.05$ the resonant amplification is reduced to 10 times the input motions.

The response of a structure to the irregular and transient excitation of an earthquake will obviously be much more complex than in the simple harmonic steady-state motion discussed above. Consider the ground motion at El Centro in 1940, the accelerogram for which is shown in Figure 5.19. If we apply this motion to a series of single-degree-of-freedom structures with different natural periods for damping, we can plot the maximum acceleration response of each of these structures as in Figure 5.20.

As with simple harmonic ground motion, the natural period and degree of damping are again evident in Figure 5.19. While no simple periodicity occurs in the ground motion of

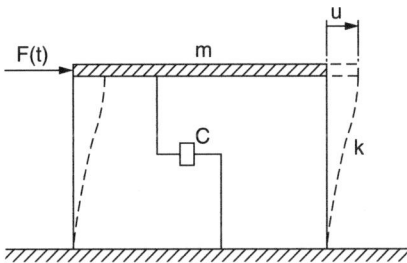

Figure 5.17 Idealized single-degree-of-freedom system

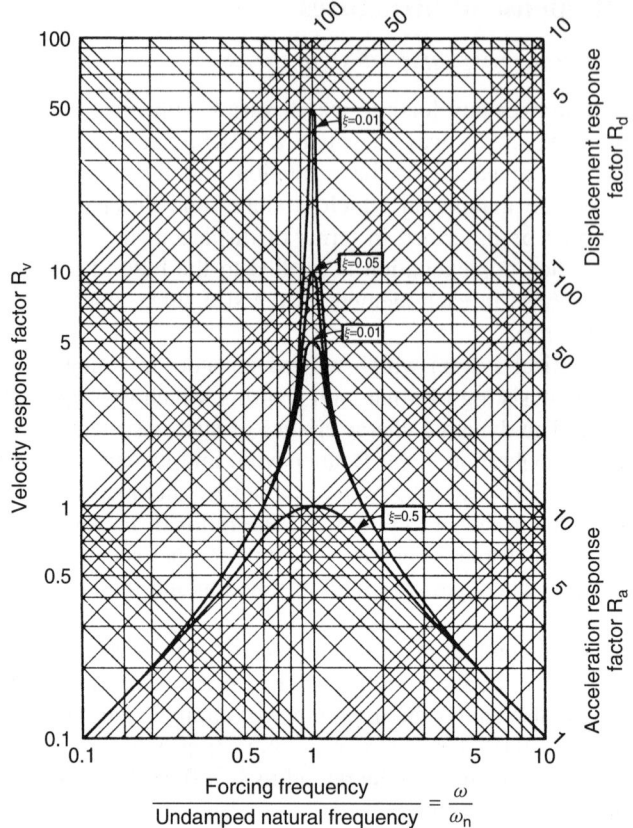

Figure 5.18 Response of linear elastic single-degree-of-freedom system to a harmonic forcing function. (Reprinted from Blake RE (1961), Basic vibration theory. In: *Shock and Vibration Handbook* (Eds CM Harris and CE Crede), Vol. 1. Reproduced with permission of the McGraw-Hill Companies)

Figure 5.19, the dominance of the shorter periods is seen from the region of acceleration responses on the left of Figure 5.20.

For example, a single-degree-of-freedom structure with a period of 0.8 s and damping $\xi = 0.02$ has a maximum acceleration of approximately 0.9 g compared with a peak input ground motion of about 0.33 g. This represents an amplification of 2.7 at $\xi = 0.02$, whereas if the damping is $\xi = 0.05$ the amplification can be seen to reduce to 1.8.

Most structures are more complex dynamically than the single-degree-of-freedom system discussed above. Multi-storey buildings, for example, are better represented as multi-degree-of-freedom structures, with one degree of freedom for each storey, and one natural mode and period of vibration for each storey (Figure 5.21). The response history of any element of such a structure is a function of all the modes of vibration, as well as of its position within the overall structural configuration.

For many multi-degree-of-freedom structures, the linear elastic responses can be computed with a high degree of mathematical accuracy. For example, assuming linear elastic

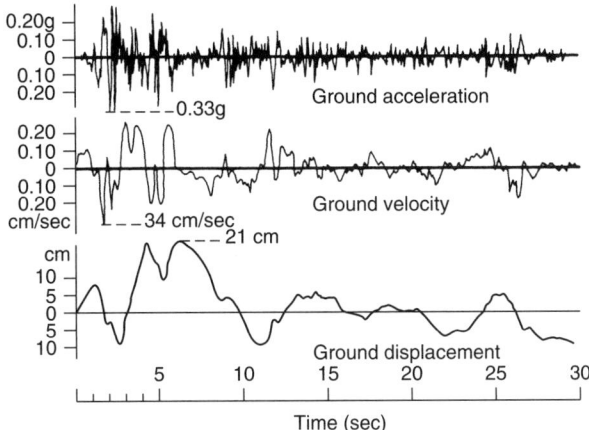

Figure 5.19 The 1940 Imperial Valley earthquake ground motion at El Centro: acceleration, velocity, and displacement in the north–south direction

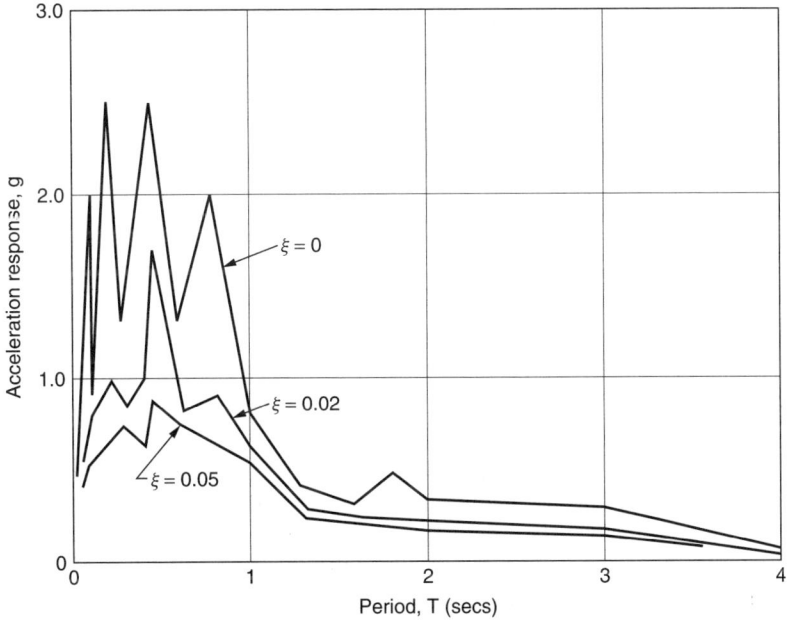

Figure 5.20 Elastic acceleration response spectra of the north–south component of the 1940 El Centro recording

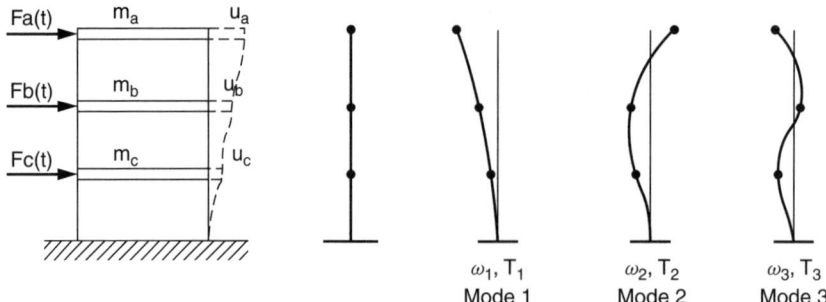

Figure 5.21 Multi-degree-of-freedom system subjected to dynamic loading

behaviour, in the dynamic analysis of a 30-storey building subjected to a ground motion 1.5 times that of Figure 5.19, the maximum horizontal shears at each floor level were computed to be as shown in Figure 5.22. Notice the considerable difference in response between the elastic case assuming 2% damping (curve 1) and that for 5% damping (curve 3). Further discussion of damping follows in Sections 5.4.2 and 5.4.4.

5.4.2 Non-linear seismic response of structures

For economical resistance against strong earthquakes, most structures must behave inelastically. In contrast to the simple linear elastic response model examined in the previous section, the pattern of inelastic stress–strain behaviour is not constant, varying with the member size and shape, the materials used, and the nature of the loading.

The typical stress–strain curves for various materials under repeated and reversed direct loading shown in Figure 5.23 illustrate the chief characteristics of inelastic dynamic behaviour, namely:

- plasticity;
- strain hardening and strain softening;
- stiffness degradation;
- ductility;
- energy dissipation.

Plasticity, as exhibited by mild steel (Figure 5.23(a)), is a desirable property in that it is easy to simulate mathematically and provides a convenient control on the load developed by a member. Unfortunately, the higher the grade of steel, the shorter the plastic plateau, and the sooner the *strain hardening* effect shown in Figure 5.23(a) sets in. *Strain softening* is the opposite of strain hardening, involving a loss of stress or strength with increasing strain as seen in Figure 5.23(c).

In the reversed loading of steel, the *Bauschinger effect* occurs, i.e. after loading past the yield point in one direction the yield stress in the opposite direction is reduced. Another characteristic of the cyclic loading of steel is the increased non-linearity in the elastic range which occurs with load reversal (Figure 5.23(b)). *Stiffness degradation* is an important feature of inelastic cyclic loading of concrete and masonry materials. The stiffness as

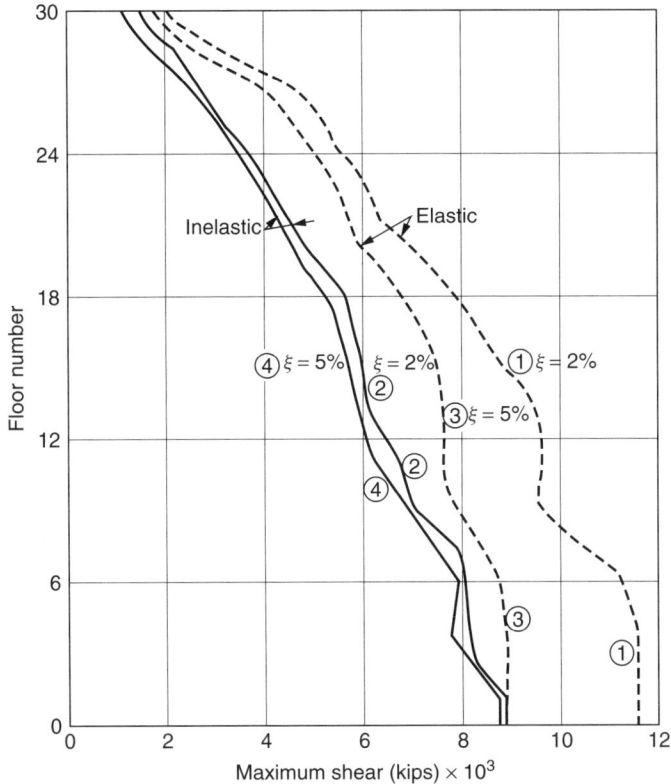

Figure 5.22 Maximum horizontal shear response for Bank of New Zealand Building, Wellington, subjected to 1.5 times the El Centro (1940) north–south component (after Smith *et al.*, 1975)

measured by the overall stress–strain ratio of each hysteresis loop of Figure 5.23(c)–(f) is clearly reducing with each successive loading cycle.

The *ductility* of a member or structure may be defined in general terms by the ratio

$$\text{ductility} = \frac{\text{deformation at failure}}{\text{deformation at yield}}.$$

In various uses of this definition, 'deformation' may be measured in terms of deflection, rotation or curvature. The numerical value of ductility will also vary depending on the particular combination of applied forces and moments under which the deformations are measured. Ductility is generally desirable in structures because of the gentler and less explosive onset of failure than that occurring in brittle materials. The favourable ductility of mild steel may be seen in Figure 5.23(a) by the large value of ductility in direct tension measured by the ratio $\varepsilon_{su}/\varepsilon_{sy}$. This ductility is particularly useful in seismic problems because it is accompanied by an increase in strength in the inelastic range. By comparison the high value of compressive ductility for plain concrete expressed by the ratio $\varepsilon_{cu}/\varepsilon_{cy}$ in Figure 5.23(c) is far less useful because of the inelastic loss of strength. Steel has the best ductility properties of normal building materials, while concrete can

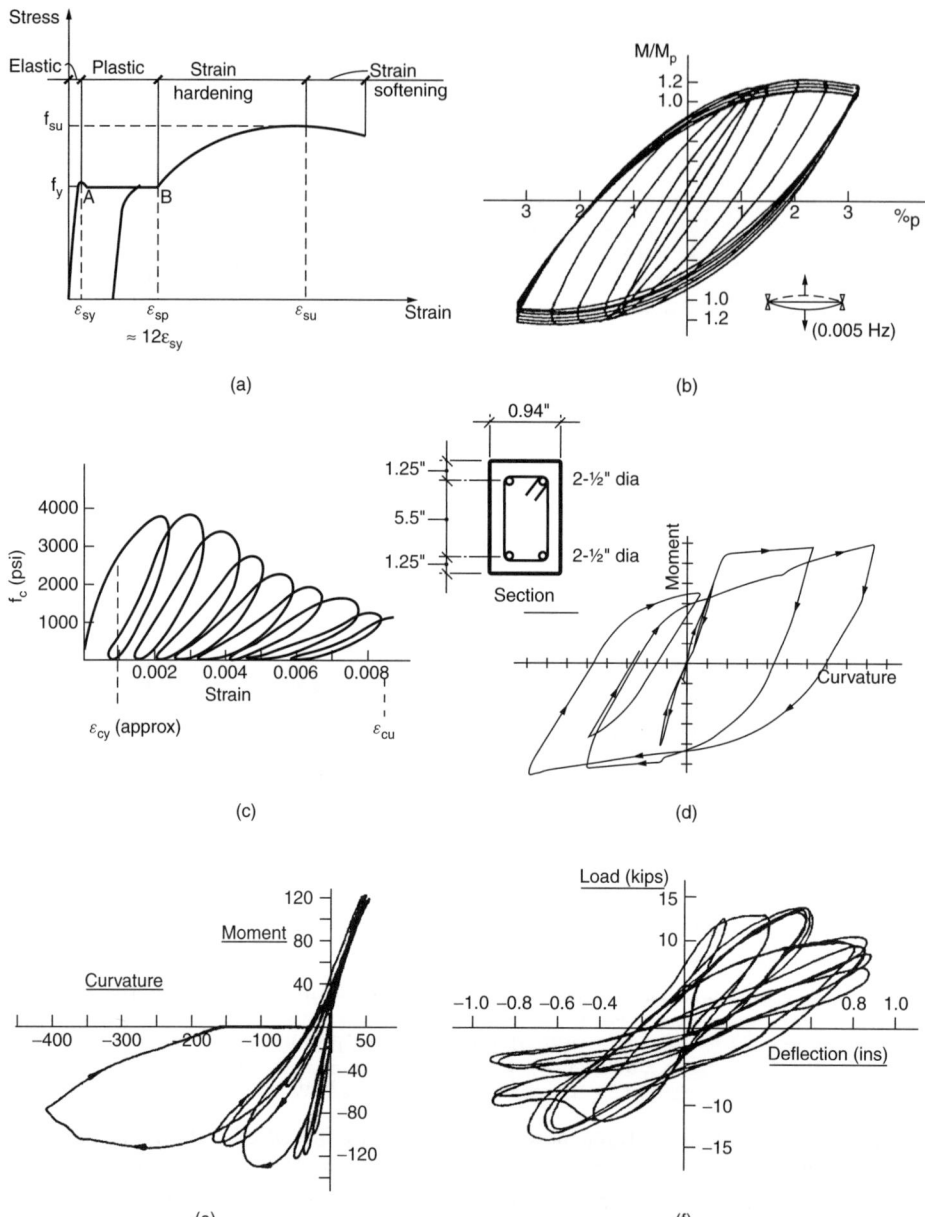

Figure 5.23 Elastic and inelastic stress–strain behaviour of various materials under repeated and reversed loading: (a) mild steel, monotonic (or repeated axial) loading; (b) structural steel under cyclic bending (after Takanashi, 1973); (c) unconfined concrete, repeated loading (after Sinha et al., 1964) (reprinted by permission of ACI International); (d) doubly reinforced concrete beam, cyclic loading (after Park et al., 1972) (reproduced by permission of the American Society of Civil Engineers); (e) prestressed concrete column, cyclic bending (after Blakeley, 1973); (f) masonry wall, cyclic lateral loading (after Williams and Scrivener, 1973)

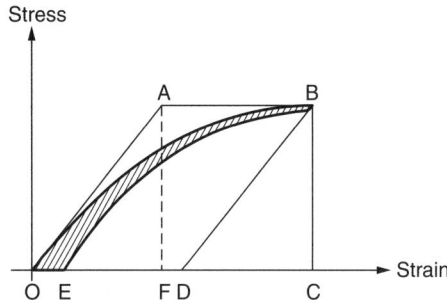

Figure 5.24 Energy stored and dissipated in idealized systems

be made ductile with appropriate reinforcement. The ductility of masonry, even when reinforced, is more dubious. Further discussion of the ductility of the various materials is found elsewhere in this book, particularly in Chapter 10.

A high *energy dissipation* capacity is often mentioned rather loosely as a desirable property of earthquake resistant construction. Strictly speaking, a distinction should be made between *temporary* dissipation and *permanent* dissipation of energy.

Compare the simple elastoplastic system represented by OABD in Figure 5.24 with the hypothetical non-linear mainly elastic system of curves OB and BE; after loading each system to B the total energy dissipated by each system is nearly equal, as represented by the area OABC and OBC, respectively. However, the ratio between temporarily stored strain energy and permanently dissipated energy for the two systems is far from equal. After unloading to zero stress it can be seen that the energy dissipated by the elastoplastic system is equal to the hysteretic area OABD, while the energy dissipated by the non-linear system is equal to the much smaller hysteretic area OBE (shaded in Figure 5.24). The energy dissipated in *hysteresis* may be expressed as an *equivalent viscous damping*, as described for soils in equation (5.1).

By way of further elucidation of the seismic energy dissipation of structures, consider Figure 5.25 derived from the inelastic seismic analysis of the Bank of New Zealand Building, Wellington. A substantial part of the energy is temporarily stored by the structure in elastic strain energy and kinetic energy. After 3 seconds the earthquake motion is so strong that the yield point is exceeded in parts of the structure and permanent energy dissipation in the form of inelastic (or hysteretic) strain energy begins. Throughout the whole of the earthquake, energy is also dissipated by viscous damping (from effects other than hysteresis), which is of course the means by which the elastic energy is dissipated once the forcing ground motion ceases. The value of damping for reducing the energy available for causing damage is suggested by Figure 5.25. It has been shown by Housner and Jennings (1977) that 2% of critical damping reduces the energy available for damage to about 50 per cent of the total energy input, while 5% damping reduces it to about 30 per cent.

It is evident from the large proportion of hysteretic energy dissipated by this building that considerable ductility without excessive strength loss is required. A brittle building with the same yield strength, but with no inelastic behaviour ($\varepsilon_u/\varepsilon_y = 1$), would have begun to fail after 3 seconds of the earthquake. In other words, stronger members would

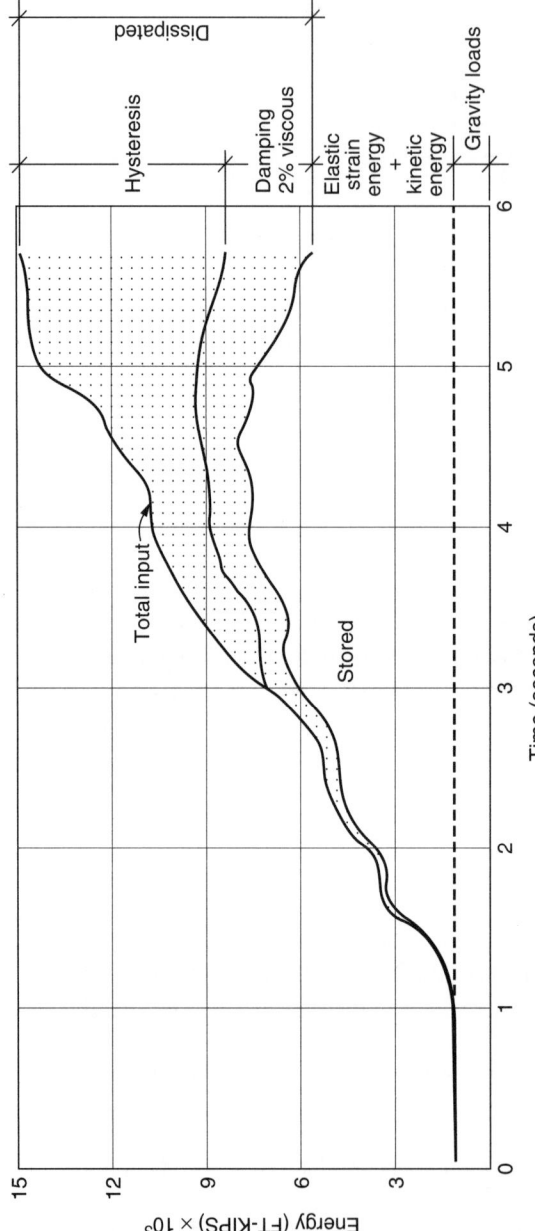

Figure 5.25 Energy dissipation in Bank of New Zealand Building, Wellington, computed for the first part of an earthquake equal to 1.5 times the El Centro (1940) north–south component (after Smith et al., 1975)

have been necessary in a purely elastic design. This can be seen in another way in Figure 5.22, showing the reduction in storey shears achieved when assuming inelastic behaviour (curve 2) as compared with the elastic case (curve 1).

We note that as energy dissipation is achieved from post-elastic deformations it means that the structure is being damaged, and a good earthquake resistant structure designed to yield exhibits high *damageability* without collapsing. Alternatively, as discussed in Section 8.5, energy-dissipating devices may be used to protect the structure from damage (e.g. yielding), by absorbing energy in elements which are replaceable.

5.4.3 Mathematical models of non-linear seismic behaviour

When examining the range and complexity of the hysteretic behaviour shown in Figure 5.23, the problems involved in establishing usable mathematical stress–strain models are obvious. It follows that many hysteresis models have been developed, such as:

(1) elastoplastic;
(2) bilinear;
(3) trilinear;
(4) multilinear;
(5) Ramberg–Osgood;
(6) degrading stiffness;
(7) pinched loops;
(8) slackness developing.

Three of the simplest of these hysteresis models are illustrated in Figure 5.26, namely elastoplastic, bilinear (non-degrading) and Ramberg–Osgood (modified). The elastoplastic and bilinear models are the most commonly used because of their simplicity and relative computational efficiency, and the bilinear form in Figure 5.26(b) is found to be suitable for most response analyses of reinforced concrete and steel structures. The Ramberg–Osgood model shown in Figure 5.26(c) was developed by Thompson and Park (1978) for closely modelling the hysteretic behaviour of prestressing steel, using the Ramberg–Osgood function for obtaining any desired shape of curve by varying the parameter r, as shown in Figure 5.26(d).

The best type of model depends heavily on the structural materials. However, compromises have to be made in selecting a suitable model, partly because the choice may be governed by the models incorporated into the available computer programs. For further discussion on methods of analysis, see Section 5.4.7.

5.4.4 Level of damping in different structures

The general influence of damping upon seismic response is discussed in Sections 5.4.1 and 5.4.2, but when choosing the level of damping for use in the dynamic analysis of a structure, the following factors should be considered.

Damping varies with the materials used, the form of the structure, the nature of the subsoil and the nature of the vibration. Large-amplitude post-elastic vibration is more heavily damped than small-amplitude vibration, while buildings with heavy shear walls and heavy cladding and partitions have greater damping than lightly clad skeletal

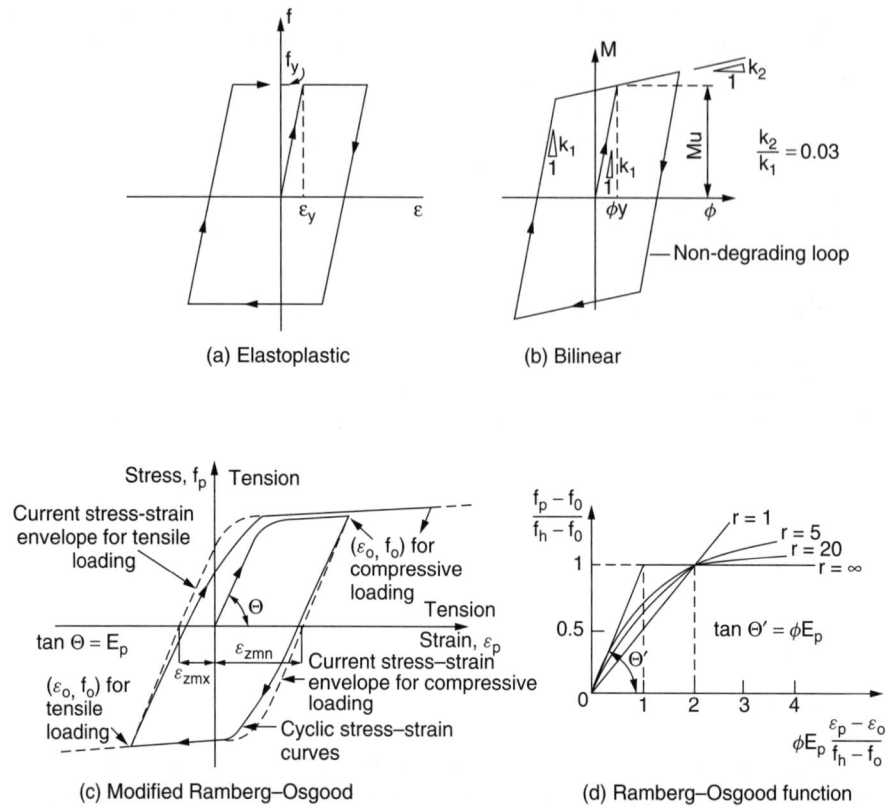

Figure 5.26 Three of the simpler idealized models of hysteretic behaviour under reversed cyclic loading. The Ramberg–Osgood model is from Thompson and Park (1978)

structures. The overall damping of a structure is clearly also related to the damping characteristics of the subsoil as discussed in Sections 5.2 and 5.3, and if the soil damping is to be incorporated into the effective damping of the structure the method given in Section 5.3.4 may be appropriate.

The many experimentally determined values of damping reported in the literature are generally derived either for individual structural components or for low-amplitude vibration of structures. Hence, for whole structures subject to strong ground motion some extrapolation of such damping data is necessary.

Table 5.11 indicates representative values of damping for a range of construction. These values are suitable for normal response spectrum or modal analysis in which viscous damping, equal in all modes, is assumed. These damping values also assume that the structure is a normal-risk structure expected to yield in the design earthquake (i.e. it conforms to item 2(a) of Table 8.1), and hence the damping due to hysteresis is included, but no allowance for radiation damping has been made. Except for a few forms of construction which have been specifically tested, insufficient evidence exists to warrant any more detailed allowance for differences in structural and non-structural form, and designers will need to use their own judgement to interpret the table.

Table 5.11 Typical damping ratios for structures

Type of construction	Damping ξ, percentage of critical
Steel frame, welded, with all walls of flexible construction	2
Steel frame, welded, with normal floors and cladding	5
Steel frame, bolted, with normal floors and cladding	10
Concrete frame, with all walls of flexible construction	5
Concrete frame, with stiff cladding and all internal walls flexible	7
Concrete frame, with concrete or masonry shear walls	10
Concrete and/or masonry shear wall buildings	10
Timber shear wall construction	15

Notes: The term 'frame' indicates beam and column bending structures as distinct from shear structures. The term 'concrete' includes both reinforced and prestressed concrete in buildings. For isolate prestressed concrete members such as in bridge decks damping values less than 5% may be appropriate, e.g. 1–2% if the structure remains substantially uncracked.

For steel and concrete buildings, Hart and Vasudevan (1975) offer a systematic method for estimating damping as a function of spectral velocity and modal frequency, based on an analysis of the response of buildings in the San Fernando earthquake. For use with dynamic analysis programs which do not permit differences in modal damping, a weighted average based on modal contributions to base shear would be appropriate. Some further data on damping of different types of structure are given in Chapter 10 by Newmark and Rosenblueth (1971).

5.4.5 Periods of vibration of structures

As indicated in various parts of this book, the periods of vibration of structures are primary tools in determining seismic response. While the periods of vibration are found during dynamic analyses by the solution of the eigenvalue equation, it is often desirable to make a quick estimate of the fundamental period T of a structure using approximate formulae. Various such formulae exist, one of the simplest being

$$T = 1.0\, C_t h_n^{0.75}, \quad \text{for the serviceability limit state,} \quad (5.45)$$

$$T = 1.25\, C_t h_n^{0.75}, \quad \text{for the ultimate limit state,} \quad (5.46)$$

where $C_t = 0.085$ for moment resisting steel frames, 0.075 for moment resisting concrete frames and eccentrically braced steel frames, and 0.050 for stiffer (walled) structures; and h_n is the height in metres to the uppermost principal mass. For further discussion of similar formulae, together with the data on which they are based, see Chopra and Goel (2000) and Goel and Chopra (1998).

The above empirical formulae have the advantage that no structural analyses are required for their use, and they are thus suitable for preliminary design purposes. However, they are obviously insensitive to the actual mass and stiffness distributions of a given building and thus are subject to significant error. Also the above formulae are not reliable for flexible construction such as portal frames, and are not applicable to timber structures which are very varied in form, e.g. timber portal frame structures may have very long periods of 1.0 s and more. Hence, in cases where these formulae are likely to be insufficiently accurate, once a static linear elastic load analysis which calculates deflections has been done, T may be conveniently estimated by a more reliable formula based on Rayleigh's method for a structure with masses of N levels, as follows:

$$T = 2\pi \sqrt{\frac{\sum_{i=1}^{N} m_i u_i^2}{\sum_{i=1}^{N} F_i u_i}} \qquad (5.47)$$

where F_i is the seismic lateral force at level i; m_i is the mass assigned to level i; and u_i is the static lateral deflection at level i due to the forces F_i.

Finally, it is noted that equations (5.45) and (5.47) are for the initial elastic state. With the stiffness degradation that occurs in the post-elastic range, the period progressively increases as shown in equation (5.46), and may become twice the elastic value (or more) prior to failure. This may increase or decrease the response, depending on where the elastic period is on the response spectrum, and such effects may deserve inclusion in the analysis.

5.4.6 Interaction of frames and infill panels

Introduction

Walls are often created in buildings by infilling parts of the frame with stiff construction such as bricks or concrete blocks. Unless adequately separated from the frame (Section 12.2), the structural interaction of the frame and infill panels must be allowed for in the design. This interaction has a considerable effect on the overall seismic response of the structure and on the response of the individual members. Many instances of earthquake damage to both the frame members and infill panels have been recorded (e.g. Stratta and Feldman, 1971).

Analytical techniques used in studying frame–panel interaction are briefly discussed below.

The effect of infill panels on overall seismic response

The principal effects of infill panels on the overall seismic response of structural frames are:

- to increase the stiffness and hence increase the base shear response in most earthquakes;
- to increase the overall energy dissipation capacity of the building; and
- to alter the shear distribution throughout the structure.

The more flexible the basic structural frame, the greater will be the above-mentioned effects. As infill is often made of brittle and relatively weak materials, in strong earthquakes the response of such a structure will be strongly influenced by the damage sustained by the infill and its stiffness-degradation characteristics.

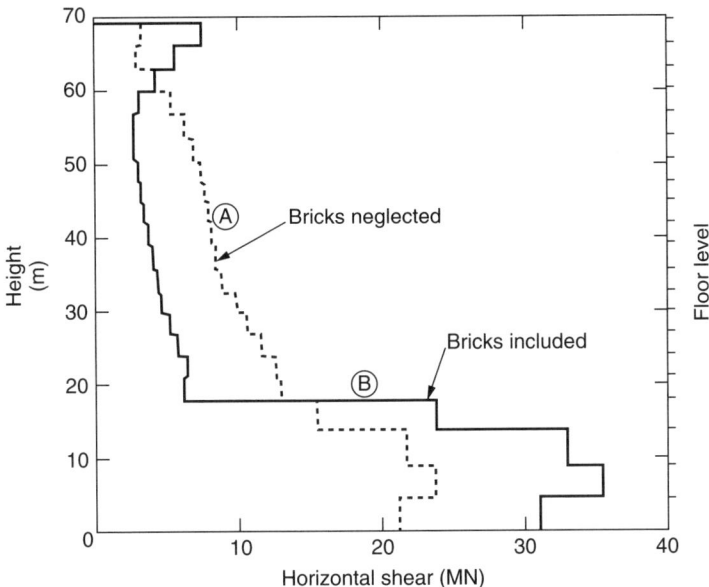

Figure 5.27 Horizontal seismic shear diagram for lift core of 20-storey hotel building showing effect of brick partitions above the fourth floor

To fully simulate the earthquake response of an infilled frame, a complex non-linear time-dependent finite-element dynamic analysis would be necessary, as provided by computer codes such as DRAIN-2DX. Such are the problems and effort involved in using such software, e.g. in modelling the structural behaviour of normal masonry infill, that more rudimentary dynamic analyses are more commonly used.

For many structures a response spectrum analysis in which the infill panels are simulated by simple finite elements will be very revealing. Figure 5.27 shows the results of such an analysis of a multi-storey hotel building, in which all of the bedroom floors (fourth to twentieth) have alternate partitions in brickwork. Curve A shows the horizontal earthquake shear distribution up the shear core ignoring the brickwork, while curve B shows the shears when an approximate allowance for the brick walls is made.

Allowing for the brick reduces the fundamental period of the structure from 1.96 s to 1.2 s, and correspondingly increases the base shear on the shear core from 21.0 MN to 31.0 MN. The effect on the distribution of shear is particularly dramatic; it can be seen how the brick walls carry a large portion of the shear until they terminate at the fourth-floor level; below this level the shear walls of the core must, of course, take the total load (see also Section 8.3.5).

In carrying out this simple type of dynamic analysis difficulty may be experienced in selecting a suitable value of shear modulus G for the infill material. Not only is the G value notoriously variable for bricks, but the infill material may not even have been chosen at the time of the analysis. Either a single representative value may have to be assumed, or it may be desirable to take a lower and a higher likely value of G in two separate analyses for purposes of comparison.

Further examples of the effect of infill on mode shapes and periods of vibration of structural frames are reported by Lamar and Fortoul (1969) and Sritharan and Dowrick (1994). In their examples the first mode period was reduced by factors of about 3–6 when comparing the infill-included with the infill-excluded cases. Mode shapes of some structures also depend upon the distribution of effect infill.

For further insights into analytical modelling of infilled frames, see the review paper by Christafulli *et al*. (2000).

The effect of infill panels on member forces

As mentioned above, considerable uncertainties are involved in estimation of the seismic interaction between infill panels and structural frames. In a study of response of low-rise buildings to moderate-strength ground shaking, modelling non-linear response with DRAIN-2DX, Sritharan and Dowrick (1994) found that:

- infill-excluded frames had highly non-linear response;
- infill-included frames responded elastically, despite the base shears being up to four times larger than without infill;
- the above results varied quantitatively but not qualitatively when element stiffness were varied by factors of up to 2.

As an easier alternative to non-linear, time-history analyses, from a straightforward finite-element response spectrum analysis some basic design information may be derived. While to take such data as definitive seismic design criteria would be misleading, sensible use of the computed forces in design would nevertheless be much better than ignoring the presence of the infill, as has often been the case in the past. By carrying out comparative analyses with and without the infill panels, at least a qualitative idea of the effect of the infill can be obtained.

(1) Infill panels. The shear stresses computed in the infill panels should give a reasonable indication whether or not the infill will survive the design earthquake. Despite being very approximately determined, the shear stress level will also help in determining what reinforcement to use in the panel and whether to tie the panel to the frame (Section 10.4.5).

(2) Frame members. The design of the beams and columns abutting the infill is generally the least satisfactory aspect of this form of seismic construction. Because of the approximations in the analytical model, the stresses in the frame members are ill defined. Failures tend to occur at the tops and bottoms of columns, due to shears arising from interaction with the compression diagonal which exists in the infill panel during the earthquake (Stratta and Feldman, 1971). Unfortunately, no comprehensive design criteria for this problem have yet been established, and further research examining the frame rather than the panel stresses is required. In simple analyses, if the analysis indicates the failure of the infill panels, the frame should be analysed with any failed panels deleted, so that appropriate frame stresses may be taken into consideration.

Because such structures often have highly non-linear behaviour arising from the interaction between infill and frame, analytical modelling is a complex problem. Christafulli et al. (2000) have reviewed the different analytical procedures, discussing how best to implement them.

5.4.7 Methods of seismic analysis for structures

The many methods for determining seismic forces in structures fall into two distinct categories;

(1) equivalent-static force analysis.
(2) dynamic analysis.

Equivalent static force analysis

These are approximate methods which have been evolved because of the difficulties involved in carrying out realistic dynamic analysis. Up until the 1990s, codes of practice worldwide relied mainly on the simpler static force approach, incorporating varying degrees of refinement in an attempt to simulate the real behaviour of the structure. Basically, they give a crude means of determining the 'total' horizontal force (base shear) V, on a structure:

$$V = ma,$$

where m is the mass of the structure and a is the seismic horizontal acceleration. V is applied to the structure by a simple rule describing its vertical distribution. In a building this generally consists of horizontal point loads at each concentration of mass, most typically at floor levels (Figure 5.28). The seismic forces and moments in the structures are then determined by any suitable statical analysis and the results added to those for the gravity load cases.

In the subsequent design of structural sections an increase in permissible elastic stresses of 33–50% is usually permitted, or a smaller load factor than normal is required for ultimate limit state design. In regions of high winds and moderate earthquake requirements, the worst design loads of taller structures may well arise from wind rather than earthquake forces. Even so, the form and detail of the structure should still be governed by seismic considerations.

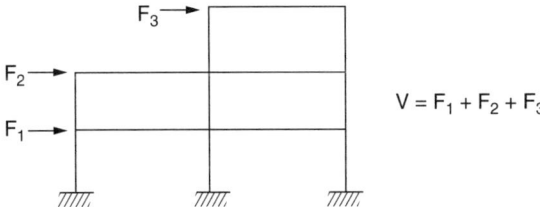

Figure 5.28 Example of frame with equivalent-static forces applied at floor levels

An important feature of equivalent-static load requirements in most codes of practice is the fact that the calculated seismic forces are considerably less than those which would actually occur in the larger earthquakes likely in the area concerned. The forces calculated in more rigorous dynamic analyses based on a realistic earthquake excitation can be as much as ten times greater than those arising from the static load provisions of some codes. This state of affairs has been justified by arguing that the force discrepancy will be taken up by inelastic behaviour of the structure, which should therefore be detailed to be appropriately ductile, and more advanced codes do have specific ductility requirements (see Section 10.1).

Dynamic analysis

For large or complex structures, static methods of seismic analysis are often deemed to be not accurate enough and many authorities demand dynamic analyses for certain types an size of structure. Various methods of differing complexity have been developed for the dynamic seismic analysis of structures. They all have in common the solution of the equations of motion, as well as the usual statical relationships of equilibrium and stiffness.

The three main techniques used for dynamic analysis are:

(1) response spectrum techniques;
(2) normal mode analysis;
(3) direct integration of the equations of motion by step-by-step procedures.

Response spectrum analysis is the simplest form of dynamic analysis. First it is necessary to find the displacement history of the structure by solving the equations of motion of the system. There is one equation of dynamic equilibrium for each degree of freedom. A single-degree-of-freedom system is shown in Figure 5.17. In the figure $F(t)$ is a force varying with time, k is the total spring constant of resisting elements, c is the damping coefficient (the damping force is usually taken as proportional to the velocity of the mass for ease of computation), and u is the displacement.

Generally, the equation of dynamic equilibrium is

$$F_\text{I} + F_\text{D} + F_\text{S} = F(t), \tag{5.48}$$

where the inertia force is

$$F_\text{I} = m\ddot{u},$$

the damping force is

$$F_\text{D} = c\dot{u},$$

and the elastic force is

$$F_\text{S} = ku.$$

Thus

$$m\ddot{u} + c\dot{u} + ku = F(t). \tag{5.49}$$

For the case of earthquake excitation (Figure 5.29), the only external loading is in the form of an applied motion at ground level, $u_g(t)$, i.e. the total acceleration of the mass m is

$$\ddot{u}_t = \ddot{u} + \ddot{u}_g.$$

Therefore

$$m\ddot{u} + c\dot{u} + ku = F_{\text{eff}}(t), \quad (5.50)$$

where

$$F_{\text{eff}}(t) = -m\ddot{u}_g$$

is the effective load resulting from the ground motion, and is equivalent to the product of the mass of the structure and the ground motion.

Response to arbitrary loading (undamped)

To find the response of a single-degree-of-freedom system to an arbitrary loading, the latter can be treated as series of short impulses (Figure 5.30). The total response to the arbitrary loading is the sum of all the impulses of duration $d\tau$, i.e.

$$u(t) = \int_0^t \frac{F(\tau)}{m\omega} \sin\omega(t-\tau)\, d\tau. \quad (5.51)$$

This is an exact expression called the Duhamel integral. Because it depends upon the principle of superposition it is applicable to linear structures only.

Response to arbitrary loading (damped)

The damped form of the Duhamel integral is

$$u(t) = \int_0^t \frac{F(\tau)}{m\omega_D} e^{-\xi\omega(t-\tau)} \sin\omega(t-\tau)\, d\tau. \quad (5.52)$$

To evaluate this response to an arbitrary loading history, such as would occur in an earthquake, many numerical integration processes are available.

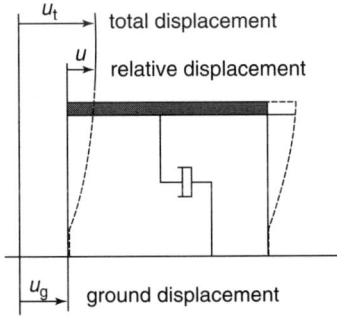

Figure 5.29 Single-degree-of-freedom system subjected to ground motion

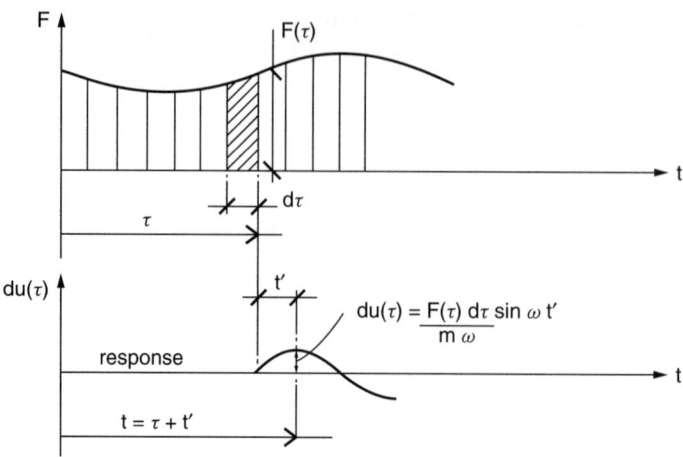

Figure 5.30 Response of single-degree-of-freedom system to impulsive undamped loading

Earthquake response

The response of damped single-degree-of-freedom structures to earthquake motion comes from above as follows. Equation (5.52) can be written in terms of the ground acceleration:

$$u(t) = \frac{1}{\omega} \int_0^t \ddot{u}_g(\tau) e^{-\xi\omega(t-\tau)} \sin\omega(t-\tau)\, d\tau. \tag{5.53}$$

Denoting the integral in the above equation by the response function, $V(t)$,

$$V(t) = \int_0^t \ddot{u}_g(\tau) e^{-\xi\omega(t-\tau)} \sin\omega(t-\tau)\, d\tau, \tag{5.54}$$

the earthquake deflection response of a lumped mass system becomes

$$u(t) = \frac{1}{\omega} V(t). \tag{5.55}$$

The effective earthquake force on the structure follows simply as

$$Q(t) = m\omega V(t). \tag{5.56}$$

Thus the effective earthquake force (or base shear) is found in terms of the mass of the structure, its circular frequency, and the response function $V(t)$ expressed in equation (5.54). Equations (5.55) and (5.56) describe the earthquake response at any time t for a single-degree-of-freedom structure, and solutions to these earthquakes depend upon the evaluation of equation (5.54).

Response spectra

To obtain the entire history of forces and displacements during an earthquake using the above equations is a tedious procedure. For many structures it will suffice to evaluate only the maximum responses. From equations (5.55) and (5.56) this means finding the maximum value of response function $V(t)$. This maximum value is called the spectral velocity, which is

$$S_v = \left\{ \int_0^t \ddot{u}_g(\tau) e^{-\xi \omega (t-\tau)} \sin \omega (t - \tau) \, d\tau \right\}_{max}. \qquad (5.57)$$

It follows that the maximum displacement or spectral displacement is

$$S_d = \frac{S_v}{\omega}, \qquad (5.58)$$

and the spectral acceleration (or spectral pseudo-acceleration) is

$$S_a = \omega S_v. \qquad (5.59)$$

From these relationships, the maximum earthquake displacement response is

$$u_{max} = S_d, \qquad (5.60)$$

and the maximum effective earthquake force or base shear is

$$Q_{max} = M S_a. \qquad (5.61)$$

If equation (5.59) is evaluated for single-degree-of-freedom structures of varying natural periods, a maximum velocity response curve (called a response spectrum) can be plotted for any given damping. See, for example, the ground acceleration history shown in Figure 5.31 and the acceleration response spectra shown in Figure 5.20. The maximum responses of a single-degree-of-freedom structure may be found directly from the spectra and equations (5.60) and (5.61).

The velocity spectrum of Figure 5.31 is for the ground motion of a specific earthquake recorded at a specific site, and the sharp discontinuities in the spectral curves indicate local resonances and characteristics of the ground motion. In any case the period of vibration of a structure cannot be known with enough certainty to design with either a peak spectral value or an adjacent trough. For general design purposes an averaged spectrum as shown in Figure 5.32 will therefore be more appropriate. To obtain a realistic result from a response from a response spectrum analysis, it is necessary to use a spectrum derived from an appropriate ground motion (see Section 4.7.4).

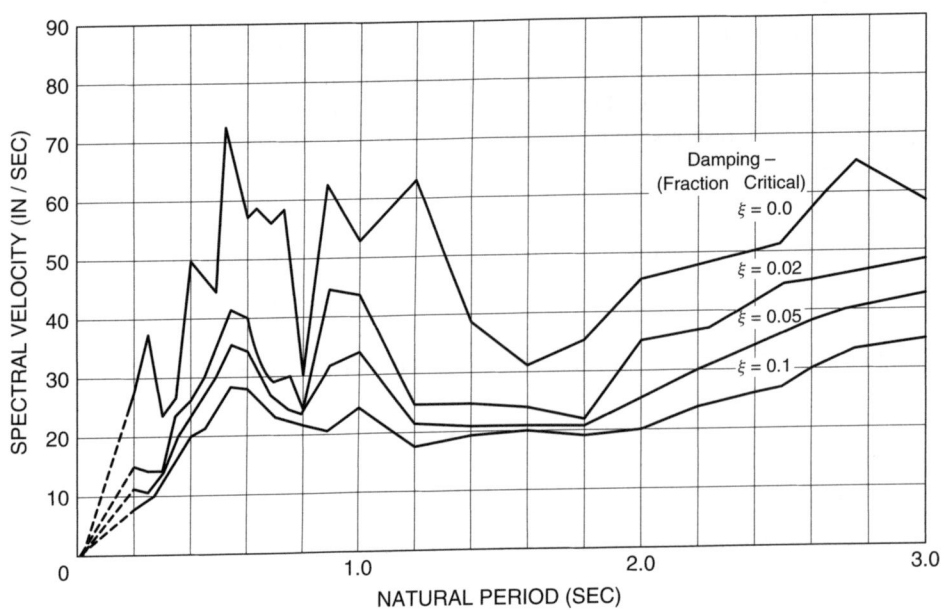

Figure 5.31 Velocity spectrum for the ground motion recorded at El Centro in the Imperial Valley earthquake of May 1940 (north–south component)

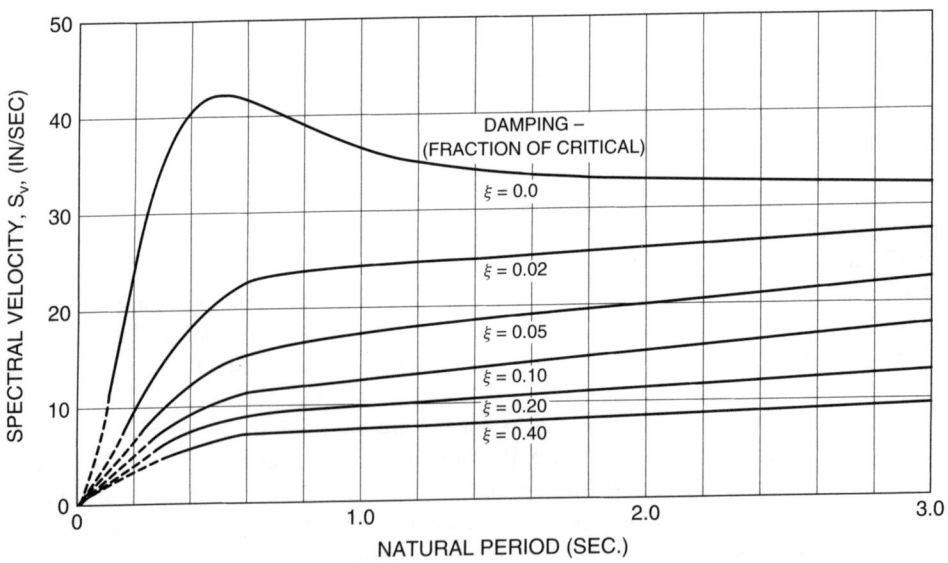

Figure 5.32 Averaged velocity response spectra, based on the spectral intensity of the 1940 El Centro record (cf. Figure 5.31)

Multi-degree-of freedom systems

In the dynamic analysis of most structures it is necessary to assume that the mass is distributed in more than one discrete lump. For most buildings the mass is assumed to be concentrated at for levels, and to be subjected to lateral displacements only. To illustrate the corresponding multi-degree-of-freedom analysis, consider a three-storey building (Figure 5.21). Each storey mass represents one degree of freedom, each with an equation of dynamic equilibrium

$$F_{Ia} + F_{Da} + F_{Sa} = F_a(t),$$
$$F_{Ib} + F_{Db} + F_{Sb} = F_b(t), \quad (5.62)$$
$$F_{Ic} + F_{Dc} + F_{Sc} = F_c(t).$$

The inertia forces are given in matrix form by

$$\begin{Bmatrix} F_{Ia} \\ F_{Ib} \\ F_{Ic} \end{Bmatrix} = \begin{Bmatrix} m_a & 0 & 0 \\ 0 & m_b & 0 \\ 0 & 0 & m_c \end{Bmatrix} \begin{Bmatrix} \ddot{u}_a \\ \ddot{u}_b \\ \ddot{u}_c \end{Bmatrix} \quad (5.63)$$

or, more generally,

$$\mathbf{F} = M\ddot{\mathbf{u}} \quad (5.64)$$

where $\mathbf{F}_I$ is the inertia force vector, $\mathbf{M}$ is the mass matrix and $\ddot{\mathbf{u}}$ is the acceleration vector. The mass matrix is of diagonal form for a lumped-mass system, giving no coupling between masses, which is a prime reason for using the lumped-mass method.

The elastic forces in equation (5.62) depend on the displacements, and using stiffness influence coefficients they may be expressed in matrix form as

$$\begin{Bmatrix} F_{Sa} \\ F_{Sb} \\ F_{Sc} \end{Bmatrix} = \begin{Bmatrix} k_{aa} & k_{ab} & k_{ac} \\ k_{ba} & k_{bb} & k_{bc} \\ k_{ca} & k_{cb} & k_{cc} \end{Bmatrix} \begin{Bmatrix} u_a \\ u_b \\ u_c \end{Bmatrix} \quad (5.65)$$

or, more generally,

$$\mathbf{F_S} = \mathbf{Ku} \quad (5.66)$$

where $\mathbf{F_S}$ is the elastic force vector, $\mathbf{K}$ is the stiffness matrix, and $\mathbf{u}$ is the displacement vector. The stiffness matrix $\mathbf{K}$ generally exhibits coupling and will best be handled by a standard computerized matrix analysis.

By analogy with the expressions (5.65) and (5.66) the damping forces may be expressed

$$\mathbf{F_D} = \mathbf{C\dot{u}} \quad (5.67)$$

where $\mathbf{F_D}$ is the damping force vector, $\mathbf{C}$ is the damping matrix and $\dot{\mathbf{u}}$ is the velocity vector. In general it is not practicable to evaluate $\mathbf{C}$, and damping is usually expressed in terms of damping coefficients.

Using equations (5.64), (5.66) and (5.67), the equations of dynamic equilibrium (5.62) may be written generally as

$$\mathbf{F_I} + \mathbf{F_D} + \mathbf{F_S} = \mathbf{F}(t), \qquad (5.68)$$

which is equivalent to

$$\mathbf{M\ddot{u}} + \mathbf{C\dot{u}} + \mathbf{Ku} = \mathbf{F}(t) \qquad (5.69)$$

The matrix equation (5.69) for a multi-degree-of-freedom system is identical in form to the single-degree-of-freedom equation (5.49).

Vibration frequencies and mode shapes

The first step in the analysis of a multi-degree-freedom system is to find its free vibration frequencies and mode shapes. In free vibration there is no external force and damping is taken as zero. The equations of motion become

$$\mathbf{M\ddot{u}} + \mathbf{Ku} = \mathbf{0}. \qquad (5.70)$$

But in free vibration the motion is simple harmonic:

$$\mathbf{u} = \hat{\mathbf{u}} \sin \omega t.$$

Therefore

$$\ddot{\mathbf{u}} = -\omega^2 \hat{\mathbf{u}} \sin \omega t, \qquad (5.71)$$

where $\hat{\mathbf{u}}$ represents the amplitude of vibration.

Substituting in equation (5.70),

$$\mathbf{K\hat{u}} - \omega^2 \mathbf{M\hat{u}} = \mathbf{0}. \qquad (5.72)$$

Equation (5.72) is an eigenvalue equation and is readily solved for ω by standard computer programs. Its solution for a system of N degrees of freedom yields a vibration frequency ω_n and a mode shape vector $\boldsymbol{\phi}_n$ which represents the *relative* amplitudes of motion for each of the displacement components in mode n. Figure 5.21 shows the shapes of the three normal modes of a typical three-storey building.

Because of the orthogonality properties of mode shapes, equation (5.69) can be written

$$\ddot{Y}_n + 2\xi_n \omega_n \dot{Y}_n + \omega^2 Y_n = \frac{\boldsymbol{\phi}_n^T \mathbf{F}(t)}{\boldsymbol{\phi}_n^T \mathbf{M} \boldsymbol{\phi}_n} \qquad (5.73)$$

where Y_n is a generalized displacement in mode n, leading to the actual displacement (see equation (5.75)) and $\boldsymbol{\phi}_n^T$ is the row mode shape vector corresponding to the column vector $\boldsymbol{\phi}_n$.

Earthquake response analysis by mode superposition

The dynamic analysis of multi-degree-of-freedom systems can therefore be simplified to the solution of equation (5.72) for each mode, and the total response is then obtained by superimposing the modal effects. The response of the nth mode at any time t may be found by evaluating the Duhamel integral (equation (5.52)):

$$Y_n(t) = \frac{L_n}{\phi_n^T \mathbf{M} \phi_n} \cdot \frac{1}{\omega_n} \int_0^t \ddot{u}_g(\tau) e^{-\xi \omega_n (t-\tau)} \sin \omega_n (t - \tau) \, dt. \tag{5.74}$$

This displacement of floor (or mass) i at time t is then obtained by superimposing the response of all modes evaluated at this time t:

$$u_i = \sum_{n=1}^{N} \phi_{in} Y_n(t) \tag{5.75}$$

where ϕ_{in} is the relative amplitude of the displacement of mass i in mode n.

Superimposing all the modal contributions, the earthquake forces in the total structure may be expressed in matrix form as

$$\mathbf{q}(t) = \mathbf{M} \phi \omega^2 Y_n(t), \tag{5.76}$$

where ϕ is the square matrix of relative amplitude distributions in each mode and ω^2 is the diagonal matrix of ω^2 for each of the n modes.

From equations (5.72) and (5.73) the entire history of displacement and force response can be defined for any multi-degree-of-freedom system, having first determined the modal response amplitudes of equation (5.75).

Response spectrum analysis for multi-degree systems

As with single-degree-of-freedom structures considerable simplification of the analysis is achieved if only the maximum response to each mode is considered rather than the whole response history. If the maximum value $Y_{n,\max}$ of the Duhamel equation (5.71) is calculated, the distribution of maximum displacements in that mode is

$$\mathbf{u}_{n,\max} = \phi_n Y_{n,\max} = \phi_n \frac{L_n}{\phi_n^T \mathbf{M} \phi_n} \frac{S_{vn}}{\omega_n} \tag{5.77}$$

and the distribution of maximum earthquake forces in that mode is

$$\mathbf{q}_{n,\max} = \mathbf{M} \phi_n \omega_n^2 Y_{n,\max} = \mathbf{M} \phi_n \frac{L_n}{\phi_n^T \mathbf{M} \phi_n} S_{an}. \tag{5.78}$$

In equations (5.77) and (5.78), S_{vn} and S_{an} are the spectral velocity and spectral acceleration for mode n, and are as defined in equations (5.54) and (5.56).

Equations (5.77) and (5.78) enable the maximum response in each mode to be determined. As the modal maxima do not necessarily occur at the same time, nor necessarily have the same sign, they are best combined on a probability basis. Various approximate formulae for superposition are used, the most common being the square root of sum of squares procedure. As an example, the maximum deflection at the top of a three-storey structure (three masses) would be

$$u_{a,\max} = \sqrt{u_{a1,\max}^2 + u_{a2,\max}^2 + u_{a3,\max}^2}. \tag{5.79}$$

This approximation is usually, but not always, conservative. An often preferred alternative with reduced errors in the modal combinations is the complete quadratic combination (CQC) method (Wilson *et al.*, 1981), in which a typical displacement component is

$$u_k = \sqrt{\sum_i \sum_j u_{ki} \rho_{ij} u_{kj}} \tag{5.80}$$

and a typical force component is

$$f_k = \sqrt{\sum_i \sum_j f_{ki} \rho_{ij} f_{kj}}, \tag{5.81}$$

where u_{ki} is a typical component of the modal displacement vector, $U_{i,\max}$. The cross modal coefficients, ρ_{ij}, are functions of the duration and frequency content of the loading and of the modal frequencies and damping ratios of the structure.

Normal mode analysis is a more complex and accurate form of dynamic analysis than the response spectrum technique and a more limited technique than direct integration (discussed next), as it depends on artificially separating the normal modes of vibration and combining the forces and displacements associated with a chosen number of them by superposition. As with direct integration techniques, actual earthquake accelerograms can be applied to the structure and a stress history determined, but because of the use of superposition the technique is limited to linear material behaviour. Although modal analysis can provide any desired order of accuracy for linear behaviour by incorporating all the modal responses, some approximation is usually made by using only the first few modes in order to save computation time. Problems are encountered in dealing with systems where the modes cannot be validly separated, i.e. where mode coupling occurs.

The most serious shortcoming of linear analyses is that they do not accurately indicate all the members requiring maximum ductility. In other words, the pattern of highest elastic stresses is not necessarily the same as the pattern of plastic deformation in an earthquake structure. For important structures in zones of high seismic risk, non-linear dynamic analysis is sometimes called for.

Direct integration provides the most powerful and informative analysis for any given earthquake motion. A time-dependent forcing function (earthquake accelerogram) is applied and the corresponding response history of the structure during the earthquake is computed. That is, the moment and force diagrams at each of a series of prescribed intervals throughout the applied motion can be found. Computer programs have been written for both linear elastic and non-linear inelastic material behaviour, using step-by-step integration procedures. Linear behaviour is seldom analysed by direct integration, unless mode coupling is involved, as normal mode techniques are easier, cheaper, and nearly as accurate. Three-dimensional non-linear analyses take the three orthogonal accelerogram components from a given earthquake, and apply them simultaneously to the structure. In principle, this is the most complete dynamic analysis technique so far devised, and is unfortunately correspondingly expensive and laborious to carry out.

For detailed information on dynamic analysis, readers are referred to textbooks such as those of Chopra (2005, 2007) and Clough and Penzien (1993), the convenient overview by Carr (1994), or the discussion of issues relating to non-linear analysis by Elnashai (2002).

Selection of method of analysis for structures

It is difficult to give clear general advice on selecting the means of analysis, as each structure will have its own technical, statutory, economic and sometimes political requirements. Broadly speaking, however, the larger and/or more complex the structure, the more sophisticated the dynamic analysis used. Table 5.12 gives a very simple indication of the applicability of the main methods of analysis.

Except in special projects, designers are unlikely to do a three-dimensional dynamic analysis, whether elastic or inelastic, which allows for two orthogonal horizontal components of ground motion simultaneously. To make some allowance for the resultant diagonal response, some codes stipulate arbitrary means of adding the separately computed orthogonal components.

It is important to note that the methods of analysis in Table 5.12 become successively more realistic *only* if the appropriate seismic loadings and a suitable model of material behaviour are used. Even with the best input, research shows (Tang and Clough, 1979) that non-linear analysis is likely to be accurate in global terms, but not at local member behaviour level. Response spectrum analysis is not only far simpler and cheaper to use than non-linear analysis but the use of smoothed spectra by definition gives more reliable allowance for applied loads than do response history methods. In general, it is better to use simpler rather than more complex methods of analysis, putting most effort into the design concept (Chapter 8) and good detailing (Chapter 10).

Method of analysis and material behaviour

Dynamic analysis techniques relating to the behaviour of soils in the soil–structure interaction problem have been discussed in Section 5.3.3. The following discussion is complementary to that for soils, having structures more specifically in mind. The problem

Table 5.12 Methods of analysis and types of structure

Type of structure	Method of analysis (two- or three-dimensional)
Small simple structures	(1) Equivalent static forces
↓	(2) Response spectra
Progressively more demanding structures	(3) Modal analysis
↓	(4) Direct integration
Large complex structures	(5) Non-linear soil–structure

of selecting an analysis method, of course, depends largely upon whether the materials are intended to be elastic or inelastic during the design earthquake. The usual methods of analysis for these two states are set out in Table 5.13.

Elastic behaviour during the design earthquake obviously has the advantage of making linear analysis entirely appropriate. It may arise because:

(1) the material is brittle, i.e. has zero ductility (e.g. porcelain, mass concrete, some masonry);
(2) the material has limited ductility (e.g. timber, ill-confined concrete, most reinforced masonry);
(3) the designer chooses to keep a ductile material within the elastic range (e.g. steel, reinforced concrete) where greater stiffness is required for functional reasons or greater safety is desired (items (2b) or (4) of Table 8.1).

Materials in the brittle category, such as masonry or porcelain, may be realistically analysed by method (1a) in Table 5.13, because their linear elastic behaviour matches the analytical assumptions. The chief problems lie in choosing an adequate safety margin (load factor) within the elastic range, to cover the normal errors involved in assessing the loading, the geometric modelling, and the ultimate strength. For cases (2) and (3) above, designed to be elastic in the design earthquake (methods 1(b) and 3(b) of Table 5.13), it is common practice to enhance the safety factor by developing as much inelastic deformation prior to collapse as practicable with nominal ductility reinforcement. This increases

Table 5.13 Seismic analysis and design procedures

Method of analysis	Material behaviour		Design provisions	Seismic loading
	Elastic	(1a)	Permissible stress or factored ultimate design	
Static linear				Full
		(1b)	Sometimes detailed for nominal ductility	
	Inelastic	(2)	Permissible stress or factored ultimate design Detailed for ductility	Reduced by factor R (Figure 5.33)
	Elastic	(3a)	Permissible stress or factored ultimate design	
Dynamic linear		(3b)	Sometimes detailed for nominal ductility	Full
	Inelastic	(4)	Permissible stress or factored ultimate design Detailed for ductility	Reduced by factor R (Figure 5.33)
Dynamic non-linear	Inelastic	(5)	Hysteresis loops required Ultimate strength design Ductility demands found from plastic deformations	Full

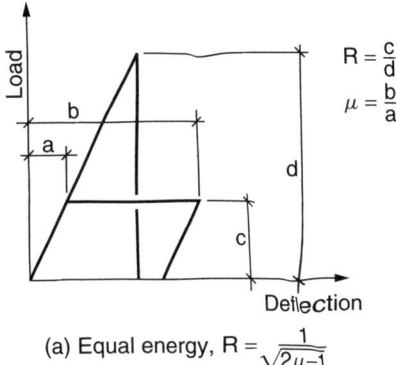
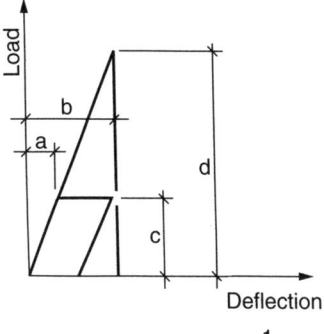

Figure 5.33 Multi-degree-of-freedom system subjected to dynamic loading. Reduction factors R for seismic loading equating elastic and inelastic response in terms of (a) energy and (b) deflection

the safety against strong shaking, particularly of longer duration. In porcelain structures it is good practice to increase the earthquake protection by deliberately enhancing the damping. For further discussion of masonry and porcelain, see Sections 10.4 and 11.2, respectively.

Materials which become inelastic in the design earthquake are more satisfactory for earthquake resistance than brittle ones because of their inelastic deformability, but are less convenient and more expensive to analyse for the same reason. Of the methods used to analyse inelastic behaviour, only method (5) attempts to model the hysteretic stress–strain behaviour directly. This cannot always be done reliably, because suitable hysteresis models are available for only limited types of element and stress condition in each construction material.

For steel structures which can be realistically modelled as regular plane frames, such as that referred to in Figures 5.22 and 5.25, method (5) may be economic from an analytical as well as from a construction point of view. It follows that most structures which will behave inelastically in the design earthquake are designed by methods (2) and (4) of Table 5.13, assuming linear elastic material behaviour in the analysis. Both methods imply the application of an artificially reduced earthquake loading within the elastic capacity of the structure, and an approximate allowance for the inelastic deformations by ensuring certain ductility levels in highly stressed zones.

This design process, which attempts to do inelastic design by an 'equivalent elastic' method, has difficulties in.

(1) allowing for inelastic deflection;
(2) allowing for stiffness degradation;
(3) determining the distribution of ductility demands; and
(4) allowing for the duration of strong shaking.

These four factors are not independent and vary with the nature of the material, the structural form, and the loading, as discussed below.

In endeavouring to define an equivalent-elastic loading, two alternative methods of equating elastic and inelastic response are commonly considered (Figure 5.32).

For simple elastoplastic systems, two alternative reduction factors for the loading, $R = c/d$, are obtained in terms of the deflection ductility factor $\mu = b/a$, as shown. Here $\mu = b/a$ is the ratio of the total deflection to the elastic deflection of the system. In the case of buildings or other structures, μ is generally referred to as the structure deflection ductility factor, where μ is the ratio of the deflection at ultimate to the deflection at first yield, measured at the top of the structure, as shown in Figure 8.9. Despite being known to be unreliable at lower periods of vibration T, the equal deflection reduction factor $R = 1/\mu$ is widely used because of its simplicity. Although R is relatively insensitive to the shape of the hysteresis loop, it is strongly dependent on the nature of the accelerogram used as input, and it appears from the work of Moss et al. (1986) that the use of $R = 1/\mu$ may not be appropriate for T less than about 1.5 s. A scheme which makes some allowance for this is discussed in Section 4.7.4.

The dependence of R on the individual accelerogram can be seen from Figure 5.34, which shows the near-source accelerograms and response spectra from El Centro in the $M_W = 7.0$ Imperial Valley earthquake of 1940 and that from Parkfield in the $M_W = 6.2$ Parkfield earthquake of 1966. Despite its smaller PGA and spectral amplitudes, damage was much greater around El Centro than that associated with the Parkfield record. This awkward fact arises largely from the greater duration of strong shaking in the larger-magnitude Imperial Valley event, and unfortunately response spectra do not account much for duration.

Clearly, great care is necessary in selecting design earthquakes of different magnitude to ensure an appropriate relationship between elastic and inelastic response. A much greater reduction factor R for Parkfield would be required than for the El Centro response. However, the use of averaged or uniform hazard spectra, together with reduction factors R for given ductility levels, is an internationally accepted earthquake analysis tool.

The determination of the distribution of ductility demand throughout a multi-redundant structure using an equivalent-elastic analysis is also unreliable. The positions of maximum moment in a frame, determined elastically, will not necessarily indicate the order of plastic hinge formation. However, in ideally regular plane frames this approximation may be reasonable, and is often taken in practice.

P-delta effect

When structures sway horizontally in earthquakes an overturning moment $M = P\Delta$ exists equal to the weight P multiplied by the horizontal displacement Δ. While usually negligible globally, the P-delta moment at individual vertical members is sometimes significant, particularly in moment resisting frames supporting substantial gravity loads, and where fully ductile design permits substantial inelastic horizontal displacements, and for longer durations of strong shaking.

A simple approximate method of allowing for P-delta effects is to increase the structural actions found from an equivalent-static analysis or a modal response spectrum analysis for the ultimate limit state by multiplying them by the ratio

$$V/(C_p W_t + V),$$

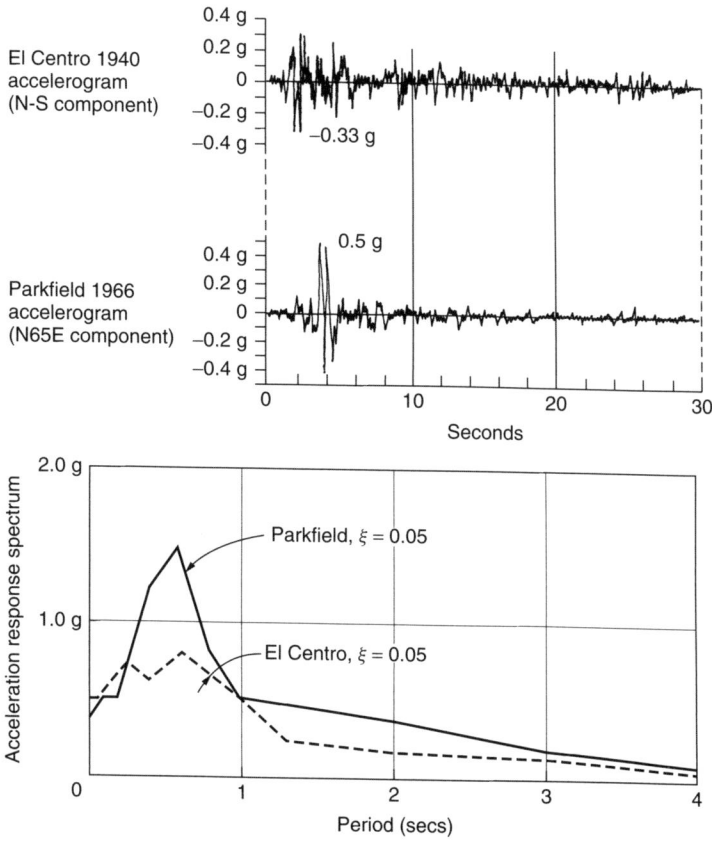

Figure 5.34 Accelerograms and elastic response spectra from El Centro in the $M_W = 7.0$ Imperial Valley earthquake of 1940 and from the $M_W = 6.2$ Parkfield earthquake of 1966

where V is the base shear, W_t is the seismic weight of the building, and C_p is given by the expression

$$C_p = 0.015 + 0.0025(\mu - 1), \tag{5.82}$$

with the limits of $0.015 < C_p < 0.03$.

P-delta effects are small enough to be ignored in buildings for which the structural ductility factor $\mu \leq 1.5$, or in low-rise buildings such that $T_1 \leq 0.6$ s and the building is of not more than three or four storeys in high-hazard zones, or five or six storeys in low-hazard zones.

References

Ambraseys NN (1973) Dynamics and response of foundation materials in epicentral regions of strong earthquakes. *Proc. 5th World Conf on Earthq Eng*, Rome **1**: CXXVI–CXLVIII.

Ambraseys NN (1988) Engineering seismology. *Earthq Eng Struc Dyn* **17**: 1–105.

Ang AH and Newmark NM (1971) Development of a transmitting boundary for numerical wave motion calculations. Report to Defence Atomic Support Agency, Contract DASA-01-0040, Washington, DC.

Applied Technology Council (1978) Tentative provisions for the development of seismic regulations for buildings. ATC 3-06, National Bureau of Standards, USA.

Beuch F, Davies TR, Pettinga JR, Finnemore M and Berrill J (2007) The Little Red Hill field experiment: seismic response of an edifice. *Proc. 2007 Conf NZ Soc Earthq Eng, Wairakei*. Paper 48 (CD-ROM).

Bielak J (1975) Dynamic behaviour of structures with embedded foundations. *Earthq Eng Struc Dyn* **3**(3): 259–274.

Blake RE (1961) Basic vibration theory. In *Shock and Vibration Handbook*. (eds CM Harris and CE Crede), Vol. **1**. McGraw-Hill, New York.

Blakeley RWG (1973) Prestressed concrete seismic design. *Bull NZ Soc Earthq Eng* **6**(1): 2–21.

Boore DM (1973) The effect of simple topography on seismic waves: Implications for the recorded accelerations at Pacoima Dam. *Bull Seis Soc Amer* **63**: 1603–1609.

Borcherdt R (1994) Estimates of site-dependent response spectra for design (methodology and justification). *Earthq Spectra* **10**(4): 617–653.

Brune JN (1984) Preliminary results on topographic seismic amplification effect on a foam rubber model of the topography near Pacoima Dam. *Proc. 8th World Conf on Earthq Eng*, San Francisco **II**: 663–670.

Building Seismic Safety Council (2000) *NEHRP Recommended Provisions for Seismic Regulations for Buildings and other Structures* (FEMA 368). Federal Emergency Management Agency, Washington, DC.

Building Seismic Safety Council (2001) *NEHRP Recommended Provisions for Seismic Regulations for New Buildings and Other Structures*. Federal Emergency Management Agency, Washington, DC.

Bycroft GN (1956) Forced vibrations of a rigid circular plate on a semi-infinite elastic space and on an elastic stratum. *Phil Trans Roy Soc London A* **248**: 327–368.

Carr AJ (1994) Dynamic analysis of structures. *Bull NZ Nat Soc Earthq Eng* **27**: 129–146.

Chopra AK (2005) *Earthquake dynamics of structures, A primer*. EERI Monograph MNO-11, 2nd edn. Earthquake Engineering Research Institute, Oakland, CA.

Chopra AK (2007) *Dynamics of Structures – Theory and Applications to Earthquake Engineering*, 3rd edn. Pearson Prentice Hall, Upper Saddle River, NJ.

Chopra AK and Goel RK (2000) Building period formulas for estimating seismic displacements. *Earthq Spectra* **16**(2): 533–536.

Christafulli FJ, Carr A and Park R (2000) Analytical modelling of infilled frame structures – a general review. *Bull NZ Soc Earthq Eng* **33**: 30–47.

Clough RW and Penzien J (1993) *Dynamics of Structures*, 2nd edn. McGraw-Hill, New York.

Danay A (1977) Vibrations of rigid foundations. *Arup J*, London **12**(1): 19–27.

Dobry R and Vucetic M (1987) Dynamic properties and seismic response of soft clay deposits. *Proc. Int Symp Geotech Eng Soft Soils*, Mexico City **2**: 51–87.

Dowrick DJ (1985) Preliminary field observations on the Chilean earthquake of 3 March, 1985. *Bull NZ Nat Soc Earthq Eng* **18**(2): 119–127.

Elnashai AS (2002) Do we really need inelastic dynamic analysis? *J Earthq Eng* **6**(Special Issue 1): 123–130.

Gazetas G and Dobry R (1984) Simple radiation damping model for piles and footings. *J Eng Mech* **110**(6): 937–956.

Geli L, Bard P-Y and Jullien B (1988) The effect of topography on earthquake ground motion: A review and new results. *Bull Seis Soc Amer* **78**: 42–63.

Goel RK and Chopra AK (1998) Period formulas for concrete shear wall buildings. *J Struc Eng* **124**(4): 426–433.

Hart GC and Vasudevan R (1975) Earthquake design of buildings: damping. *J Struc Divn* **101**(ST1): 11–30.

Housner GW and Jennings PC (1977) The capacity of extreme earthquake motions to damage structures. In *Structural and Geotechnical Mechanics – A Volume Honoring NM Newmark*. (ed. WJ Hall). Prentice-Hall, Englewood Cliffs, NJ, pp. 112–116.

Idriss IM and Boulanger RW (2008) *Soil liquefaction during earthquakes*, MNO-12. Earthquake Engineering Research Institute, Oakland, CA.

Ishibashi I (1992) Discussion to: Effect of soil plasticity on cyclic response, by M Vucetic and R Dobry (1991), *J Geotech Eng* **118**: 830–832.

Kausel E (1974) Forced vibrations of circular foundations on layered media. Research Report R74-11, Dept of Civil Eng, Massachusetts Institute of Technology.

Kausel E (1976) Nonlinear behaviour of soil-structure interaction. *J Geotech Eng Divn* **102**(GT11): 1159–1170.

Kramer SL (1996) *Geotechnical Earthquake Engineering*. Prentice Hall, Upper Saddle River, NJ.

Kramer SL and Elgamal AW (2001) Modeling soil liquefaction hazards for performance-based earthquake engineering. Pacific Center for Earthquake Engineering Report PEER 2001/13, Berkeley, CA.

Lam N and Wilson J (1999) Estimation of site natural period from a borehole record. *Australian J Struc Eng* **SE1**(3): 179–199.

Lamar S and Fortoul C (1969) Brick masonry effect in vibration of frames. *Proc. 4th World Conf Earthq Eng*, Chile **II**(A3): 91–98.

Liao SSC, Veneziano D and Whitman RV (1988) Regression models for evaluating liquefaction probability. *J Geotech Eng* **114**: 389–411.

Luco JE (1974) Impedance functions for a rigid foundation on a layered medium. *Nuclear Eng and Design* **31**(2): 204–217.

Luco JE and Westmann RA (1971) Dynamic response of circular footings. *J Eng Mech Divn* **97**: 1381–1395.

Luco JE, Wong HL and Trifunac MD (1975) A note on dynamic response of rigid embedded foundations. *Earthq Eng Struc Dyn* **4**(2): 119–127.

Lysmer J and Kuhlemeyer RL (1969) Finite dynamic model for infinite media. *J Eng Mech Divn* **95**(EM4): 859–877.

McVerry GH, Hodder SB, Hefford RT and Heine AJ (1984) Records of engineering significance from the New Zealand strong-motion network. *Proc. 8th World Conf Earthq Eng*, San Francisco **II**: 199–206.

Moss PW, Carr AJ and Buchanan AH (1986) Seismic design loads for low-rise steel buildings. *Proc. Pacific Struc Steel Conf*, Auckland **1**: 149–165.

Mylonakis G and Gazetas G (2000) Seismic soil-structure interaction: beneficial or detrimental? *J Earthq Eng* **4**: 277–302.

Newmark NM and Rosenblueth E (1971) *Fundamentals of Earthquake Engineering*. Prentice-Hall, Englewood Cliffs, NJ.

Park R, Kent DC and Sampson RA (1972) Reinforced concrete members with cyclic loading. *J Struc Divn* **98**(ST7): 1341–1360.

Penzien J and Tseng WS (1976) Seismic analysis of gravity platforms under soil-structure interaction effects. *Proc. Offshore Technology Conf*, Houston, TX, Paper No 2674.

Poulos HG and Davis EH (1974) *Elastic Solutions for Soil and Rock Mechanics*. John Wiley & Sons, Inc., New York.

Roesset JM (1972) Fundamentals of soil amplification. In *Seismic Design of Nuclear Reactors* (ed. RJ Hansen), MIT Press, Cambridge, MA, pp. 183–244.

Schnabel PB, Seed HB and Lysmer J (1971) Modification of seismograph records for effects of local soil conditions. Report No EERC 71-8, Earthquake Engineering Research Center, University of California, Berkeley.

Seed HB (1975) Earthquakes effects of soil-foundation systems. In *Foundation Engineering Handbook* (eds HF Winterborn and HY Fang). Van Nostrand Reinhold Company, New York, pp. 700–732.

Seed HB and Idriss IM (1969) Influence of soil conditions on ground motions during earthquakes. *J Soil Mech and Found Divn* **95**(SM1): 99–137.

Seed HB and Idriss IM (1970) Soil moduli and damping factors for dynamic response analysis. Report No EERC 70-10, Earthquake Engineering Research Center, University of California, Berkeley.

Seed HB and Romo MP (1986) Analytical modelling of dynamic soil response. *Proc. Int Conf on the 1985 Mexico Earthquakes*. Mexico City, American Society of Civil Engineers.

Seed HB, Whitman RV, Dezfulian H, Dobry R and Idriss IM (1972) Soil conditions and building damage in 1967 Caracas earthquake. *J Soil Mech and Found Divn* **98**(SM8): 787–806.

Seed HB, Wong RT, Idriss IM and Tokimatsu K (1984) Moduli and damping factors for dynamic analyses of cohesionless soils. Report No UCB/EERC-84/14, Earthquake Engineering Research Center, University of California, Berkeley.

Seed HB, Wong RT, Idriss IM and Tokimatsu K (1986) Moduli and damping factors for dynamic analyses of cohesionless soils. *J Geotech Eng* **112**: 1016–1032.

Silver ML and Seed HB (1969) The behaviour of sands under seismic loading conditions. Report No EERC 69-16, Earthquake Engineering Research Center, University of California, Berkeley.

Sinha BP, Gerstle KH and Tulin LG (1964) Stress-strain relationships for concrete under cyclic loading. *J Amer Conc Inst* **61**(2): 195–211.

Smith IC, Spring KCF and Sidwell GK (1975) The design and construction of the Bank of New Zealand, Wellington, New Zealand. *Bull NZ Nat Soc Earthq Eng* **8**(2): 142–168.

Sritharan S and Dowrick DJ (1994) Response of low-rise buildings to moderate ground shaking, particularly the May 1990 Weber earthquake. *Bull NZ Nat Soc Earthq Eng* **27**(3): 205–221.

Stratta JL and Feldman J (1971) Interaction of infill walls and concrete frames during earthquakes. *Bull Seis Soc Amer* **61**(3): 609–612.

Structural Engineers Association of California (1980) *Recommended lateral force requirements and commentary*. SEAOC.

Takanashi K (1973) Inelastic lateral buckling of steel beams subjected to repeated and reversed loadings. *Proc. 5th World Conf Earthq Eng*, Rome **1**: 795–798.

Tang DT and Clough RW (1979) Shaking table earthquake response of steel frames. *J Struc Divn* **105**(ST1): 221–243.

Tani K (1995) General report: Measurement of shear deformation of geomaterials – field tests. *Proc. Int Symp Pre-Failure Def Charac Geomaterials*, Sapporo **2**: 1115–1135.

Tazoh T, Nakahi S, Shimizu K and Yokata H (1984) Vibration characteristics of soil and their applicability to the field. *Proc. 8th World Conf on Earthq Eng*, San Francisco **II**: 679–686.

Thompson KJ and Park R (1978) Stress-strain model for prestressing steel with cyclic loading. *Bull NZ Nat Soc Earthq Eng* **11**(4): 209–218.

Vaish AK and Chopra AK (1974) Earthquake finite element analysis of structure-foundation systems. *J Eng Mechanics Divn* **100**(EM6): 1101–1116.

Veletsos AS and Meek JW (1974) Dynamic behaviour of building-foundation systems. *Earthq Eng Struc Dyn* **3**(2): 121–138.

Veletsos AS and Nair VVD (1975) Seismic interaction of structures on hysteretic foundations. *J Struc Divn* **101**(ST1): 109–129.

Veletsos AS and Verbic B (1973) Vibration of viscoelastic foundations. *Earthq Eng Struc Dyn* **2**(1): 87–102.

Veletsos AS and Wei YT (1971) Lateral and rocking vibrations of footings. *J Soil Mech and Found Divn* **97**(SM9): 1227–1248.

Watt BJ, Boaz IB and Dowrick DJ (1976) Response of concrete gravity platforms to earthquake excitations. *Proc. Offshore Tech Conf*, Houston, TX, Paper No 2673.

Watt BJ, Boaz IB, Ruhl JA, Shipley SA, Dowrick DJ and Ghose A (1978) Earthquake survivability of concrete platforms. *Proc. Offshore Techn Conf*, Houston, TX, Paper No 3159, 957–973.

Werkle H and Waas G (1986) Dynamic stiffness of foundations in inhomogeneous soils. *Proc. 8th Euro Conf Earthq Eng*, Lisbon.

Whitman RV and Richart FE (1967) Design procedures for dynamically loaded foundations. *J Soil Mech and Found Divn* **93**(SM6): 169–191.

Whitman RV, Roesset JM, Dobry R and Ayestaran L (1972) Accuracy of modal superposition for one-dimensional soil amplification analysis. *Proc. Int Conf on Microzonation*, Seattle **2**: 483–498.

Williams D and Scrivener JE (1973) Response of reinforced masonry shear walls to static and dynamic cyclic loading. *Proc. 5th World Conf Earthq Eng*, Rome **2**: 1491–1494.

Wilson EL, Der Kiureghian A and Bayo EP (1981) A replacement for the SSRS method of seismic analysis. *Earthq Eng Struct Dyn* **9**: 187–194.

Wolf JP (1985) *Dynamic Soil-Structure Interaction*. Prentice Hall, Englewood Cliffs, NJ.

Wolf JP (1988) *Soil-Structure Interaction in the Time Domain*. Prentice Hall, Englewood Cliffs, NJ.

Wolf JP and Song C (1996) *Finite-Element Modelling of Unbounded Media*. John Wiley & Sons, Ltd, Chichester.

Wolf JP and Song C (1999) The guts of soil-structure interaction. *Int Symp Earthq Eng*, Budva, Montenegro.

Yasuda S, Nagase H, Oda S, Masuda T and Morimoto I (1994) A study of appropriate number of cyclic shear tests for seismic response analyses. *Proc. Int Symp Pre-Failure Def Charac Geomaterials*, Sapporo **1**: 197–202.

Youd TL (1991) Mapping of earthquake-induced liquefaction for seismic zonation. *Proc. 4th Int Conf Seismic Zonation*, **1**: 111–147.

Zaslavsky Y and Shapira A (2000) Experimental study of topographic amplification using the Israel seismic network. *J Earthq Eng* **4**: 43–65.

Zhao Z (1990) Seismic soil-structure interaction. PhD thesis, Dept of Civil Engineering, University of Canterbury, New Zealand.

6

Earthquake Vulnerability of the Built Environment

6.1 Introduction

To manage and minimize risk in future earthquakes, by design, planning and retrofitting, we need to understand and evaluate the earthquake vulnerability of the built and natural environments. This is best done by developing models by studying damage in past earthquakes, and quantifying the data on damage to a much greater degree than is possible in earthquake reconnaissance reports, or is given in the excellent book by Steinbrugge (1982).

Earthquake damage to the built environment is caused by a number of factors beyond the principal cause, ground shaking – for example, landslides (as listed in Section 3.1). In this chapter, our discussion of vulnerability is restricted mainly to that relating directly to ground shaking. All other effects, such as subsidence, landslides, liquefaction and earthquake induced fires (Section 7.2.2), are supplementary phenomena.

The vulnerability of items of the built environment to damage in earthquakes varies enormously. It depends on the robustness of the item, which may be inherent or the result of its earthquake resistant design. *Thus, vulnerability may be defined as the degree of damage of a given item of the built environment to a given strength of shaking.* It is helpful to describe the degree of damage both qualitatively and quantitatively, as discussed below.

6.2 Qualitative Measures of Vulnerability

Vulnerability of different classes of construction has long been described in words in the subjective intensity scales. This is illustrated in Table 6.1 where the degree of damage to six classes of construction is described for six Modified Mercalli intensities (MM6–MM11). An aspect of this table that is worth noting is the speculative nature of intensity MM11.The author knows of no verified instances of intensities higher than MM10 worldwide. It appears that the upper bound on intensity is MM10, as shown

---

Earthquake Resistant Design and Risk Reduction   D. Dowrick
© 2009, John Wiley & Sons, Ltd

Table 6.1 Intensity versus construction class performance (slightly modified[1] from Dowrick, 1996)

Intensity	Construction classes					
	Pre-code			Post-code 'brittle' era	Capacity design era	Special low damage
	I[2]	II	III	IV	V	VI
MM11	*All* destroyed	Many destroyed				
MM10	*All* destroyed	Most destroyed	Heavily damaged, some collapse	Heavily damaged, some collapse	Damaged, some with partial collapse	Minor[3] damage, a few moderate damages
MM9	Many destroyed	Heavily damaged, some collapse	Damaged, some with partial collapse	Damaged, some with partial collapse	Moderately damaged, a few with partial collapse[4]	A few instances of damage
MM8	Heavily damaged, some collapse	Damaged, some with partial collapse	Damaged in some cases	Damaged in some cases, some flexible frames seriously	Damaged in some cases, some flexible frames moderately	
MM7	Cracked, some minor masonry falls	A few damaged		A few instances of damage		
MM6	Slight damage may occur					

Notes:
[1] MM12 deleted and changes to MM10 and MM11 shown in italics.
[2] Construction classes paraphrased below are given in full in Appendix A. Type I, poor quality unreinforced masonry (URM), or pre-code reinforced concrete (RC) with a weak storey. Type II, average-quality URM. Type III, pre-code reinforced masonry or concrete (subdivided here as follows): III(1), brick walled with RC ring beams; III(2), RC beams, columns, floors, plus brick infill. III(3), RC walled (no weak storeys). Type IV, post-code (c. 1935–1975).
[3] This is speculative, on the optimistic side.
[4] Allows for structures of this era not having capacity design or being 'below average'.

by Dowrick and Rhoades (2005a) from data from near-surface-rupturing New Zealand earthquakes, in their plot of intensity at the centre (I_0) and the innermost isoseismal (I_{ii}) against magnitude (Figure 6.1). Also, more recently it has been found using the two-dimensional source model of Dowrick and Rhoades (in preparation) that MM intensities do not exceed MM10 in the most powerful surface-rupturing earthquakes (e.g. see Figure 4.39). The possibility remains that while the while the maximum intensity for an isoseismal of substantial area is MM10, that topographical amplification within such a zone could result in a local area of MM11.

Another way of making qualitative measures of vulnerability is in terms of *damage states*. It is useful to compare the damage states of non-domestic buildings in the Napier/Hastings area resulting from the intensity of MM10 induced by the pre-code 1931 Hawke's Bay earthquake in New Zealand, as described by Dowrick (1998). For this purpose the buildings were divided into five subsets: (a) types I and II, (b) type III(1), (c) type III(2), (d) type III(3), and (e) timber (as defined in the footnote to Table 6.1). The

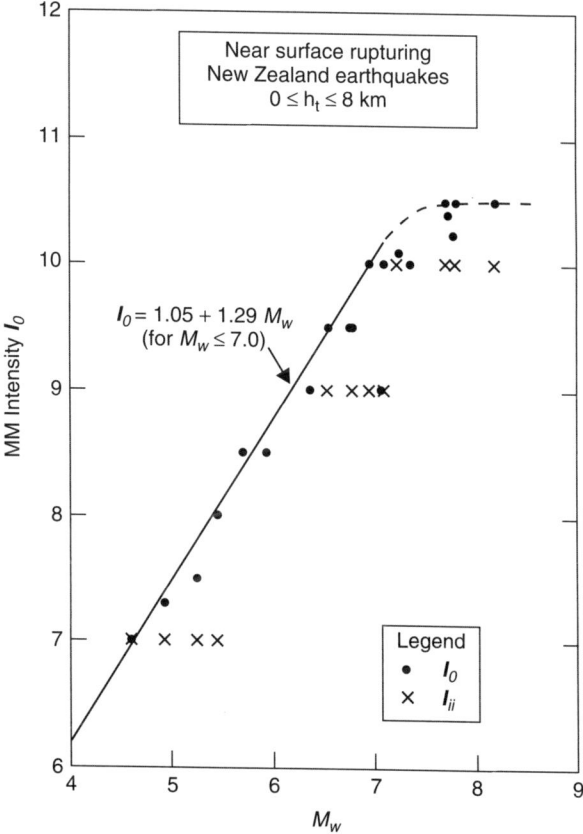

Figure 6.1 Plot of data for intensity at centre (I_0) and innermost isoseismal (I_{ii}) for near-surface-rupturing earthquakes, with upper bound line for I_0 versus M_W (modified from Dowrick and Rhoades, 2005a)

degree of damage to each building has been assessed according to a scale of four damage states:

(1) OK – damage zero or slight;
(2) damaged – cracked or moderately cracked, parapets and gables fall (no volume loss);
(3) partial collapse – volume loss less than 50%;
(4) collapse – volume loss 50% or greater.

The data from this assessment are plotted in histogram form in Figure 6.2. These plots show very clearly that the buildings of subset (a) are much more severely damaged than any of the other four subsets. The percentages of buildings suffering some degree of collapse (damage states 3 and 4) are 71%, 23%, 0%, 0% and 1% for building subsets (a) to (e), respectively.

As damage states 3 and 4 cause most casualties, these figures confirm the appropriateness of past and present priorities to reduce the risk of collapse of URM buildings. While this is well understood by engineers, as is the safety of timber construction (e), what has not been widely recognized is the almost collapse-free performance at MM10 of pre-code low-rise concrete buildings with walls (van de Vorstenbosch et al., 2002). This is true not only for buildings with concrete walls (d) but also, remarkably, for concrete beam and column buildings with brick infill in this data set. However, it should be borne in mind that brick infill panels are not always as reliable as they were in Hawke's Bay.

6.3 Quantitative Measures of Vulnerability

6.3.1 Introduction

For most purposes in risk assessment, it is necessary to have quantitative measures of *vulnerability* (also named *fragility* in some contexts) of the classes of property under consideration. This is conveniently done in terms of a damage ratio D_r, defined as

$$D_r = \frac{\text{Cost of damage to an item}}{\text{Value of that item}}, \qquad (6.1)$$

where value is best expressed in terms of replacement value, and D_r is a function of the strength of shaking and the physical nature of the item considered. It follows that D_r would most helpfully be modelled in an attenuation function in terms of magnitude, distance and scatter. With the small number of good D_r data sets yet available, we are mostly limited to describing D_r as a function of intensity, but are able to examine the distribution (scatter) of D_r well in those terms. The population of property items for any given distribution of D_r is drawn from the area between two adjacent isoseismals, so that the MM7 intensity zone (for example) is defined as the area between the MM7 and the MM8 isoseismals (Figure 6.3).

The New Zealand studies of damage ratios referred to below, all lead by the author of this book, are of four earthquakes:

- Hawke's Bay, 1931, $M_W = 7.8$;
- Wairarapa, 1942, $M_W = 7.1$;

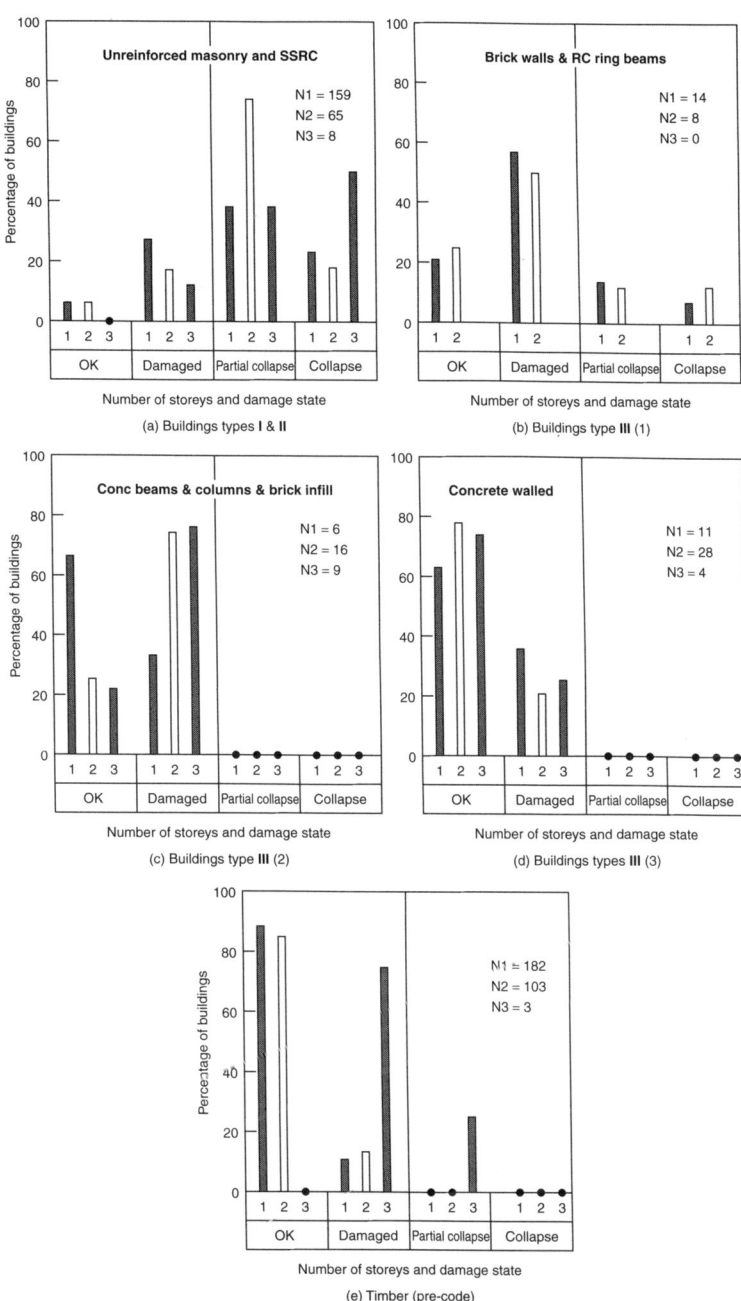

Figure 6.2 Damage distributions for non-domestic buildings in intensity MM10 in Napier and Hastings in the $M_W = 7.8$ Hawke's Bay, New Zealand earthquake of 1931. N1, N2, N3 are the numbers of buildings of 1, 2 and 3 storeys respectively in each subset. Building types are defined in the footnotes to Table 6.1 (from Dowrick, 1998)

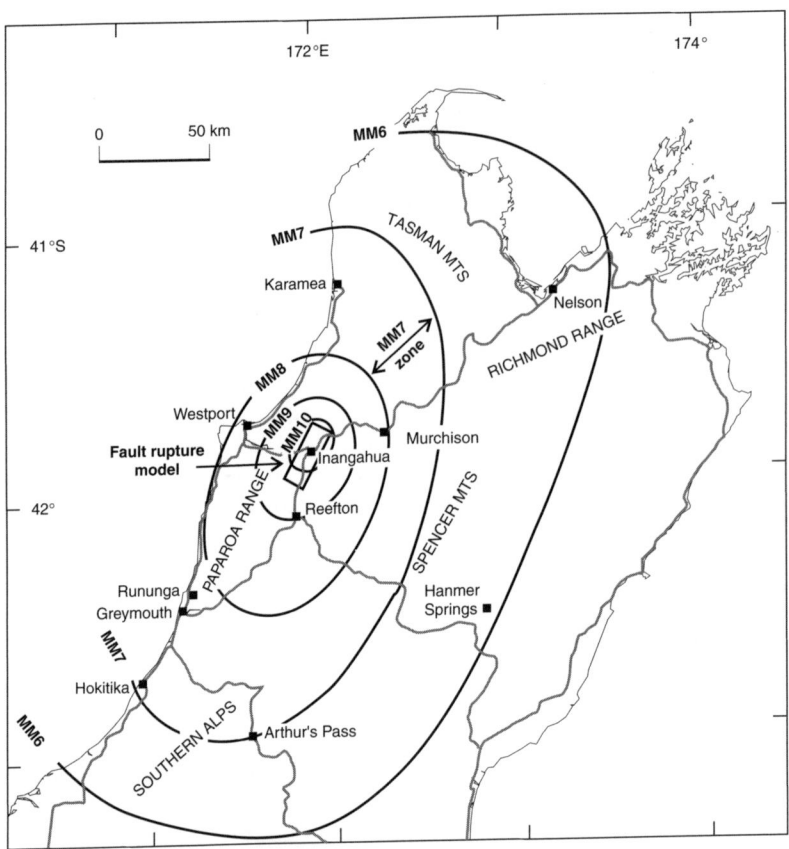

Figure 6.3 Map showing inner isoseismals, state highways and key place names for the 1968 Inangahua, New Zealand, earthquake. The fault rupture model dips at 45° with the top on the east side (from Dowrick et al., 2001)

- Inangahua, 1968, $M_W = 7.2$;
- Edgecumbe, 1987, $M_W = 6.5$.

In all of these studies, the robustness of the results was maximized by three procedures:

- accounting for all property items, damaged and undamaged, in each area considered;
- use of the actual repair costs in all cases (including the cost of the insurance deductible);
- use of the replacement value in all cases except household contents.

The use of these three procedures make these studies the only ones in the world (to the knowledge of the author) to have been so thoroughly conducted up to the time of writing (2009).

A typical distribution of damage ratios is that for non-domestic buildings shown in histogram form in Figure 6.4, from a study of damage ratios in the Edgecumbe earthquake by Dowrick and Rhoades (1993). This distribution fits the truncated lognormal form well,

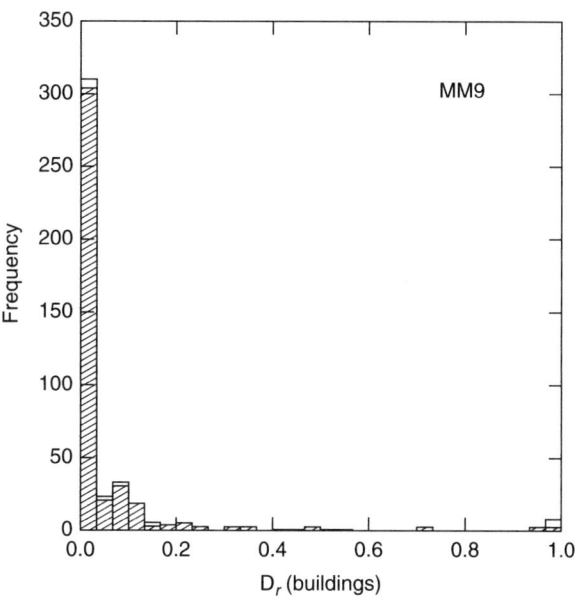

Figure 6.4 Histogram of damage ratios for non-domestic commercial and industrial buildings with non-zero damage in the MM9 zone of the 1987 Edgecumbe, New Zealand, earthquake (from Dowrick and Rhoades, 1993)

as do all the other distributions studied by those authors (e.g. Figure 6.5(a)). The lognormal distribution has the density function

$$f(x) = \frac{1}{\sigma x\sqrt{2\pi}} \exp\left[-\frac{1}{2}(\log_e x - \mu)^2/\sigma^2\right], \quad x > 0. \tag{6.2}$$

Here the parameters μ and σ are estimated by the sample mean and standard deviation of the natural log of the damage ratio, example values being given in Table 6.2. From this table it is seen that the scatter within distributions varies considerably, and even for larger populations ($n > 100$) the normal variability parameter σ lies within a wide range 0.7–1.77.

The mean damage ratio for all buildings in a given MM intensity zone is a useful parameter for various purposes, e.g. for comparing the earthquake resistance of different classes of property. Considering all N items (damaged and undamaged) in an MM intensity zone, we give here two principal ways of defining the mean D_r: first,

$$\overline{D}_r = \frac{\sum_{i=1}^{n} [\text{cost of damage to item } i]}{\sum_{i=1}^{n} [\text{value of item } i]} \tag{6.3}$$

where n is the number of damaged items; secondly,

$$D_{rm} = \frac{\sum_{i=1}^{n} D_{r_i}}{N}. \tag{6.4}$$

In general, D_{rm} with its associated confidence limits is a more reliable and useful tool than $\overline{D}_r$. If derived from large, homogeneous populations, $\overline{D}_r$ and D_{rm} tend to be similar in value, while for more inhomogeneous populations (with large ranges of replacement values and vulnerabilities) $\overline{D}_r$ and D_{rm} may differ widely. The values of $\overline{D}_r$ and D_{rm} for the various classes of property are presented in Table 6.2.

It has been found that the damage ratio is sometimes related to property value (Rhoades and Dowrick, 1999). If it is, $\overline{D}_r$ and D_{rm} tend to differ quite markedly. For example, if higher-valued properties tend to have higher damage ratios, then $\overline{D}_r$ tends to exceed D_{rm}. In some studies (Dowrick et al., 1995, 2001) the tendency is for $\overline{D}_r$ for houses to be less than D_{rm}, for most subsets. This indicates that lower-valued houses tend to have higher damage ratios. Such a trend could arise from a number of causes, including underestimation of the replacement values of low-valued properties, and/or by the costs associated with some of the main types of damage being independent of replacement value.

In general, D_{rm} is a much more robust statistic than $\overline{D}_r$, and is therefore preferred for modelling future events. This is illustrated in Figure 6.6, where values of D_{rm} and $\overline{D}_r$ are plotted for subsets of the data, together with their associated uncertainty intervals. The uncertainty intervals were determined by resampling many times from the empirical distribution of damage ratios and property values (Rhoades and Dowrick, 1999). They represent the variability that can be expected if similar populations of property are subjected to the same level of shaking in future earthquakes. It is seen that the uncertainty intervals for D_{rm} are much narrower than for $\overline{D}_r$, and that the location of $\overline{D}_r$ within its uncertainty interval is erratic, being highly asymmetrically placed in the 1935–1964 data set. The relationship between D_{rm} and $\overline{D}_r$ is similarly erratic as seen by comparing the values presented for D_{rm} and $\overline{D}_r$ for a wide range of property items in Table 6.2. These effects arise because of the wide range of property values within a single data set. $\overline{D}_r$ is sensitive to damage ratios for high-value property items.

Plots of cumulative probability of damage ratios in the Inangahua earthquake are shown for all houses by MM intensity in Figure 6.5(a). While very small amounts of damage occur at MM5, with a near-zero probability of damage occurring (Figure 6.5(c)), Figure 6.6 shows that the amount of damage is very small and the practical threshold of damage is at MM6 for houses, especially those with brittle chimneys. This is consistent with the definitions of the MM intensity scale (Dowrick et al., 2008) – see Appendix A – and confirms that the outer isoseismals of Figure 6.3 have been appropriately located.

When considering mean damage ratios, parallel effects are observed to those discussed above in terms of damage ratio distributions. Figure 6.5(b) shows plots of D_{rm} for six intensity zones. The values of D_{rm} (including all damage) for the Inangahua earthquake range up to 0.34 at MM10.

Next, consider mean damage ratio as affected by brittle chimney damage. The influence of chimneys on D_{rm} is very apparent in Figure 6.5(b), where D_{rm} is seen to range from 2.0×10^{-5} (excluding chimneys) at MM5 to 0.048 (including chimneys) at MM8, and then flattening off to rise only slightly to 0.050 at MM9. This plateau is a result of chimney damage reaching a near-maximum at MM8. The dominance of brittle chimneys as an indicator of vulnerability is also illustrated by the ratio of D_{rm} including chimneys with that excluding chimneys, which in round figures is 1.8, 6, 9, 9, 5 and 1.2 for MM5 to MM10, respectively. The figure of 1.2 for the MM10 zone is close to that of 1.3 obtained

Table 6.2 Basic statistics of the distribution of damage ratio for some subclasses of property in the intensity MM9 zones of two New Zealand earthquakes

Property class		n	N	μ	σ	D_{rm}	$\overline{D}_r$
Houses							
All 1- and 2-storeys[1]		2040	2800	−3.25	1.38	0.091	0.070
one-storey, incl. chimney costs[2]		368	455	−3.24	1.01	0.050	0.036
one-storey, excl. chimney costs[2]		268	455	−5.08	1.30	0.0096	0.0071
Household contents							
Edgecumbe earthquake[1]		2210	2800	−2.80	1.18	0.092	0.079
Inangahua earthquake[2]		280	321	−4.04	1.02	0.020	0.022
One-storey non-domestic buildings[3]							
Code era built: 1935–1964		72	154	−3.29	1.69	0.063	0.034
1965–1969/79		60	118	−3.28	1.62	0.054	0.085
1969/79–1987		57	133	−3.71	1.67	0.033	0.063
Equipment[4]							
Vulnerability class:	Robust	80	197	−3.64	1.34	0.023	0.006
	Medium	116	247	−3.13	1.59	0.052	0.031
	Fragile	11	16	−0.90	0.80	0.32	0.48
Stock[4]							
Vulnerability class:	Robust	23	82	−3.41	1.04	0.015	0.022
	Medium	35	53	−2.54	1.19	0.091	0.053
	Fragile	52	77	−1.73	1.07	0.22	0.48

Notes:
[1] Dowrick (1991).
[2] Dowrick et al. (2001).
[3] Dowrick and Rhoades (1997b).
[4] Dowrick and Rhoades (1995).

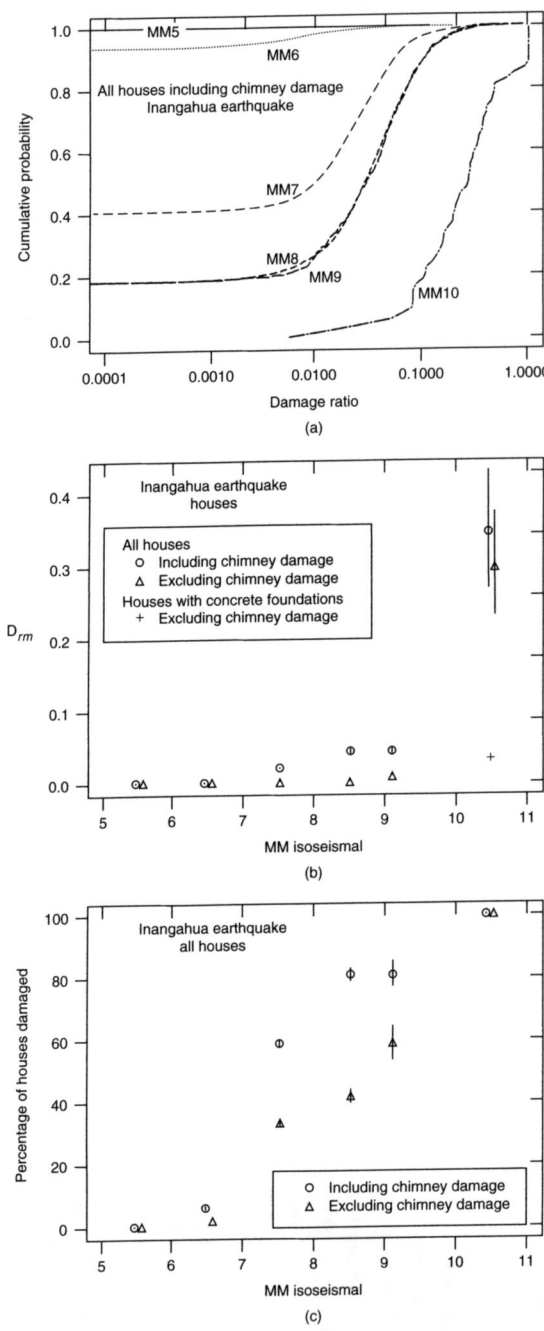

Figure 6.5 Three measures of vulnerability plotted as functions of MM intensity for all houses in the Inangahua earthquake: (a) cumulative probability distributions of damage ratio; (b) D_{rm} and its 95% confidence limits; (c) percentage of houses damaged with 95% confidence limits (from Dowrick et al., 2001)

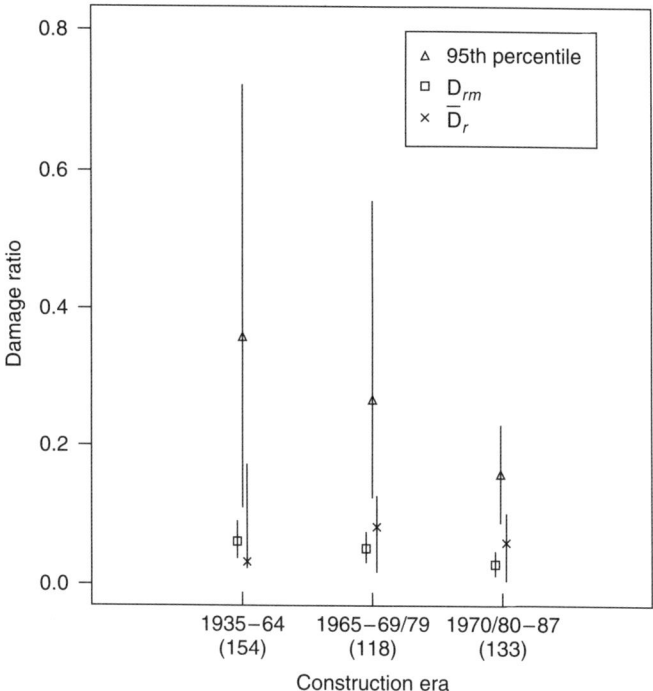

Figure 6.6 D_{rm}, $\overline{D}_r$ and the 95th percentile of $\overline{D}_r$ for non-domestic buildings of different code eras in the MM9 zone of the Edgecumbe earthquake. The uncertainty limits are the 2.5% and 97.5% quantiles of the distributions of D_{rm} and $\overline{D}_r$ determined by resampling (from Dowrick and Rhoades, 1997a)

for the MM10 zone in Napier in 1931 (Dowrick *et al.*, 1995) where all houses also had brittle chimneys, and were of similar construction (weatherboarded and piled) to those of the Inangahua area.

In addition to distributions of damage ratios and mean damage ratios, a third measure of vulnerability is the percentage of property items that are damaged. This is derived from the fraction n/N, examples of which are given in Table 6.2, and as plotted in Figure 6.5(c). From the latter, it is seen that at most intensities a much greater percentage of houses with brittle chimneys are damaged than those without chimneys.

6.3.2 Vulnerability of different classes of buildings

In Figure 6.6, it can be seen that non-domestic buildings built in New Zealand's first two earthquake code eras had much the same vulnerability, but buildings built in the most recent era for which data are available (1970–1987) were found to be statistically significantly better than pre-1970 buildings, and the maximum damage levels as reflected in the 95th percentile had decreased. This improvement is attributed to the influence of greater ductility requirements of the codes of that era.

The difference in vulnerability of different construction eras is more dramatically shown in Figure 6.7, where the performance of pure brick buildings at intensity MM8 is compared with that of three other classes of building:

- non-domestic buildings of reinforced masonry from the code era 1935–1979;
- houses with brittle chimneys;
- houses excluding chimney damage.

Considering only one-storey buildings, Dowrick and Rhoades (2002) found that the mean damage ratio for the pure brick Wairarapa buildings was (1) approximately three times that for timber framed houses with brittle chimneys in the Inangahua earthquake; (2) seven times that for 1935–1979 concrete masonry buildings in the Edgecumbe earthquake; and (3) 28 times that for timber framed houses excluding chimney damage in the Inangahua earthquake.

As well as considering the code era and materials of construction, the vulnerability of buildings needs to be measured according to the number of storeys. For example, in the study of pure brick buildings in the Wairarapa earthquake (Dowrick and Rhoades, 2002), the mean damage ratio for two-storey pure brick buildings, at 0.22, was 57% greater than that for one-storey buildings. The difference between the D_{rm} values for these two subsets fell just short of statistical significance ($p = 0.057$), but is consistent with the differences between one- and two-storey buildings found in studies of the Edgecumbe earthquake at intensity MM7 and MM9 (Figure 6.8) and the Inangahua earthquake at MM7 and MM8.

Why are two-storey buildings more vulnerable than those of one storey? This effect is presumed to arise from the fact that single-storey buildings generally have relatively more of their mass located above base level compared to buildings of two or more storeys. So what about buildings of more than two storeys? Unfortunately, the author has had only rather small subsets of data on buildings of three or more storeys (and none more than six storeys), but the data which do exist suggest that such buildings are about as vulnerable as those of two storeys. This may be largely the result of the fact that taller buildings have relatively more engineering design input than do two-storey buildings.

Another very basic feature of buildings that may influence their vulnerability is their size, and this factor has been examined in a number of cases (Dowrick and Rhoades, 1997b, 2002; Dowrick et al., 1995). In the study of the Wairarapa earthquake, the mean damage ratio for both one- and two-storey brick buildings was found to reduce substantially with increasing floor area (Figure 6.9). Such a trend was not seen elsewhere, except in houses with chimney damage, and the reason for it is not evident. However, this trend explains why $\overline{D_r}$ (at 0.12) is so much smaller than D_{rm} (at 0.17) for the brick buildings (Dowrick and Rhoades, 2002).

Further information on the vulnerability of different classes of buildings is given later in relation to Figure 10.54 (Section 10.5.1).

6.3.3 *Vulnerability of contents of buildings*

The contents of buildings considered here comprise a wide range of property items: contents of houses, and plant, equipment and stock of a non-domestic nature.

Quantitative Measures 215

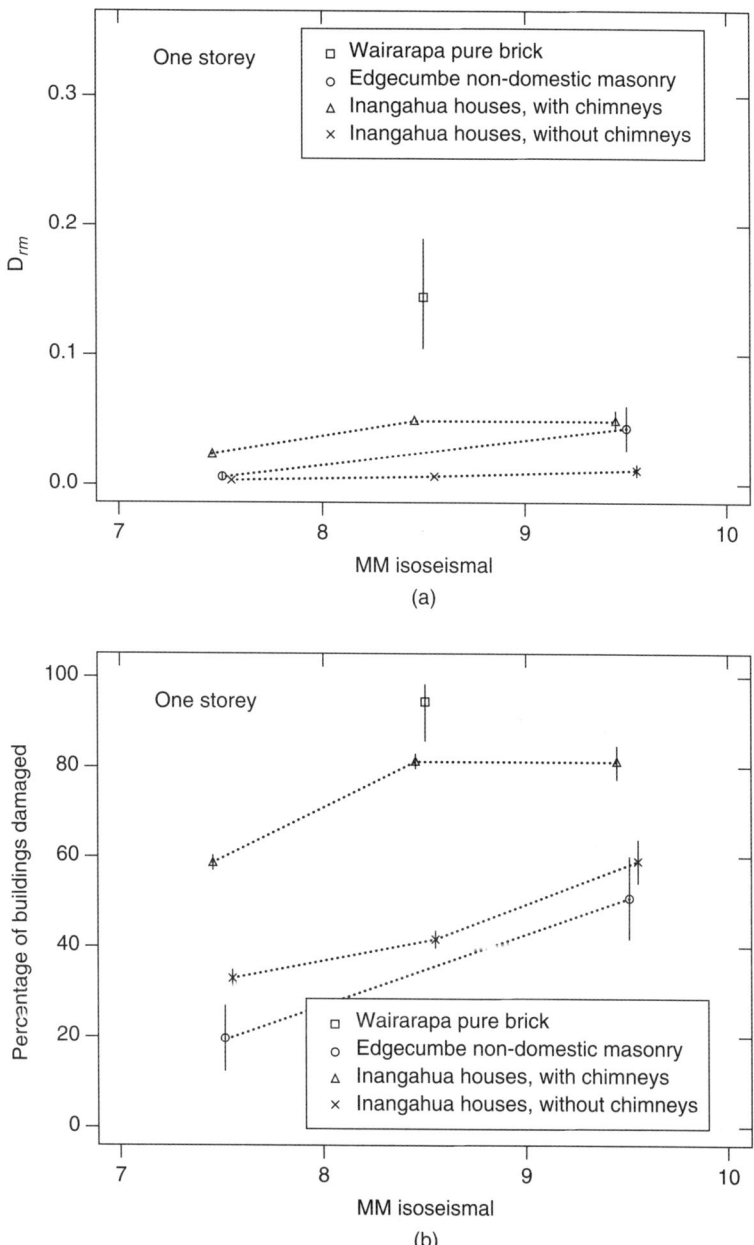

Figure 6.7 Two vulnerability measures for pure brick Wairarapa, New Zealand, non-domestic buildings compared with those for three other building classes from the Inangahua and Edgecumbe earthquakes: (a) D_{rm} with its 95% confidence limits; (b) percentage of buildings damaged with 95% confidence limits. The houses are timber framed (from Dowrick and Rhoades, 2002)

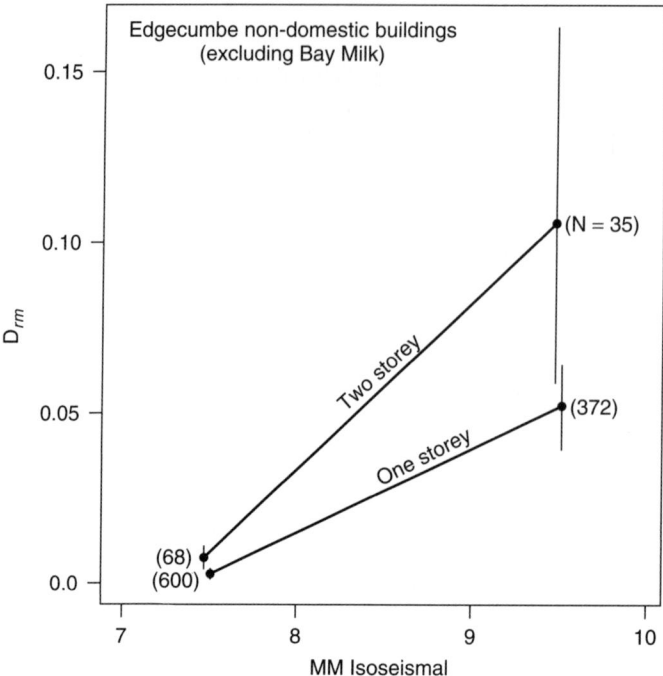

Figure 6.8 D_{rm} and its 95% confidence limits for one- and two-storey non-domestic buildings (all materials), for intensities MM7 and MM9 in the 1987 Edgecumbe earthquake (from Dowrick and Rhoades, 1997b)

Household contents

Regarding household contents, the damage ratios for these have been studied for two New Zealand earthquakes, the 1987 Edgecumbe earthquake (Dowrick, 1991) and the 1968 Inangahua earthquake (Dowrick *et al.*, 2001). From the 1987 event, basic statistics of damage ratio distributions were obtained for three intensity zones (MM7–MM9), while six intensity zones (MM5–MM10) were studied for the 1968 event. In Table 6.2, the basic statistics are presented for the MM9 intensity zone for household contents for these two events. It can be seen that the mean damage ratios differ widely, D_{rm} being 0.092 and 0.022. This difference is surprisingly large considering that (i) insurance policies were similar, (ii) the method of establishing the damage ratios was the same for the two events, (iii) the differences between the D_{rm} values of the two events are quite small for MM6–MM8 (Figure 6.10b), and (iv) the isoseismal maps were drawn in the same way by the same organization. Some of the difference may be due to the fact that the relationship between D_{rm} and MM intensity is particularly non-linear at about MM9, as seen in Figure 6.10(b).

Correlation of damage ratios for buildings and contents

A feature of damage that is of interest in predicting losses in future events is the relationship of damage to buildings and their contents. Individual cases of fragile contents

Quantitative Measures 217

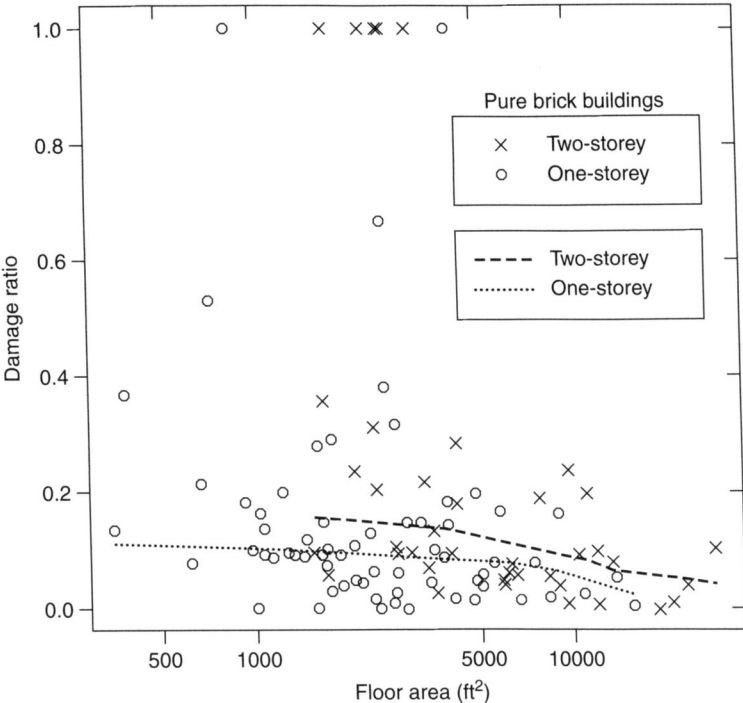

Figure 6.9 Plot of damage ratio versus floor area for one- and two-storey pure brick buildings in the MM8 intensity zone of the Wairarapa earthquake, including a robust fit of the median damage ratio curve (from Dowrick and Rhoades, 2002)

(e.g. large collections of pottery or glassware) in robust buildings, have of course suggested that the correlation might be weak between D_r (buildings) and D_r (contents) on a building-by-building basis, despite the fact that mean damage ratios for such populations might be roughly proportional, as is generally the case, and despite the fact that the shapes of the statistical distributions are similar.

For both domestic and non-domestic property, the author has found, surprisingly, that there is zero correlation between D_r (buildings) and D_r (contents), regardless of the strength of shaking. This situation is illustrated for domestic property in Figure 6.11.

Plant, equipment and stock

This section reports on a comprehensive study by Dowrick and Rhoades (1995) of non-domestic property in the Edgecumbe earthquake. The type of property studied was defined to mean *equipment, plant, stock and other contents* in and associated with commercial, industrial and institutional properties. The latter included commercial residential property, which incorporated hotels, motels, hostels and rest homes providing short-term accommodation, and long-term rental accommodation of more than one storey. Property of the above type that was government-owned at the time of the earthquake was also included

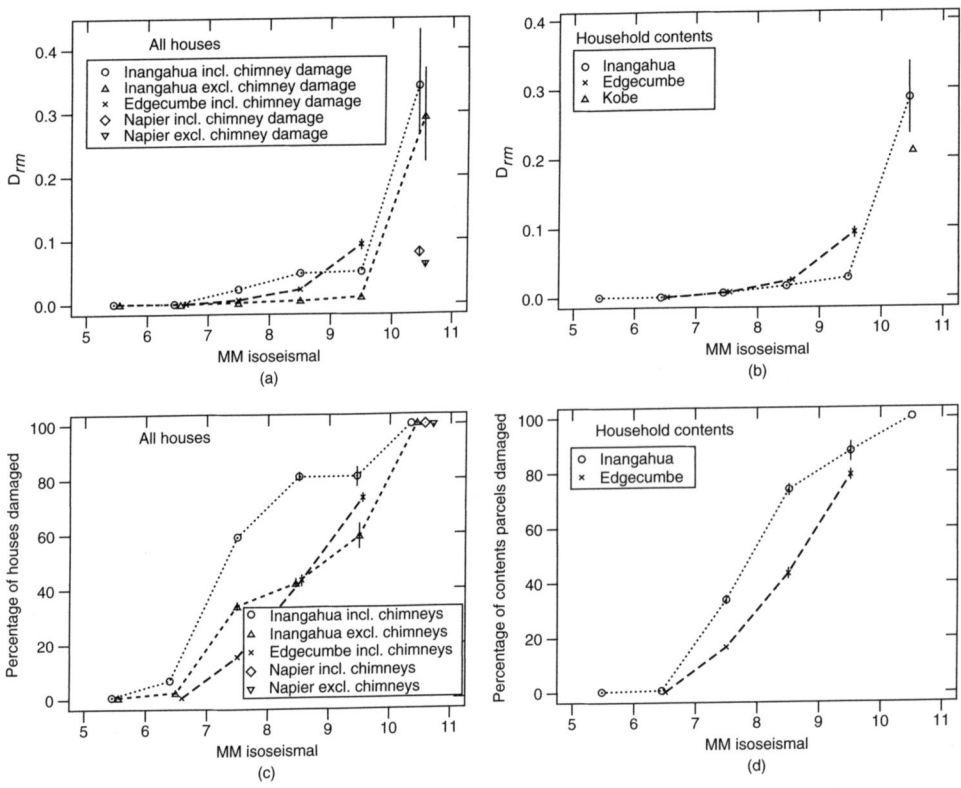

Figure 6.10 Vulnerability measures for domestic property from the Inangahua earthquake compared with those from other earthquakes: (a) D_{rm} for houses; (b) D_{rm} for contents; (c) percentage of houses damaged; (d) percentage of contents parcels damaged (from Dowrick *et al.*, 2001)

(e.g. equipment in schools, post offices, telecommunications, electricity department buildings). Most public utility equipment was included (e.g. pumps in drainage pumphouses), the only known exceptions being the equipment external to the buildings at the Kawerau and Edgecumbe electricity supply switchyards, the utility networks (power and telecommunication lines and underground water and sewer pipes), and the Matahina Hydrodam and associated equipment. As there was no standard terminology for classifying insured items, for the purposes of this study the following definitions were used:

- *Equipment* – 'fixed' equipment or plant for tasks such as manufacturing, measuring, computing and storage; contents of non-domestic buildings, except stock. (Mobile plant was excluded, e.g. vehicles and mobile cranes.)
- *Stock* – mostly manufactured items stored in bulk prior to sale or hire.

Because there are often so many individual items of equipment or stock at a given site, in general it is impracticable to find D_r for every item individually (in contrast to buildings). The available data covered groups of items rather than individual items. Any such group was called a 'parcel'. A parcel may comprise all of the equipment in a

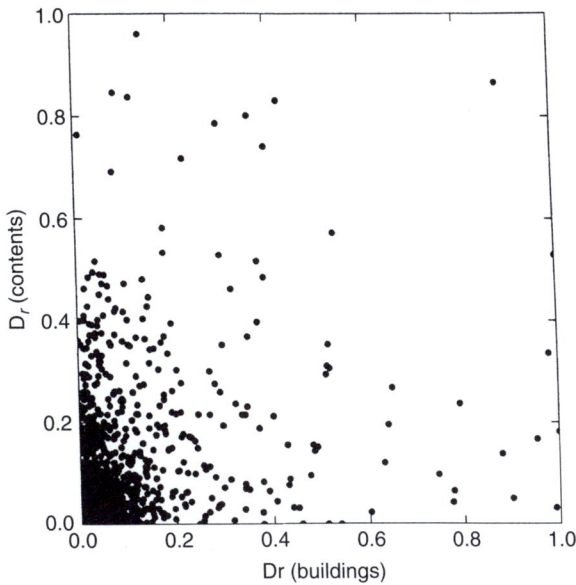

Figure 6.11 Plot of damage ratios for buildings versus contents for the MM9 zone of the Edgecumbe earthquake showing zero correlation, although D_r (buildings) $\approx D_r$ (contents) (from Dowrick, 1991). (Reprinted by permission of John Wiley & Sons, Ltd)

given building or on a given site, or any part thereof, depending on the available data on damage costs or values of the property concerned.

So that the replacement values and relative vulnerability of the total population of equipment and stock could be evaluated for each intensity zone, it was necessary to find out the number of parcels of each (damaged and undamaged) in each area of interest, and to obtain a description of the nature of each. To this end, insurance claim data were complemented with data from owner-occupiers of uninsured property (mostly government-owned), and most of the damaged properties were visited.

To examine the earthquake resistive nature reasonably closely, but still with workably simple classes, each parcel of equipment and stock was placed (*a priori*) in one of three classes of relative earthquake vulnerability: robust, medium or fragile. By *a priori* is meant that the classification was assigned from the physical nature and/or situation of the particular parcel without reference to its damage ratio in this event. Various factors were taken into account, including whether the parcel was designed for earthquake resistance or not, e.g. stainless steel tanks in the food processing industry were not designed for earthquakes, and were clearly in the fragile class. Inherently tough items such as rolls of newsprint and pumps and some non-brittle items such as clothing were classed as robust. Unsecured brittle items such as most ceramics and glassware were obviously fragile. The *situation* of some items may affect their vulnerability, such as whether they are secured against falling, e.g. glassware, mechanical equipment and desktop computers. It was encouraging to find that by 1995 a number of laboratories in New Zealand had devised 'operator-friendly' means of securing glassware items which are kept on shelves or benches, but which need to be moved quite frequently. A range of examples of the vulnerability classes that were adopted for this study is given in Table 6.3, and the statistics of the damage ratios are given in Table 6.2 and Figure 6.12.

While the criteria for the vulnerability of many items were reasonably definitive, subjective judgements variously arise, such as for items which seem to be near a class boundary, e.g. whether robust or medium. The subjectivity is greatest when there are many disparate items of equipment or stock within a given parcel. Such cases can formally be dealt with by weighting the contribution of each item according to its replacement value. In this study, with so many parcels to be considered, this could only be done very roughly. Because of the wide variety of items of equipment used in shops and offices, the vulnerability of parcels of equipment in shops and offices was taken to be medium in most instances.

Comparing D_{rm} for the fragile and robust subsets (Figure 6.12 and Table 6.2), the ratio of $D_{rm}(F)$ to $D_{rm}(R)$ for all equipment in the MM9 zone is approximately 14 (or 8 if the contentious Bay Milk factory is included). Although $D_{rm}(F)$ is based on a small subset, a large difference between the seismic resistance of these two extreme vulnerability classes is of course to be expected. This result suggests that at intensity MM9 earthquake protection of equipment could reduce the damage levels by about an order of magnitude (on average) for a wide range of unprotected highly vulnerable equipment (where 'operator-friendly' protection systems can be found).

A similar comparison for all stock at MM9 (Figure 6.12 and Table 6.2) shows that the ratio of $D_{rm}(F)$ to $D_{rm}(R)$ is 15, a similar difference to that found for equipment. Unfortunately it is not practicable to protect most items of fragile stock (generally in shops), but this result is of value for insurance purposes.

The foregoing discussion has concentrated on the results for the strongest intensity zone (MM9), but it is of interest also to consider the relative damage levels in the MM7 zone. It was found that the ratio of $D_{rm}(F)$ to $D_{rm}(R)$ for all equipment was 23 and for all stock was 160. These ratios are larger than the equivalent ones for the MM9 zone, as is to be expected – the threshold intensity of shaking for damage to occur to fragile items is lower than for robust ones. This fact is fundamental to the intensity scale, as shown in Table 6.1.

As noted above, in most cases office equipment was classified as medium vulnerability, and was considered likely to be of a similar nature overall to shop equipment. As seen in Figure 6.12, D_{rm} for office equipment was much less than D_{rm} for the other two medium subsets (industrial and shops), being similar in value to D_{rm} for the robust equipment subsets. This result is probably explained by the fact that the office subset is quite small (only 32 parcels, compared with 103 shop equipment parcels) so that the estimate of the mean D_{rm} is not very robust. Also, most of the offices in the MM9 zone were located in Kawerau at the fringe of the MM9 zone, while a bigger percentage of the shops and industrial sites were nearer than Kawerau to the fault rupture at the centre of the MM9 zone.

6.3.4 Damage models as functions of ground-motion measures

Damage ratios

The damage ratios given herein from studies of New Zealand are all evaluated as functions of Modified Mercalli intensity in the form of isoseismals (e.g. Figure 6.3). It has

Table 6.3 General characteristics and examples of vulnerability classification of equipment and stock (from Dowrick and Rhoades, 1995)

	Robust	Medium	Fragile
General characteristics	Designed for earthquakes Inherently tough Well restrained against falling, or overturning; low aspect ratio, H/B[1] Thick members, anti-buckling	Designed for earthquakes Intermediate between robust and fragile	Not designed for earthquake Inherently brittle Not restrained against overturning and of high aspect ratio, H/B Thin members, likely to buckle
Equipment	Heavy fixed equipment: Pumps Generators Turbines Transformers Handling equipment for heavy/bulky materials Most engineering workshops Some furniture Motels, kindergartens, churches	Some computers Some transformers Some stainless steel food processing equipment Some furniture Most offices and shops Mix of robust, medium and fragile	Loose equipment, such as Computers (some) Microscopes Laboratory glassware Refractory materials (e.g. kiln linings) Glassware Ceramics Stainless steel food processing equipment Some furniture
Stock	Timber logs Heavy processed timber Paper products Clothing Motor vehicles Chainsaws	Some food Some pharmacy stock Mix of robust, medium and fragile	Most food and drink Chilled food[2] Most pharmacy stock Paint

Notes:
[1] H/B = Height/Breadth (for centre of mass).
[2] Vulnerable to power failure.

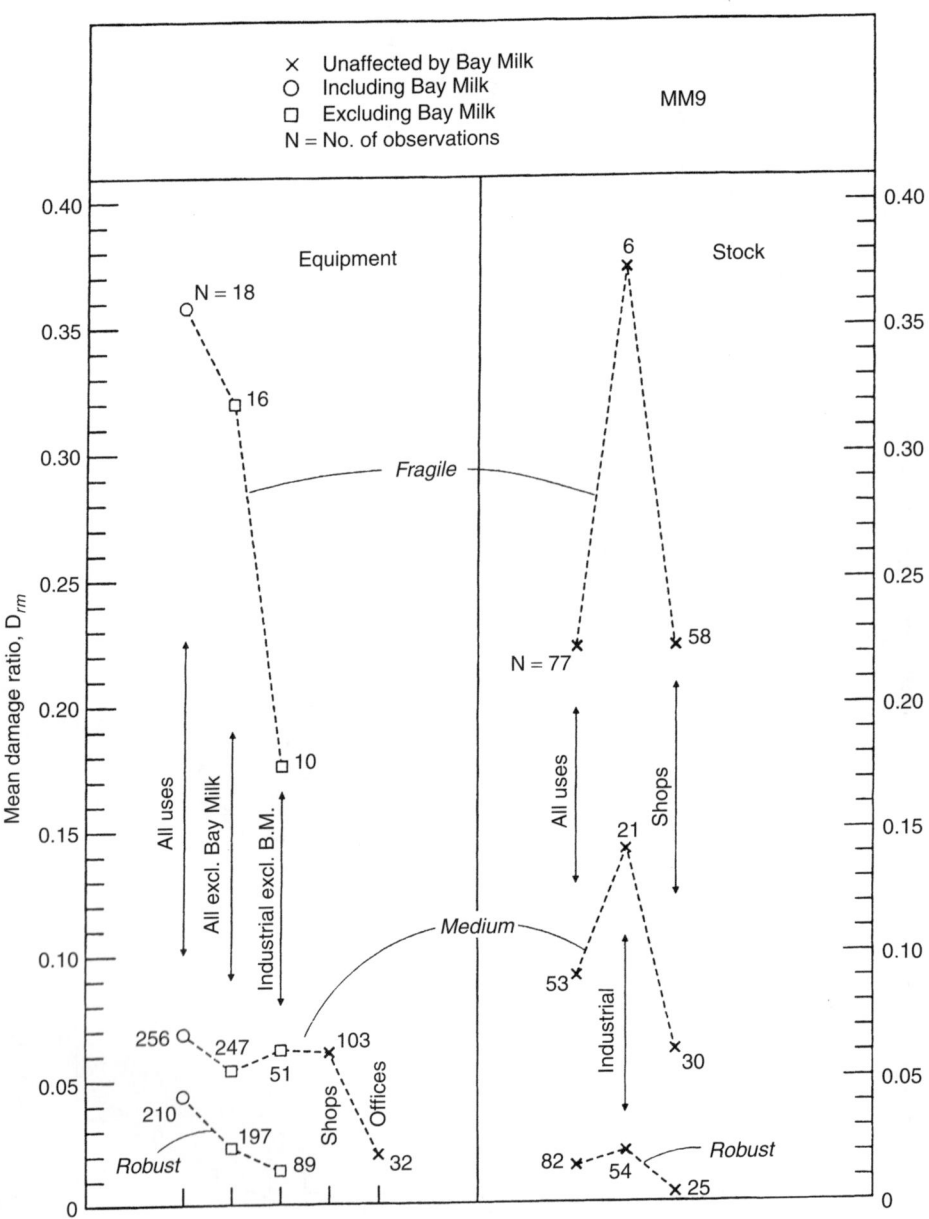

Figure 6.12 Mean damage ratios for equipment and stock in three vulnerability classes for various use classes in the MM9 zone of the Edgecumbe earthquake (from Dowrick and Rhoades, 1995)

not been possible to reliably relate D_r to ground-motion measures as too few strong motion recordings have been made in the relevant earthquakes. Specifically, of the four earthquakes studied (listed in Section 6.3.1), no instrumental records are available for the 1931 and 1942 events, 16 peak ground accelerations were recorded in the 1968 event, and two PGAs were recorded in the 1987 event. Only one spectral record was available. On the other hand, reasonably good isoseismal maps were available for all four events.

It has thus been considered appropriate to relate the damage ratios to MM intensity. And so, to apply the damage ratios to future events for risk assessment purposes, a reliable predictive model for MM intensity is necessary. To that end, a great effort has been made to develop such a model for New Zealand (Dowrick and Rhoades, 2005a). McGuire and Toro (2000) have shown that non-linear instrumental ground motions have the potential to be better than MM intensity for predicting future losses, but that the difference is not great.

Semi-theoretical building damage functions

An alternative approach to the above empirical method of modelling vulnerability from damage costs, bypassing the use of MM intensities, has been developed for the USA (Kircher *et al.*, 1997a). Their approach involves building damage functions that were developed for the Federal Emergency Management Agency (FEMA/NIBS) earthquake loss estimation method as formulated by Whitman *et al.* (1997). Unlike the models given above, which are based on MM intensity, the new functions use quantitative measures of ground shaking (and ground failure) and analyse model building types in a similar manner to the engineering analysis of a single structure. These functions estimate the probability of discrete states of structural and non-structural building damage that are used as inputs to the estimation of building losses, including economic loss, casualties and loss of function (Kircher *et al.*, 1997b).

Since 1997 substantial developments have taken place in this methodology, especially with the pivotal *fragility functions* ratio curves which are equivalent to the damage ratio cumulative probability curves of Figure 6.5(a). Figure 6.13 shows fragility curves for three different design philosophies (in order of decreasing damage): (a) ductile design, (b) design for control and repairability, and (c) damage avoidance design. Ductile design is has been the standard approach to the earthquake resistant design of structures, but increasingly engineers are seeking ways of reducing damage through the other two design philosophies. An example of the ultimate approach, damage avoidance design, is discussed in Section 8.6.

Ongoing work on the creation of fragility functions is described by Porter *et al.* (2007), Applied Technology Council (2005) and Wen and Ellingwood (2005). It will be interesting to see how these potentially powerful models develop with time, and as they are calibrated by comparing their results with actual losses in future earthquakes. Calibration of fragility curves of relevant structural classes could be done using the damage ratio curves discussed in this chapter, allowing for the fact that the damage ratios are for total damage, that is, they include non-structural as well as structural damage, and are related to MM intensity rather than ground-motion parameters. As noted in Section 12.1, damage costs to non-structure can be larger than for structure.

Another semi-theoretical damage function is that based on inter-storey drift, as developed for Mexico City, and discussed in Section 7.2.3.

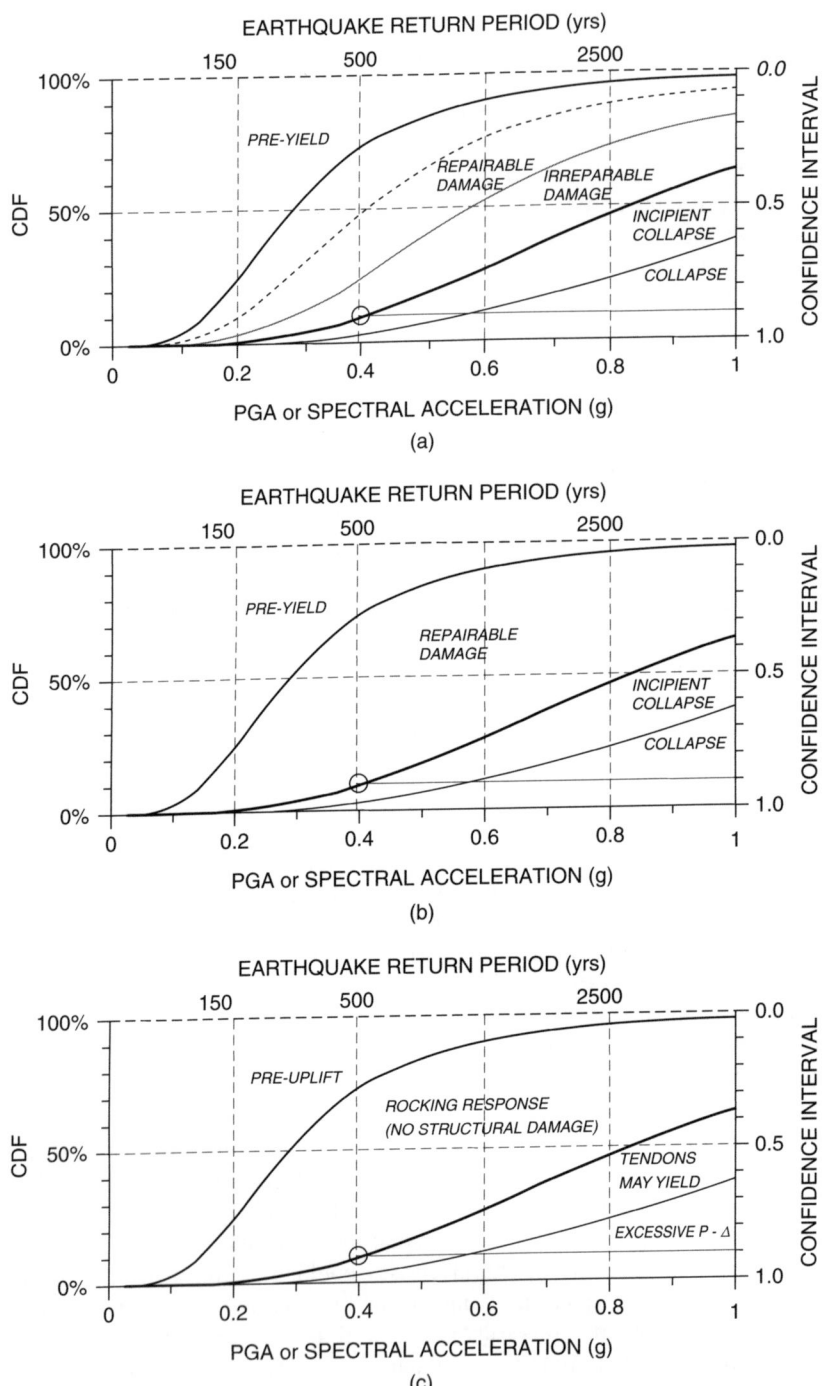

Figure 6.13 Fragility curves for different design philosophies: (a) ductile design; (b) design for control and repairability; and (c) damage avoidance design (after Mander, 2004)

6.3.5 Microzoning effects on vulnerability functions

Microzoning effects in very strong shaking (MM10)

The effects of different ground conditions on the response of structures, referred to as microzoning effects, are functions of both frequency and amplitude of vibration. The dependency on amplitude has been shown in peak ground acceleration terms by Idriss (1990), such that PGAs on soils are greater than PGAs on rock at low amplitudes, while the reverse is true at high amplitudes. This occurs because the weaker the soil is, the lower is the acceleration that it can transmit.

An opportunity to evaluate this phenomenon was provided in a study of damage ratios for houses (single-storey timber) in Napier in the Hawke's Bay earthquake (Dowrick *et al.*, 1995). Napier was located over the source, so the shaking was obviously of high amplitude. As shown in Figure 6.14, three ground classes were mapped throughout the town,

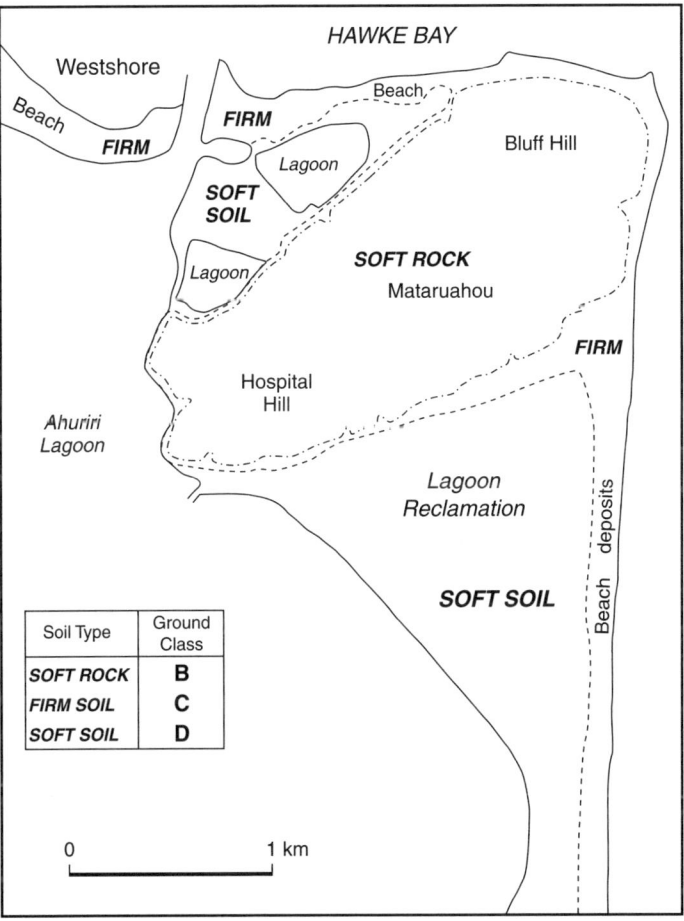

Figure 6.14 Simplified ground class map for Napier, New Zealand, 1931 (Dowrick *et al.*, 1995)

namely (B) soft rock, (C) beach gravels and sands (firm), and (D) harbour reclamation (soft). The geological descriptions of the ground classes are as follows:

B Early Quaternary marine sediments (very compact silts, sands, and limestones) of Mataruahou (Scinde Island) and the other hills and former islands to the west. Mataruahou itself is predominantly made up of cemented limestone. These marine sediments are effectively 'bedrock' in this area and are likely to act as a stiff, dense rock material during strong earthquake shaking.
C Dense sands and fine–medium gravels of the sand spits. These materials are classed as 'firm ground'. The top 5–10 m at least, are likely to comprise very dense sands and gravels deposited in a high energy beach environment. Experience with similar materials indicates that they will exhibit SPT N values of 50 or greater, and are unlikely to show ground damage due to high-intensity (MM10) shaking.
D Reclaimed swamp and lagoon areas. These are classed as 'soft ground' and are likely to vary both laterally and with depth, and to consist predominantly of interlayered mixes of poorly consolidated, saturated, fine grained soils (muds) and organic material with peat horizons to moderate depths (possibly up to 30 m or more).

The mean damage ratio for the houses (short-period structures) in each of these microzones is given in Table 6.4 and plotted in Figure 6.15, where it is seen that the weaker the surface layer the lower the average damage level (except when ground damage occurred). For example, D_{rm} on the 'firm' ground was 54% higher than on the 'soft' ground without ground damage, and is clearly statistically significantly different. This trend is consistent with Idriss's observations regarding PGAs. When ground damage occurred, however, the damage was worst. The reader should note that while the relative values of D_{rm} in Table 6.4 and Figure 6.15 are valid, the absolute values are much lower (a third to a half) than would prevail today, because of the exceptionally low costs of repairs that were obtained in the economic climate of the slump in 1931.

It is noted that New Zealand single storey timber houses are short-period structures. The effect on longer-period structures is likely to be opposite to that on short period structures, so microzoning rules should be different for low-rise and high-rise buildings on soft/weak ground in high-strength shaking. Another issue that arises is that it has often been reported that the worst damage in strong shaking has occurred on soft ground, sometimes regardless

Table 6.4 Basic statistics of the distributions of damage ratio for single-storey weatherboard houses by ground class in the Hawke's Bay earthquake (from Dowrick et al., 1995)

Subclass		n	N	μ	σ	D_{rm}	$\overline{D_r}$
MM10 Zone							
One-storey timber houses (excl. drains, incl. extra decorating):							
Ground class:	Rock[1]	417	417	−3.01	0.70	0.062	0.057
	Firm (beach)	281	281	−3.14	0.77	0.056	0.048
	Soft (reclaim) No grd dam[2]	844	844	−3.56	0.82	0.036	0.032
	Soft (reclaim) Grd dam only	101	101	−2.76	0.72	0.077	0.069

[1] Predominantly cemented limestone.
[2] grd dam = Ground damage (i.e. liquefaction) sites.

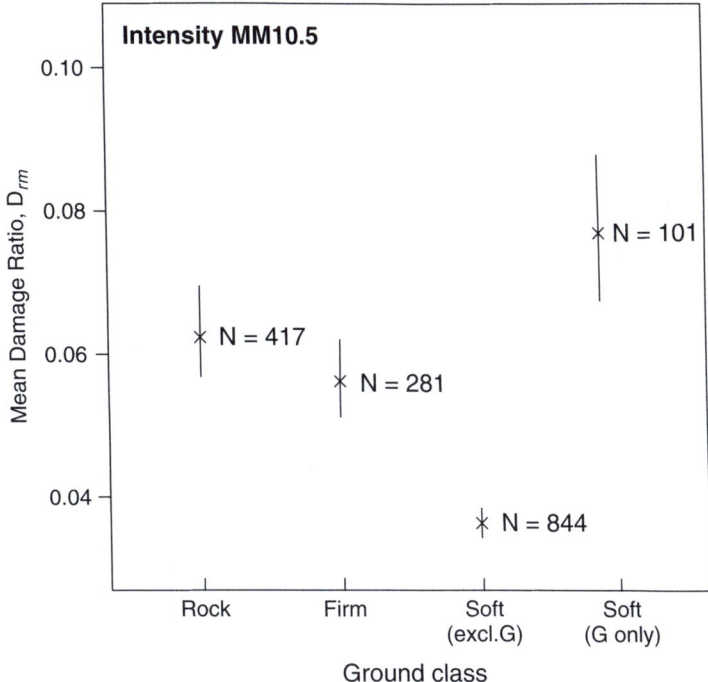

Figure 6.15 D_{rm} with its 95% confidence limits for single-storey timber houses in three different ground classes (microzones) in Napier in the 1931 Hawke's Bay earthquake. N is the total number of houses in each subset (from Dowrick *et al.*, 1995)

of the periods of vibration of the buildings. This can certainly be the case for brittle construction, or poorly founded buildings, or when ground damage such as liquefaction occurs (as occurred in Napier in 1931, as discussed above in relation to Table 6.4). In the above case of houses in Napier, they were all timber-framed weatherboarded houses, which are very tough. Clearly, in analysing microzoning effects on damage it is very important to be careful in defining the structural classes of the buildings that are to be differentiated between.

Microzoning and foundation effects

We next consider the microzoning effects measured through damage ratios obtained for the Inangahua earthquake in studies by Dowrick *et al.* (2003, 2005). In this case, damage ratios were calculated from the insured costs of damage to houses and household contents in seven towns: Inangahua, Reefton, Westport, Greymouth, Runanga, Hokitika and Nelson. As seen from the isoseismal map (Figure 6.3), these towns cover the intensity range MM6.4 to a little over MM10.

The basis of the microzones was the geology of any deposits overlying bedrock, as mapped in the earlier microzoning study of Suggate and Wood (1979). Their map for Greymouth is reproduced here, slightly annotated, in Figure 6.16. Inangahua and Reefton

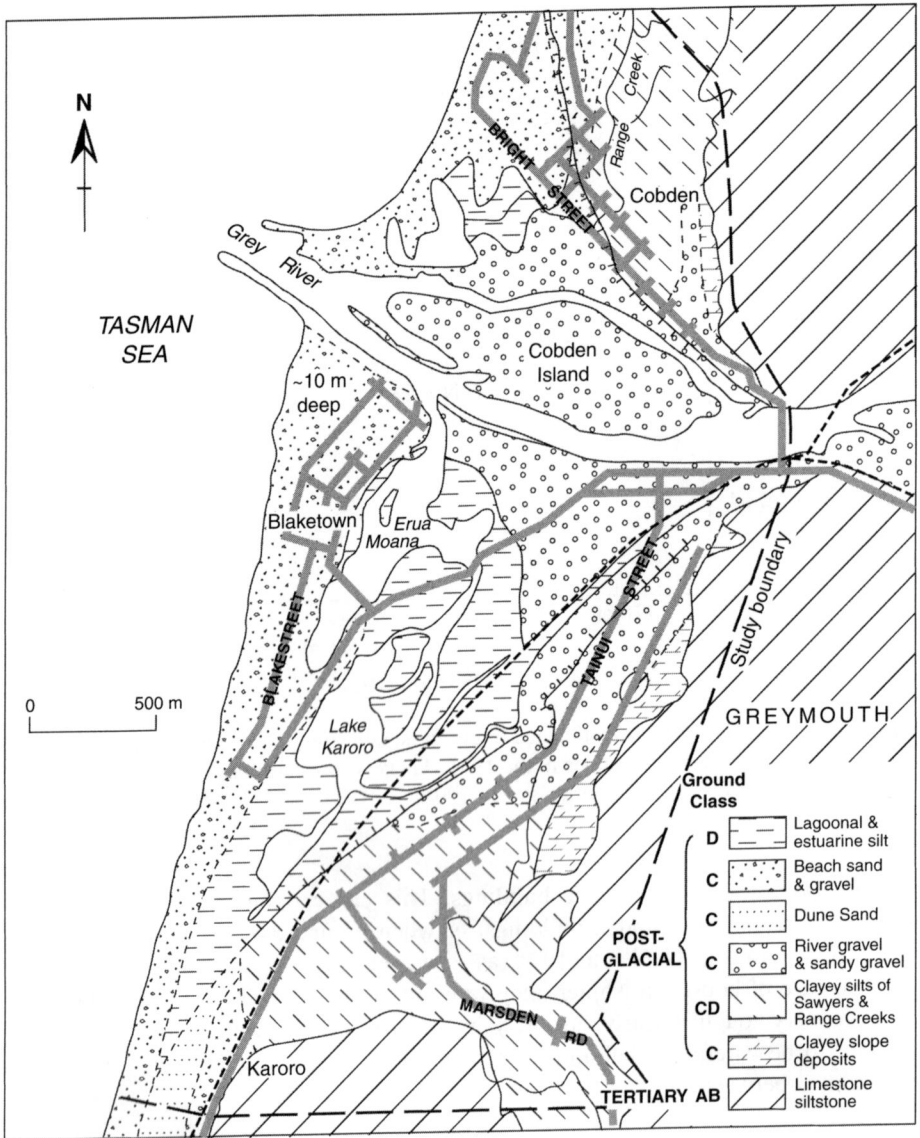

Figure 6.16 Microzoning map of Greymouth, New Zealand, with ground classes derived from geology, based on Suggate and Wood (1979) (from Dowrick *et al.*, 2003)

are entirely ground class C. The ground classes AB, C and D conform to the definitions used in the 2004 New Zealand loadings standard. These definitions are as follows:

Ground class AB: rock. Rock with less than 3 m thickness of stiff overburden. (Classes A and B in the 2004 New Zealand loadings standard.)

Ground class C. shallow soil. Sites where the low-amplitude natural period is less than or equal to 0.6 s, or sites with depths of soil not exceeding those listed in Table 3.1, but excluding very soft soil sites.

Ground class D. deep or soft soil. Sites where the low-amplitude natural period is greater than 0.6 s, or sites with depths of soils exceeding those listed in Table 3.1, but excluding very soft soil sites.

Ground class CD. Soil sites which were uncertain, but would be either class C or D.

In addition to microzoning, the effects on vulnerability were considered for foundations, which were classified into two types:

(1) fully piled, with piles generally unbraced (Figure 6.17(a));
(2) concrete foundation wall around perimeter (Figure 6.17(b)).

Effects of foundation construction on damage to houses

The effects of foundation construction type on mean damage ratio over a wide range of intensities are shown in Figure 6.18. The comparisons are made for one-storey houses on ground class C, with weatherboard wall cladding, and excluding and including chimney damage. It can be seen that houses with concrete perimeter wall foundations perform better than those on unbraced piled foundations right through the intensity range MM6.4 to a little over MM10. At most intensities, D_{rm} for houses on piled foundations is several times greater than those for houses with concrete foundations.

Effects of microzoning on damage to houses as a function of MM intensity

Figure 6.19 plots the mean damage ratios (from Table 6.5) for one-storey weatherboard houses, (excluding chimney damage) in Greymouth at intensity MM7.5, with the two types of foundation, and on the four different ground classes described above. Two very different patterns are seen. First, houses with concrete foundations have steadily increasing damage levels as the ground becomes more flexible. This pattern follows the well-established trends of peak ground acceleration and peak spectral acceleration, both of which can be expected to increase (at this intensity) as the ground becomes more flexible. Second, houses with unbraced piled foundations respond very differently, with those on ground class CD being the most damaged.

The behaviour of the houses on piles in Figure 6.19 is surprising, but is presumably explained by dynamic response effects. The peak response at ground class CD suggests that some resonance is occurring. This possibility is supported by the likelihood that the natural period of vibration for piled weatherboard houses and ground class CD are both about 0.4 seconds.

Figure 6.20 plots the mean damage ratios for one-storey weatherboard houses in Westport at intensity MM8.5, with two types of foundation, and on ground classes C and D. In both the cases of chimney damage excluded (Figure 6.21(a)) and chimney damage included (Figure 6.20(b)), D_{rm} is greater on ground class D than on ground class C, and in both the piled foundation cases this difference is statistically significant.

Figure 6.21 shows plots of the mean damage ratios for one-storey weatherboard houses in Hokitika at intensity MM7.0, with two types of foundation and on ground classes C and

(a)

(b)

Figure 6.17 (a) The most common type of pre-1968 West Coast, New Zealand, house, one storey with a corrugated iron roof weatherboard wall cladding and piled (unbraced) foundations. (b) A pre-1968 one-storey West Coast house with tiled roof, stucco veneer walls and concrete perimeter wall foundations

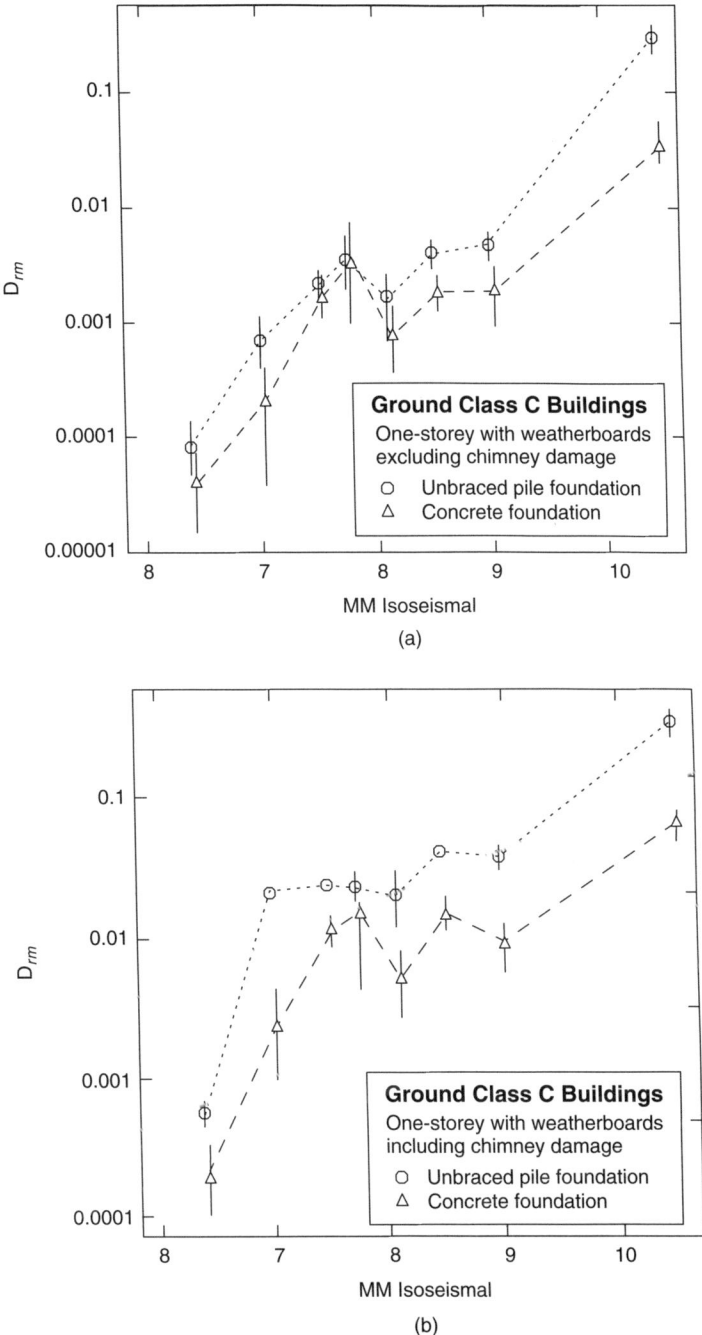

Figure 6.18 Mean damage ratio and its 95% confidence limits for one-storey weatherboard houses on ground class C, as a function of MM intensity and two types of foundation, in the Inangahua earthquake: (a) excluding chimney damage; (b) including chimney damage (from Dowrick *et al.*, 2005)

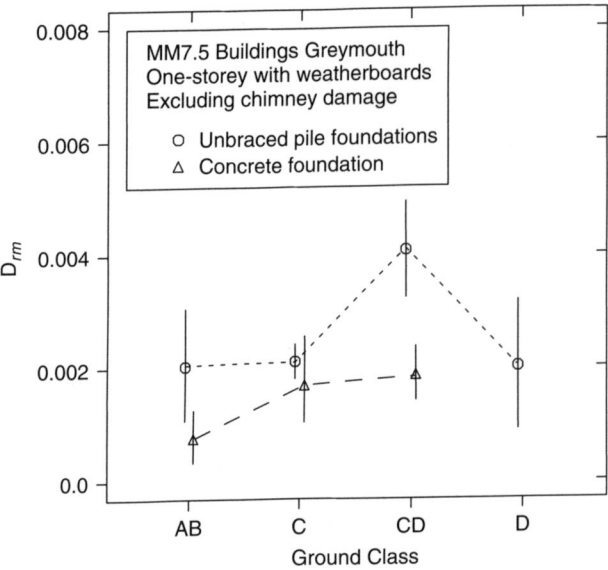

Figure 6.19 Mean damage ratio and its 95% confidence limits for single-storey weatherboard houses, excluding chimney damage, at intensity MM7.5 (Greymouth, New Zealand) for two types of foundation and four ground classes, in the Inangahua earthquake (from Dowrick *et al.*, 2005)

Table 6.5 Sample of basic statistics of the distributions of damage ratio by class of domestic property and ground class in the Inangahua earthquake (from Dowrick *et al.*, 2005)

Property class	n	N	μ	σ	$\overline{D}_{rm}$	$\overline{D}_r$
MM7.5 Greymouth excl. chimney damage						
One-storey houses, weatherboard, unbraced piles						
GC AB	22	62	−5.51	1.11	0.0021	0.0019
GC C	358	971	−5.73	1.12	0.0021	0.0018
GC CD	201	396	−5.44	1.15	0.0041	0.0040
GC D	14	41	−5.45	1.12	0.0020	0.0017
One-storey houses, weatherboard, concrete foundations						
GC AB	14	60	−6.16	1.41	0.00076	0.0010
GC C	70	216	−5.84	1.26	0.0018	0.0018
GC CD	84	242	−5.67	1.08	0.0018	0.0021
GC D	0	0				
Household contents, unbraced piles						
GC AB	21	66	−5.03	1.01	0.0030	0.0031
GC C	238	985	−4.78	1.07	0.0036	0.0031
GC CD	151	399	−4.45	1.10	0.0075	0.0073
GC D	16	41	−4.32	1.68	0.0034	0.0020

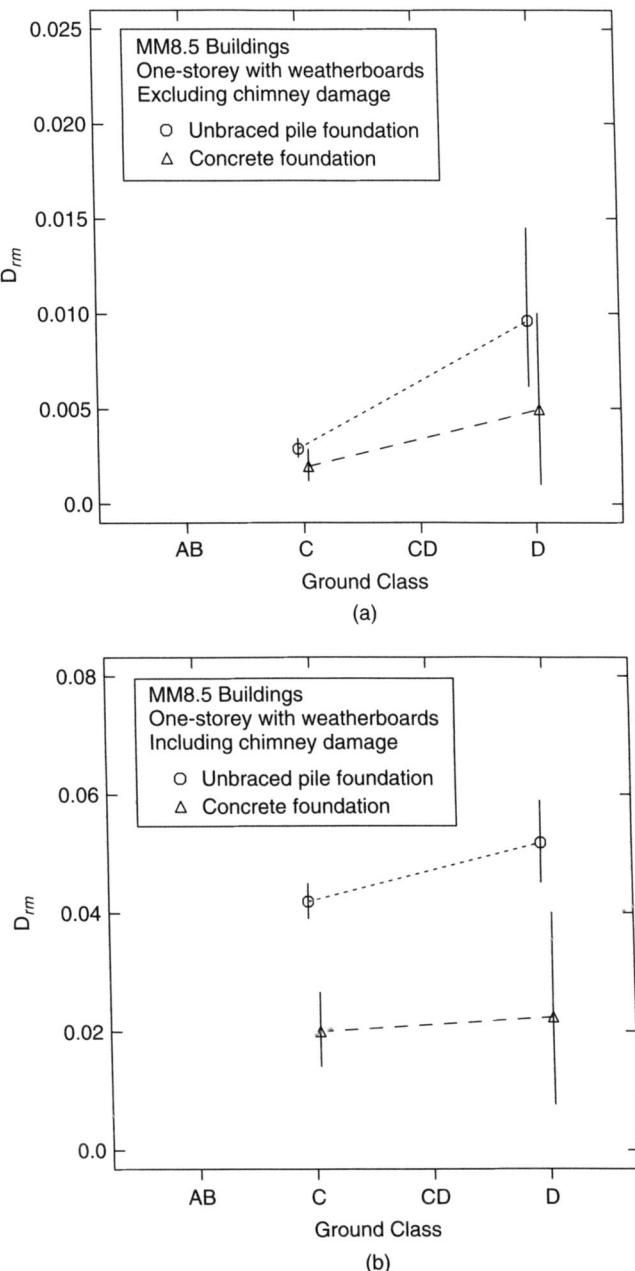

Figure 6.20 Mean damage ratio and its 95% confidence limits for single-storey weatherboard houses, at intensity MM8.5 (Westport), for two types of foundation and two ground classes, in the Inangahua earthquake: (a) excluding chimney damage; (b) including chimney damage (from Dowrick *et al.*, 2003)

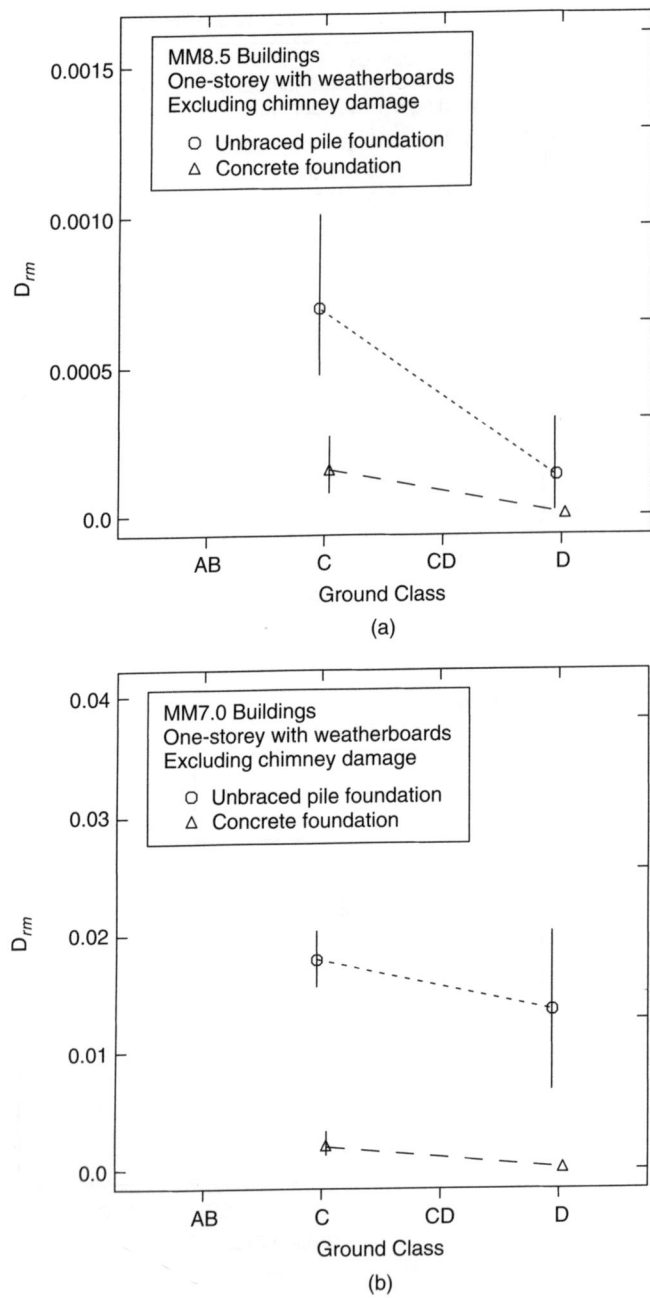

Figure 6.21 Mean damage ratio and its 95% confidence limits for single-storey weatherboard houses, at intensity MM7.0 (Hokitika), for two types of foundation and two ground classes, in the Inangahua earthquake: (a) excluding chimney damage (b) including chimney damage (from Dowrick *et al.*, 2003)

D. In both the cases of chimney damage excluded (Figure 6.21(a)) and chimney damage included (Figure 6.21(b)), D_{rm} is less on ground class D than on ground class C, and in three of the four cases plotted this difference is statistically significant. This result is the opposite of that found at intensity MM8.5, shown in Figure 6.20. The effect shown in Figure 6.21 is surprising, as it might be expected that the response at the low amplitudes of shaking at MM7.0 would produce the same responses on soft soils as those observed for the moderately strong shaking of MM8.5.

It is of interest to compare the results for MM8.5 and MM7.0 (Figures 6.20 and 6.21) with the results for ground classes C and D, at the intermediate intensity of MM7.5 as plotted in Figure 6.19. It can be seen that at intensity MM7.5 the results are equivocal, there being no significant statistical difference between D_{rm} for these two ground classes for either type of foundation. Thus the MM7.5 responses are transitional between the opposing responses at MM7.0 and MM8.5. This suggests that the apparently anomalous results for MM7.0 are real, and are presumably caused by dynamical soil–structure interaction effects.

Effects of microzoning and house foundation type on damage to household contents

Figure 6.22 plots the mean damage ratios for contents of one-storey houses at intensity MM7.5 (Greymouth), with two types of foundation and on four different ground classes. Here we observe the same resonance effect on ground class CD discussed for houses in relation to Figure 6.19.

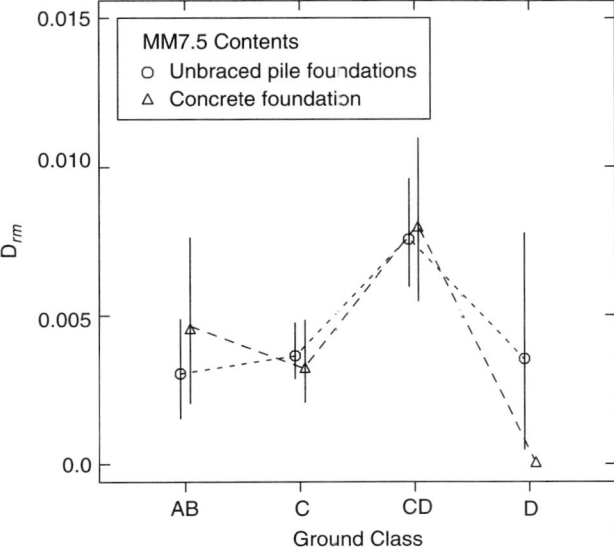

Figure 6.22 Mean damage ratio and its 95% confidence limits for household contents in houses on piled and concrete foundations and on four ground classes at intensity MM7.5 (Greymouth) (from Dowrick et al., 2005). Compare with damage to houses in Figure 6.19

Microzoning and risk assessment methodology

It is evident from the surprising and complex results of the Inangahua earthquake microzoning study discussed above that much remains to be understood about microzoning effects on earthquake damage levels. Foremost amongst other things that should be done, it appears that microzoning maps need to be based on more information than do surface geology maps, and the required extra criteria (such as engineering properties of the soil) need to be better understood. In addition, vulnerability models used in estimating earthquake losses should be functions not only of structural type and strength of shaking, but also of ground class.

6.3.6 Upper and lower bounds on vulnerability

The mean damage ratio data for buildings and equipment from the various studies referred to above are plotted in Figure 6.23, and lines representing approximate upper and lower

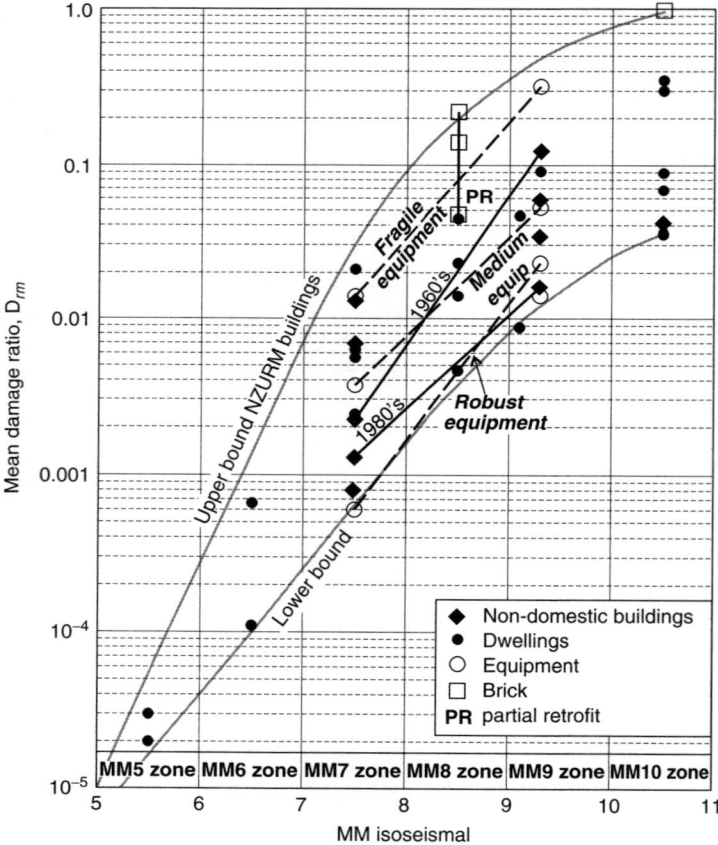

Figure 6.23 Mean damage ratio data from New Zealand earthquakes, for buildings and equipment as a function of intensity, with approximate upper and lower bounds (from Dowrick, 2003)

bounds to the data have been drawn. It is interesting to note that the most robust buildings and equipment have similar levels of vulnerability, and the same is true for fragile equipment and brittle buildings. The improvement in vulnerability of buildings from the 1960s to the 1980s is very apparent in this plot.

The upper-bound curve is relevant to well-built URM as built in New Zealand (see Buildings Type II in Appendix A), and indicates less vulnerability than that of the poorest forms of URM, say adobe. In a similar manner, the lower-bound curve does not represent the true least vulnerability of buildings, which of course would have $D_r = 0$ for individual perfect structures, and the mean damage ratio $D_{rm} \to 0$. However, the lower bound as drawn involves low values of D_{rm}, which are realistically attainable without unreasonable cost, at least for low-rise buildings. At the time of writing no good quantitative data on D_r for high-rise buildings exist, though it is known that post-1980 buildings in Kobe sustained little damage (Park et al., 1995) at intensities of c. MM10. In this regard it is noted that at intensity MM10.5, a data point of $D_{rm} = 0.04$ has been plotted for non-domestic buildings. This is an estimate made qualitatively by the author for the virtually undamaged performance of pre-code low-rise reinforced concrete walled buildings in the Hawke's Bay earthquake of 1931 (van de Vorstenbosch et al., 2002). This excellent performance is consistent with that of timber dwellings on concrete foundations (with $D_{rm} = 0.035$) in the Inangahua earthquake (Dowrick et al., 2001).

Curiously, these upper and lower bounds for buildings come close to enclosing all of the data for contents of both houses and non-domestic buildings (i.e. stock), as can be seen in the plot on Figure 6.24.

6.3.7 Earthquake risk reduction potential

The potential for earthquake risk reduction for buildings and equipment is illustrated by Figure 6.23. Here are plotted the mean damage ratios, D_{rm}, over a range of intensities from MM5 to MM10, as found for New Zealand buildings and equipment in various earthquakes. It is seen that the lower bound D_{rm} is about one-thirtieth of the upper-bound value over the range of damaging intensities MM7–MM10. This suggests that in New Zealand there is the potential for about an order of magnitude reduction in earthquake losses, if the whole built environment were to be converted near to the lower bound of vulnerability. This would save billions of dollars as well as many hundreds of casualties in a Wellington fault earthquake.

6.3.8 Human vulnerability to casualties

To quantify the vulnerability of people in earthquakes we shall consider the historical case of New Zealand. A summary of the known earthquake deaths and hospitalized injured people in New Zealand is presented in Table 6.6. It can be seen that such casualties have occurred in 16 earthquakes, which is an average rate of one casualty-causing earthquake per decade. The estimated total numbers of death and hospitalized injured in Table 6.6 are 298 and 653, respectively. These numbers should be considered as best estimates. For example, there are considerable uncertainties of 10% or so in the numbers of casualties in the 1931 Hawke's Bay earthquake which gave rise to over 90% of the casualties.

From Table 6.6, it can be seen that casualties have occurred in earthquakes of a wide range of magnitude (5.6–8.2), and from Modified Mercalli intensity MM7 to MM10. Thus

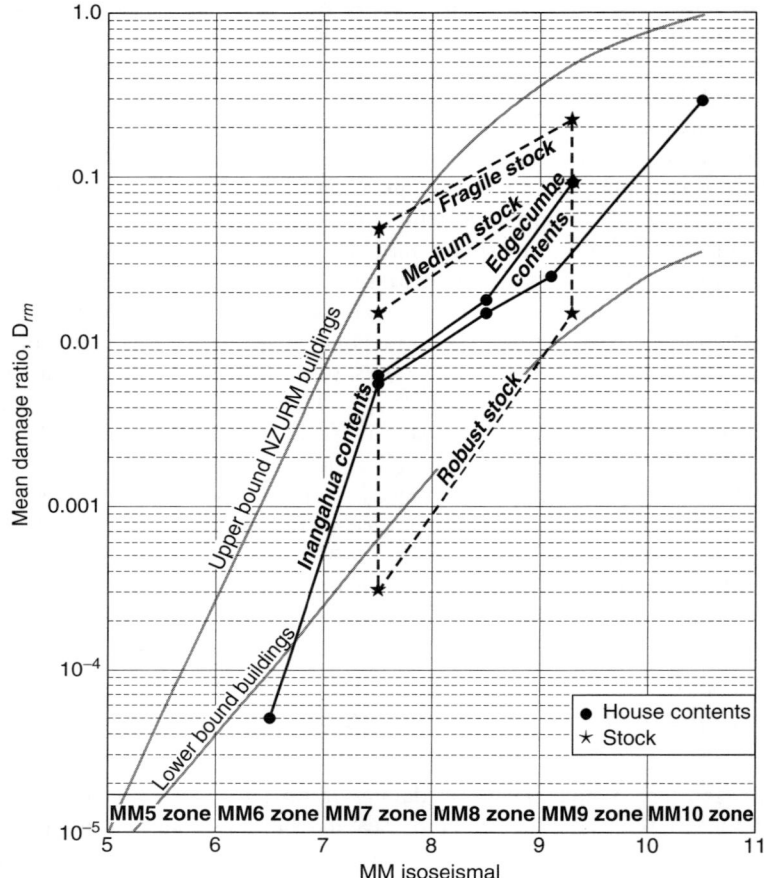

Figure 6.24 Mean damage ratio data from New Zealand earthquakes, for household contents and non-domestic contents (i.e. stock), as a function of intensity, with approximate upper and lower bounds for buildings

intensity MM7, which is the effective threshold of structural damage for New Zealand construction, is correspondingly the threshold for hospitalized injured. It follows that the search for historical data on casualties has been concentrated on (but not limited to) earthquakes which have caused intensities of MM7 or greater in populated areas. In all, 71 such earthquakes were identified as having occurred, with magnitudes in the range 4.9–8.2. The maximum intensity was MM10, MM9, MM8 and MM7 in 4, 10, 18 and 39 events, respectively.

An important feature of Table 6.6 is what it says about the vulnerability of people to death or injury as a function of intensity. It can be seen that 1% of casualties have occurred at intensity MM7, 4% at each of MM8 and MM9, and the overwhelming majority (91%) have occurred at MM10.

Quantitative Measures   239

Table 6.6 Numbers of deaths and hospitalized injured as a function of Modified Mercalli intensity in New Zealand earthquakes, 1840–2001, excluding indirect casualties (from Dowrick and Rhoades, 2005b)

Local date and time		M_W	MM7		MM8		MM9		MM10	
			Dth	Inj	Dth	Inj	Dth	Inj	Dth	Inj
1843	Jul 8	1645	7.5	–	–	2	–	–	–	–	–
1848	Oct 17	1540	6?	3	–	–	–	–	–	–	–
1855	Jan 23	2102	8.2	–	1	2	–	6[3]	4?	–	–
1882	([1])	Day	5–6	–	–	3[2]	–	–	–	–	–
1897	Dec 7	0240	6.5	–	1	–	–	1	–	–	–
1901	Nov 15	0745	6.8	–	0	–	–	–	–	–	–
1913	Apr 12	1912	5.6	1	–	1	–	–	–	–	–
1914	Oct 6	0646	6.6	–	–	–	1	–	–	–	–
1922	Dec 25	1503	6.4	–	–	3	1	12	–	–	–
1929	Jun 16	1017	7.7	–	–	2	18	5	8	–	–
1931	Feb 3	1047	7.8	–	1	–	3	–	–	254[4]	594[4]
1932	Sep 15	0125	6.8	–	–	–	–	–	–	–	–
1934	Mar 5	2346	7.4	–	–	–	2	–	–	–	–
1942	Jun 24	2316	7.1	–	–	–	–	–	–	–	–
1968	May 23	0424	7.2	–	–	–	–	–	1	2[5]	2[5]
1987	Mar 2	1342	6.5	–	–	–	–	–	–	–	–
Totals				4	3	13	28	24	13	256[6]	596[6]

Notes:
[1]Date and hour not known.
[2]Could have been MM8.
[3]5–7 deaths.
[4]Includes 27 injured who died.
[5]Includes one injured who died.
[6]Includes 28 injured who died.

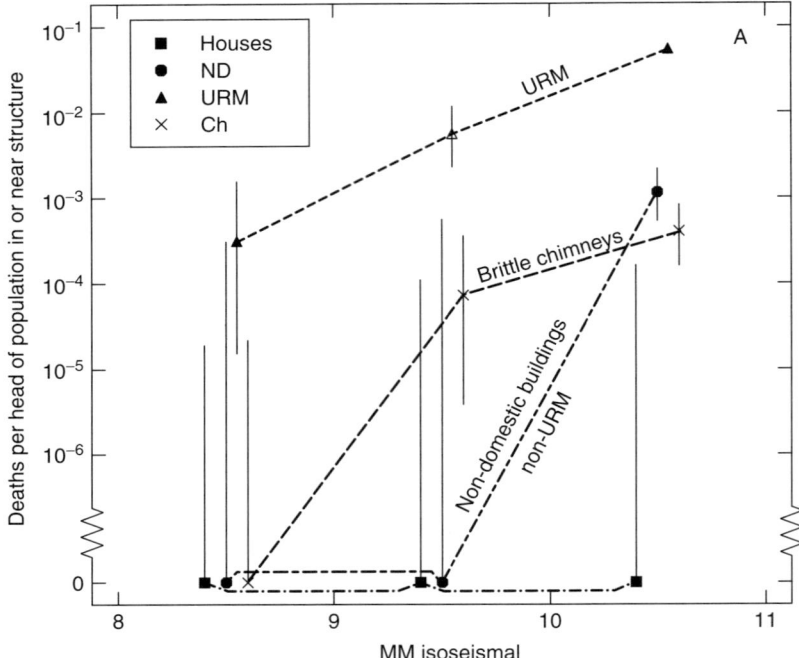

Figure 6.25 Death rates in New Zealand earthquakes in the period 1840–2001, based on the population in or near four types of structure (from Dowrick and Rhoades, 2005b)

Using the statistics of deaths compared to exposure for people (in and beside) four types of structure implied by Table 6.6, the death rates for such people as a function of intensity are plotted in Figure 6.25. It can be seen that URM buildings are much the most dangerous type of structure. To obtain the risk of death in any such building, the death rate in Figure 6.25 has to be multiplied by the hazard, i.e. the probability of occurrence of the intensity under consideration at the building's location. More discussion of Figure 6.25 has been given by Dowrick and Rhoades (2005b).

6.3.9 Inter-earthquake effects

Different earthquakes produce different observed levels of vulnerability of any given class of property, similar to differences that are observed in other phenomena, such as attenuation of ground-motion parameters. There are considerable inter-earthquake differences in D_{rm} and percentage of items damaged, as seen in the plots for the Inangahua and Edgecumbe earthquakes in Figure 6.10. (For comments on the Napier results, see Section 6.3.5.) Such differences arise for a number of reasons, notably:

- intrinsic earthquake differences;
- subtle differences in classes of property, nominally labelled the same in different studies;
- samage costs measured in different insurance regimes, including differences in insurance deductibles;

- differences of various types between studies in different countries, including representativeness of samples of data when whole populations are not studied, or methods of mapping intensities – such differences are likely to exist between the otherwise similar studies of similar houses, as reported here for New Zealand and by Steinbrugge and Algermissen (1990) for California.

6.3.10 Damage due to liquefaction-induced differential ground deformations

Bird et al. (2005) have developed a straightforward analytical method of predicting building response to liquefaction-induced differential settlement and differential lateral deformation of foundations of buildings. An outcome of the method is sets of vulnerability (fragility) curves of the type shown in Figure 6.26. The vertical line across the probability plots on the figure shows that for buildings subjected to a different settlement $\Delta_{FV} = 200$ mm, a single building would have a 35% probability of experiencing slight damage, a 6% probability of moderate damage, an 8% probability of extensive damage and a 51% probability of complete damage. While this method is based on sound principles, at present it is not backed up by any indication of the uncertainties associated with estimating either the differential settlement or damage levels.

Unfortunately the improvement of the foundation structure and/or soils (see Section 9.1.6) to prevent the above damage to the superstructure may well be too expensive.

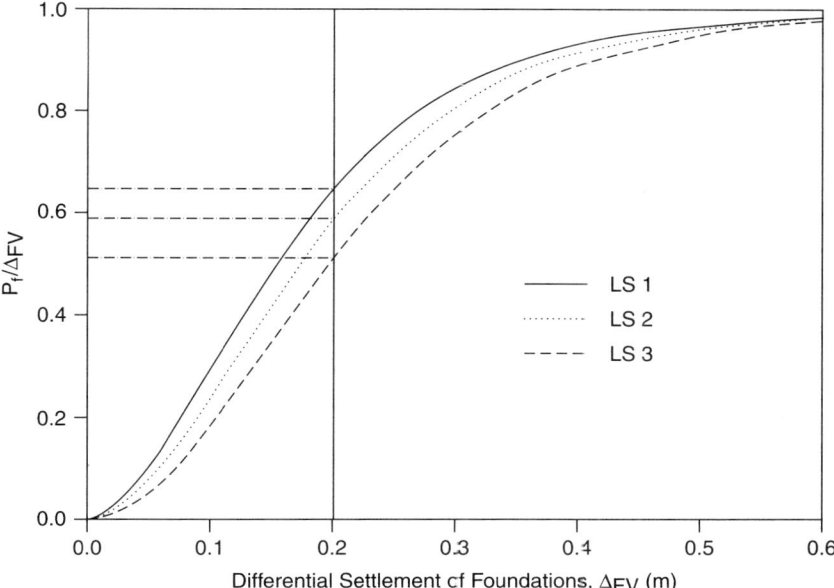

Figure 6.26 Illustrative vulnerability curves for multi- and single-bay, poor quality, reinforced concrete frame buildings characteristic of European building stock, subjected to differential settlement (Limit State 1 (LS1) = slight damage, LS3 = failure) (from Bird et al., 2005)

Furthermore, supposing that measures were taken to avoid liquefaction damage, the superstructure of this type of (brittle) building would of course also need retrofitting to make it survive the earthquake. This would further increase the cost of reducing the vulnerability of the building to the desired level.

References

Applied Technology Council (2005) *Guidelines for seismic performance assessment of buildings 25% complete draft*. Report prepared by the ATC for the Department of Homeland Security, Washington, DC.

Bird JF, Crowley H, Pinho R and Bommer JJ (2005) Assessment of building response to liquefaction-induced differential ground deformation. *Bull NZ Soc Earthq Eng* **38**(4): 215–234.

Dowrick DJ (1991) Damage costs for houses and farms as a function of intensity in the 1987 Edgecumbe earthquake. *Earthq Eng and Struc Dyn* **20**: 455–469.

Dowrick DJ (1996) The Modified Mercalli earthquake intensity scale: revisions arising from recent studies of New Zealand earthquakes. *Bull NZ Nat Soc Earthq Eng* **29**(2): 92–106.

Dowrick DJ (1998) Earthquake risk for property and people in New Zealand. *Proc. NZ Nat Soc Earthq Eng Tech Conf*, Wairakei, 43–50.

Dowrick DJ (2003) Earthquake risk reduction actions for New Zealand. *Bull NZ Soc Earthq Eng* **36**(4): 1–24.

Dowrick DJ and Rhoades DA (1993) Damage costs for commercial and industrial property as a function of intensity in the 1987 Edgecumbe earthquake. *Earthq Eng Struc Dyn* **22**: 869–884.

Dowrick DJ and Rhoades DA (1995) Damage ratios for plant, equipment and stock in the 1987 Edgecumbe, New Zealand earthquake. *Bull NZ Nat Soc Earthq Eng* **28**(4): 265–278.

Dowrick DJ and Rhoades DA (1997a) Inferences for design, insurance and planning from damage evaluation in past New Zealand earthquakes. *J Earthq Eng* **1**(1): 77–91.

Dowrick DJ and Rhoades DA (1997b) Vulnerability of different classes of low-rise buildings in the 1987 Edgecumbe, New Zealand, earthquake. *Bull NZ Nat Soc Earthq Eng* **30**(3): 227–241.

Dowrick DJ and Rhoades DA (2002) Damage ratios for low-rise non-domestic brick buildings in the magnitude 7.1 Wairarapa, New Zealand, earthquake of 24 June 1942. *Bull NZ Soc Earthq Eng* **35**(3): 135–148.

Dowrick DJ and Rhoades DA (2005a) Revised models for attenuation of Modified Mercalli intensity in New Zealand earthquakes. *Bull NZ Soc Earthq Eng* **38**(4): 185–214.

Dowrick DJ and Rhoades DA (2005b) Risk of casualties in New Zealand earthquakes. *Bull NZ Soc Earthq Eng* **38**(2): 53–72.

Dowrick DJ and Rhoades DA (in preparation) A distributed source approach to modelling the spatial distribution of MM intensities resulting from large crustal new Zealand earthquakes. *Bull NZ Soc Earthq Eng*.

Dowrick DJ, Rhoades DA, Babor J and Beetham RD (1995) Damage ratios and microzoning effects for houses in Napier at the centre of the magnitude 7.8 Hawke's Bay, New Zealand, earthquake of 1931. *Bull NZ Nat Soc Earthq Eng* **28**(2): 134–145.

Dowrick DJ, Rhoades DA and Davenport PN (2001) Damage ratios for domestic property in the magnitude 7.2 1968 Inangahua, New Zealand, earthquake. *Bull NZ Soc Earthq Eng* **34**(3): 191–213.

Dowrick DJ, Rhoades DA and Davenport PN (2003) Effects of microzoning and foundations on damage ratios for domestic property in the magnitude 7.2 1968 Inangahua, New Zealand, earthquake. *Bull NZ Soc Earthq Eng* **36**(1): 25–46.

Dowrick DJ, Rhoades DA and Davenport DA (2005) Microzoning effects on damage in two major New Zealand earthquakes. *Bull NZ Soc Earthq Eng* **38**(1): 1–18.

Dowrick DJ, Hancox GT, Perrin ND and Dellow GA (2008) The Modified Mercalli intensity scale – Revisions arising from New Zealand experience. *Bull NZ Soc Earthq Eng* **41**(3): 193–205.

Idriss IM (1990) Response of soft soil sites during earthquakes. *Proc. H Bolton Seed Memorial Symp*, Berkeley, 273–289.

Kircher CA, Nassar AA, Kustu O and Holmes WT (1997a) Development of building damage functions for earthquake loss estimation. *Earthq Spectra* **13**: 663–682.

Kircher CA, Reitherman RK, Whitman RV and Arnold C (1997b) Estimation of earthquake losses to buildings. *Earthq Spectra* **13**: 703–720.

Mander J (2004) Beyond ductility: the quest goes on. *Bull NZ Soc Earthq Eng* **37**(1): 35–44.

McGuire RK and Toro GR (2000) Models of residential loss based on non-linear instrumental ground motion. Paper 1141, *Proc. 12th World Conf Earthq Eng*, Auckland NZGS 85, NZ Geol Surv, Dept Scientific and Indust Res, New Zealand.

Park R, Billings IJ, Clifton GC *et al.* (1995) The Hyogo-ken Nanbu earthquake of 17 January 1995. *Bull NZ Nat Soc Earthq Eng* **28**(1): 1–98.

Porter K, Kennedy R and Bachman R (2007) Creating fragility functions for performance-based engineering. *Earthq Spectra* **23**(2): 471–489.

Rhoades DA and Dowrick DJ (1999) Variability of damage ratios for property in earthquakes. *Earthq Spectra* **15**(2): 297–316.

Steinbrugge KV (1982) *Earthquakes, Volcanoes, and Tsunamis – An Anatomy of Hazards*. Skandia America Group, New York.

Steinbrugge KV and Algermissen ST (1990) *Earthquake Losses to Single-Family Dwellings: California Experience*, US Geological Survey Bulletin 1939. US Government Printing Office, Washington, DC.

Suggate RP and Wood PR (1979) Inangahua earthquake – damage to houses. Report NZGS 85, New Zealand Geol Surv, Dept Scientific and Ind Res.

van de Vorstenbosch G, Charleson AW and Dowrick DJ (2002). Performance of reinforced concrete buildings in the 1931 Hawke's Bay, New Zealand, earthquake. *Bull NZ Soc Earthq Eng* **35**(3): 149–164.

Wen YK and Ellingwood BR (2005) The role of fragility assessment in consequence-based engineering. *Earthq Spectra* **21**(3): 861–877.

Whitman RV, Anagnos T, Kircher CA, Lagorio HJ, Lawson RS and Schneider P (1997) Development of a national earthquake loss estimation methodology. *Earthq Spectra* **13**(4): 643–661.

7

Earthquake Risk Modelling and Management

7.1 Earthquake Risk Modelling

To manage earthquake risk, it is essential to know the size of the risk and the amount by which the risk is reduced by taking some particular action or set of actions. We thus need to be able to quantify one or more of the nine socio-economic consequences of earthquakes listed in Section 1.3.1. Earthquake risk is generally quantified by computer modelling, which takes account of the hazard and the quantity and nature of the people and/or property at risk, as expressed in equation (1.1).

To fully account for the seismic hazard, not only the ground shaking has to be considered, but also associated earthquake induced hazards which include the nine mainly geological consequences such as liquefaction and landslides listed in Section 3.1, and earthquake induced fires.

Earthquake risk modelling is carried out for a wide range of elements of the built environment:

- buildings;
- contents of buildings;
- fixed and mobile plant and equipment;
- lifelines.

Such modelling may be carried out in its own right, and of course is also a necessary first step in modelling risk of death or injury to people. Another step in risk modelling and management is to identify all the contributing factors which could be improved in order to reduce risk, as discussed for Wellington, New Zealand, in Section 1.4.

7.2 Material Damage Costs

The most common type of earthquake risk modelling is the estimation of the direct financial cost due to material damage to a subset of the built environment. This is done for widely disparate sets of property ranging from a single item or parcel of items, such as the contents of a particular building, to all (or selections of all) the property in a city, or for the total losses for a given earthquake.

7.2.1 Damage costs directly due to ground shaking using empirical damage ratios

Consider the case of a large earthquake occurring on a surface-rupturing fault. The scenario map of such an event is given in Figure 7.1, which shows the Modified Mercalli intensity isoseismals of a magnitude $M_W = 7.5$ earthquake on the Wellington fault, which runs through the centre of the capital city of New Zealand. The intensities depicted range from heavy damage at intensity MM10 in the near-source region to MM7,

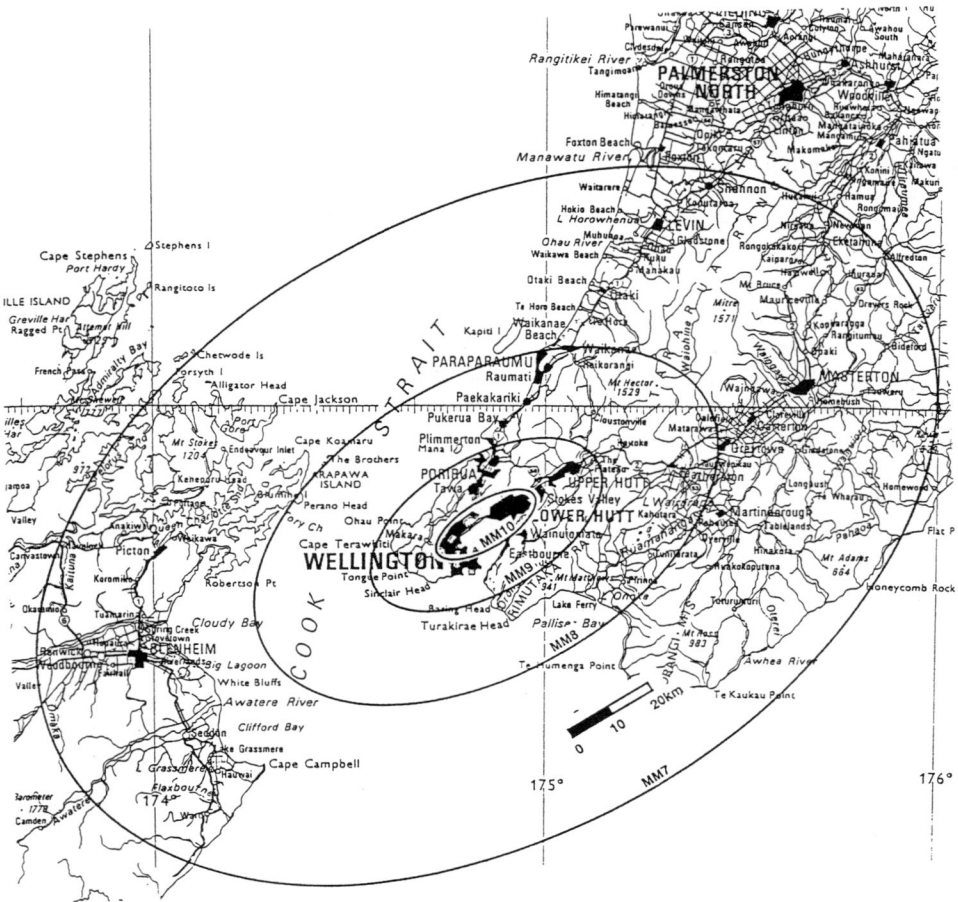

Figure 7.1 Isoseismal map of a magnitude 7.5 earthquake on the Wellington fault in New Zealand (average recurrence interval 600 years), prepared using the attenuation model of Dowrick and Rhoades (1999)

Table 7.1 Replacement values of non-domestic buildings at risk and estimated damage costs due to ground shaking and earthquake induced fires for the earthquake scenario shown in Figure 7.1

Intensity zone	Replacement value of non-domestic buildings at risk ($ million)	D_{rm}	Damage cost ($ million)	
			Shaking	Earthquake fire
MM10	8000	0.2	1600.0	175
MM9	1500	0.12	180.0	20
MM8	400	0.03	12.0	0
MM7	1800	0.007	12.6	0
MM7–MM10	11700		1804.6	195

which is the approximate damage threshold to all but very vulnerable structures (e.g. see Table 6.1).

In modelling the risk for non-domestic buildings in the above scenario, the property at risk is lumped into intensity zones, such that the MM7 zone is the area between the MM7 and MM8 isoseismals. The total replacement value of such buildings in the four intensity zones is given in Table 7.1, as are the average values adopted for the mean damage ratio D_{rm} for each zone. It is seen that the MM10 zone contains two-thirds of the total replacement value of $11.7 billion in the affected area, i.e. within the MM7 isoseismal. The centre of the isoseismal pattern has been deterministically positioned so as to maximize the losses estimated from an earthquake on this fault.

Table 7.1 represents a simple spreadsheet calculation, in which the damage cost due to ground shaking is

$$\text{Damage cost} = D_{rm} \times \text{Replacement value.} \qquad (7.1)$$

The values of D_{rm} adopted in calculations such as those summarized in Table 7.1 need to take into account the mix of buildings of various vulnerabilities in a given zone, using the D_{rm} data such as those given in Chapter 6. Because good-quality damage ratio data are available for only a few earthquakes and a few classes of buildings, subjectively assigned safety margins may need to be added. Buildings in different countries and of different construction styles and dates have different vulnerabilities, and how much better or worse they will perform than the classes discussed in Chapter 6 in general has not yet been objectively quantified.

It should be noted that the above estimate is now considered to require revision, involving replacing the isoseismal map in Figure 7.1 with one derived using the two-dimensional source model of Dowrick and Rhoades (in preparation); see, for example, see Figure 4.33. This would make the MM10 zone three times as long, enclosing the urban area of Upper Hutt and thus increasing the estimated losses.

7.2.2 Damage costs due to earthquake induced fires

As well as the costs of damage directly due to ground shaking, Table 7.1 gives estimates of the losses due to earthquake induced fires. These estimates were made using the mean percentage of non-domestic buildings burnt out arising from a study by Dowrick *et al.*

Table 7.2 Mean percentage of buildings burnt out by earthquake induced fires in central New Zealand (from Dowrick *et al.*, 1990)

Intensity zone	Mean percentage of buildings burnt out	
	Houses	Non-domestic buildings
MM7, MM8	0	0
MM9	0.23	1.8
MM10	0.35	2.4

(1990) of the environment in central New Zealand, the results of which are given in Table 7.1. As with mean damage ratios D_{rm} due to shaking, fire damage models are region-specific, depending on many factors which include the separation of buildings and other flammable materials, incidence of ignition sources, terrain and vegetation, the climate and wind regime, and post-earthquake operational fire-fighting resources. These factors were first developed for Californian studies by Scawthorn (1987). Table 7.2 shows that the fire risk for non-domestic buildings is 7–8 times that for houses. This is a natural consequence in fire spread situations of the mostly smaller separations between non-domestic buildings in the main commercial districts, than the average separations between houses.

The incidence and extent of earthquake induced fires varies not only with the strength of shaking, but also from earthquake to earthquake. While occasional single buildings may catch fire even at modest intensities in many earthquakes, major conflagrations with fire spreading between adjacent buildings are not the norm, even at high intensities in large earthquakes. Nevertheless, major fires that are well known have occurred in large earthquakes, such as San Francisco (1906), Tokyo (1923), Napier (1931) and Kobe (1995). In the case of the 1906 earthquake, it was estimated that the total cost of shaking damage was $80 million, while the cost of the fire damage was $320 million.

7.2.3 Damage cost estimation using structural response parameters

In the above methods of estimating damage costs (Sections 7.2.1 and 7.2.2), earthquake hazard is defined in terms of MM intensity and the risk is related to the intensity directly through empirical damage ratios. For obtaining reliable results this requires a robust attenuation model for MM intensity and robust damage ratio models for the relevant property types. In many parts of the world, either or both of these models are not available, thus prompting earthquake engineers to develop models using strong ground-motion attenuation and damage ratios linked to structural response parameters, notably the fragility functions discussed in Section 6.3.4. A leading example of such an approach is that developed specifically for estimating regional losses in the USA in the form of a software package, known as HAZUS-MP. This software was developed by the National Institute for Building Standards for the Federal Emergency Management Agency, and is available as a free download from http://www.fema.gov/hazus. A further example comes from Mexico, with a computer model first generated for estimating regional losses from buildings in Mexico City, and then generalized to deal with the rest of the country (Ordaz *et al.*, 2000).

In the Mexico method, the damage cost for a building on a given ground class is related to its maximum inter-storey drift, γ_i, estimated from its spectral acceleration, $S_a(T)$, as

$$\gamma_i = \frac{\beta_1 \beta_2 \beta_3 \beta_4 (\eta N^\rho)^2}{4\pi^2 Nh} S_a(T) \tag{7.2}$$

where β_1 is the ratio between the maximum lateral displacement at the upper level of the structure and the spectral displacement, considering linear elastic behaviour; β_2 is the ratio between the maximum storey drift and the global distortion of the structure, which is defined as the maximum lateral displacement at the upper level divided by its total height; β_3 is the ratio between the maximum lateral displacement with inelastic behaviour and the maximum lateral elastic displacement; β_4 is the ratio between elastic and inelastic β_2 factors; N is the number of storeys; h is the storey height; and η and ρ are factors used in estimating the fundamental period of the structure, $T = \eta N \rho$.

The expected gross damage of a structure, given a maximum storey distortion, is calculated with

$$E(\beta|\gamma_i) = 1 - \exp\left[\left(\frac{\gamma_i}{\gamma_0}\right)^\varepsilon\right] \tag{7.3}$$

where β is the gross damage conditioned on the maximum storey distortion γ_i, and ε and γ_0 are parameters of structural vulnerability that depend upon the structural system and the date of construction, and $E(\cdot)$ stands for expected value. β is the damage ratio, as defined in equation (6.1), and lies between 0 and 1. Equation (7.3) was supplied by E. Miranda (pers.comm.), as a correction of that in Ordaz, et al.

The Mexico model described above incorporates other features which permit probabilistic loss modelling, and for the purposes of estimating insurance losses has provisions for limiting the losses to the maximum loss for a given building and the insurance deductible (Ordaz et al., 2000).

7.3 Estimating Casualties

As demonstrated forcefully by Table 1.1 and Figure 1.2, human casualties are a major problem in earthquakes. For managing the risk from a future strong earthquake centred in a populated area, realistic estimates of the number and nature of casualties are needed by various bodies, particularly emergency services, healthcare providers and insurers. An example of such an event is provided by the case of the large earthquake on the Wellington fault for which one component of the damage costs has been discussed in Section 7.2. As most casualties are generally caused by damage to the built environment, it can be seen from Figure 7.1 and Table 7.1 that in this event the risk of casualties in the MM10 intensity zone is much higher than elsewhere.

For modelling earthquake casualties, the numbers and locations of people in relation to the various life-threatening hazards need to be modelled, as well as the level of hazard to people in each situation. Building-related casualties arise for falling parts of buildings and their contents, as well as collapse of all or part of the structure. The latter effect is directly related to the loss of volume of the building. Different classes of building suffer different volume losses, sometimes for much the same damage cost. The main reason for the development of ductile structures has been to minimize casualties rather than damage, and buildings which are subjected to high ductility demands and suffer little loss of volume may nevertheless be unrepairable. Thus, damage cost alone is an inadequate criterion for casualty prediction.

When estimating injuries it is important to know the distribution of degree of injury, often considered in three classes: seriously injured, moderately injured and uninjured or lightly injured (i.e. not requiring admission to hospital).

Considering the case of a magnitude 7.5 earthquake on the Wellington fault (Figure 7.1), Cousins et al. (2006) estimated the casualties from a workday and a nighttime event for the Accident Compensation Commission in New Zealand for reinsurance purposes. Using the following equation (from Spence et al. 2008), they estimated the number of killed, in any building in any zone, as

$$N_5 = P.NB.(M_{51}.DS_1 + M_{52}.DS_2 + M_{53}.DS_3 + M_{54}.DS_4 + M_{55}.DS_5)$$
$$= P.NB.\left(\sum M_{5j}.DS_j\right) \text{ with } j = 1, \ldots, 5 \text{ over five damage levels} \quad (7.4)$$

where P = the total number of occupants per building in that class and zone, NB is the number of buildings in that class and zone, M_{ij} is the expected proportion of occupants suffering injury class i given that the building suffers damage state j, and $DS_1, \ldots, DS_5$ are the probabilities of the building being in each damage state as a result of the given scenario. Similar formulae give the numbers in each of the injury categories, i.e. for five injury states defined by $i = 1, \ldots, 5$.

The overall results of this partly probabilistic study are given in Table 7.3, where it is seen that the best estimate (mean of the model) for deaths at 11 a.m. (620) is over three times as great as that for the 2 a.m. event (180). This large difference is caused mainly by the fact that at night far more people in the Wellington region are in their mostly timber-framed houses than during the working day when many more people are in non-domestic buildings which suffer much more volume loss than do timber-framed New Zealand houses.

Table 7.3 Summary of deaths and hospitalized injured people estimated for two $M_W = 7.5$ scenario earthquakes on the Wellington fault as depicted in Figure 7.1

		Workday 11 a.m. event			Night-time 2 a.m. event		
Cause of casualties		Deaths	Seriously injured	Moder. injured	Deaths	Seriously injured	Moder. injured
Earthquake shaking damage to buildings		455	348	3903	114	206	3130
Buildings sheared by fault		76	17	94	22	9	63
Misc. other causes		93	64	205	43	35	102
Totals	Best estimate[1]	620	410	4200	180	250	3300
	90th percentile[1]	1000	700	7000	500	500	6000
	10th percentile[1]	300	200	2000	100	100	1500

Note: [1]Totals have been rounded by Dowrick

Of the casualties caused by building collapse (part or complete), it was found that the majority occurred in a relatively small number of buildings designed prior to the introduction of 'capacity design' and ductility (Chapter 10) in the 1970s.

As well as listing casualties due to ground shaking, Table 7.3 also gives the casualties estimated from other causes, including those associated with buildings straddling the causative fault, mostly (but not all!) built before the fault was identified. A field survey found that 74 houses and 70 other buildings had been built across this strike-slip fault, which is expected to have shear displacements of about 5 m horizontally and up to about a metre vertically. The casualties attributed to 'miscellaneous other causes' in Table 7.3 include causes such as localized earthquake induced fires, landslides, liquefaction, damage to bridges and falling contents of buildings.

When considering how the daytime casualties on a work day in Table 7.3 are geographically distributed throughout the intensity zones of the scenario event shown in Figure 7.1, according to the best estimate 94% of the deaths occur in the MM10 zone and the remaining 6% occur in the MM9 zone. The total daytime populations at risk are 238,300 in the MM10 zone and 91,100 in the MM9 zone. This means that mortality rates of 0.29% and 0.05% per capita are expected for the MM10 and MM9 zones, respectively. Put another way, the best estimate of the chance of a person being killed in the intensity MM10 zone is 1 in 350, at MM9 it is 1 in 2000, while at MM8 it is zero. These casualty rates are about an order of magnitude less than those found for people in or near URM buildings in New Zealand (Figure 6.23).

When uncertainties in the model are taken into account, as shown in Table 7.3, the 90th percentile estimates of casualties are as much as twice those of the median values, while the 10th percentile estimates are about half of their respective medians. It is also noted that the above estimate is now considered to require revision, involving replacing the isoseismal map in Figure 7.1 with one derived using the two-dimensional source model of Dowrick and Rhoades (in preparation); see, for example, Figure 4.33). This would make the MM10 zone three times as long, enclosing the urban area of Upper Hutt and increasing the estimated losses.

Another example of the causal distribution of deaths in earthquakes comes from the $M_W = 6.7$ Northridge, California earthquake, as given in Table 7.4 compiled by Durkin (1996). In this event 30% of the deaths were caused by structural failures, 10% by damage to non-structural elements and contents, and 60% were from other causes. Surprisingly, the biggest single cause of deaths was heart attacks, which were classified by the Coroner's office as indirectly earthquake related. This suggests that the estimate of deaths for future earthquakes should include allowances for heart attacks.

An issue that arises when considering numbers of casualties is the ratio of injured to deaths. Table 7.5 gives such ratios for several real large earthquakes (plus two models) in three different countries. Surprisingly, it can be seen that the ratios vary greatly from much less than unity to much greater than unity. Clearly the relationship between numbers killed and numbers injured is very dependent on construction materials, time of day and the situation of people. Considering the right-hand column of Table 7.5, the smallest injured/death ratio (0.06) is for the Buller earthquake in which all the deaths (16) were caused by landslides, and only one non-fatal injury occurred, caused by falling bricks. The Northridge earthquake provides the other extreme with a ratio of total injured/death of 25 (Durkin, 1996).

Table 7.4 Deaths by structural and other causes in the 1994 Northridge, California, earthquake. (Reprinted from Durkin, ME (1996) Casualties patterns in the 1994 Northridge, California earthquake. *11th* World Conference a Earthquake Engineering, Paper No. 979, with permission from Elsevier Science)

Cause of death	Deaths	Percentage of total
Structural failure		
Buildings		
Wood frame		
Apartment building collapse	16	22
Single family residential	4	5
Mobile home		
Mobile home collapse	1	1
Other structures		
Freeways		
Collapsed freeway overpass	1	1
Total related to structural failure	22	30
Non-structural elements/contents		
Microwave oven/Heart attack	1	1
Collectibles	2	3
Respirator failure	3	4
Electrocution	1	1
Total related to non-structural elements/Building contents	7	10
Other causes		
Falls	5	7
Automobile accidents	3	4
Fire	1	1
Suicide	1	1
Exposure	2	3
Heart attacks	30	42
Total related to other causes	43	60
Total fatalities	72	100

The rates of mortality and serious injury increase with increasing volume loss, i.e. with fewer voids in structures in which trapped people can survive. Thus, landslides have no voids and hence high death rates, as does unreinforced masonry construction. The latter was the cause of most of the casualties in the Hawke's Bay earthquake, which had a lower injured/death ratio of 1.56. As the Wellington built environment is similar in terms of earthquake resistant construction standards to that of Los Angeles (Northridge), the modelling of the hypothetical 2006 Wellington earthquake appropriately results in the prediction of a ratio of total injured/deaths quite similar at 20 to the actual ratio of 25 for the 1994 Northridge event.

In addition to the injured people needing hospitalization, casualties of course occur for which the injured need only outpatient treatment. In modern-day Californian earthquakes,

Table 7.5 Ratio of numbers injured per death in some large earthquakes (deaths from induced heart attacks excluded where known)

No.	Earthquake	Time	M_W	$I_m{}^1$	Hospitalized injured per death		
					Seriously injured	Moderately injured	Serious + moderate
1	1929 Buller, NZ	10.17 a.m.	7.7	9	0.06	0	0.06
2	1931 Hawke's Bay, NZ	10.47 a.m.	7.8	10	0.34	1.22	1.56
3	1945 Mikawa, Japan	3.38 a.m.	6.8	9			0.65
4	1968 Inangahua, NZ	4.24 a.m.	7.2	10	0	0.5	0.5
5	1994 Northridge, USA	4.31 a.m.	6.7	9			25
6	1995 Kobe, Japan	5.46 a.m.	6.9	10	0.33		
7	1999 Chi Chi, Taiwan	1.47 a.m.	7.6	10	0.29		4.5^3
8	Wellington model^2	11 a.m.	7.5	10	0.66	6.8	7.4
9	Wellington model	2 a.m.	7.5	10	1.4	18	20

Notes: $^1 I_m$ is maximum MM intensity in populated areas.
^{2}See Table 7.3.
^{3}Unclear if this includes unhospitalized injured (Tsai *et al.* 2001).

Durkin (1996) gives high ratios of hospitalized to total injured, i.e. 8%, 17.5% and 10% for the 1971 San Fernando, 1989 Loma Prieta and 1994 Northridge earthquakes, respectively. Not nearly so many light injuries have occurred in New Zealand earthquakes. This is because the last New Zealand earthquake causing many casualties occurred in 1931, when pre-code brickwork caused mainly deaths and serious injuries. This again emphasizes the point that the incidence of casualties is very situation-dependent, and much still needs to be learned about casualty modelling.

7.4 Business Interruption

By the term *business* we here refer to any organization, such as shops, factories, schools, clubs, hospitals, and governmental bodies. *Business interruption* is the name often given to the costs of loss of business arising from any cause, in our case from an earthquake. In Section 1.3.1 business interruption is one of a list of nine socio-economic consequences of earthquakes and is one of the hardest to model. This is illustrated graphically by the flowchart in Figure 7.2. It is seen that there are three general areas which may cause business loss, i.e.

- upstream effects;
- direct material damage; and
- downstream effects.

Upstream effects are those to do with supplies of anything that a business uses or consumes, such as power, raw materials or components. *Downstream effects* comprise damage to dispatch routes or loss of a market, for example, a customer's business. As seen in

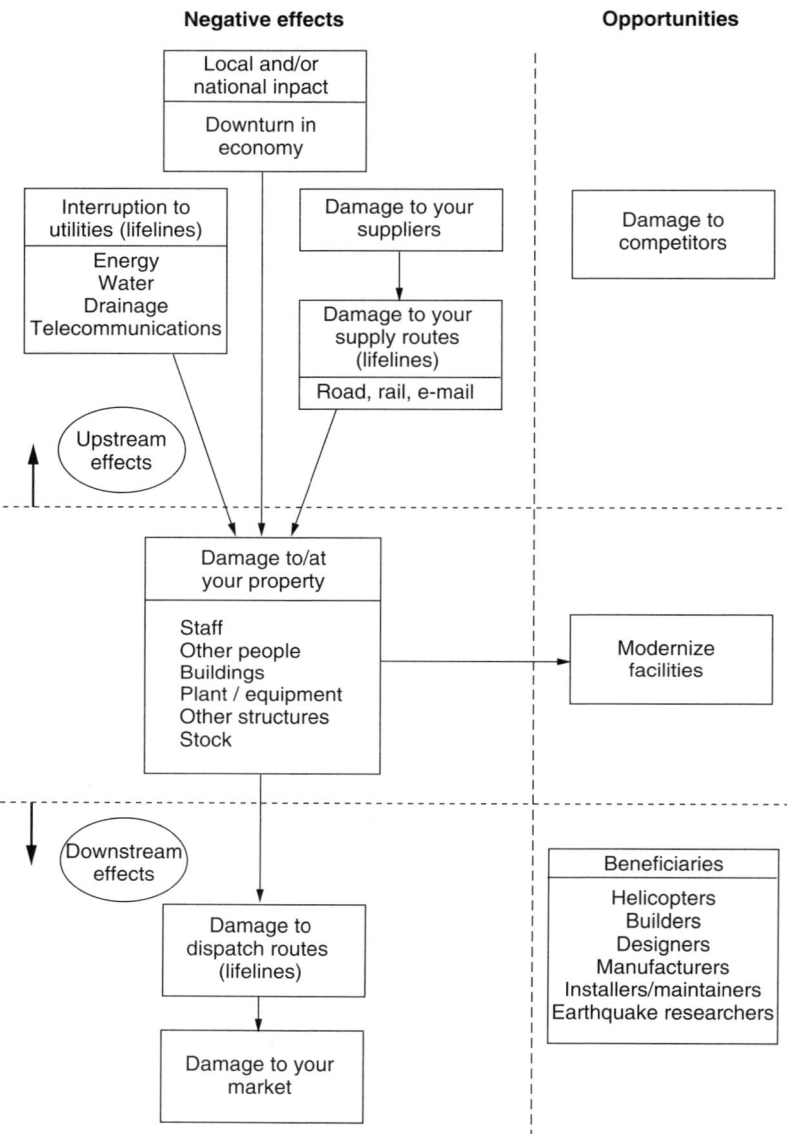

Figure 7.2 Earthquake business interruption/opportunity flowchart

Figure 7.2, the consequences of an earthquake can be both negative and/or positive for any given business. An example of related negative and positive effects comes from the 1987 Edgecumbe, New Zealand, earthquake, in which the public hospital in the largest affected town was put out of commission for some time. This caused a serious decrease in business to the local undertaker (who was not insured against such a loss), because fatally

Table 7.6 Examples of effects of an earthquake on a given business or other organization

Business	Material damage		Business response	
	Damage ratio[1]	Cost ($ million)	Loss ($ million)	Gain ($ million)
A	0	0	Total	
B	0	0	Large	
C	0	0		Large
D	0.1	10	60	
E	0.1	10	5	
F	0.1	10		20
G	1.0	RV[2]	Large	

Notes: [1] Damage ratio, $D_r =$ (Damage cost) $\div$ (RV).
[2] RV = Replacement value.

ill patients were being sent to hospitals elsewhere. This of course led to a corresponding increase in business for undertakers in other places. A further example of downstream effects comes from the destruction of the port at Kobe in the 1995 Hyogo-ken Nanbu earthquake: this caused serious negative impacts on shipping worldwide (Tom Holzer, personal communication, 2008).

As shown in Figure 7.2, business interruption can be caused by local or distant earthquakes, even those in other countries. The effects on businesses are clearly very variable and often unpredictable. A range of very different possible outcomes are listed in Table 7.6, which can be considered as those for either different businesses, or alternative negative and positive outcomes for the same business in different scenarios. Because of this inherent variability, estimation of effects of earthquakes on businesses is best studied by considering various likely scenarios, modelling a range of possible upstream and downstream effects to supplier and customer bases for each scenario. Such modelling involves estimating the length of time after the earthquake that each consequence lasts. An interesting example of such a study is that of modelling time delays caused by damage to the transportation system in the San Francisco area in a magnitude 7.5 earthquake on the San Andreas fault (Kiremidjian et al., 2001), using GIS-based methodology to tackle a complex problem. More recently, Comerio (2006) has developed modelling of downtime taking account of various factors including irrational situation-specific components affecting the time needed to mobilize for repairs.

7.5 Reduction of Business Interruption

A range of measures for mitigating business interruption may be appropriate depending on the nature and location(s) of the business. These will depend upon cost-effectiveness, and include the following:

(1) Create low-damage built environment (e.g. use more structural walls in buildings; hold down key equipment/plant; create stable storage); see Section 8.6.
(2) Prioritize: give greatest protection to key functions (critical residual operations).
(3) Control earthquake induced fire hazard.

(4) Safe-shutdown systems for plant.
(5) Develop emergency and recovery plans (minimize cost and time of interruption to business).
(6) Establish operational flexibility, duplication, spare capacity (e.g. on separate sites).
(7) Establish alternative sources of supply.
(8) Establish alternative sources of energy.
(9) Establish early warning, early action plan.
(10) Purchase business interruption insurance (e.g. consider insurance of profits as discussed by Fawcet, 1988).

Some of the key questions that need to be answered in relation to the fitness of the business to respond to an earthquake disaster are the following:

- In what ways will the disaster:
    (a) damage the business?
    (b) be a business opportunity?
- Is the risk of business interruption big enough to worry about?
- Can you afford:
    (a) to do nothing?
    (b) not to do something?
- Is business interruption likely to be permanent for you (e.g. land permanently flooded by regional land downthrow)?
- In what respects is prior mitigation more important to you? For example, fewer casualties, less trauma, less downtime, less unemployment, less national impact.
- How much insurance (material damage and business interruption) is necessary and/or prudent business practice?

7.6 Management of and Planning for Earthquakes

Management of and planning for earthquakes involve a wide range of activities, with generally complex issues, in particular:

- planning of land use;
- planning of disaster emergency response;
- planning of economic response;
- planning of social response.

Damage scenarios based on damage ratios, such as in Figure 7.1, provide information on the potential outcomes of future earthquakes which is highly relevant to planning of land use, and of economic and social response to earthquakes. Such damage maps highlight the existence and extent of high-risk zones within existing urban areas, and the potential development of future black spots if extensive development is proposed in the vicinity of frequently rupturing major faults. As illustrated in Figure 6.10 and Table 7.1, damage levels increase rapidly with intensity, so that development in regions which are certain to experience intensity MM9 or greater at 'unacceptable intervals' should be discouraged or at least carefully regulated.

Considering the case of Wellington (Figure 7.1), much of its urban area lies in a zone which is expected to experience intensity MM10 when the Wellington fault ruptures, that is, at an average recurrence interval of about 600 years. Within the MM9 and MM10 zones, there are areas subject to a range of earthquake induced hazards, which include microzoning for ground-motion amplification, liquefaction, landslides, inundation and tsunami. In simplistic terms, if most of the development currently located within 10 km of the Wellington fault was relocated to a zone 10–20 km from the fault, then the damage and casualties in the scenario earthquake would be enormously reduced. For example, over 100 buildings are located astride the fault. Also, based on the study of casualties discussed in Section 7.3, casualties would reduced by 80–90% if such a relocation was made.

The risk associated with other possible major hazard sources in any given region would also have to be considered when making such risk mitigation decisions. In the case of Wellington, other major faults running nearly parallel to the Wellington fault also exist in the region. One of these lies about 4 km to the north-west of the Wellington fault and another lies about 18 km to the south-east over a mountain range. Both of these faults are less active than the Wellington fault, generating large earthquakes but less frequently, with approximate average recurrence intervals of 3000 and 1500 years respectively, compared to 600 years for the Wellington fault. For the centre of the city to be damaged severely once in two or three thousand years is clearly better than once in 600 years, but preferably it should be planned not to happen at all, so these faults should be avoided as well. There are regions further away where no major faults exist, but other issues, such as possible flooding and what to do about Wellington's safe deep-water port and the enormous costs of moving a city, would be prohibitive. Obviously, the threat of a major fault is best dealt with before cities are allowed to grow across them, rather than afterwards.

Issues arise as to what kind of development should be permitted and at what distances from major faults/ Also, for existing cities, how do we compare the socio-economic cost of leaving it where it is, with or without strengthening/retrofitting?

A societal limitation on the effectiveness of land-use planning controls, even when they are put in place, is that they are not always properly enforced. An example of this comes from the city of Anchorage, in the zone of the very destructive Turnagain Heights landslide (Figure 3.2) which occurred in the magnitude 8.4 Alaska earthquake of 1964. Following the earthquake, planning controls forbade new building on the affected land, but subsequently this restriction has not been rigorously enforced.

The planning of social responses to earthquakes addx an extra, largely subjective, dimension to the already complex problem of estimating the physical and financial consequences. In order to advance the social side of the management of risks from a range of hazards, the Emergency Management Office of the local government of Wellington City in New Zealand commissioned the development of a GIS-based software package (Cousins et al., 2000). In addition to modelling the geographic distribution of damage losses, other economic losses and casualties, this software incorporates an interactive module so that parametric analyses and other sensitivity tests can be performed to evaluate the effect that various mitigation strategies, options, and policies may have on risk reduction. The results are in a format that enables risk management policy recommendations to be formulated and verified. Through its availability on the Internet interactive use by members of the community enables them to gain increased awareness of the relationships between hazards and risk to themselves or others.

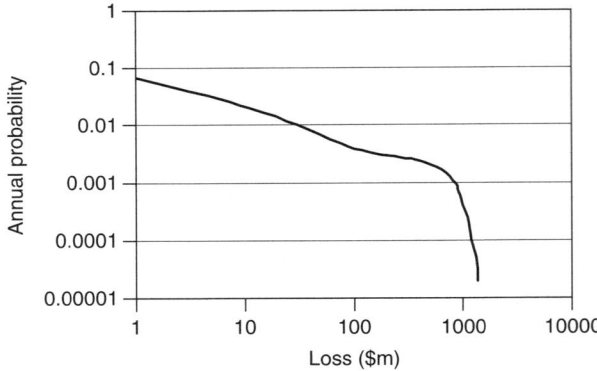

Figure 7.3 Annual probability of exceedence for earthquake losses to houses in Hutt City, New Zealand (from Smith and Cousins, 2002)

An important tool in management of risk is to model the earthquake losses for a given part of the built environment as a function of probability of occurrence. This has been done by Smith and Cousins (2002) using a synthetic earthquake catalogue, and a Monte Carlo procedure for calculating the probability of occurrence of ground motions.

For each fault source affecting a given site or region, the return period T relates to a probability $1/T$ that an earthquake will occur in any given year. If a random number in the range (0,1) is generated for each year of the synthetic catalogue, each time it is less than $1/T$ indicates an earthquake. If it is greater than $1/T$, no earthquake is recognized. Examining all the faults in this way determines which of them will generate earthquakes in any one year, and the process is repeated for each year of the synthetic catalogue.

Using an attenuation model, and a vulnerability function and the Monte Carlo process, Smith and Cousins compiled the statistics of losses for houses in Hutt City (part of Greater Wellington; see Figure 7.1). They found that it was necessary to consider a synthetic catalogue of 100,000 years' duration in order to obtain a stable set of statistics. Their cumulative probability distribution is shown in Figure 7.3.

Smith (2006) points out that no single measure of risk is adequate for making well-informed decisions in earthquake risk management. In particular, the annualized loss is a poor indicator of likely losses. The conditional expected values of the loss, expressed in terms comparable to scenario losses, are useful for conveying to decision-makers the likely extent of losses. The net present value (i.e. annualized loss) gives a measure of total future losses, taking into account appropriate discounting for events far into the future, but is inadequate for use on its own. Utility theory introduces the subjective value judgements of the decision-maker to indicate the level of expenditure with which he should be comfortable in order to mitigate the risk.

Smith and Cousins comment as follows:

> The data used in [Figure 7.3] were all the domestic buildings in Hutt City (total value $4871 million). Their locations have been grouped by suburb. More detailed modelling could use a GIS formulation, specifying the actual location of each building, though the effort is substantially greater and the advantage may not be significant in this case. The methodology is the same.

The total damage due to earthquake will be much greater than [Figure 7.3] indicates, when other factors such as damage to commercial and public buildings and to infrastructure are added, and especially business interruption. These can all be modelled, albeit with some difficulty, but the present paper is not affected by its use of only domestic buildings in its damage portfolio.

As seen in Figure 7.3, the worst loss is $1350 million, which is 28% of the total value of houses at risk. Also seen in the figure is that the worst loss is caused by an event with an annual probability of occurrence of about 10^{-4}, i.e. having a return period of nearly 10,000 years.

7.7 Earthquake Insurance

From its humble beginnings in the early part of the twentieth century, earthquake insurance has grown to a stage where it is a major factor in the management of earthquake risk. In the decade of the 1960s, when earthquake insurance began to be appreciable, total world insurance losses were US$55 million (in 1999 values), while for the decade of the 1990s the losses had increased about 430 times to US$23.3 billion (Smolka, 2000). Obviously, only a small part of this increase arises from the doubling of world population over those years. In the same period, the world's total economic losses due to earthquakes increased by a factor of 11, from US$19 billion to US$209 billion. The losses in the 1990s were dominated by three major earthquakes: Northridge (1994), Kobe (1995) and Izmit (1999). As global seismic energy release per decade is fairly constant, it appears that this increase has more to do with the location of the earthquakes, hitting major urban areas, than with anything else.

The role of insurance in earthquake risk management has been discussed in simplistic terms in Section 1.3.1, and methods of modelling losses have been discussed in earlier parts of this chapter. More information on earthquake insurance is available in Earthquake Engineering Research Institute (2000), Walker (1997) and Smolka (2000).

7.8 Earthquake Risk Management in Developing Countries

The management of earthquake risk in developing countries deals with the same problems in principle as are faced in developed countries. But the scale of the problems is worsened considerably because developing countries have less available resources of all kinds (physical, financial, educational and administrative) per capita. The problems are magnified by rapidly increasing populations (especially in cities), uncontrolled urban development spreading into marginal and more hazardous areas, and inappropriate construction materials and practices. Poverty, social and economic marginalization and inadequate education greatly limit the choices of increasing numbers of the population. As well as these problems, governmental agencies in developing countries have to give greater priority to dealing with the day-to-day problems, such as pollution of water and air, inadequate sanitation, healthcare, droughts and floods, rather than the less frequent visitations of damaging earthquakes.

Valuable contributions to earthquake risk management and mitigation in some developing countries have resulted from initiatives taken in the 1990s, as part of the International

Decade for Natural Disaster Reduction (IDNDR). In some cities at risk, there was a welcome shift from largely reactive response and relief measures to proactive risk mitigation measures. In parallel with some networks in developed countries, multidisciplinary teams of engineers, scientists, public officials, journalists, community leaders and members of the general public worked together. The efforts of the local people were supplemented by people from various international organizations and national organizations from developed countries.

One such project which made a lot of progress comes from Ecuador – the Quito Project. According to Fernandez and Yepes (1997), some of the most beneficial strategies were:

- working in multidisciplinary teams;
- building partnerships among different countries and institutions, and
- seeking locally specific solutions.

To continue making progress, Fernandez and Yepes comment that 'the capacity and desire of the local community to reduce their own vulnerabilities is the keystone for building sustainable solutions'.

Another developing country which has made good progress in earthquake risk management for its capital city is Colombia. In Bogotá in the 1990s, according to Mattingly (2000), the following substantial actions were put in hand:

- updating seismic code standards;
- evaluation of seismic vulnerability of hospitals and design of their rehabilitation and structural and non-structural reinforcement;
- detailed evaluation of vulnerability of critical points of lifelines and emergency response plans;
- development of standards for design of urban gas networks.

An important insight into this project comes from Cardona (1999): 'This project is an example of a study in a developing area, where the political will and the agreement of the different institutions involved, constitutes the basis to get effective results; [importantly] without a huge amount of financial resources as is usually required.'

A productive action of the IDNDR secretariat, sponsored by the United Nations, is the Risk Assessment Tools for Diagnosis of Urban Areas against Seismic Disaster (RADIUS) initiative. Starting with risk assessment case studies of nine cities, practical tools were published by IDNDR (1999), including:

- a manual to prepare earthquake damage scenarios for urban areas;
- a graphic software package for computer simulation as easier application of the manual;
- case studies;
- a guide for simple assessment of buildings and houses.

This 23-page document contains a CD-ROM with RADIUS tools and final reports. Details of how to obtain it are available through Google by typing in the title, as given in the references. The document can be downloaded free of charge, and the CD-ROM can be viewed. The RADIUS projects in the case study cities were designed around a core partnership between local government and local academic or scientific institutions. Clearly, the next stages, implementation and ongoing work, are crucial to the success of the initiative for the case study cities themselves, and for other cities. Fortunately, such extensions of RADIUS-like collaborative projects on earthquake risk reduction are seen in the Earthquakes and Megacities Initiative (EMI), which started in 1997, and which focuses on megacities with significant earthquake hazard in developing countries. It thus appears that the excellent platform built by the IDNDR has stimulated some of the desired longer-term risk reduction activities.

7.9 Impediments to Earthquake Risk Reduction

There are many impediments to earthquake risk reduction (as discussed in various parts of this book, e.g. Section 1.4), which may be divided into two types: finding out what to do physically to the environment; and implementation. While steady progress is being made in research on what to do, the most intractable impediments arise in getting that knowledge effectively implemented. Even in relatively wealthy communities with well-developed bureaucracies implementation meets impediments, particularly in relation to the existing built environment.

This problem can be partly attributed to inadequate promotion of research solutions by the researchers, partly due to inadequate language, as discussed by people such as Killip (2001), a psychologist with the Building Officials Institute of New Zealand. Killip points out that too much of what researchers or technologists of various kinds write or say is not in language readily understood by many potential implementers.

The problem is also contributed to by those receiving the advice, who may be either unwilling or unable to enforce the adoption of improved risk mitigation measures. As an example, deficiencies were found in this respect in various parts of the USA, in studies by political scientists as summarized by May and Feeley (1999). In one initiative to overcome this aspect of the problem in the United States, the Federal Emergency Management Agency (1998) produced a guidebook (FEMA 313) for state earthquake and mitigation managers, and called for delivery agents who would use and disseminate this book to help (a) generate increased understanding and support for seismic safety, and (b) promote the adoption and enforcement of seismic codes.

While different countries have impediments to earthquake risk reduction that differ from each other in detail and degree, they are nevertheless the same in principle. Thus, when planning procedures to overcome any particular impediment, we can learn from each other.

A disturbing non-technical impediment to reducing earthquake risk is that caused by corruption at the design and construction stages of projects in many countries. This takes the form of bribery of people such as designers, builders, inspectors and other officials by people who stand to make money if less building materials are used than required and specified to meet the prevailing earthquake design regulations. The author has come across this type of behaviour in a number of countries, and no doubt this phenomenon is much more widespread worldwide than is openly acknowledged. Sadly, the results of such behaviour have severe social and financial consequences. A recent widely publicized example comes from China, in the magnitude 7.9 earthquake in Sichuan province in 2008, where thousands of children were killed in the collapse of defective school buildings. This was attributed to systematic non-enforcement of regulations in return for bribes. The offices housing the corrupt officials had conformed to the regulations and withstood the earthquake. Similar greed-inspired tragedies have occurred in many other countries, and similar unnecessary tragedies are waiting to happen. The Chinese government initially promised to punish those responsible, but sadly later imprisoned a school teacher, Liu Shaokun, who had posted photographs on the Internet of collapsed schools for inciting families of victims to petition and for disseminating anti-government rumours (Guardian, 2008). At the time of writing, Amnesty International was preparing to petition for Shaokun's release.

See also Section 1.4.

7.10 Further Reading and Software

In addition to the references cited throughout the foregoing text, books which elucidate the more formal aspects of earthquake risk modelling and planning are those of Haimes (1998), Woo (1999), Jones (2000), the special volume on loss estimation in the USA (EERI, 2000), and McGuire (2004). Geoscience Australia has developed a program (EQRM) for estimating seismic hazard and risk, suitable for any part of the world, which is available for free download at https://source-forge.net/projects/eqrm. A general book on earthquake risk and its mitigation is that of Coburn and Spence (2002).

References

Cardona DM (1999) Multidisciplinary Risk Mitigation Project of Bogota, Columbia. Paper prepared for Earthquakes and Megacities Initiative (EMI) Twin Cities Meeting, Seeheim, Germany, March 14–16.

Coburn A and Spence R (2002) *Earthquake Protection*, 2nd edn. John Wiley & Sons, Ltd, Chichester.

Comerio MC (2006) Estimating downtime in loss modeling. *Earthq Spectra* **22**(2): 349–365.

Cousins J, Spence R and So E (2006) Wellington area casualty estimation – 2006 update. GNS Consultancy Report 2006/88 to the Accident Compensation Corporation, Wellington, New Zealand.

Cousins WJ, Heron D, Jensen S, Kozuch M and Savage J (2000) Integrated and Interactive Risk Assessment Platform for Wellington, New Zealand. Paper No. 1624, *Proc. 12th World Conf on Earthq Eng*, Auckland.

Dowrick DJ and Rhoades DA (1999) Attenuation of Modified Mercalli Intensity in New Zealand earthquakes. *Bull NZ Soc Earthq Eng* **32**(2): 55–89.

Dowrick DJ and Rhoades DA (in preparation) A distributed source approach to modelling the spatial distribution of MM intensities resulting from large crustal New Zealand earthquakes. *Bull NZ Soc Earthq Eng*.

Dowrick DJ, Cousins WJ and Sritharan S (1990) Report on Potential Losses to the EQC due to Fire Following Large Earthquakes in Central New Zealand. Report for Marsh & McLennan Ltd for Treasury, DSIR Physical Sciences, DSIR, Lower Hutt.

Durkin ME (1996) Casualties patterns in the 1994 Northridge, California Earthquake. *Eleventh World Conference on Earthquake Engineering*. Paper No. 979, Elsevier.

Earthquake Engineering Research Institute (2000) *Financial Management of Earthquake Risk*. EERI Endowment Fund White Paper. EERI, Oakland, CA.

Fawcett BA (1988) *The Insurance of Profits*. Robins MBS (NZ) Ltd, Auckland.

Federal Emergency Management Agency (1998) *Promoting the Adoption and Enforcement of Seismic Building Codes: A Guidebook for State Earthquake and Mitigation Managers*, FEMA 313. FEMA, Washington, DC.

Fernandez J and Yepes H (1997) Knowledge and technology transfer: Improving the urban seismic risk management in Quito-Ecuador. *First International Earthquakes and Megacities Workshop*, Seeheim, Germany, Release II, United Nations University: 295–300.

Guardian (2008) Sichuan earthquake activist sent to labour camp for year. *Guardian Weekly*, 8 August, p. 5.

Haimes YY (1998) *Risk Modelling, Assessment and Management*. John Wiley & Sons, Inc., New York.

International Decade for Natural Disaster Reduction (1999) *RADIUS: Risk Assessment Tools for Diagnosis of Urban Areas against Seismic Disaster*. IDNDR, United Nations, Geneva.

Jones BG (ed.) (2000) Economic Consequences of Earthquakes: Preparing for the Unexpected. NCEER SP-0001. National Center for Earthquake Engineering Research, Buffalo, NY.

Killip RE (2001) 'Tell It Like It Is': How to Communicate Your Future Vision for Earthquake Engineering to Clients and Building Owners. Paper No. 7.03.01, *Proc. Tech Conf of New Zealand Earthquake Engineering*, Wairakei, New Zealand.

Kiremidjian AS, Moore J, Fan Y, Hortacsu A, Burnell K and Legnie J (2001) Earthquake risk assessment for transportation systems: analysis of pre-retrofitted system. *PEER Center News* **4**(1): 1–5.

Mattingly S (2000) Advances in Seismic Risk Management in Developing Countries. Paper No. 2829. *Proc. 12th World Conf on Earthq Eng*, Auckland.

May PJ and Feeley TJ (1999) Earthquake risk reduction in the western United States. *PEER Center News* **2**(4): 7–9.

McGuire RK (2004) *Seismic Hazard and Risk Analysis*, MNO-10. EERI, Oakland, CA.

Ordaz M, Miranda E, Reinose E and Perez-Rocha LE (2000) Seismic Loss Estimation for Mexico City. Paper No. 1902, *Proc. 12th World Conf Earthq Eng*, Auckland.

Scawthorn C (1987) *Fire Following Earthquake: Estimates of the conflagration risk to insured party in greater Los Angeles and San Francisco*. All Industry Research Council, Oak Brook, IL.

Smith WD (2006) Decision tools for earthquake risk management, including net present value and expected utility. *Bull NZ Soc Earthq Eng* **39**(3): 170–175.

Smith WD and Cousins WJ (2002) Effective ways to model earthquake risk. *Technical Conf Soc Earthq Eng*, Napier, Paper No 2.2.

Smolka A (2000) Earthquake research and the insurance industry. *NZ Geoph Soc Newsletter*, No. **55**: 22–25.

Spence R, So E, Jenny S, Castella H, Ewald M and Booth E (2008) The global earthquake vulnerability estimation system (GEVES): an approach for earthquake risk assessment for insurance applications. *Bull Earthq Eng* **6**(3): 463–483.

Tsai Y-B, Yu T-M, Chao H-L and Lee C-P (2001) Spatial distribution and age dependence of human-fatality rates from the Chi-Chi, Taiwan, earthquake of 21 September 1999. *Bull Seism Soc Am* **91**: 1298–1309.

Walker GR (1997) Prediction of Insurance Loss from Earthquakes. *Bull NZ Nat Soc Earthq Eng* **30**(1): 40–45.

Woo G (1999) *The Mathematics of Natural Catastrophes*. Imperial College Press, London.

8

The Design and Construction Process – Choice of Form and Materials

8.1 The Design and Construction Process – Performance-Based Seismic Design

Earthquakes provide architects, engineers, constructors and enforcers with a number of important considerations foreign to the non-seismic design and construction process. As some of these criteria are fundamental in determining the form of the 'structure', it is crucial that adequate attention is given to earthquake considerations at the correct stages in the process. To this end, a simplified flowchart of the design and construction process for earthquake resistant infrastructure is shown in Figure 8.1.

Although the real interrelationships between all the factors shown in the diagram are obviously much more complex than indicated, the overall sequence is correct. All factors 1–8 are related when evaluating the level of seismic risk, as the risk depends not only on the possible earthquake loadings, but also on the capacity of construction to avoid damage.

The *design brief* (box 1 of Figure 8.1) for different projects is developed by the designers with varying amounts of input from the owner, and varying aspects and degrees of detail of the brief are subject to the owner's agreement.

Few owners wish to be involved in deciding the acceptable level of risk, but in any case it is important that the owner should be informed of the risks consequent to the available options. Even when the design is done according to a good local code, high risks may still exist.

It is emphasized that *poor design concepts cannot be made to perform well in strong earthquakes*, whereas good design concepts often perform well despite major shortcomings in analysis and detailing. This observation is addressed to the whole design team, which, depending on the nature of the project, will include some or all of the following: engineers of all disciplines, architects and builders.

Earthquake Resistant Design and Risk Reduction D. Dowrick
© 2009, John Wiley & Sons, Ltd

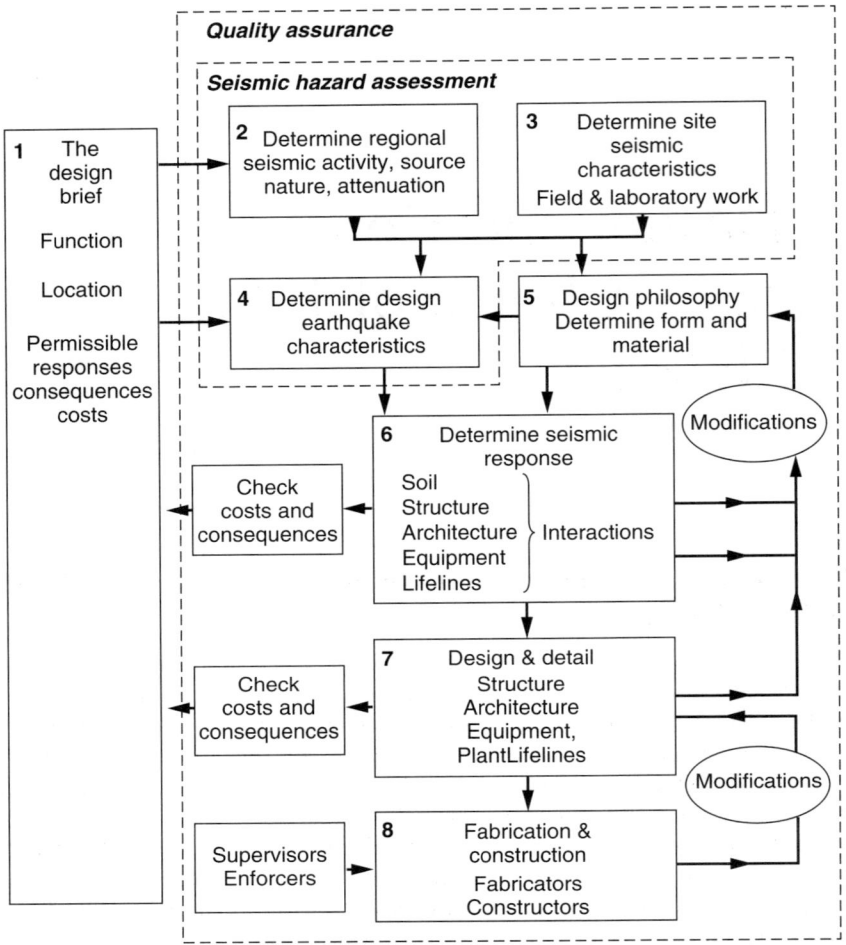

Figure 8.1 Simplified flowchart for the design and construction of earthquake resistant infrastructure (non-seismic factors omitted for clarity).

The need for cooperation between the various members of a design team is, of course, not restricted to earthquake resistant design, but the need for good conceptual design is especially important in this context. As well as its need for earthquake resistance *per se*, good design philosophy has become increasingly important in recent years for countering the dangers of errors and loss of design direction arising from rapidly growing complexities in analytical techniques and detailing requirements. It is thus apparent that decisions made by the design and construction team at this conceptual stage will generally be more important than any other aspect of the design process.

It follows from the above that this chapter is the most important in this book.

8.2 Criteria for Earthquake Resistant Design

8.2.1 Performance-based seismic design

From around the 1980s there was a growing awareness that building codes, in general, provided a good level of life-safety protection, but were significantly less reliable in minimizing property damage in moderate and even small earthquakes. This concern led to the birth of *performance-based seismic engineering*, or *performance-based seismic design*. Work by the Structural Engineers Association of California (SEAOC) resulted in the publication of tentative design guidelines on this subject as an Appendix to the SEAOC (1999) Blue Book. The following brief discussion is based on that publication.

The first stage in this design process is the *selection of performance objectives*. This selection is made by the client, in consultation with the design professional, based on considerations of the client's expectations, the seismic hazard exposure, economic analysis and acceptable risk. A performance objective is a coupling of the expected performance level with expected levels of earthquake ground motions. The options available are given in the matrix shown in Figure 8.2. Here, for example, for a normal-use building the basic objective would generally be chosen, for which up to four scenarios would be designed for, ranging from the frequent earthquake design level (43-year return period) with the fully operational earthquake performance level to the building being near collapse in the very rare earthquake design level (970-year return period). In other regions different return periods may be required, as discussed later in relation to Table 8.1.

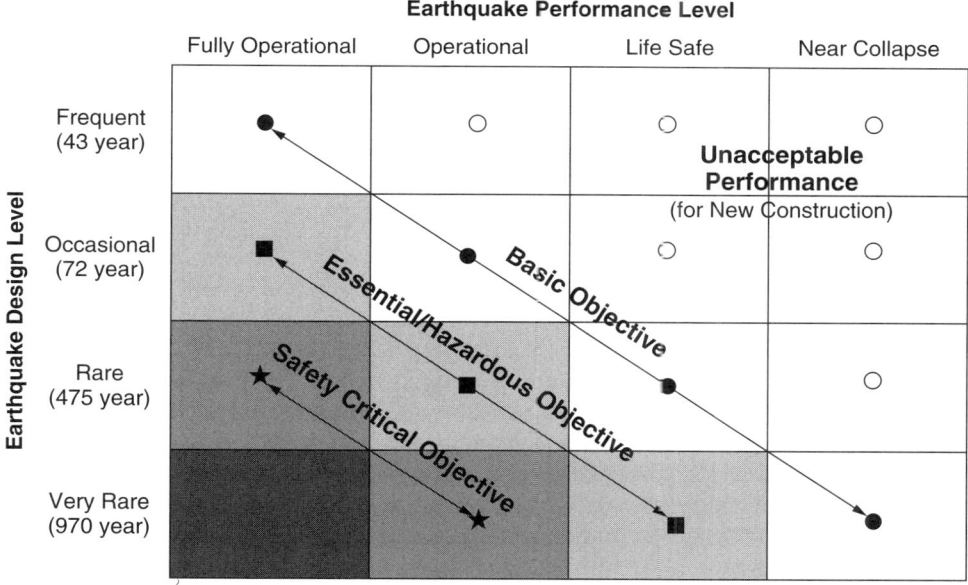

Figure 8.2 Seismic performance objectives for buildings recommended by SEAOC (1999)

Table 8.1 Hierarchy of limit-state design criteria for different levels of earthquake hazard

		(A) Serviceability limit state	(B) Ultimate limit state	(C) Survivability limit state Maximum considered earthquake (or MCE)
Normal-use structures or equipment	Response condition	(1) Undamaged Elastic	(2a) No collapse Post-yield cycling Limited deformation (Repairable) (2b) Pre-yield	(3) Pre-collapse
	Typical return period (yr)	(4) 25–72	(5) 100–500	(6) 1000–2500
Critical structures or equipment	Response condition	(7) Pre-yield		(8) As (2a) or Pre-collapse
	Typical return period (yr)	(9) 100–1000		(10) 1000–2500

Proceeding to the design stage, we follow the SEAOC direct displacement-based design approach (discussed in Section 10.1.1), rather than their strength design approach.

Step 1: Estimate target displacements. The target displacement for each selected performance level (limit state) is estimated based on the inter-storey drift limits for the selected structural system.

Step 2: Determine effective period. The effective period at each limit state under consideration can be found from Figure 10.2(d) using the target displacement and the appropriate damping level for the structure and limit state considered.

Step 3A: Determine effective stiffness. The effective stiffness at each selected performance level can be found from the equation

$$K_e = 4\pi^2 m_e / T_e^2 \tag{8.1}$$

(see also equation (10.1)), where m_e and T_e are the effective mass and effective period, respectively.

Step 3B: Determine required initial stiffness. A conservative estimate of the required initial stiffness K_i at the effective yield point can be found from

$$K_i = \mu K_e. \tag{8.2}$$

Step 4: Determine required strength. The required strength can be found from the base shear

$$V_{\text{Base}} = K_e \Delta_T, \tag{8.3}$$

where Δ_T is the target displacement at the effective height.

Step 4A: Determine required yield strength. The required yield strength V_y can be found from

$$V_y = V_{\text{Base}} / \Omega, \tag{8.4}$$

where Ω is the system overstrength factor.

Step 5: Prepare preliminary design/system element sizes. A simplified design procedure is given by SEAOC.

Step 6: Verify preliminary design using pushover analysis procedures. In the pushover analysis, the structure is pushed to reach the target displacement at the effective height, to check that the inter-storey drift limits are not exceeded and that the element ductility demands are within the achievable capacities.

Step 7: Modify design if required. If either of the above two requirements is not met, the design should be modified and step 6 repeated.

Step 8: Prepare final design and detailing. Having successfully completed steps 6 and 7, capacity design procedures should be followed to ensure that the yielding elements possess the design ductility and that the remaining elements have sufficient strength to develop the overstrength of the yielded elements.

Finally, it is noted that at the time of writing (2009) performance-based seismic design is mainly used in research and in some retrofitting projects, and is as yet little used for the design of new structures. A drawback to its use is its inability to incorporate torsional effects.

8.2.2 Function, cost and reliability

The basic principle of any design is that the product should meet the owner's requirements. The owner's requirements may be reduced to just three criteria: function, cost and reliability. While the terms *function* and *cost* are simple in principle, *reliability* concerns various technical factors relating to serviceability and safety. As the above three criteria are interrelated, and because of the normal constraints on cost, compromises with function and reliability generally have to be made. In considering the means of achieving the above requirements, it is necessary to take into account both the limitations and the opportunities arising from the availability of construction materials and components and of construction skills. The criteria governing reliability in earthquake resistance are discussed below.

8.2.3 Criteria for reliability of performance

General serviceability and safety criteria

The term *reliability* is used here in its normal language qualitative sense, and in its technical sense, where it is a quantitative measure of *performance* at given limit states, as stated in terms of probabilities (of failure or survival). Aspects of the probabilistic ingredients of reliability control are discussed in other parts of this book, notably the evaluation of seismic hazards and the question of acceptable risks. The required reliability is achieved if enough of the elements of the design behave satisfactorily under the design earthquake (ground shaking and other geological hazards, listed in Section 3.1). The elements that may be required to behave in agreed ways during earthquakes include structure, architectural elements, equipment and contents.

The design criteria governing the satisfactory behaviour or reliability of the above elements relate to one or more levels of loading which in some codes are referred to as *limit-state design criteria*. These criteria vary widely for different elements. For structures or equipment a typical hierarchy of earthquake limit states is shown in Table 8.1, where column A is for *serviceability* criteria and columns B and C are principally concerned with *safety* through damage control. The choice of the terminology used to identify the hazard levels B and C is problematical, as usage varies in the literature.

Regarding hazard levels for the ultimate limit-state design of normal-use structures in Table 8.1 (item 5) there is a trend for advanced earthquake loadings codes to adopt criteria based on 10% exceedance in 50 years, i.e. a return period of 475 years, rather than the lower values of 100 years or so that were formerly more common.

Up to the end of the twentieth century, it was mostly considered sufficient to design normal-use structures and equipment to meet two criteria:

(1) In moderate, relatively frequent earthquakes the structure or equipment should be undamaged (serviceability limit state, item 1, Table 8.1).

(2) In strong, rarer earthquakes the structure or equipment could be damaged but should not collapse ('usual' design earthquake limit state, item 2a, Table 8.1).

The main intention of the second of these criteria was to save human lives, while the definitions of the terms 'strong', 'rare', 'moderate' and 'frequent' have varied from place to place, and have tended to be rather imprecise because of the uncertainties in the state of the art. With the growing concern about the cost of repairs for earthquake damage, and the improvement in our ability to specify loadings and carry out structural analyses, more specific attention to the serviceability limit state has become appropriate, and calls for new *low-damage* structures are common (see Section 8.6).

Table 8.1 gives the hierarchy of design criteria for two different classes of construction with acceptable risks near the two ends of the risk spectrum for engineered structures or equipment, i.e. what we have called *normal-use* and *critical*, respectively. For normal-use structures designed for the ultimate limit state with what we have called the 'usual' design earthquake (e.g. usual code design loads) there is a range of response condition criteria which need to be satisfied as represented by items 2a and 2b, depending on the degree of post-yield behaviour (i.e. ductility) that may be called upon. Item 2a, where some degree of ductility can be demanded, is the most common condition, while item 2b, where the structural materials must stay in the elastic state, is generally reserved for brittle materials. However, the latter design criterion is sometimes invoked for construction in an intermediate risk class between normal use and critical. As examples, some offshore oil platforms are designed to remain elastic in the 'design' earthquake, limit state B in Table 8.1 (American Petroleum Institute, various years).

Regarding the *serviceability limit state*, consider the case of New Zealand where the return period for serviceability is 25 years. Dowrick (2006) found that for that hazard level the highest intensity anywhere in the country is MM7. This affects only two small areas comprising about 10% of the country, shown in Figure 8.3. At that intensity, *no loss of function* (predictable by a serviceability design check) had been reported in any structures classified as Buildings Type III (brittle) or better, since the introduction of reinforced concrete construction. It thus appeared that it was unnecessary to carry out design checks for serviceability of normal-use non-domestic buildings throughout New Zealand. But as there had been little local experience of intensities greater than MM7 affecting fully ductile flexible buildings, it was considered that for the time being in the two zones where the 25-year intensity is MM7 such buildings should be checked, until further appropriate field experience has been obtained. However, it may be more appropriate for New Zealand to adopta longer return period for serviceability, for example 72 years as for the operational performance level in Table 8.2.

Criteria for post-yield behaviour

The following remarks refer particularly to item 2a of Table 8.1, where normal-risk structures are being designed to the normal degree of seismic hazard and where the post-yield behaviour of the construction materials is being utilized, and some (unspecified) strength capacity is in reserve for larger earthquake loading. Traditionally, design criteria have concentrated on the important strength-related properties of ductility and energy dissipation, which comprise a subject requiring further elucidation, as discussed in Section 5.4.2.

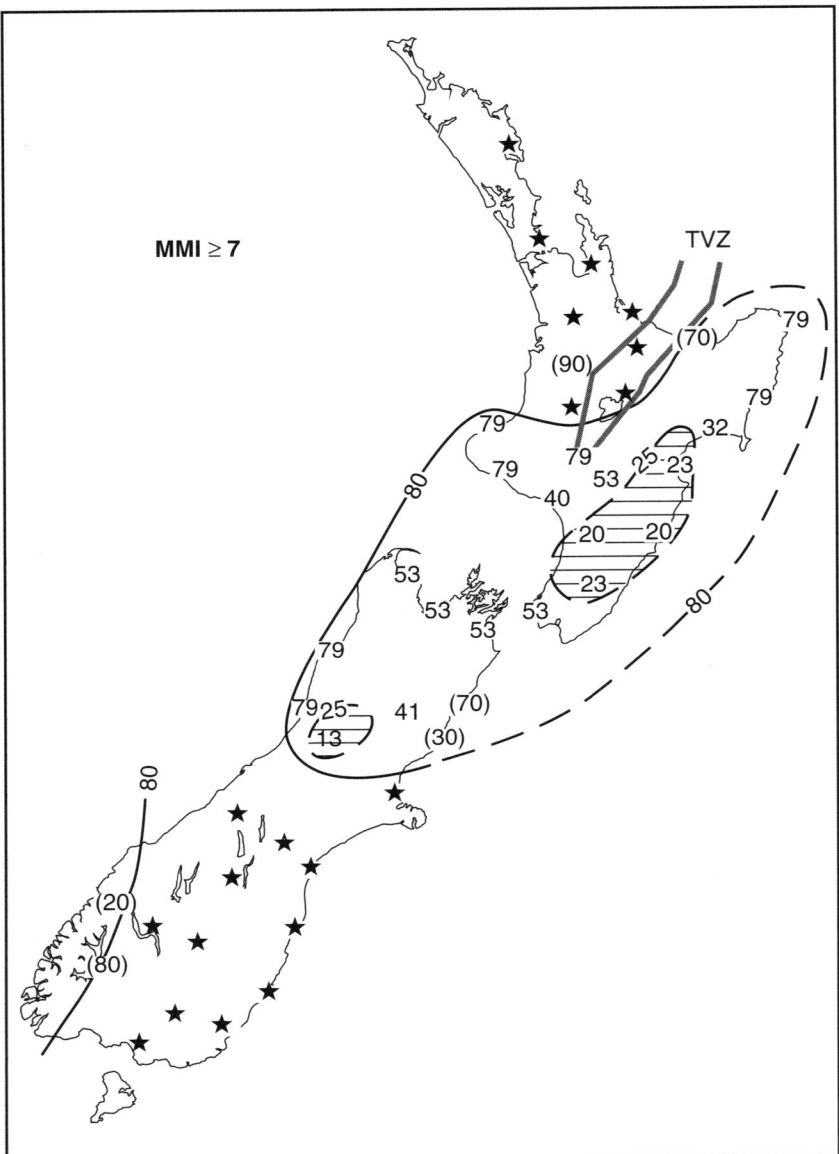

Figure 8.3 Map of New Zealand showing average return periods for historical isoseismal intensities for MMI ≥ 7. In the hatched areas the return period is 25 years, New Zealand's hazard level for serviceability (after Dowrick, 2006)

Table 8.2 Comparative merits of stiff and flexible construction (which is not seismically isolated)

	Advantages	Disadvantages
Flexible structures	(1) Specially suitable for short-period sites, for buildings with long periods	(1) Higher response on long-period sites
	(2) Ductility arguably easier to achieve	(2) Flexible framed reinforced concrete is difficult to reinforce
	(3) Non-structure may invalidate analysis	(3) More amenable to analysis
		(4) Non-structure difficult to detail
Stiff structures	(1) Suitable for long-period sites	(1) Higher response on short-period sites
	(2) Easier to reinforce stiff reinforced concrete (i.e. with shear wall)	(2) Appropriate ductility not easy to knowingly achieve
	(3) Non-structure easier to detail	(3) Less amenable to analysis

This creates a complex physical situation where damage is, by definition, occurring, and limits on deformation and hence damage are both very important and difficult to ensure. It has long been recognized that deformations rather than stresses often control the design of structures, and design methods based on deformation rather than strength criteria are being developed in various countries (e.g. Priestley *et al.*, 2007; see Section 10.1.1 below).

Conflicts between requirements of strength and deformation sometimes arise, because increasing the stiffness to reduce deformations often increases the earthquake response of the system, and may involve the use of more brittle components, reducing the amount of post-yield capacity available. A discussion of stiff versus flexible structures is given in Section 8.3.6. Some means of reducing the strength demands on structures or equipment, such as base isolation and rocking foundations (Section 8.4), also introduce special deformation considerations.

In item 2a of Table 8.1 is the criterion *repairable*, implying that normal-risk construction should be repairable after the occurrence of the design earthquake. It appears in brackets in the table because it was not a widespread requirement of earthquake codes at the time of writing. However, for bridge design in New Zealand, if the potential plastic hinge zones in piers are expected to form in inaccessible locations for repairs (e.g. underground), the bridge has to be designed to withstand higher loads. Also some code requirements implicitly reduce damage (e.g. limitations on drift). However, as noted above regarding the serviceability limit state, the growing concern over the costs of earthquake damage and the difficulty of repairing much post-yield damage, especially to non-structural elements, demands that more attention should be given to repairability at the design stage. This, of course, could also help reduce the danger to life, which is the traditional fundamental design criterion, as noted above. However, the subject of repairability can be a very difficult one, requiring consideration of the interplay between all components (e.g.

structure, architecture and equipment), and much research is still required to provide more definitive design criteria.

It can be seen from the above discussion that the principles for reliable design for post-yield behaviour require that the system conforms to criteria for strength, deformation, and repairability, which in some cases are complex, conflicting, and/or ill defined.

Survivability in extreme events

In addition to the two earthquake limit states discussed above, we sometimes need to consider that of *survivability* in more extreme events. While this has long been a required design condition for critical facilities (notably nuclear power plants and large dams), there has been a growing tendency to consider the survivability of other types of construction with acceptable risk levels much nearer to the norm. This practice has developed for a variety of reasons, in particular the following: (1) earthquake loads can be much stronger in rarer events than the 'usual' design earthquake of even 500 years' return period; (2) the growing desire of property owners to reduce their earthquake risk especially to strategic interest; (3) a growth in our ability to reliably model extreme earthquake hazard in various parts of the world; and (4) a growth in our ability to design for extreme shaking.

The above trend is so strong that advanced earthquake codes such as those of the USA (International Building Code), Europe (EC8) and New Zealand (AS/NZS 1170.0, 2002) have incorporated requirements to design for survivability in extreme events, with a growing consensus that the latter should be defined as shaking with a return period of 2500 years for normal-use structures. This level of hazard is often referred to as the *maximum considered earthquake*. Figure 8.4 shows that the response spectrum acceleration at a period $T = 0.2$ s in the high-hazard zone of San Francisco is about 1.65 times stronger for 2500 years shaking than for 500 years, while in low-hazard Charleston the ratio is much greater at about 4.3. This is typical of the difference between high- and low-hazard regions.

Although the loadings in such rare events may be very high, they can be designed against easily in most cases by appropriate choice of structural form (Section 8.3).

Criteria for critical structures or equipment

The design criteria for critical-risk construction vary widely, depending on the nature of the facility concerned. In general, they are similar in principle to those discussed above for normal-risk construction, the main difference being that the various response conditions are required to occur at much lower probabilities, as indicated by the typical criteria given in Table 8.1.

Here again, terminology is varied. For instance, in the nuclear power industry the survivability event is generally referred to as the safe shutdown earthquake, and the operating basis earthquake may perhaps be best related to the serviceability limit state despite the nuclear industry's necessary preoccupation with safety.

Effect of workmanship and buildability on reliability

Good workmanship, complying with design requirements, obviously is fundamental to reliability. Designs which are easy to build are more likely to conform to specification than

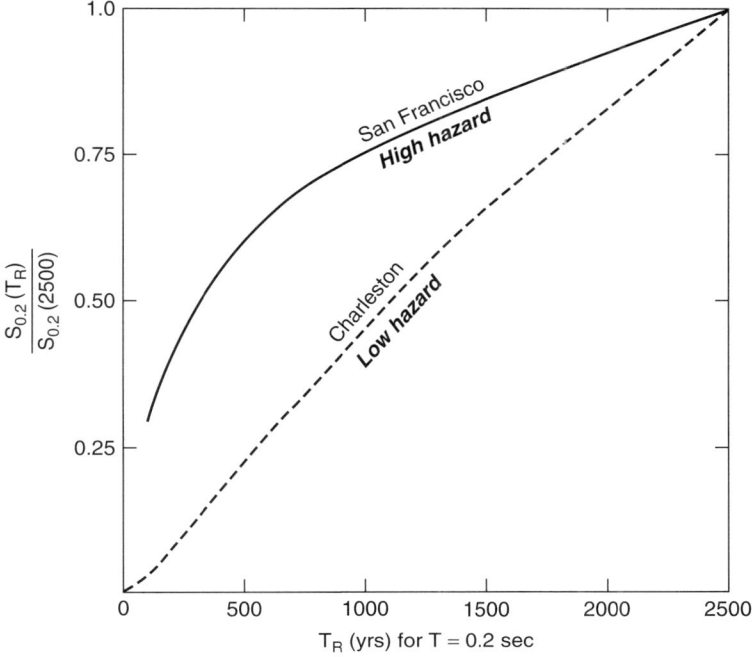

Figure 8.4 Acceleration response spectral ratio $S_{0.2}(T_R)/S_{0.2}$ (2500 years) versus mean return period for a high-hazard and a low-hazard location (derived from Leyendecker *et al.*, 2000)

less buildable construction. While these factors are difficult to quantify, the importance of buildable design in creating reliable earthquake resistance should be borne in mind throughout the design process.

8.3 Principles of Reliable Seismic Behaviour – Form, Material and Failure Modes

8.3.1 Introduction

In seeking the optimum of the proposed construction, designers should choose forms and materials that give the best failure modes in earthquakes within functional and cost requirements. The form or configuration of the construction is the geometrical arrangement of all of the elements, i.e. structure, architecture, equipment and contents. The importance of structural form was first highlighted by the author (Dowrick, 1977), and the baton has been taken up from the architectural point of view by Arnold and Reitherman (1982), and more recently by others such as Giuliani (2000), the Federal Emergency Management Agency (FEMA, 2006) and Charleson (2008). Following studies of the performance of buildings in earthquakes, the Applied Technology Council (1982) concluded that configuration and detailing may play the key roles in providing earthquake resistance, while further confirmation was provided by the 1985 Chilean earthquake (Dowrick, 1985). Obviously, similar principles apply not only to buildings but also to other forms of construction.

Widespread professional recognition of the importance of structural form is reflected in the incorporation of requirements for structural regularity in earthquake code provisions.

To achieve reliable earthquake resistance, the form of construction should be decided from consideration of the following factors:

- simplicity and symmetry;
- length in plan;
- shape in elevation;
- uniformity and continuity;
- stiffness;
- failure modes; and
- foundation conditions.

These topics are discussed below, together with the influence of construction materials and failure mode control on reliability. Foundation conditions are discussed in relation to stiffness and failure modes. In addition to the discussion below, see also Section 8.6 on low-damage structures and Section 13.2, including Figure 13.2.

8.3.2 Simplicity and symmetry

Earthquakes repeatedly demonstrate that the simplest structures have the greatest chance of survival. There are three main reasons for this. First, our ability to understand the overall behaviour of a simple structure is markedly greater than it is for a complex one – for example, torsional effects are particularly hard to predict on an irregular structure. Secondly, our ability to understand simple structural details is considerably greater than it is for complicated ones. Thirdly, simple structures are likely to be more buildable than complex ones.

Symmetry is desirable for much the same reasons. It is worth pointing out that symmetry is important in both directions in plan (Figure 8.5), and helps in elevation as well. Lack of symmetry produces torsional effects which are sometimes difficult to assess, and can be very destructive.

The introduction of deep re-entrant angles into the façades of buildings introduces complexities into the analysis which makes them potentially less reliable than simple forms. Buildings of H-, L-, T- and Y-shape in plan have often been severely damaged in earthquakes, such as the Hanga Roa Building in Viña del Mar in the 1985 San Antonio, Chile, earthquake. This 1970, 15-storey, Y-shaped reinforced concrete building failed at the junction between one of the wings and central core area. Such plan forms should only be adopted if an appropriate three-dimensional earthquake analysis is used in the design.

An asymmetrical shape that can be readily made to work in strong earthquakes is where there are structural walls on three perimeter façades (i.e. in a U-shape) which is common in shops. The torsional moments in the horizontal plane are resisted by the pair of parallel walls. Many such buildings have been subjected to strong shaking in New Zealand earthquakes without collapse (see Section 13.2).

An asymmetrical structural form not shown in Figure 8.5 is that with structural walls on only two (adjacent) perimeter façades. These buildings are referred to as 'corner buildings' because they are usually built on street corners. This form of asymmetry is

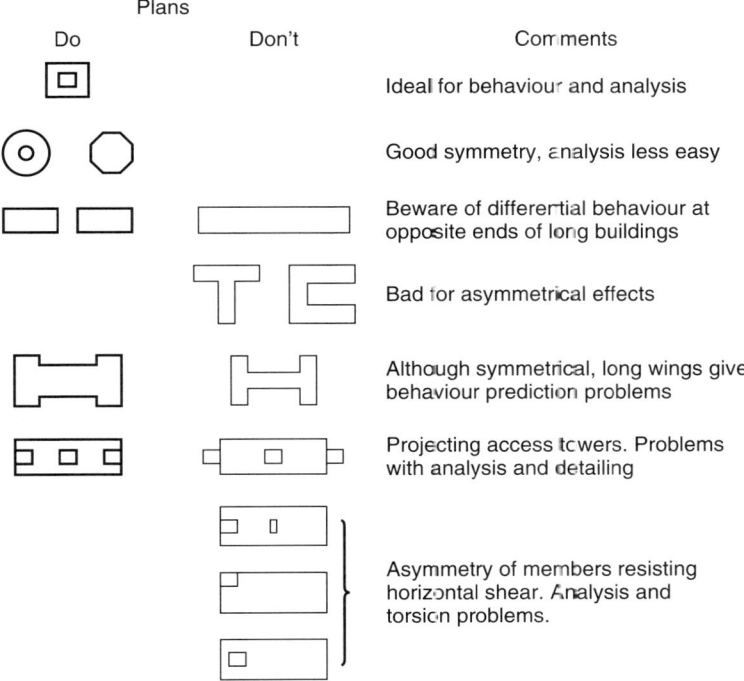

Figure 8.5 Simple rules for plan layouts of seismic buildings (*only* with dynamic analysis and careful detailing should these rules be broken)

to be avoided in high-rise buildings, some of which have collapsed (e.g. in Mexico City in 1985). However, low-rise corner buildings seem not to be especially vulnerable, as all such buildings of up to three storeys have performed well in strong shaking in New Zealand earthquakes (Dowrick and Rhoades, 2000).

External lifts and stairwells provide similar dangers, and should be used with the appropriate attention to analysis and design. In the 1971 San Fernando, California, earthquake external access towers at the Olive View Hospital were not tied into the buildings they were meant to serve, and either collapsed or rotated so far as to be useless.

8.3.3 Length in plan

Structures which are long in plan naturally experience greater variations in ground movement and soil conditions over their length than short ones. These variations may be due to out-of-phase effects or to differences in geological conditions, which are likely to be most pronounced along long bridges where depth to bedrock may change from zero to very large. The effects on structure will differ greatly, depending on whether the foundation structure is continuous, or a series of isolated footings, and whether the superstructure is continuous or not. Continuous foundations may reduce the horizontal response of the superstructure at the expense of push-pull forces in the foundation itself. Such effects should be allowed for in design, either by designing for the stresses induced in the structure or by permitting the differential movements to occur by incorporating movement gaps.

Movement gaps are relatively easy to design in bridge structures, but tend to be unreliable in buildings because of design, workmanship or cost difficulties. Insufficient gap width is often provided, perhaps because the true deformations in the post-elastic state were underestimated. Where adequate gap width is provided, in practice the gaps often become ineffective because of solids such as dirt or builder's rubble blocking them, and hammering between adjacent structures occurs. For example, there were many examples of damage from improper articulation in buildings in the 1985 San Antonio, Chile, earthquake (Dowrick, 1985). Also, Wada *et al.* (1984) have studied examples of collapse of buildings due to battering across movement gaps, analysing the dynamics of battering.

8.3.4 Shape in elevation

As indicated in Figure 8.6, very slender structures and those with sudden changes in width should be avoided in strong earthquake areas. Very slender buildings have high column forces, and foundation stability may be difficult to achieve. Also higher mode contributions may add significantly to the seismic response of the superstructure. Height–width ratios in excess of about 4 lead to less economical structures and require dynamic analysis for proper evaluation of seismic responses. For comparison, in the design of latticed towers for wind loadings, aspect ratios in excess of about 6 become uneconomical.

Sudden changes in width of a structure, such as setbacks in the façades of buildings, generally imply a step in the dynamic response characteristics of the structure at that height, and modern earthquake codes have special requirements for them. If such a shape is required in a structure, it is best designed using dynamic earthquake analysis, in order to determine the stress concentrations at the notch and the shear transfer through the horizontal diaphragm below the notch.

8.3.5 Uniform and continuous distribution of strength, stiffness and mass

This concept is closely related to that of simplicity and symmetry. The structure will have the maximum chance of surviving an earthquake if the following conditions are satisfied:

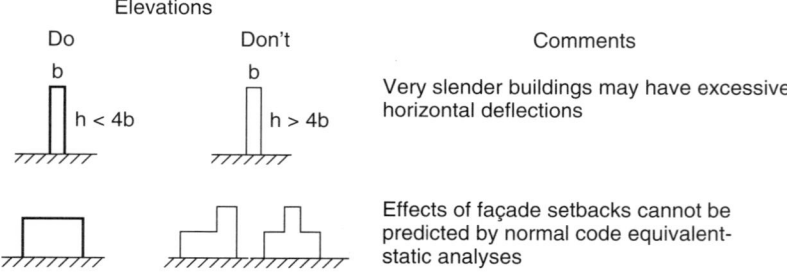

Figure 8.6 Simple rules for elevation shapes of seismic buildings (*only* with dynamic analysis and careful detailing should these rules be broken)

(1) The load-bearing members are uniformly distributed.
(2) All columns and walls are continuous and without offsets from roof to foundation.
(3) All beams are free of offsets.
(4) Columns and beams are coaxial.
(5) Reinforced concrete columns and beams are nearly the same width.
(6) No principal members change section suddenly.
(7) The structure is as continuous (redundant) and monolithic as possible.
(8) There are no irregular or asymmetric large concentrations of mass.

In qualification of the above recommendations, it can be said that while they are not rigidly mandatory they are well proven, and the less they are followed the more vulnerable and expensive the structure will become.

While it can readily be seen how these recommendations make structures more easily analysed and avoid undesirable stress concentrations and torsions, some further explanation may be warranted. The restrictions on architectural freedom implied by the above sometimes make their acceptance difficult. Perhaps the most contentious is that of uninterrupted vertical structure, especially where cantilevered façades and columns supporting shear walls are fashionable. But sudden changes in lateral stiffness up a building are *not* wise (Figure 8.7), first because even with the most sophisticated and expensive computer analysis the earthquake stresses cannot be determined well, and secondly because the demands on effective structural detailing become very high. Severe damage and collapse of buildings with sudden big changes in vertical structure have occurred in many earthquakes, most dramatically in Kobe in 1995 (Park *et al.*, 1995). Sometimes such severe effects are caused by the failure of infill in framed structures, leading to the unintended creation of a soft (weak) storey (Section 12.2.1). During the 1999 Kocaeli earthquake ($M_W = 7.4$) and other recent earthquakes in Turkey it has been observed that buildings with heavy overhangs and heavy balconies sustained heavier damage than those with regular elevations.

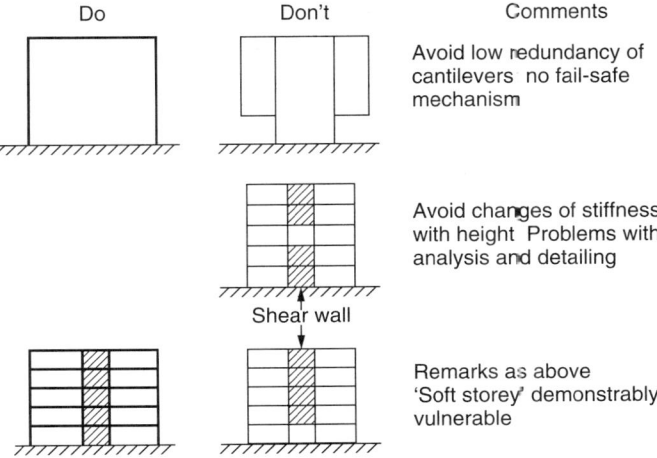

Figure 8.7 Simple rules for vertical frames in seismic buildings

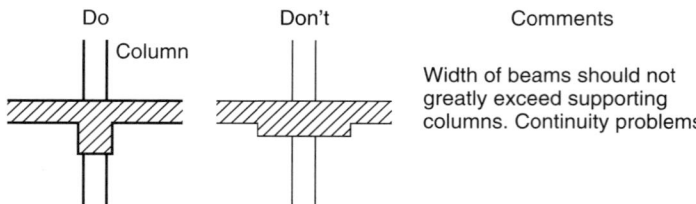

Figure 8.8 Simple rules for widths of beams and columns in seismic reinforced concrete moment resisting frames

Item (5) above recommends that in reinforced concrete structures, contiguous beams and columns should be of similar width. This promotes good detailing and aids the transfer of moments and shears through the junctions of the members concerned. Very wide, shallow beams have been found to fail near junctions with normal-sized columns (Figure 8.8).

The remaining main point worth elaborating is item (7) above, which says that a structure should be as redundant as possible. The earthquake resistance of an economically designed structure depends on its capacity to absorb apparently excessive energy input, mainly in repeated plastic deformations of its members. Hence, the more continuous and monolithic a structure is made, the more plastic hinges and shear and thrust routes are available for energy dissipation. This is why it is so difficult to make precast concrete structures work well for strong earthquake motions.

Making joints monolithic and fully continuous is not only important for energy dissipation; it also eliminates a frequent source of serious local failure due to high local stresses engendered solely by the large movements and rotations caused by earthquakes. This problem can arise in such places as the connection of major beams to slabs or minor beams, and of beams to columns or corbels.

8.3.6 *Appropriate stiffness*

In designing construction to have reliable seismic behaviour, the design of structures to have appropriate stiffness is an important task, often made difficult because so many criteria, often conflicting, may need to be satisfied. The criteria for the stiffness of a structure fall into three categories, i.e. the stiffness is required:

(1) to create desired vibrational characteristics of the structure (to reduce seismic response, or to suit equipment or function);
(2) to control deformations (to protect structure, cladding, partitions, services);
(3) to influence failure modes.

Stiffness to suit required vibrational characteristics

With regard to vibrational characteristics, we note first that it would be desirable in general to avoid resonance of the structure with the dominant period of the site as indicated by the peak in the response spectrum (Figure 3.3). This is particularly true for flexible longer-period structures, while shorter-period structures with ample structural walls can be made to work on any kind of site.

In the case of sites where the soil is soft and deep enough to amplify the lower frequencies, resonance with longer-period structures may occur, and high frequencies may be largely filtered out. The extreme case of this is in the Lake Bed Zone of Mexico City, where in the powerful 1957 and 1985 earthquakes old low-rise unreinforced masonry buildings were undamaged, although adjacent to heavily damaged modern high-rise buildings. A survey of buildings in central Mexico City conducted by the National University (UNAM, 1985; Scawthorn et al., 1986) after the 1985 earthquake found for example that 1% of one- to five-storey buildings were damaged, compared with 14% of nine- to twelve-storey buildings. The partial inverse of this situation is that taller, more flexible structures are suited to rock sites.

Unfortunately, in terms of conventional construction, often it will not be possible to arrange the structure to benefit in this respect. In industrial installations it may be necessary to have very stiff structures for functional reasons, or to suit the equipment mounted thereon, and this will of course override any preference for seismic performance.

However, if we turn to unconventional techniques, notably the use of base isolation (Section 8.5), it is often possible to greatly modify the horizontal vibrational characteristics of a structure whether it is inherently stiff or flexible above the isolating layer. This not only allows the horizontal seismic responses to be greatly reduced, but does not conflict with some functional needs for high stiffness (e.g. nuclear reactors or containment structures are inherently stiff).

Stiffness to control deformation

The importance of deformation control in enhancing safety and reducing damage and thus improving the reliability of construction in earthquakes is now well recognized (Section 8.2). The stiffness levels required to control damaging interaction between structure, cladding, partitions and equipment vary widely, depending upon the nature of components and the function of the construction, but stiff construction is obviously better than flexible in this regard. The seismic deformations of conventional construction can be greatly reduced by the use of seismic isolation (Section 8.5), so that relatively flexible moment resisting frames may be able to satisfy the design deformation criteria, and P-delta column moments will be greatly reduced.

Stiffness affects failure modes

Different levels of stiffness can be created by such widely differing structural configurations that wide differences in potential failure modes arise. In general, stiffer construction implies the existence of less favourable failure modes from an earthquake design point of view, and this needs special design attention, as discussed in Section 8.3.8.

Stiff structures versus flexible

The terms 'stiff' and 'flexible' are relative ones, and must be interpreted with care. Some of their effects depend in part on the height of structure concerned. Table 8.2 summarizes some of the comparative merits of stiff and flexible construction, some of which have been discussed in the earlier parts of this section. A few further points of comparison are highlighted by discussing fully flexible structures.

Fully flexible structures may be exemplified by many modern beam and column buildings, where non-structure has been carefully separated from the frame. No significant shear elements exist, actual or potential: all partitioning and infill walls are isolated from frame movements, even the lift and stair shaft walls are completely separated. The cladding is mounted on rocker and roller brackets (of non-corrosive material). Apart from the points listed in Table 8.2 it has further disadvantages. Floor-to-floor lateral drift and permanent set may be excessive after a moderate earthquake. In reinforced concrete the joint detailing is very difficult. There is no hidden redundancy (extra safety margin) provided by non-structure as in traditional construction.

To overcome the difficulties imposed by the deformability of more flexible construction over the years, there has been a trend to avoid using traditional moment resisting frames by various means such as shear walls (various forms), bracing (various forms), base isolation, and energy absorbing devices. These will:

- reduce lateral drift;
- reduce reinforced concrete joint detailing problems;
- help to ensure that plasticity develops uniformly over the structure;
- prevent column failure in sway due to the P-delta effect (i.e. secondary bending resulting from the product of the vertical load and the lateral deflection).

In conclusion, it can be said that in many situations either a stiff or a flexible structure can be made to work, but the advantages of the two forms need careful consideration when choosing between them.

8.3.7 Choice of construction materials

Reliability of construction in earthquakes is greatly affected by the materials used for the constituent elements of structure, architecture, and equipment. It is seldom possible to use the ideal materials for all elements, as the choice may be dictated by local availability of local construction skills, cost constraints, or political decisions.

Purely in terms of earthquake resistance, the best materials have the following properties:

(1) high ductility;
(2) high strength–weight ratio;
(3) homogeneity;
(4) orthotropy;
(5) ease in making full strength connections.

Generally, the larger the structure, the more important the above properties are. By way of illustration, the applicability of the major structural materials to buildings is given in Table 8.3. The term 'good reinforced masonry' refers to properly detailed hollow concrete blocks as discussed in Section 10.4.4.

Table 8.3 Suitable construction materials for moderate to high earthquake loading

	Type of building		
	High-rise	Medium-rise	Low-rise
Best	(1) Steel	(1) Steel	(1) Timber
Structural materials in approximate order of suitability	(2) *In situ* reinforced concrete	(2) *In situ* reinforced concrete	(2) *In situ* reinforced concrete
		(3) Good precast concrete[1]	(3) Steel
		(4) Prestressed concrete	(4) Prestressed concrete
		(5) Good reinforced masonry[1]	(5) Good reinforced masonry
			(6) Precast concrete
Worst			(7) Primitive reinforced masonry

[1] These two materials only just qualify for inclusion in the medium-rise bracket. Indeed, some earthquake engineers would not use either material in these circumstances.

Most fully precast concrete systems are not well suited for highly ductile earthquake resistance, because of the difficulty of achieving a monolithic and continuous structure.

The order of suitability shown in Table 8.3 is, of course, far from fixed, as it will depend on many things such as the qualities of materials as locally available, the type of structure, and the skill of the local labour in using them.

All these factors being equal, there is arguably little to choose between steel and *in situ* reinforced concrete for medium-rise buildings, as long as they are both well designed and detailed. For tall buildings steelwork is generally preferable, though each case must be considered on its merits. Timber performs well in low-rise buildings, partly because of its high strength–weight ratio, but must be detailed with great care. Further discussion of the use of different materials is given in Chapter 10. Developing countries have special problems in selecting building materials, from the points of view of cost, availability, and technology. Further discussion of these factors is given by Flores (1969).

The choice of construction material is important in relation to the desirable stiffness (Section 8.3.6). It is worth bearing in mind while choosing materials that if a flexible structure is required then some materials, such as masonry, are not suitable. On the other hand, steelwork is used essentially to obtain flexible structures, although if greater stiffness is desired diagonal bracing or reinforced concrete shear panels may sometimes be incorporated into steel frames. Concrete, of course, can readily be used to achieve almost any degree of stiffness.

A word of warning should be given here about the effect of non-structural materials on the structural response of buildings. The non-structure, mainly in the form of partitions, may greatly stiffen an otherwise flexible structure and hence must be allowed for in the structural analysis. This subject is discussed in more detail in Section 8.3.8.

8.3.8 Failure mode control

Failure modes of a complete system

Underlying the principle of failure mode control is the assumption that structural elements of a certain minimum strength will be provided, as required by the strength limit states of codes of practice (Section 8.2). This means that some overall probability of failure should not be exceeded. For structures which are required to be stiff, or for those which are inherently brittle, it may suffice simply to design the structure to remain elastic in the design earthquake, i.e. to conform with item (2b) in Table 8.1.

However, in general, good design not only seeks to keep the overall probability of failure below a given level but also arranges the system such that less desirable modes of failure are less likely to happen than others. This increases the reliability of the design by decreasing the potential for damage and increasing the overall safety. The less desirable modes of failure for structures are:

(1) those resulting in total collapse of the structure (notably through failure of vertical load-carrying members); and
(2) those involving sudden failure (e.g. brittle or buckling modes).

While the above principle is good practice for *any* type of loading, it is particularly important for moderate to strong earthquake loading, because such loading is often so much more demanding on structures than other environmental loadings, and generally involves stress incursions well into the post-elastic range in the parts of the structure. It is therefore highly desirable to control both the location and the manner of the post-elastic behaviour, i.e. to design for failure mode control.

To reduce the probability of occurrence of failure modes (1) and (2) above, earthquake codes commonly have requirements that give added strength (i) to vertical load-carrying elements and (ii) to members carrying significant shear or compressive loads. Figure 8.9 illustrates alternative failure modes for a multi-storey moment resisting frame. Clearly, the column sidesway mechanism is less desirable than the beam sidesway mechanism, as the former will lead to earlier total collapse than the latter. However, while it is possible and desirable to design so that plastic hinges form in beams rather than columns, it is not possible to eliminate plastic hinges from vertical structure completely. A number of potential plastic hinge zones are generally required in the lowest level of columns or walls even in the preferred failure mode, as in Figure 8.9(b).

While beam-hinging failure mechanisms are obviously preferable, the desired configuration for a structure sometimes dictates that a column-hinging failure mode cannot be avoided. In this case, in line with the above philosophy, some earthquake codes require that the structure be designed for a higher level of loading.

In the general case, the number of possible failure modes increases with increasing number of elements, and plastic hinges are likely to form at different locations in different earthquakes (Sharpe and Carr, 1975). Detailing for plastic hinge control may not be sufficient based on solely linear frame analysis, because hinge positions do not necessarily occur only at the locations of maximum moment indicated in a linear analysis.

The number of possible failure modes is substantially reduced by suppressing the chances of occurrence of undesirable failure mechanisms, as discussed above, but some

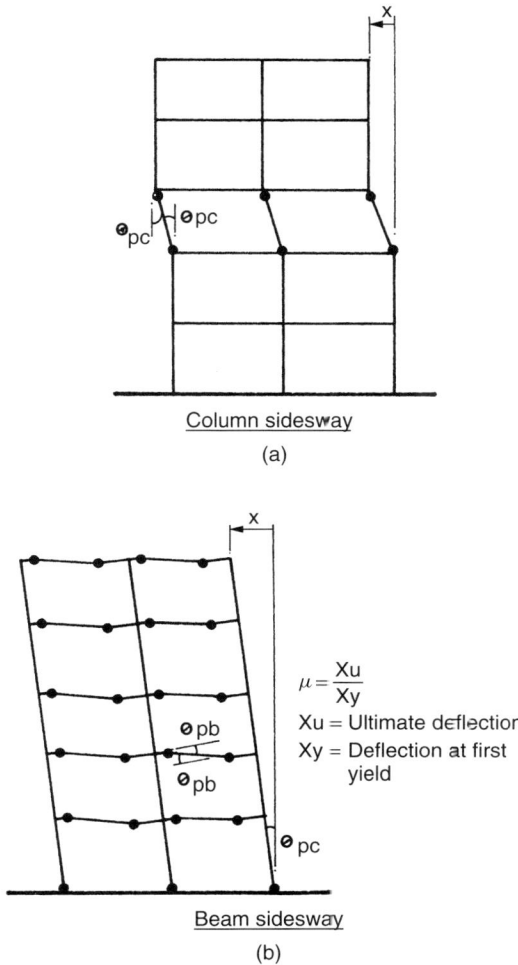

Figure 8.9 Alternative plastic hinge mechanisms for a typical multi-storey frame

uncertainty over the manner of overall failure remains unless failure mode control is systematically carried out for all elements of the construction. In the 1970s a method of doing this began to be developed in New Zealand (Paulay, 1979), where it was called *capacity design*. This procedure requires the designer to impose a mode of overall failure on the structure, which demands that the parts of the structure (mainly beams) that yield in the chosen failure mode are detailed for high energy dissipation, and that the remainder of the structure (mainly columns) has the strength capacity to ensure that no other yielding zones are likely to occur. This principle is straightforward to apply to most structures with few members, but otherwise may be problematical. In theory, it not only helps to maximize safety but, by dictating where the damage will occur, it enables designers to improve the repairability of the structure and interacting elements.

Failure mode control is generally unnecessary for single-storey structures (unless they have abnormally heavy roofs), and is considered to be unnecessary for most structures in regions of low seismic hazard. However, it is being implemented in some form of the capacity design procedure in design codes for moderate and high seismic hazard regions of countries in various parts of the world, including New Zealand, the USA and Europe.

In the foregoing discussion, we have considered how to control failure modes by structuring a system in certain ways. However, these good intentions are often frustrated if elements other than the superstructure, i.e. the part normally analysed for seismic response, are not also appropriately designed and constructed. Thus, it is essential that *non-structure* and *substructure* have suitable forms, as discussed below.

Finally, it is noted that failure mode control will be implemented through effective *workmanship*, and that the *buildability* of the design plays a crucial role.

Non-structure and failure mode control

Non-structural elements have an important role in the reliability or predictability of seismic response of any given type of construction. In considering the form of a structure, it is important to be aware that some items which are normally non-structural become structurally very responsive in earthquakes. This means anything which will interfere with the free deformations of the structure during an earthquake. In buildings the principal elements concerned are cladding, perimeter infill walls and internal partitions. Where these elements are made of very flexible materials, they will not affect the structure significantly. However, very often it will be desirable for non-structural reasons to construct them of stiff materials such as precast concrete, or concrete blocks, or bricks. Such elements can have a significant effect on the behaviour and safety of the structure. Although these elements may be carrying little vertical load, they can act as shear walls in an earthquake with the following important negative or positive effects. They may:

(1) reduce the natural period of vibration of the structure, hence changing the intake of seismic energy and changing the seismic stresses of the 'official' structure;
(2) redistribute the lateral stiffness of the structure, hence changing the stress distribution, sometimes creating large asymmetries;
(3) cause premature failure of the structure usually in shear or by pounding;
(4) suffer excessive damage themselves, due to shear forces or pounding;
(5) prevent failure of otherwise inadequate moment resisting frames.

First, let us consider the negative effects of infill construction. The more flexible the basic structure is, the worse the effects can be; and they will be particularly dangerous when the distribution of such 'non-structural' elements is asymmetric or not the same on successive floors. Stratta and Feldman (1971) discussed some of the effects of infill walls during the Peruvian earthquake of May 1970.

In attempting to deal with the above problems, either of two opposite approaches may be adopted. The first is knowingly to include those extra shear elements into the official structure as analysed, and to detail accordingly. This method is appropriate if the building is essentially stiff anyway, or if a stiff structure is desirable for low seismic response on the site concerned. It means that the shear elements themselves will probably require seismic reinforcement. Thus, 'non-structure' is made into real structure. For notes on the analysis of such composite structures (see Section 5.4.6).

The second approach is to prevent the non-structural elements from contributing their shear stiffness to the structure. This method is appropriate particularly when a flexible structure is required for low seismic response. It can be effected by making a gap against the structure, up the sides and along the top of the element. The non-structural element will need restraint at the top (with dowels, say) against overturning by out-of-plane forces. If the gap has to be filled, a really flexible material must be used. Some advice on the detailing of infill walls is given in Sections 10.4.5 and 12.2

Unfortunately, neither of the above solutions is very satisfactory, as the fixing of the necessary ties, reinforcement, dowels, or gap treatments is time-consuming, expensive, and hard to supervise properly. Also, flexible gap fillers will not be good for sound insulation.

It can be seen from the above discussion that in regions of moderate to high seismic hazard, solid infill walls should not be added or subtracted from existing buildings without checking the earthquake resistance consequences. This may happen by accident in earthquake shaking, as noted in Section 8.3.5.

Finally, the positive side of infill walls (item (5) above) should not be neglected. Many buildings would have had their performance improved by infill in past earthquakes. For example, low-rise pre-code reinforced concrete buildings in the intensity MM10 zone of the 1931 Hawke's Bay, New Zealand, earthquake were evidently saved from much more serious damage by brick infill (van de Vorstenbosch et al., 2002). Many examples of this behaviour were also observed in the 2001 Bhuj, India, earthquake, and in New Zealand earthquakes (Section 13.2).

While the positive stress redistribution of infill panels is readily explained (Sritharan and Dowrick, 1994), all the pros and cons are not easily predicted. However, it appears that simple unreinforced infill, disposed symmetrically or U-shaped in plan, and built full wall height is more likely to be beneficial than not, if it does not fall out prematurely.

Substructure and failure mode control

Although the form of the substructure must have a strong influence upon the seismic response of structures, little comparative work has been done on this subject. The following notes briefly summarize what appears to be good practice at the present time.

The basic rule regarding the earthquake resistance of substructure is that *integral action* in earthquakes should be obtained. This requires adequate consideration of the dynamic response characteristics of the superstructure and of the subsoil. If a good seismic resistant form has been chosen for the superstructure (Sections 8.3.1–8.3.6) then at least the plan form of the substructure is likely to be sound, i.e.:

(1) vertical loading will be symmetrical;
(2) overturning effects will not be too large;
(3) the structure will not be too long in plan.

As with non-seismic design, the nature of the subsoil will determine the minimum depth of foundations. In earthquake areas this will involve consideration of the following factors:

(a) transmission of horizontal base shears from the structure to the soil;
(b) provision for earthquake overturning moments (e.g. tension piles);
(c) differential settlements (Figure 8.10);
(d) liquefaction of the subsoil;
(e) the effects of embedment on seismic response.

The effects of depth of embedment are not easy to evaluate reliably, but some allowance for this effect can be made in soil–structure interaction analyses (Section 5.3), or when determining at what level to apply the earthquake loading input for the superstructure analysis.

Three basic types of foundation may be listed as:

- discrete pads;
- continuous rafts;
- piled foundations.

Piles, of course, may be used in conjunction with either pads or rafts. Continuous rafts or box foundations are good seismic forms only requiring adequate depth and stiffness. Piles and discrete pads require more detailed consideration in order to ensure satisfactory integral action which deals with so many of the structural requirements implied in (1)–(3) and (a)–(e) above. Integral action should provide sufficient reserves of strength to deal with some of the differential ground movements which are not explicitly designed for at present. Where a change of soil type occurs under a structure (Figure 8.10), particular care may be necessary to ensure integral substructure action.

This discussion of substructure form is applicable to structures on softer soils only, as structures on rock are naturally integral per media of the rock itself. For a more detailed discussion of foundation design, see Section 9.1.

Finally, it is noted that piled foundations offer a special opportunity for failure mode control through base isolation, as discussed in Section 8.5.4.

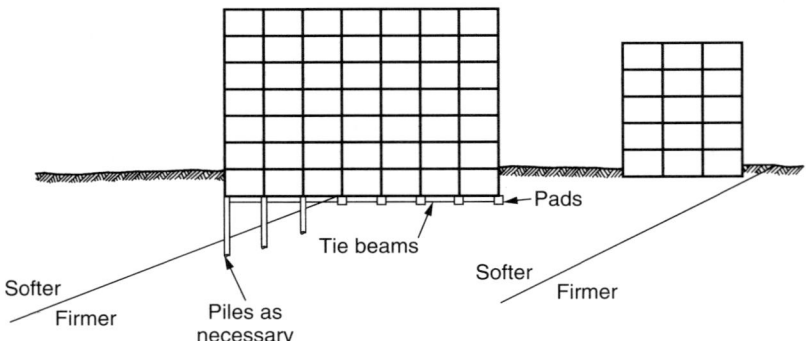

Figure 8.10 Typical structures founded on two types of soil, requiring precautions against differential seismic movements

8.4 Specific Structural Forms for Earthquake Resistance

In the preceding sections of this chapter, we have considered the principles underlying good earthquake resistant design, which should be applied to the specific structural forms utilized for a given project. The various structural forms in use around the world all have their strong and weak points, conforming better to some of the above principles than others. The main structural forms suitable for earthquake resistance are:

(1) moment resisting frames;
(2) framed tube structures;
(3) structural walls (shear walls);
(4) concentrically braced frames;
(5) eccentrically braced frames;
(6) hybrid structural systems.

Design details of various aspects of the above forms are discussed under the relevant construction materials in Chapter 10, while a general overview of the seismic resistant attributes of these forms is given below.

8.4.1 Moment resisting frames

Moment resisting frames comprise one of the commonest forms of modern structure, in widespread use in buildings and industrial structures. Their great advantage for seismic resistance is that, by definition, they avoid potentially brittle shear failure modes, but they tend to sway excessively. They are made from steelwork, concrete and timber.

8.4.2 Framed tube structures

The framed tube system is a special case of the moment resisting frame, which usually consists of closely spaced wide steel columns combined with relatively deep beams. These frames are usually, but not only, located on the perimeter of the structure, and introduce more stiffness to overcome the problems of excessive horizontal deflection of orthodox moment resisting frames, at the expense of a reduction in ductility. They have been widely used for tall buildings in high wind regions since the 1960s, and more recently have been used in earthquake zones (Amin and Louie, 1984). The framed tube system may be seen as a compromise between 'pure' moment resisting and shear structures.

8.4.3 Structural walls (shear walls)

The term *structural walls* or *shear walls* refers to structures in which the resistance to horizontal forces is principally provided by walls. These walls are usually constructed of concrete, masonry, timber or steel, while other lesser structural materials such as gypsum, or composites are also encountered.

As mentioned earlier in this chapter, the great advantage of structural walls is the protection their natural stiffness offers to non-structure through limiting inter-storey deflections.

Earlier designs of concrete structural walls exhibited classical brittle shear failure modes in some earthquakes, particularly the 1964 Alaska event. Subsequent research has shown how these walls should be designed to overcome this problem, through appropriate reinforcing of ordinary *cantilever walls* or through the use of concrete *coupled walls* (Section 10.4.4).

Coupled walls make special use of lintel beams between adjacent walls (Figure 10.30), such that these coupling beams have ductility and energy-dissipating characteristics, which help to protect the walls from excessive damage.

8.4.4 Concentrically braced frames

Concentrically braced frames are here defined as those where the centre-lines of all intersecting members meet at a point (Figure 10.11). This traditional form of bracing is, of course, widely used for all kinds of construction such as towers, bridges, and buildings, creating stiffness with great economy of materials in two-dimensional trusses or three-dimensional space frames. Concentrically braced frames are constructed from steel, timber and concrete, and composite forms are frequently met such as timber beams and columns with steel diagonals (Figure 10.67).

The bracing may take the form of either a single diagonal in a bay or a double bracing in an X shape (Figure 10.11). Braced frames have the advantage over moment resisting frames of having smaller horizontal deflections in moderate earthquakes, but are more inclined to undesirable buckling modes and have less reliable ductility.

If the diagonals are very slender and hence capable of tensile resistance only, as is often the case in steel construction, the seismic resistance is not as good as when the bracing is capable of compressive as well as tensile resistance. This is partly because in tension-only bracing there is a greater tendency for incremental permanent deflections to occur in one direction only. Also, shock loadings tend to occur as bracings straighten from the buckled zero-load state to the tensile load-carrying state.

8.4.5 Eccentrically braced frames

Traditional design of trussed structures lays great importance on keeping the forces in the structure to axial only, avoiding moments by ensuring that the centre-lines of all intersecting members meet at a point, i.e. concentrically (Section 8.4.4). However, starting in the late 1970s, the concept of using deliberately eccentric bracing for earthquake resistance purposes has been found to have certain advantages, so far principally for steel structures, with major structures being designed this way (Figure 10.12).

In eccentrically braced frames the axial forces in the braces are transmitted to the columns through bending and shear in the beams, and, if designed correctly, the system possesses more ductility than concentrically braced frames while retaining the advantage of reduced horizontal deflections which braced systems have over moment resisting frames. This system conforms in part to the requirement for good earthquake design of *failure mode control* (Section 8.3.8), in so far as post-elastic behaviour of the frame is largely confined to selected portions of the beams and sudden failure modes are suppressed. However, the issues of the degree of damage that will be incurred in the floors, and the repairability of the floors and the beams, need to be considered for any given building. As shown in Figure 8.11, eccentrically braced frames subjected to lateral sway

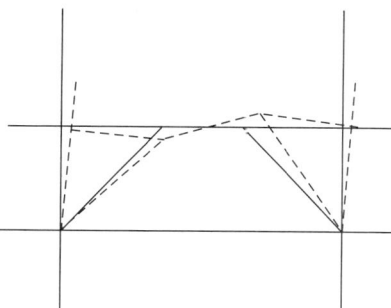

Figure 8.11 Deformed shape of eccentrically braced frame subjected to lateral sway, showing potential for substantial secondary damage to floors and non-structure. Reproduced by permission of the Earthquake Engineering Reseach Institute

deform in such a way as to cause distress in the floors and non-structural elements. The 2001 Nisqually, USA, earthquake pointed out some such drawbacks. Even though the maximum intensity was only about MM7, the secondary damage to external wall claddings, internal walls and ceilings was very expensive to repair.

8.4.6 Hybrid structural systems

Structures are often built in which the lateral resistance is provided by more than one of the above methods. The most common of these hybrid systems are those in which moment resisting frames are combined with either structural walls or diagonally braced frames.

While hybrid systems are often unavoidable and can provide good seismic resistance, care must be taken to ensure that the structural behaviour is correctly modelled in the analysis. Interaction between the different components can be large, and is not necessarily obvious, and many papers have been written on this subject. For example, for low-rise buildings it may be reasonable in many cases to assume that the walls or the braced bays resist the entire horizontal earthquake load, and the moment resisting frame is not required to resist horizontal earthquake forces. However, deformations are still imposed on the moment resisting members, requiring some seismic design consideration such as detailing for ductility.

For a given plan layout, the contribution of the moment resisting frame to lateral load resistance increases with height of structure, so that while the walls may take most of the horizontal shear in low-rise buildings, the moment resisting frame becomes the dominant partner for very tall buildings (Ghoubhir, 1984).

8.5 Passive Control of Structures – Seismic Isolation and Energy-Dissipating Devices

8.5.1 Introduction

Earthquake ground motions impart kinetic energy into structures, and the principles outlined above seek to control the location and extent of the damage caused by this energy.

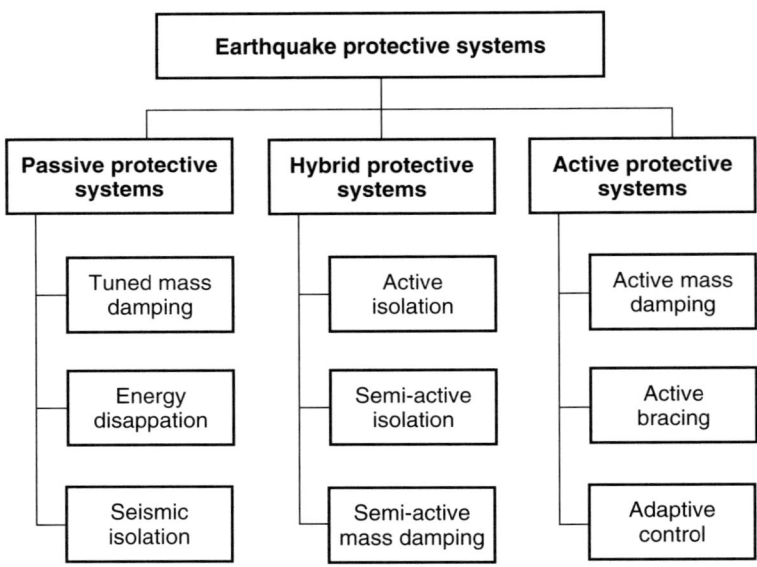

Figure 8.12 Family of earthquake protective systems (from Buckle, 2000)

This philosophy can be extended beyond the structural forms described in Section 8.4 to any means which may further protect the structure by reducing the amount of energy which enters it.

The family of earthquake protective systems has grown to include passive, active and hybrid (semi-active) systems as shown in Figure 8.12. Passive systems are the best known and these include seismic (base) isolation and passive (mechanical) energy dissipation. Seismic isolation is the most developed member of the family at the present time with continuing developments in hardware, applications, design codes (e.g. FEMA, 1997) and retrofit manuals (see Chapter 13).

The innovative research work on devices for the passive control of structures seismic loads was done in the New Zealand Physics and Engineering Laboratory starting in the 1970s, as described by Skinner *et al.* (1993). The latest book on the seismic isolation and energy absorbing devices is by Kelly *et al.* (2008). Applications are to now be found in almost all of the seismically active countries in the world, but mainly in New Zealand, USA, Japan and Italy.

According to Robinson (1998, p. 63):

> Very strong support for the principles of seismic isolation is given by the results of the January 1994 Los Angeles earthquake. The fact that of the ten hospitals affected by the Los Angeles earthquake, only the hospital seismically isolated by a lead-rubber bearing system was able to continue to operate. This seven-storey hospital (the University of Southern California Teaching Hospital) underwent ground accelerations of 0.49g, while the rooftop acceleration was 0.21g, that is attenuation by a factor of 1.8. The Olive View Hospital, nearer to the epicentre of the earthquake, underwent a top floor acceleration of 2.31g compared with its base acceleration of 0.82g, a magnification by a factor of 2.8.

It is interesting to note that the Kobe (Hyogo-ken Nanbu) earthquake of January 17, 1995 led to a sudden and significant change in application of passive control technologies for seismic design in Japan. In the three-year period prior to the 1995 earthquake, 15 seismically isolated buildings were licensed for construction. In the three years following the earthquake, 450 isolated buildings were approved, and this trend continued.

8.5.2 *Isolation from seismic motion*

The principle of isolation is simply to provide a discontinuity between two bodies in contact so that the motion of either body, in the direction of the discontinuity, cannot be fully transmitted. The discontinuity consists of a layer between the bodies which has low resistance to shear compared with the bodies themselves. Such discontinuities may be used for isolation from horizontal seismic motions of whole structures, parts of structures, or items of equipment mounted on structures. Because they are generally located at or near the base of the item concerned, such systems are often referred to as *base isolation* (Figure 8.13), although the generic term *seismic isolation* is preferable.

The layer providing the discontinuity may take various forms, ranging from very thin sliding surfaces (e.g. PTFE bearings), through rubber bearings a few centimetres thick, to flexible or lifting structural members of any height. To control the seismic deformations which occur at the discontinuity, and to provide a reasonable minimum level of damping to the structure as a whole, the discontinuity must be associated with energy-dissipating devices. The latter are also usually used for providing the required rigidity under serviceability loads, such as wind or minor earthquakes. Because substantial vertical stiffness is generally required for gravity loads, seismic isolation is only appropriate for horizontal motions.

The soft layer providing discontinuity against horizontal motions cannot completely isolate a structure. Its effect is to increase the natural periods of vibration of the structure, and to be effective the periods must be shifted so as to reduce substantially the response of the structure. For example, a three-storey building might typically have its fundamental period shifted from 0.3 s to 2.0 s by being changed from fixed base to isolated. If the structure were located on a rock site, and had a design response spectrum as for rock in Figure 3.3(a), this would reduce the elastic response of the structure by a factor of about 10. However, a similar period shift for a structure on soft soil might not achieve a worthwhile reduction in response, or could even result in an increased response, as may be inferred from the spectra for softer soil sites in Figure 3.3. Clearly, the shape of the design spectrum, the fixed base period and the period shift are the three factors which determine whether base isolation has any force-reducing effect (or, indeed, the opposite!).

The *location of the isolating devices* should obviously be as low as possible to protect as much of the structure as possible. However, cost and practical considerations influence the choice of location. On bridges it is generally appropriate to isolate only the deck where isolation from thermal movements is required anyway. In buildings the choice may lie between isolating at ground level, or below the basement, or at some point up a column. Each of these locations has its advantages and disadvantages relating to accessibility and to the very important design considerations of dealing with the effects of the shear displacements on building services, partitions, and cladding, as described in Figure 8.13, which was derived from Mayes *et al.* (1984).

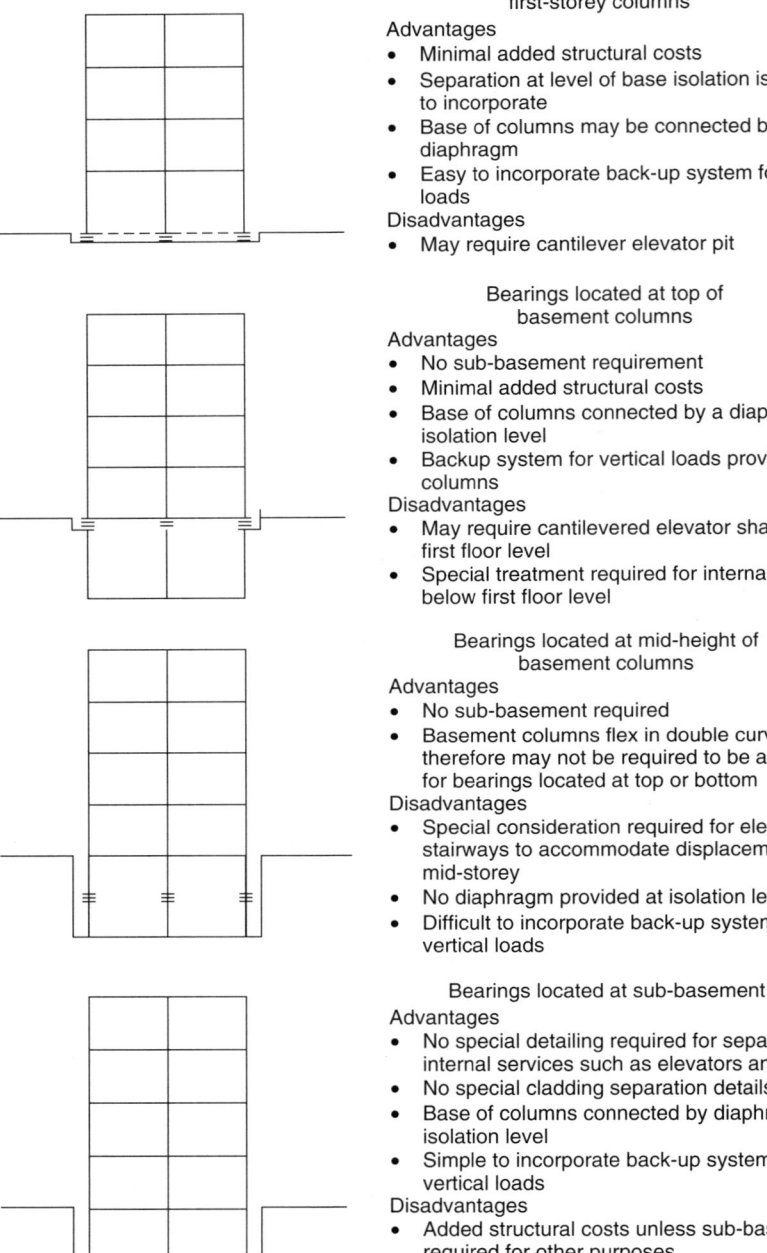

Figure 8.13 Different locations for base isolation of buildings (from Mayes *et al.*, 1984). (Reproduced by permission of the Earthquake Engineering Research Institute)

8.5.3 Seismic isolation using flexible bearings

The most commonly used method of introducing the added flexibility for seismic isolation is to seat the item concerned on either rubber or sliding bearings. The energy dissipators (dampers) that must be provided may come in various forms. For use with standard bridge-type bearings made of rubber or sliding plates, any of the energy dissipators mentioned in Section 8.5.6 may be suitable. In addition, all-in-one devices, incorporating both isolation and damping, are used, namely lead-rubber and high damping rubber bearings.

The most effective device, the *lead-rubber bearing* (Robinson and Tucker, 1977), is conceptually and practically very attractive for seismic isolation, as it combines all of the required design features of flexibility and deflection control into a single component. As shown in Figure 8.14, it is similar to the laminated steel and rubber bearings used for temperature effects on bridges, but with the addition of a lead plug energy dissipator. Under cyclic shear loading the lead plug causes the bearing to have high hysteretic damping behaviour, of almost pure bilinear form (Figure 8.15). The high initial stiffness is likely to satisfy the deflection criteria for serviceability limit-state loadings, while the low post-elastic stiffness gives the potential for a large increase in period of vibration desired for the ultimate limit state design earthquake.

As shown by Tyler and Robinson (1984), hysteretic behaviour is very stable under increasing cyclic displacements. In dynamic tests on bearings $280 \times 230 \times 113$ mm in size, shear displacements of up to ± 140 mm at frequencies of 0.1–0.3 Hz were applied,

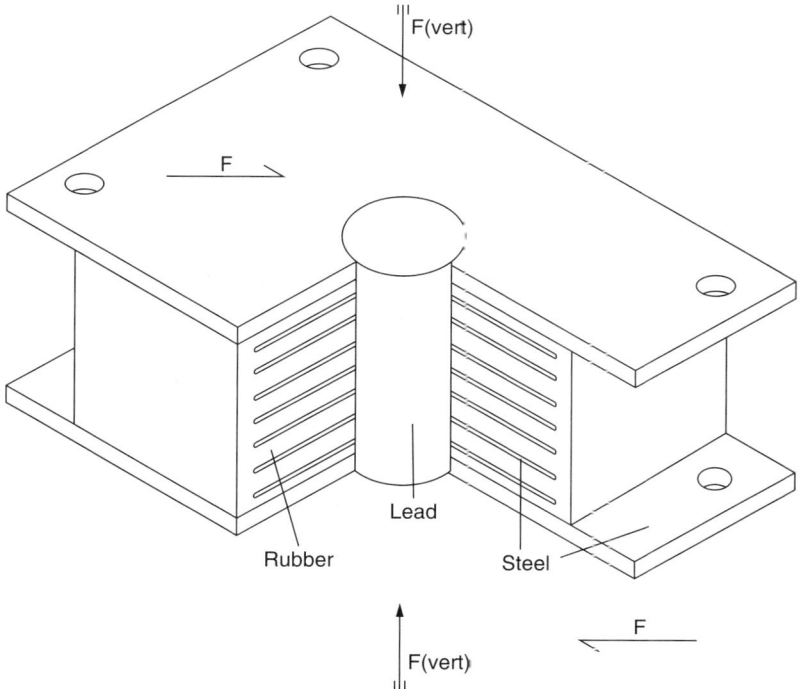

Figure 8.14 Construction of a patented lead-rubber bearing (after Robinson and Tucker, 1977)

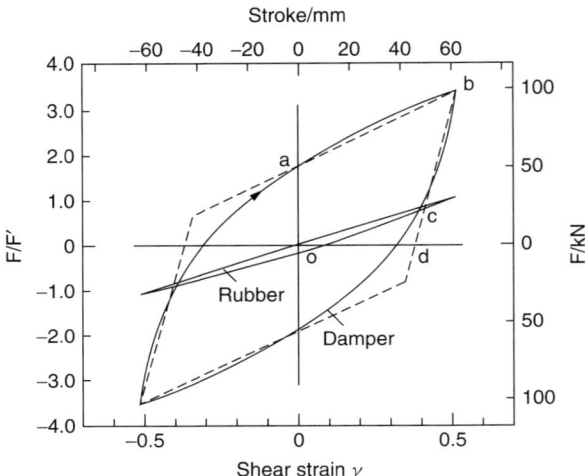

Figure 8.15 Typical hysteretic behaviour of a lead-rubber bearing (after Robinson and Tucker, 1977)

giving shear strains in the rubber of up to ±200%. The weight of the structure on the bearings ranged from 35 to 455 kN. It was concluded that with peak strains in the rubber in excess of 100%, the bearings would continue to function satisfactorily for a sequence of very large earthquakes.

The first building in the world to be built using lead-rubber bearings for seismic isolation was the William Clayton Building (Megget, 1978) in Wellington, New Zealand, designed c. 1978. It has a four-storey ductile moment resisting frame, a section through the building being shown in Figure 8.16. The inter-storey drifts calculated for the isolated building were about 10 mm, and were uniform over the building's height. For comparison, the maximum drift for the non-isolated model was 52 mm per storey for the top two storeys. An overall deflection ductility factor of only $\mu = 1.6$ (μ is defined in Section 5.4.7) was required for the isolated building, whereas $\mu = 7.6$ would have been required for the non-isolated condition. A more recent example of a structure protected by lead-rubber bearings is discussed in Section 11.3.3.

Lead-rubber bearings have also been used in a rapidly growing number of bridges in New Zealand, Italy, Japan, the USA and elsewhere. These bearings have a wide range of applications where they are likely to lead not only to less damaged structures in earthquakes, but also sometimes to cheaper construction. Design procedures are well established, including methods using design graphs (e.g. Skinner et al., 1993) and design codes (e.g. FEMA, 1997).

Further developments in seismic isolation are going on, as described, for example, by Robinson (2000). One such development is an improvement to both the rubber isolation bearing and to the lead-rubber bearing, consisting of a centre-drive to the top and bottom of the bearings. The 'centre-drive' approach has two advantages: first, it allows a greater displacement to height ratio; and second, it provides increased damping capacity at large displacements, thus providing some of the additional damping needed for resisting 'near-fault fling'.

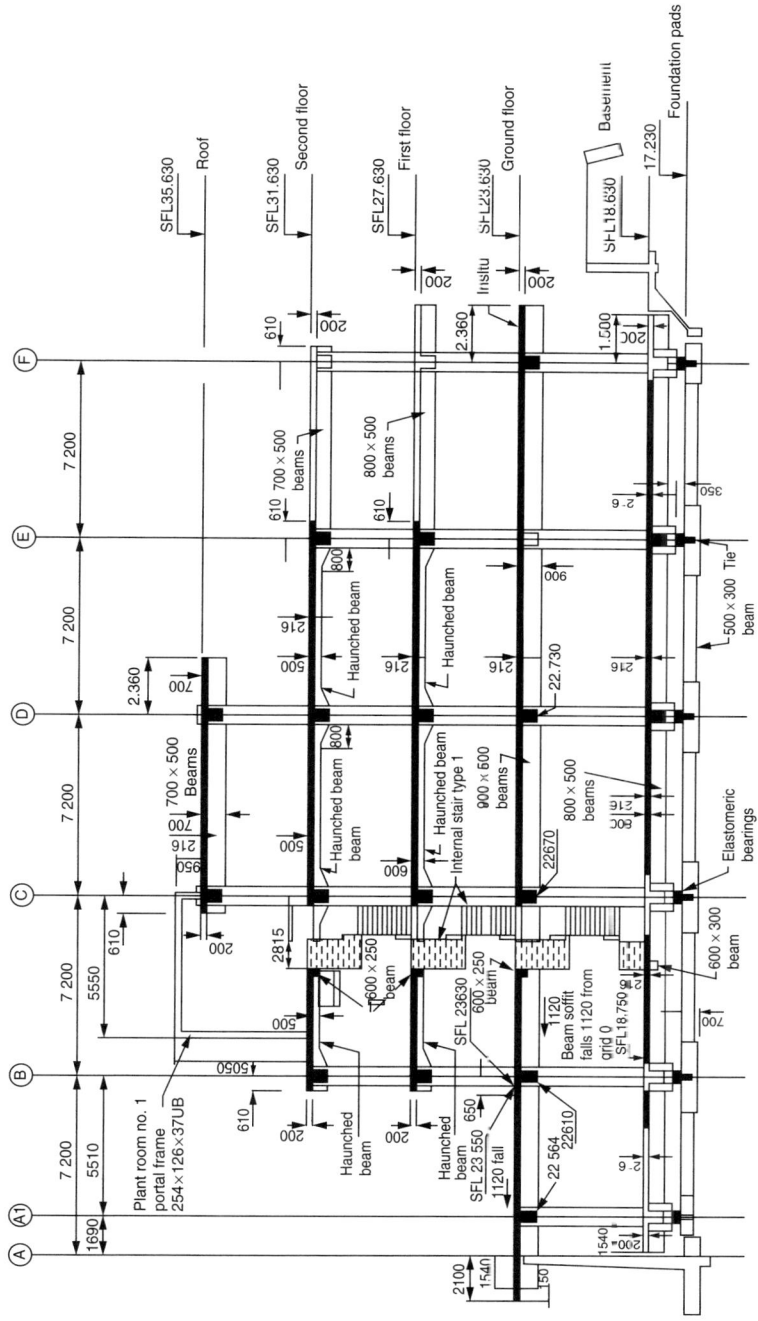

Figure 8.16 Section through the William Clayton Building, Wellington, New Zealand, the first building to be seismically isolated. The lead-rubber bearings are shown beneath the basement (after Megget, 1978)

8.5.4 Seismic isolation using flexible piles and energy dissipators

An interesting alternative to the use of lead-rubber bearings is the isolation system used first for Union House (Boardman *et al.*, 1983), a 12-storey office block in Auckland, New Zealand, completed in 1983 (Figure 8.17). As the building required end-bearing piles about 10 m long, the designers took the opportunity of making the piles flexible and separating them from lateral contact with the soft soil layer overlying bedrock by surrounding them with a hollow sleeve, thus creating the flexibility required for base isolation. Deflection control was imposed by tapered steel energy dissipators (Figure 8.18) located at ground level. The structure was built of reinforced concrete except that the superstructure was diagonally braced with steel tubes. Lateral flexibility of the piles was attained by creating hinges of low moment resistance at the top and bottom of each pile.

The earthquake analysis was carried out using non-linear dynamic analysis. Under the design earthquake loading the horizontal deflection of the first floor relative to the ground (i.e. at the dissipators) was calculated to be ±60 mm. The response of the building was also checked under a 'maximum credible earthquake' to ensure that adequate clearance was provided at the energy dissipators, and that no significant yielding would occur in the superstructure. In this survivability state the horizontal deflection at the dissipators was ±130 mm and a provision for ±150 mm was made. Because of the structural discontinuity at ground level, the lift shaft and the bottom story façade had to be supported from the first floor above ground level.

Another building protected in the same way is the 10-storey Wellington Central Police Station (Charleson *et al.*, 1987), built in the 1980s. The isolation system enables achievement of the design aim of making it fully operational after the expected next magnitude 7.5 earthquake on the Wellington fault located about a kilometre away. Cost and time comparisons of the isolated and non-isolated equivalent structure estimated a capital cost saving of $300,000 and a construction time saving of three months, representing $150,000. Together these equal a substantial saving of nearly 7% in the total construction cost.

8.5.5 Rocking structures

As well as the methods described in the preceding sections, the flexibility required to reduce seismic response may be obtained by allowing part of the structure to lift during large horizontal motions. This mechanism is referred to variously as uplift, rocking, or stepping, and involves a discontinuity of contact between part of the foundations and the soil beneath, or between a vertical member and its base. The good performance of many ordinary structures in very strong ground shaking can only be explained by rocking having beneficially occurred during the earthquake. For example, this appears to have been the case for some pre-code low-rise brittle concrete buildings in the near-fault region of the Hawke's Bay, New Zealand, earthquake of 1931 (van de Vorstenbosch *et al.*, 2002).

As early as the 1970s, various studies (e.g. Meek, 1975) were made of 'natural' rocking systems, i.e. those not incorporating displacement-controlling energy dissipators relying solely on uplifting columns or rocking of raft or local pad foundations to produce the desired effects. However, despite apparently favourable results, such structures have not yet been enthusiastically adopted in practice. This is probably due to continuing design uncertainties regarding factors such as soil behaviour under rocking foundations in the design earthquake, the possible overturning of slender structures in survivability events,

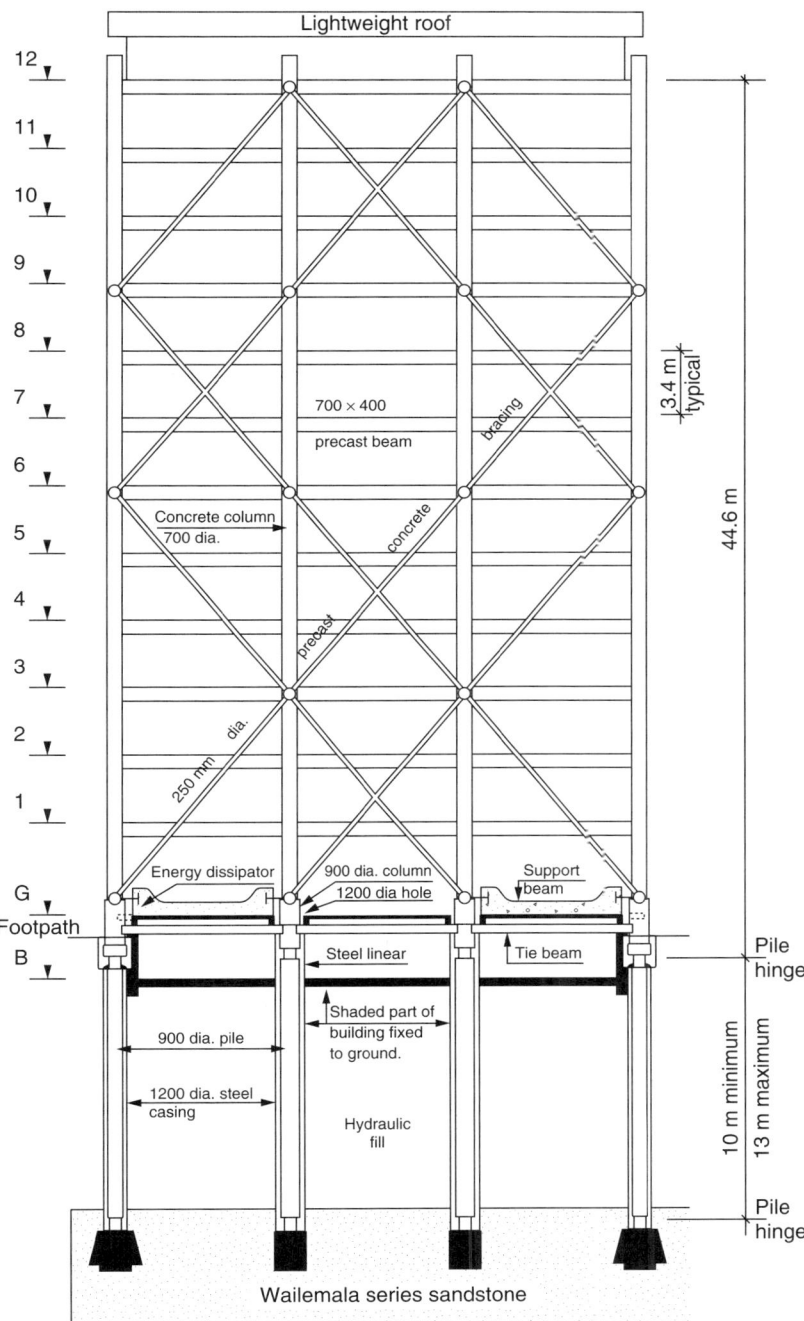

Figure 8.17 Section through Union House, Auckland, New Zealand, showing isolating piles and energy dissipators (after Boardman *et al.*, 1983)

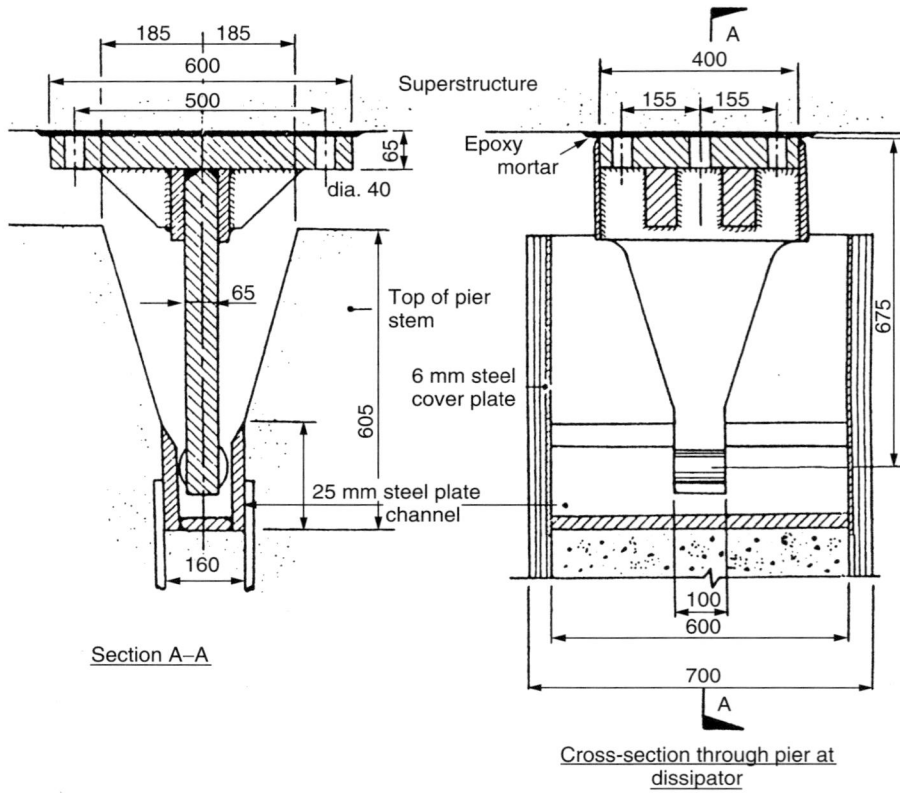

Figure 8.18 Cantilever steel plate dampers of the type used in Union House and Dunedin Motorway Bridge (Skinner *et al.*, 1980)

or possible impact effects when the separated interfaces slam back together (Meek, 1975). However, based on field experience like that of the Hawke's Bay earthquake cited above (van de Vorstenbosch *et al.*, 2002), more boldness in allowing rocking of squat structures seems to be justified.

With the addition of energy absorbers the above hazards are lessened, and utilization of the advantageous flexibility of uplift has been put to practical effect in completed constructions, a bridge and chimney being discussed below.

The first such structure to be built was the South Rangitikei Railway Bridge in New Zealand, the design of which was carried out *c.* 1971. The bridge deck is 320 m long, comprising six prestressed concrete spans, about half of which is at a height of 70 m above the riverbed. The piers consist of hollow reinforced concrete twin shafts 10.7 m apart coupled together with cross beams at three levels, so that they act as a kind of portal frame lateral to the line of the bridge. At their base, these shafts are seated on an elastomeric bearing (Figure 8.19) and lateral rocking of the portals is possible under the control of a steel torsion-beam energy dissipator of the type shown in Figure 8.19. As with other forms of base isolation, substantial reductions in earthquake stresses are possible, as

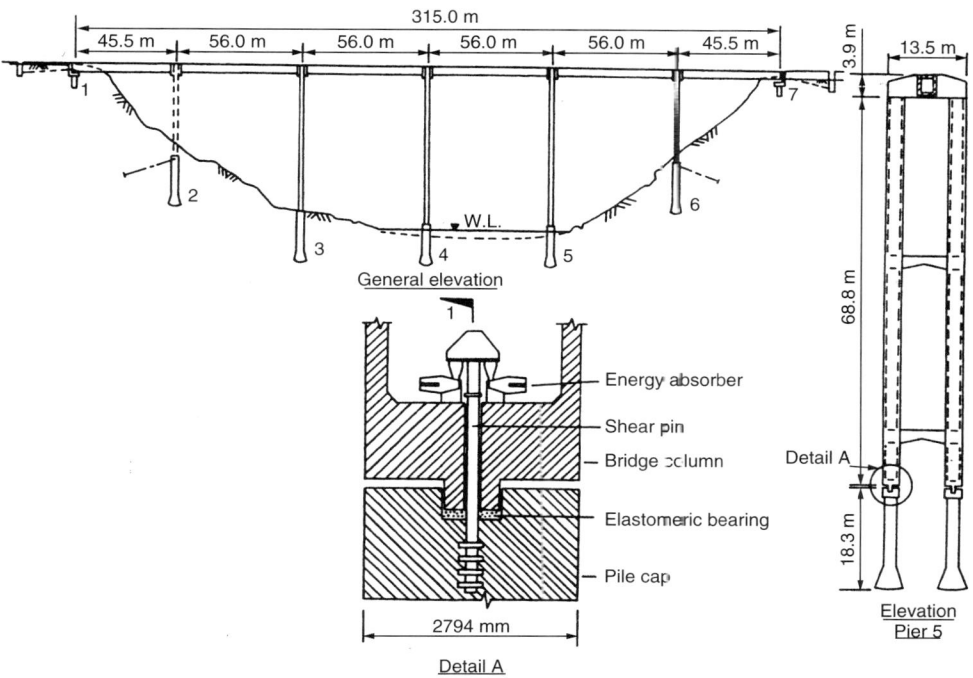

Figure 8.19 South Rangitikei railway bridge, New Zealand, showing locations of bearings and torsion-beam energy dissipators (Skinner *et al.*, 1980)

described for an early investigation of this bridge by Beck and Skinner (1974), the final configuration differing in detail but not in principle.

A further example of the use of rocking in seismic design is given in Section 8.6.

8.5.6 Energy dissipators for seismically isolated structures

In the preceding sections on isolation methods, we have discussed a number of energy dissipators (dampers) that have been used with seismically isolated structures, namely:

(1) lead plugs, in lead-rubber bearings (Figure 8.14);
(2) tapered steel plate cantilevers (Figure 8.18);
(3) steel torsion-beams (Figure 8.20).

A variety of other devices have been investigated which are also suitable in this situation:

(4) lead extrusion devices;
(5) flexural beam dampers.

A general overview of some of these energy dissipators is given by Skinner *et al.* (1993) and Hanson and Soong (2001).

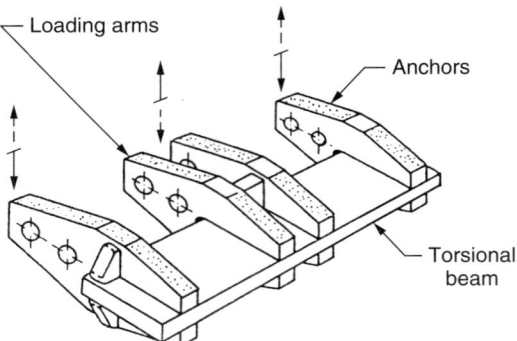

Figure 8.20 A torsion beam hysteretic damper. The arrows show the opposing actions of the structure and its support (Kelly et al., 1972)

8.5.7 Energy dissipators for non-isolated structures

As discussed above, energy-dissipating devices are an essential component of seismic isolation systems, and they also may be used to reduce seismic stresses in non-isolated structures. Various forms of energy-dissipating devices have been developed for such structures, some of which are discussed below. In addition to the discussion below, the reader is referred to the book by Skinner et al. (1993), and one specifically on energy-absorbing devices by Hanson and Soong (2001).

Energy dissipators in diagonal bracing

Diagonal bracings incorporating energy dissipators control the horizontal deflections of the frame and also the locations of damage, thus protecting both the main structure and non-structure. A practical example is provided by a six-storey government office building constructed in Wanganui, New Zealand, in 1980. This building obtains its lateral load resistance from diagonally braced precast concrete cladding panels, thus minimizing the amount of internal structure to ease architectural planning. Each diagonal brace contains a steel insert consisting of a sleeve housing a specially fabricated steel tube 90 mm diameter and 1.4 m long, which was designed to yield axially at a given load level. A movement gap was provided through the surrounding structure, and buckling was prevented by the surrounding sleeve and concrete.

A number of devices to be connected to diagonal steel bracing show high energy-absorbing capabilities, such as the lead extrusion damper (Skinner et al., 1993) also used in base isolated structures (Section 8.5.6). Pall and Marsh (1982) developed friction damped devices to suit both X- and K-bracing (but the reliability of the friction forces needs to be established).

A device developed in the late 1990s (Monti et al., 1998) is a lead-shear damper which, unlike most other seismic dampers, is suitable for damper vibrations from other sources such as wind. For example, one such device has design working displacements in the range of 50 μm to 10 mm. The effectiveness of its hysteretic damping at three levels of displacement is shown in Figure 8.21.

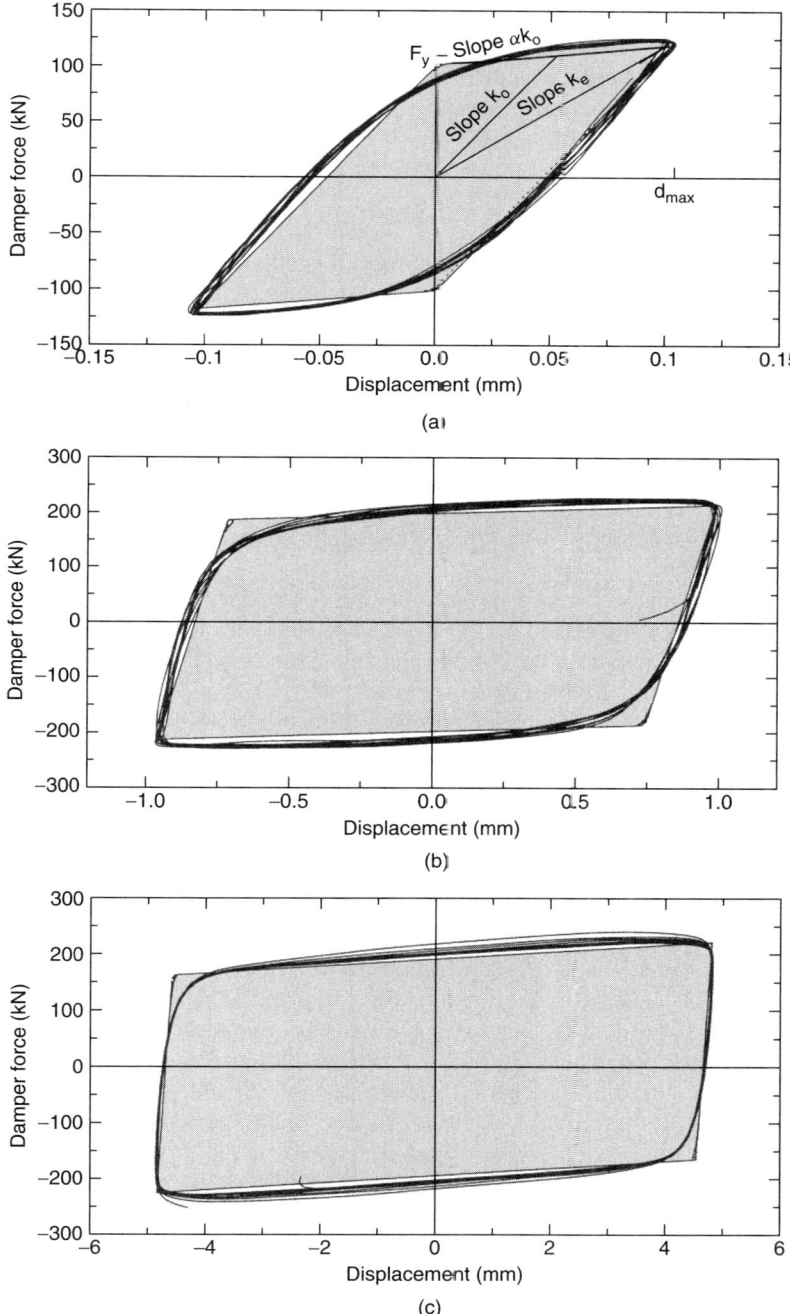

Figure 8.21 Hysteresis loops of the Penguin Vibration Damper and its bilinear-spring model (shaded) (from Monti et al., 1998): (a) $d_{max} = 0.1$ mm; (b) $d_{max} = 1.0$ mm; (c) $d_{max} = 4.9$ mm

Energy dissipation in bolted beam–column joints

If bolted joint interfaces are designed to permit controlled sliding rotations in earthquake motion, not only is energy dissipated, but also damage is limited or eliminated. The development and behaviour of promising devices of this type are described by Clifton (2005) and Clifton *et al.* (2007).

8.6 Low-Damage Structures – Damage Avoidance Design

Much work has been done in recent years to find economical ways of creating structures which suffer low damage in design-level shaking, using some of the devices discussed above as well as subsequent ideas. For example, efforts have been made to increase the numbers of ways in which to use rocking mechanisms in columns and walls, with or without use of supplementary energy absorbers and vertical prestressing to recentre the vertical member after the shaking (e.g. Mander and Cheng, 1997). The aim is to find affordable means of avoiding damage to *normal-use structures* in design- level earthquake shaking (i.e. 475-year hazard level). Achieving this would not only be of great value to owners of property subjected to very strong shaking, but could also make an important contribution to sustainability of building components, a matter of considerable importance to the world.

In the USA a major research project on this topic for buildings of various heights was carried out over the period *c.* 1999–2000. It involved the use of precast/prestressed concrete, and was known as the Precast Seismic Structural Systems (PRESSS) research program (Nakaki *et al.*, 1999; Priestley *et al.*, 1999). The system utilized unbonded prestressing tendons. Energy dissipation can be provided by non-prestressed steel (and other energy-dissipating devices have since been used). It was deemed that the structures of buildings designed according to PRESSS would suffer little damage in the design event and would be economical. Design recommendations for such structures have subsequently been incorporated into the New Zealand concrete code (NZS 3101, 2006); see its Appendix B. While buildings using PRESSS are being built in California, it has not yet been adopted by US codes, which is retarding the use of this promising building system.

A recent project (completed early in 2009) where low-damage levels were specifically sought by the client is a group of three new student accommodation buildings for the University of Victoria in Wellington, New Zealand. The buildings extend up to 11 storeys in height, and it was found that the most economical structure would consist of a structural steel frame with a lightweight façade. To meet the challenge of providing a cost-effective solution the designers (Connell Wagner) developed a new system for damage avoidance applicable to high-rise steel-framed buildings (Gledhill *et al.*, 2008). After reviewing the options available, the chosen system for the 11-storey building incorporates a transverse braced frame (Figure 8.22) comprising concentrically braced frames (see Section 10.2.6) which act as rocking steel shear walls, controlled by pretensioned Ringfeder friction springs located under all four columns (Figure 8.23). The two concentrically braced frames are connected with central coupling beams equipped with sliding hinge joints. Lateral loads are distributed to fixed head piles through large shear key plates under the ground-floor collector beam. The relevant research papers are by Clifton (2005) and Clifton *et al.* (2007). For the design manual for Ringfeder springs see Ringfeder (1991).

Damage Avoidance Design

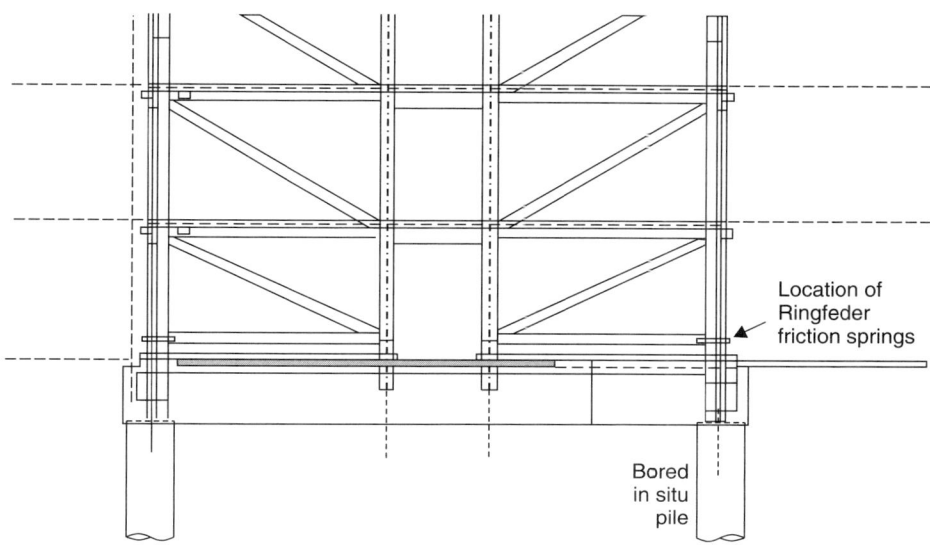

Figure 8.22 Elevation of transverse bracing frame of the low-damage Fairlie Terrace Student Accommodation Building in Wellington, New Zealand (after Gledhill *et al.*, 2008)

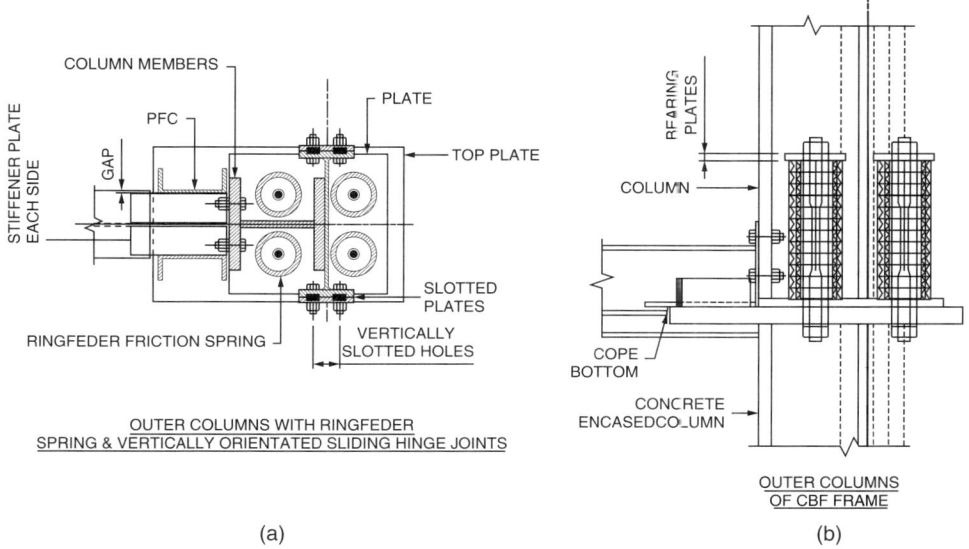

Figure 8.23 Construction details of the Fairlie Terrace Student Accommodation Building in Wellington New Zealand: (a) plan on outer column base with Ringfeder springs; (b) elevation of base of outer column base of concentrically braced frame with Ringfeder springs on holding down bolts

The fragility function approach to damage avoidance design referred to in Section 6.3.4 was not used in the above project.

In the design of low-damage buildings, thought should of course be given to the protection of the non-structure, as well as the structure. Guidance is given in Chapters 11 and 12, and also is a component of performance-based seismic design as described by SEAOC (1999). In the case of the Wellington buildings discussed above, standard detailing of non-structure was deemed generally adequate, except for those components affected by uplift at the base during rocking. (The buildings have no ceilings.)

8.7 Construction and the Enforcement of Standards

It is obvious that the standard of construction should match the standard of the design by meeting the requirements of the drawings and specifications for any given project. As described above, in relation to Figure 8.1, the design should be buildable and within the skills and experience of the constructors and fabricators. This often requires the involvement, to some degree, of the contractor in the design process. In larger and more complex projects, this may be automatic when the contractor is involved from the start, such as in turn-key or design-build contracts.

The enforcement of standards (quality assurance) is too often a vexed question, the degree of enforcement and pervasiveness varying from project to project and country to country. Even when enforcement of standards is officially required, their execution may be ineffective for various reasons, such as inadequacy of resources, inefficiency or corruption. In some cases, the standards of design may be adequately controlled, but are not achieved in the end product because of lack of control of standards of fabrication or construction. Although seldom publicly reported, inspection of property damaged in earthquakes has all too often found major differences between what was specified and what was built. For example, after the 1999 Kocaeli earthquake in Turkey, of a random sample of 25 schools it was found that in only two did the concrete strength meet the building code requirement. Thus, one of the biggest challenges in efforts to reduce earthquake risk is the enforcement of standards at all stages of the design and construction process.

A report (Hoover and Greene, 1996) prepared in the USA examines the above problems, focusing on the relationship between the education of construction tradespeople, code enforcement personnel, and the earthquake performance of structures. The focus includes strategies to improve the education of, and existing training methods for, construction workers and building inspectors.

8.8 Developing Countries

Developing countries of course have special difficulties in reducing earthquake risk, arising from inadequate knowledge and physical resources. The fact that this is the most important chapter of this book (Section 8.1) is especially true for developing countries. People from such places should take particular note of Sections 8.3 and 8.7 above. A wise seismic structural form, tied together by means of so-called *intermediate technology*, and with basic but thorough enforcement of these simple standards will do much to reduce earthquake risk at minimum cost (see also Section 10.6).

References

American Petroleum Institute (various years) *API Recommended Practice for Planning, Designing, and Constructing Fixed Offshore Platforms*. American Petroleum Institute, Washington, DC.

Amin NR and Louie JJC (1984) Design of multiple framed tube high rise steel structures in seismic regions. *Proc. 8th World Conf Earthq Eng*, San Francisco **V**: 347–354.

Applied Technology Council (1982) *An Investigation of the Correlation between Earthquake Ground Motion and Building Performance*, ATC-10. Berkeley, California.

Arnold C and Reitherman R (1982) *Building Configuration and Seismic Design*. John Wiley & Sons, Inc., New York.

AS/NZS 1170.0 (2002) *Structural Design Actions, Part 0: General Principles*. Standards Australia, Sydney; Standards New Zealand, Wellington.

Beck JL and Skinner RI (1974) The seismic response of a reinforced concrete bridge pier designed to step. *Earthq Eng Struc Dyn* **2**: 343–358.

Boardman PR, Wood BJ and Carr AJ (1983) Union House – cross braced structure with energy dissipators. *Bull NZ Nat Soc Earthq Eng* **16**(2): 83–97.

Buckle IG (2000) Passive control of structures for seismic loads. *Bull NZ Soc Earthq Eng* **33**: 209–221.

Charleson AW (2008) *Seismic Design for Architects – Outwitting the Quake*. Elsevier, Oxford.

Charleson AW, Wright PD and Skinner RI (1987) Wellington Central Police Station, base isolation of an essential facility. *Proc. Pacific Conf Earthq Eng, New Zealand* **2**: 378–388.

Clifton GC (2005) *Semi-rigid joints for moment resisting steel framed seismic-resisting systems*. Heavy Engineering Research Association, Manukau City, HERA Report R4-134.

Clifton GC, MacRae GA, Mackiven H, Pampanin S and Butterworth J (2007) Sliding hinge joints and subassemblies for steel moment frames. *Proc. NZ Soc Earthq Eng*, Wairakei, New Zealand, Paper 19 (CD-ROM).

Dowrick, DJ (1977) Structural form for earthquake resistance. *Proc. 6th World Conf Earthq Eng*, New Delhi **2**: 1826–1833.

Dowrick DJ (1985) Preliminary field observations of the Chilean earthquake of 3 March 1985. *Bull NZ Nat Soc Earthq Eng* **18**(2): 119–127.

Dowrick DJ (2006) The serviceability of normal-use, non-domestic buildings in earthquakes – Are serviceability design checks necessary? *Bull NZ Soc Earthq Eng* **39**(4): 208–214.

Dowrick DJ and Rhoades DA (2000) Earthquake damage and risk experience and modelling in New Zealand. *Proc. 12th World Conf Earth Eng, Auckland*.

FEMA (1997) *NEHRP guidelines for the seismic rehabilitation of buildings*. Federal Emergency Management Agency, Report FEMA-273, Washington, DC

FEMA (2006) *Designing for earthquakes: A manual for architects*. Federal Emergency Management Agency, Publication No 454. Washington, DC, online or CD/DVD.

Flores R (1969) An outline of earthquake protection criteria for a developing country. *Proc. 4th World Conf Earthq Eng, Chile III* **J4**: 1–14.

Ghoubhir ML (1984) Earthquake resistance of structural systems for tall buildings. *Proc. 8th World Conf Earthq Eng, San Francisco* **V**: 491–498.

Giuliani H (2000) Seismic resistant architecture: a theory for the architectural design of buildings in seismic design. *Proc. 12th World Conf Earthq Eng*, Auckland, Paper No 2456.

Gledhill SM, Sidwell GK and Bell DK (2008) The damage avoidance design of tall steel frame buildings – Fairlie Terrace Student Accommodation Project, Victoria University of Wellington. *Proc. NZ Soc Earthq Eng*, Wairakei, Paper 63 (CD).

Hanson RD and Soong TT (2001) *Seismic Design with Supplemental Energy Dissipating Devices*. Earthquake Engineering Research Institute, Oakland, CA.

Hoover CA and Greene MR (1996) *Construction Quality, Education and Seismic Safety*. EERI Publication EF 96-02, Earthquake Engineering Research Institute, Oakland, CA.

Kelly JM, Skinner RI and Heine AJ (1972) Mechanisms of energy absorption in special devices for use in earthquake resistant structures. *Bull NZ Nat Soc Earthq Eng* **5**(3): 63–88.

Kelly TE, Robinson WH and Skinner RI (2008) *Seismic isolation for designers and structural engineers*. Robinson Seismic Ltd. http://www.robinsonseismic.com.

Leyendecker EV, Hunt RJ, Frankel AD and Ruckstales KS (2000) Development of maximum considered earthquake ground motion maps. *Earthq Spectra* **16**: 21–40.

Mander JB and Cheng C-T (1997) Seismic resistance of bridges based on damage avoidance design. NCEER, Technical Report NCEER-97-0014.

Mayes RL, Jones LR, Kelly TE and Button MR (1984) Design guidelines for base-isolated buildings with energy dissipators. *Earthq Spectra* **1**(1): 41–74.

Meek JL (1975) Effects of foundation tipping on dynamic response. *J Struc Divn* **101**(ST7): 1297–1311.

Megget LM (1978) Analysis and design of a base-isolated reinforced concrete frame building. *Bull NZ Nat Soc Earthq Eng* **11**(4): 245–254.

Monti MD, Zhao JX, Gannon CR and Robinson WH (1998) Experimental results and dynamic parameters for the Penguin Vibration Damper (PVD) for wind and earthquake loading. *Bull NZ Nat Soc Earthq Eng* **31**: 177–193.

Nakaki SD, Stanton JF and Sritharan, S (1999) An overview of the PRESSS five-story precast test building. *PCI J* **44**: 26–39.

NZS 3101 (2006) *Concrete Structures Standard*. Standards New Zealand, Wellington, Parts 1 and 2.

Pall AS and Marsh C (1982) Response of friction damped braced frames. *J Struc Divn* **108**(ST6): 1313–1323.

Park R, Billings IJ, Clifton GC *et al.* (1995) The Hyogo-ken Nanbu earthquake of 17 January 1995. *Bull NZ Nat Soc Earthq Eng* **28**: 1–98.

Paulay T (1979) Capacity design of earthquake resisting ductile multi-storey reinforced concrete frames. *Proc. 3rd Canadian Conf Earthq Eng*, Montreal **2**: 917–948.

Priestley MJN, Calvi GM and Kowalsky MJ (2007) *Displacement-Based Seismic Design of Structures*. IUSS Press, Pavia, Italy.

Priestley MJN, Sritharan S, Conley JR, and Pampinin S (1999) Preliminary results and conclusions from the PRESSS five-story precast concrete test building. *PCI J* **46**(6): 42–67.

Ringfeder (1991) Ringfeder Friction Springs. Catalogue R60E, Ringfeder GmbH, Krefeld, Germany.

Robinson WH (1998) Passive control of structures, the New Zealand experience. *ISET J Earthq Technology* **35**: 63–75.

Robinson WH (2000) Two developments in seismic isolation – A 'centre drive' and 'friction ball'. *Proc. 12th World Conf Earthq Eng*, Auckland, Paper No 1178.

Robinson WH and Tucker AG (1977) A lead-rubber shear damper. *Bull NZ Nat Soc Earthq Eng* **10**(3): 151–153.

Scawthorn C, Celebi M and Prince J (1986) Performance characteristics of structures, 1985 Mexico City earthquake. In *The Mexico Earthquakes – 1985 – Factors Involved and Lessons Learned*. ASCE, 217–232.

Sharpe RD and Carr AJ (1975) The seismic resistance of inelastic structures. *Bull NZ Nat Soc Earthq Eng* **8**(3): 192–203.

References

Skinner RI, Taylor RG and Robinson WH (1980) Hysteretic dampers for protection of structures from earthquakes. *Bull NZ Nat Soc Earthq Eng* **13**(1): 22–36.

Skinner RI, Robinson WH and McVerry GH (1993) *An Introduction to Seismic Isolation*. John Wiley & Sons, Ltd, Chichester.

Sritharan S and Dowrick DJ (1994) Response of low-rise buildings to moderate earthquake shaking, particularly the May 1990 Weber earthquake. *Bull NZ Nat Soc Earthq Eng* **27**(3): 205–221.

Stratta JL and Feldman J (1971) Interaction of infill walls and concrete frames during earthquakes. *Bull Seism Soc Amer* **61**(3): 609–612.

Structural Engineers Association of California (1999) *Recommended Lateral Force Requirements and Commentary*. SEAOC.

Tyler RG and Robinson WH (1984) High strain tests on lead-rubber bearings for earthquake loadings. *Bull NZ Nat Soc Earthq Eng* **17**(2): 90–105.

UNAM (1985) *Efectos de los sismos de Sept de 1985 en las Construcciones de la Ciudad de México*. Technical report, Universidad Nacional Autónoma de México.

van de Vorstenbosch G, Charleson AW and Dowrick D (2002) Performance of reinforced concrete buildings in the 1931 Hawke's Bay, New Zealand, earthquake. *Bull NZ Nat Soc Earthq Eng* **35**: 149–164.

Wada A, Shinozaki Y and Nakamura N (1984) Collapse of building with expansion joints through collision caused by earthquake motion. *Proc. 8th World Conf Earthq Eng, San Francisco* **IV**: 855–862.

9

Seismic Design of Foundations and Soil-Retaining Structures

9.1 Foundations

9.1.1 Introduction

As discussed in Chapter 5, the properties and dynamic behaviour of soils and their relationships to structures are complex and involve large uncertainties. In view of the challenges thus imposed on designers of foundations, Pender (1996) proposed three levels of *design analysis* for foundations, in a review of earthquake resistant design of foundations where design analysis was defined as all the calculation and analysis that is a central part of the design process. He described the three levels as follows:

Level 1: with respect to displacement estimates, the soil is assumed to remain elastic during seismic loading. Capacity calculations are done using traditional methods.

Level 2: at this level, 'engineering' methods which account for the real behaviour of cyclically loaded soil are used. Approximate techniques are needed to take account of the cyclic loading on the soil strength and the expected cyclic shear strain on the soil stiffness. This is challenging, particularly for foundations embedded in saturated sand. Insight for these methods will come from Level 1 approaches, laboratory and field studies, back analysis of the observed earthquake response of foundations, and the methods of Level 3.

Level 3: in this case, a full analysis is undertaken that accounts properly for the dynamic loading, non-linear soil properties, generation of excess pore pressures during cyclic loading, strain softening, and the complexities of the soil–structure interaction. This is hardly a design analysis, not least because it is usually very difficult to match the sophistication of the numerical modelling with a comparable level of effort in determining soil parameter values. Nevertheless, it is useful for verifying the methods of Level 2.

Earthquake Resistant Design and Risk Reduction D. Dowrick
© 2009, John Wiley & Sons, Ltd

Pender went on to discuss the above levels as follows:

> In solving geotechnical problems it is probably the level 2 analysis that is potentially the most useful design tool. However, it seems at the time of writing (1995) that this level is in need of substantial further development. On the other hand, methods corresponding to level 1 are in a surprisingly good state both for pile and shallow foundations, particularly as useful results for foundation stiffness have been expressed in the form of simple formulae that can be evaluated with a spreadsheet or similar type of general purpose computational software.
>
> It is necessary to relate the above three levels of design analysis to the two levels of seismic loading discussed earlier. Serviceability limit state design requires that there is no structural damage. The foundation equivalent would be no, or at most small, permanent displacement. However, this does not mean completely elastic behaviour of the foundation soil as such behaviour is limited to a very small strain range. Thus geotechnical design needs to use level 2 methods to arrive at the appropriate soil modulus. Ultimate limit state design of structures founded on shallow foundations allows mobilization of a considerable part of the foundation capacity but usually not failure.

The horizontal interaction stresses between the soil and the foundation are arguably more problematical than the vertical stresses, as comparatively little is known about allowable seismic passive pressures and the effect of seismic active pressure in different foundation situations. Indeed, it is customary to assume even more arbitrary distributions for horizontal stress between foundations and soil than for vertical stress. The main problems (peculiar to earthquakes) of foundation design occur in transferring the base shear of the structure to the ground, and in maintaining structural integrity of the foundation during differential soil deformations. Some design guidance on these problems now follows under the headings of:

- shallow foundations;
- deep box foundations;
- caissons; and
- piled foundations.

9.1.2 Shallow foundations

The horizontal seismic shear force at the base of the structure must be transferred through the substructure to the soil. With shallow foundations it is normal to assume that most of the resistance to lateral load is provided by friction between the soil and the base of the members resisting horizontal load. Other footings and slabs in contact with the ground may also be assumed to provide shear resistance if they are suitably connected to the main resisting elements. The total available resistance to lateral movement of the structure may be taken to be equal to the product of the dead load carried by the elements considered and the coefficient of sliding friction between the soil and the substructure. Typical values of friction angles for foundations are given in Table 9.1.

In some cases, further horizontal resistance will arise from the passive soil pressures developed against subsurface elements. If this resistance is taken into account it is often deemed wise to restrict the calculated total restraint by reducing either the frictional

Table 9.1 Typical interface friction angles (NAVFAC, 1982)

	Interface materials	Interface friction angle δ
Mass concrete against:	clean sound rock	25
	clean gravel, gravel–sand mixtures, coarse sand	29–31
	clean fine to medium sand, silty medium to coarse sand, silty or clayey gravel	24–29
	clean fine sand, silty or clayey fine to medium sand	19–24
	fine sandy silt, non-plastic silt	17–19
	medium-stiff and stiff clay and silty clay	17–19
Formed concrete against:	clean gravel, gravel–sand mixture, well-graded rock fill with spalls	22–26
	clean gravel, silty sand–gravel mixture, single-size hard rock fill	17–22
	silty sand, gravel, or sand mixed with silt or clay	17
	fine sandy silt, non-plastic silt	14
Steel sheet piles against:	clean gravel, gravel–sand mixture, well-graded rock fill with spalls	22
	clean sand, silty sand–gravel mixture, single-size hard rock fill	17
	silty sand, gravel, or sand mixed with silt or clay	14
	fine sandy silt, non-plastic silt	11

force or the passive resistance force by 50%. To ensure that the passive restraint can be developed, appropriate measures must be taken on site, such as adequate compacting of backfill against sides of footings.

Shallow foundations are often of a form that is highly vulnerable to damage from differential horizontal and vertical ground movements during earthquakes. It is therefore good practice even in quite low structures, especially those founded on soft soils, to provide ties between column pads. In the absence of a more realistic method, an arbitrary design criterion for such ties is to make them capable of carrying compression and tension loads equal to 10% of the maximum vertical load in adjacent columns. However, it may be possible to resist some or all of these horizontal forces by passive action of the soil, particularly for light buildings. The designer may also have a choice between providing the tie action at the bottom floor level (in tie beams or in the slab), or at some other position in relation to the foundations.

9.1.3 Deep box foundations

In the seismic design of deep box foundations, designers must rely mainly on normal structural and geotechnical static design techniques, supplemented where appropriate by consideration of known seismic phenomena, such as seismically enhanced soil pressures. The natural stiffness and strength of box-shaped foundations should be utilized to advantage in distributing the seismic forces from the soil and the superstructure through rational load paths in the foundation, with an adequate safety factor.

Although less susceptible to damage from ground motions than isolated pad footings, deep box foundations nevertheless require proper design to withstand strong earthquakes. This was exemplified by the virtual destruction of underground water tanks in the 1971 San Fernando earthquake (Wyllie *et al.*, 1973). This failure demonstrated the importance of internal walls to provide an egg-crate type of stiffening; it also showed the valuable contribution that concrete keys could have provided to the strength of construction joints, which moved 0.6 m in shear despite the presence of steel reinforcement normal to the joint.

Before completing the design of the foundations it is assumed that the dynamic characteristics of the subsoil have been determined as discussed in Chapter 3 and Section 5.2, and a suitable form for the substructure should also have been chosen as suggested in Section 8.3.8.

It then remains to design the foundations for appropriate seismic forces which arise (1) directly from the deformation of the adjacent soil, and (2) as a result of the earthquake forces acting in the superstructure. While our ability to estimate the seismic forces from (2) is now quite advanced, there remains a great deal of uncertainty about the magnitude and effect of the forces induced directly by the ground. This is true despite the increasing attempts to elucidate the soil–structure interaction problem by sophisticated analytical and experimental techniques.

In design practice, it is often found convenient to consider two separate stress systems: the seismic vertical stresses (e.g. due to overturning moments), and the seismic horizontal stresses (e.g. due to the base shear on the structure). Overturning moments are not usually a problem for buildings as a whole, unless it is very slender, but can be difficult for individual footings, such as column pads or shear wall strip footings. The foundation should, of course, be proportioned so as to keep the maximum bearing pressures due to the overturning moments and gravity loads within the allowable seismic value for the soil concerned.

Sand may behave worse in dynamic than static loading, for example subsiding or liquefying (Section 5.2). On the other hand, some clays sustain substantially higher dynamic than static loadings (M. Pender, personal communication, 2001), although such beneficial effects are generally ignored in design. According to Seed (1960), some sensitive clays lose strength under dynamic loading.

Permissible bearing pressures need to be worked out on a case-by-case basis, using appropriate soil mechanics theory, and allowing for foundation geometry and soil properties. However, for preliminary design purposes only, the bearing pressures taken from an out-of-print publication by the New Zealand Ministry of Works (NZMOW, 1970) quoted in Table 9.2 may be helpful; here the bearing pressures are reduced by 25% for medium gravel and medium sand and increased by 50% for rock and very stiff or medium stiff clay. The latter values are given some support by Ishihara *et al.* (1974). They found that the cohesion component c of partially saturated clays (i.e. a volcanic clay (PI = 30) and a sandy clay (PI = 18)) was higher under dynamic loading (c_d) than under static loading (c) such that $c_d/c = 2.4$ and 1.86, respectively. The angle of internal friction was unchanged by the rate of loading. The values in Table 9.2 may in some cases be over-conservative, and well-informed geotechnical advice should in any case be taken for the actual soil conditions for final design of each project.

Table 9.2 Allowable bearing pressures on soils, for preliminary design only (New Zealand Ministry of Works, 1970)

Soil types		Allowable bearing pressures (kPa)					Standard penetration blow count (N)	Apparent cohesion c_u (kPa)
		Long-term loads			Total loads (including seismic loads)			
Soft or broken rock		960			1440		30	
Gravel	Dense	285–570			285–570			
	Medium	96–285			72–215			
		Well graded	Uniform	Well graded		Uniform		
Sand[1]	Dense	240–525	120–265	240–525		120–265	30	
	Medium	96–240	48–120	72–180		40–90	15–30	
	Very stiff	190–380		285–570			15–30	100–200
Clay[2]	Medium stiff	48–190		72–285			4–15	25–100
	Soft	0–48		0–48			0–4	0–25
Peat, silts made ground		To be determined after investigation						

Notes: [1]Reduce bearing pressures by half below the water table.
[2]Alternatively: allow 1.2 times c_u for round and square footings, and 1.0 times c_u for length/width ratios of more than 4.0. Interpolate for intermediate values.

9.1.4 Caissons

Caissons are similar to piles in that they are relatively slender at least in one direction, and are used as isolated foundations spaced at intervals to support structures which may be long in plan, such as bridges or large buildings. In bridge construction the terms *caissons* and *piers* are sometimes used interchangeably. Where caissons penetrate the soil deeply, they need special consideration of soil–structure interaction effects, as discussed below for piles. Caissons may differ from piles in that they may often be treated as rigid rocking structures, rather than bending structures. This typically occurs for bridge piers in the direction lateral to the axis of the bridge.

9.1.5 Piled foundations

Introduction

The reliable design of *piles* for earthquake loads is difficult because of the uncertainties involved in determining the design deformation state of the piles. This is partly due to the uncertainties involved in assessing lateral soil–pile interaction, and partly to the complexity of behaviour of pile groups. As indicated in Figures 9.1, 9.4 and 9.5, high bending moments may occur at various locations up the pile. In addition to the locations of high bending moments indicated by these idealized moment diagrams, high stresses may be induced at other depths due to local shear failure of weak layers of soil or due to liquefaction, or due to loss of lateral support from the soil because of scour in waterways or settlement of loose deposits.

The seismic design of piled foundations will include consideration of the vertical and horizontal stresses and the structural integrity of the foundation. Vertical seismic loads in

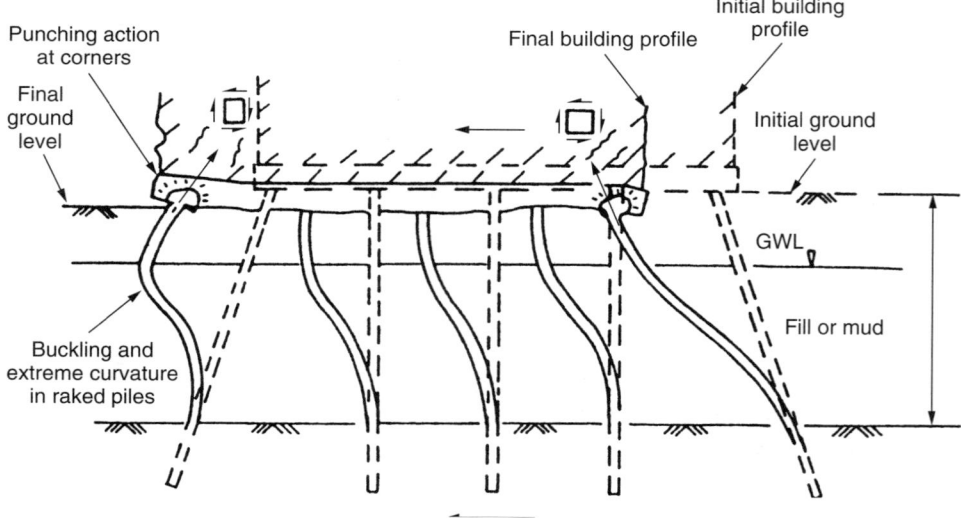

Figure 9.1 Interaction of raked piles and pilecap during an earthquake

individual piles may vary greatly, depending upon their position in relation to the rest of the pile group and to the superstructure (Figure 9.1). Some piles, particularly those at the edges or corners of pile systems, may have to carry large tensile as well as compression forces during earthquakes.

Lack of structural integrity has caused failure of piled foundations in earthquakes, such as that of San Fernando, 1971. Sufficient continuity reinforcement must be provided between the piles and the pile cap, and the piles themselves must obviously be able to develop the required tensile, compression and bending strength. Where plastic hinges are likely to form in concrete piles, suitable confinement reinforcement must be provided, as it is in columns.

As a supplement to the following discussion of piles, the reader is referred to the specialist literature, especially the extensive review of Pender (1993) and the earlier text of Poulos and Davis (1980).

Dynamic response of piles

In response to horizontal ground motions, it appears that piles generally follow the formations of the ground, and do not cut through the soil. It also seems that piles are subject to two distinct failure mode zones:

(1) In the upper part of the pile, say the top $10D$ (D = diameter), the response is affected by the presence of the free soil surface, which permits the soil adjacent to the pile to yield and move upwards in a wedge (Figure 9.2). Also, the upper part of the pile has inertia loads induced in it by the surrounding soil and the structure above.
(2) In the lower part of the pile, the surrounding soil dominates the response, and flexibility or ductility is required to permit the pile to safely conform to the curvatures imposed by the soil deformations.

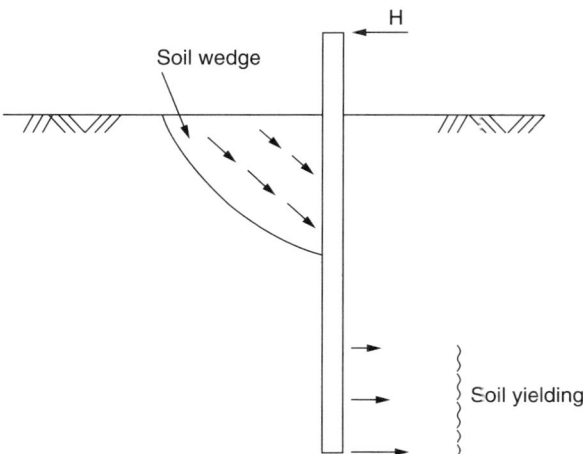

Figure 9.2 Limiting lateral load on a pile defined by two separate failure modes

In the dynamic response analysis of piled foundations for design purposes, because the soil–pile interaction is so complex, it is usual to simplify the structural modelling problem, often as much as in the following opposing options:

(1) Ignore the soil entirely, using only the stiffness of the pile, after having first defined some depth to pile fixity based on soil stiffness.
(2) Ignore the horizontal and rotational stiffness of the pile, using only the stiffness of the soil.

The development of more sophisticated, research-oriented analytical techniques, similar to the shear beam model described in Section 5.2.2, has been along two main lines:

(1) a continuous elastic model (e.g. Gazetas and Dobry, 1984);
(2) a discrete model with lumped masses, springs and dashpots (e.g. Penzien, 1970; Blaney et al., 1976).

Such techniques permit pile stresses as well as stiffness to be estimated. The linear analyses are generally conducted in the frequency domain, pile stiffnesses and damping being expressed in frequency-dependent terms, as discussed in Section 5.3.3. Significant differences between the results of such studies need to be resolved, particularly for the low-frequency properties.

Method (2) above is used for non-linear as well as linear analyses. Some of the non-linear analyses studied that have been done have not been true dynamic earthquake analyses of the shear beam type noted above, but have been of either repeated or quasi-static cyclic loading nature, in which the soil–pile system has been loaded by a horizontal load and perhaps a moment applied at the top. Clearly, such a loading model closely represents wind or wave loads, but for seismic loading it would give better estimates of stiffness than of worst deformations in the pile.

In an example of a full dynamic earthquake response analysis, some light was thrown on the likely behaviour of long piles in deep sensitive clay in a sophisticated non-linear analysis of a bridge described by Penzien (1970). In this case, it was found that if subjected to moderate earthquake loading like that of El Centro in 1940, the piles would have been deformed to their yield curvatures.

Equivalent static lateral analysis of piles

In the majority of design projects, pile design and foundation modelling for superstructure analysis will be carried out with reference to (separate) equivalent-static load analyses of the piles. The latter comprise a number of methods which may be divided into three categories:

(1) limiting (or ultimate) loads;
(2) elastic continuum;
(3) non-linear discontinuum (Winkler springs).

Only the elastic continuum method will be considered here.

Elastic continuum method

Considering an elastic pile embedded in an elastic soil and loaded at the pile head, the displacement u and the rotation θ at the ground surface are given by

$$u = f_{uH} H + f_{uM} M, \qquad (9.1)$$

$$\theta = f_{\theta H} H + f_{\theta M} M, \qquad (9.2)$$

where H is the applied horizontal load, M is the applied moment, and f_{uH}, f_{uM}, $f_{\theta H}$, $f_{\theta M}$ are flexibility coefficients. From the reciprocal theorem $f_{\theta H} = f_{uM}$.

For a long pile the flexibility coefficients are functions of the ratio of the Young's moduli of the pile and the soil, Poisson's ratio and Young's modulus of the soil, and the pile diameter. For short piles, the pile length is also required in the expressions for the flexibilities.

The flexibility coefficients for a long circular pile are expressed in terms of a modulus ratio:

$$\begin{aligned} K &= \frac{E_p}{E_s} \text{ for constant soil modulus} \\ K &= \frac{E_p}{E_{s(at\ z=D)}} \text{ for variable modulus} \end{aligned} \qquad (9.3)$$

where E_p is Young's modulus for the pile material, E_s is Young's modulus of the soil, z is the depth below ground level, and D is the pile diameter.

The pile head can be loaded with a horizontal shear force, a moment or both. When the shear force is applied to a pile shaft above the ground surface, it is convenient to express the resulting pile head moment in terms of an eccentricity defined in the following alternative ways:

$$e = \frac{M}{H}, \quad f = \frac{M}{DH}. \qquad (9.4)$$

Lateral elastic displacements of a single 'long' pile

Consider the case where the soil modulus increases linearity with depth. The equations for this case are given by Budhu and Davies (1987, 1988), except for those involving Poisson's ratio where a value of $v = 0.5$ has been adopted by Pender (1993), where resulting simplifications are given below. Young's modulus of the soil and the stiffness ratio are

$$E_s = mD, \quad K = \frac{E_p}{mD}, \qquad (9.5)$$

where m is the rate of increase in Young's modulus with depth. Budhu and Davies give values of m for sands of various densities. These values are intended for static loading of piles, so they are not appropriate for dynamic excitation of piles embedded in liquefiable sands.

The active length of the pile is

$$L_a = 1.3DK^{0.222}. \tag{9.6}$$

If the pile length is greater than that given by the above equation, then the pile is 'long' and the following equations for the flexibility coefficients can be used:

$$f_{uH} = \frac{3.2K^{-0.333}}{mD^2},$$

$$f_{uM} = f_{\theta H} = \frac{5.0K^{-0.556}}{mD^3}, \tag{9.7}$$

$$f_{\theta M} = 13.6\frac{K^{-0.778}}{mD^4}.$$

The location and maximum moment in the pile section are given by

$$L_{\text{Max}} = 0.41L_a, \tag{9.8}$$

$$M_{\text{Max}} = I_{MH}DH, \tag{9.9}$$

where $I_{MH} = aK^b$, $a = 0.6f$, and $b = 0.17f^{-0.3}$. If I_{MH} is greater than 8, a value of 8 is used.

Non-linear lateral displacements of a single 'long' pile

For estimating the effect of local soil failure at the pile–soil interface, Davies and Budhu (1986) proposed a modification factor to be applied to the elastic behaviour model. Thus for a free-head pile, the pile head displacement, rotation and maximum moment are found from

$$u_y = I_{uy}u_E,$$

$$\theta_y = I_{\theta y}\theta_E, \tag{9.10}$$

$$M_{My} = I_{My}M_{ME},$$

where I_{uy}, $I_{\theta y}$ and I_{My} are yield influence factors, u_E is the elastic pile head displacement from equations (9.1) and (9.7), θ_E is the elastic pile head rotation from equations (9.2) and (9.7), and M_{ME} is the maximum elastic pile shaft moment from equation (9.9).

For piles in cohesive soils, the yield influence factors are given by

$$I_{uy} = 1 + \frac{h - 14k^{0.32}}{40k^{0.53}},$$

$$I_{\theta y} = 1 + \frac{h - 14k^{0.32}}{54k^{0.53}}, \tag{9.11}$$

$$I_{My} = 1 + \frac{h - 8k^{0.32}}{96k^{0.48}},$$

where

$$h = \frac{H}{cD^3}, \quad k = \frac{K}{1000}, \quad (9.12)$$

c being the rate of increase of undrained shear strength with depth (kN/m³).
For piles in *cohesionless soils*, the yield influence factors are

$$I_{uy} = 1 + \frac{h - k^{0.35}}{6k^{0.65}},$$

$$I_{\theta y} = 1 + \frac{h - k^{0.35}}{11k^{0.35}}, \quad (9.13)$$

$$I_{My} = 1 + \frac{h}{20k^{0.35}},$$

where

$$h = \frac{H}{K_p \gamma D^3}, \quad k = \frac{K \exp(0.07(\phi - 30))}{1000}, \quad (9.14)$$

in which ϕ is the friction angle of the sand, $K_p = (1 + \sin\phi)/(1 - \sin\phi)$, and γ is the appropriate unit weight of the sand to give the variation of vertical effective stress with depth.

The non-linear response of a pile as given by the above expressions has been calculated by Pender (1993) for the case where the friction angle of the sand is 35°, its unit weight is 10 kN/m³, H/M is 2.3 and the yield moment of the pile is 1575 kNm. The resulting non-linear plots of pile head displacement and rotation, and the maximum moment in the pile, are plotted as functions of the horizontal shear load on the pile head in Figure 9.3.

Lateral capacity of a single 'long' pile

A much used approach for estimating the ultimate lateral capacity of a pile is that of Broms (1964a, 1964b). He proposes a simple method of estimating the maximum lateral load for two cases, that is, in cohesive and cohesionless soils. Broms considers free-head and fixed-head cases and short, intermediate and long piles. The soil reaction and force diagrams for free-head and fixed head piles in the two soil types are illustrated in Figures 9.4 and 9.5. Broms assumed that the ultimate lateral resistance of the pile is developed when the soil yields and plastic hinges develop in the piles. Charts are provided by Broms to aid the calculation. However, the method of Budhu and Davies (1987, 1988) involves simple equations from which to calculate the capacity directly, as described below for long piles.

Referring to Figure 9.4 for *cohesive soils*, let the ultimate bearing capacity of the soil be $9s_u$, where s_u is the undrained shear strength. For free-head piles the ultimate lateral capacity is

$$H_u = s_u D^2 [(2n_c + 100f^2)^{0.5} - 10f], \quad (9.15)$$

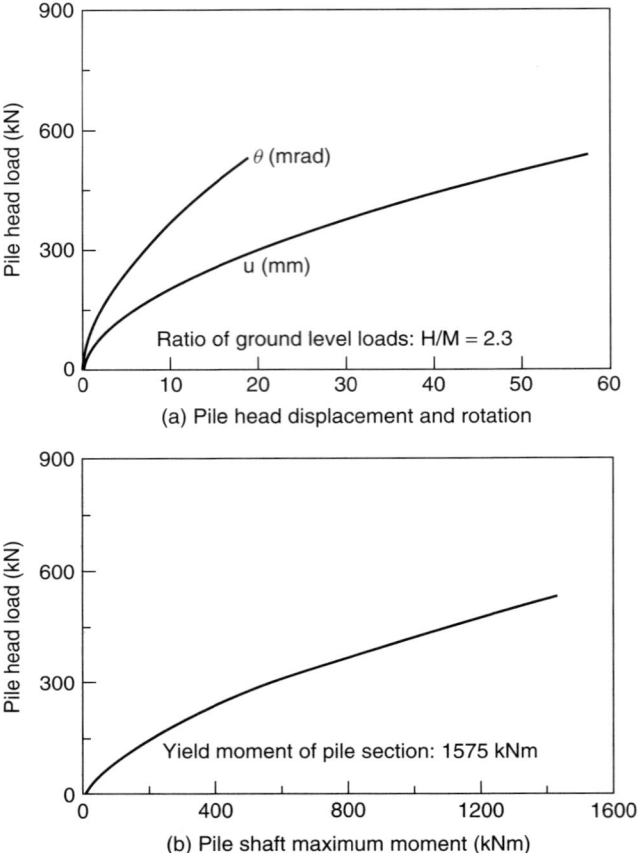

Figure 9.3 Plots of non-linear response of a free-headed pile in sand with linearly varying soil modulus, calculated with the equations of Davies and Budhu (1986) (from Pender, 1993)

where f is defined by equation (9.4), and the ratio n_c is found from

$$n_c = \frac{10 M_y}{s_u D^3}. \tag{9.16}$$

The position of the yield moment (and the length of pile shaft over which soil failure occurs) is given by

$$f_s = \frac{H_u}{9 s_u D} + e_0, \tag{9.17}$$

where e_0 and f_s define the simplified lateral stress regime adopted for the top of the pile (Figure 9.4).

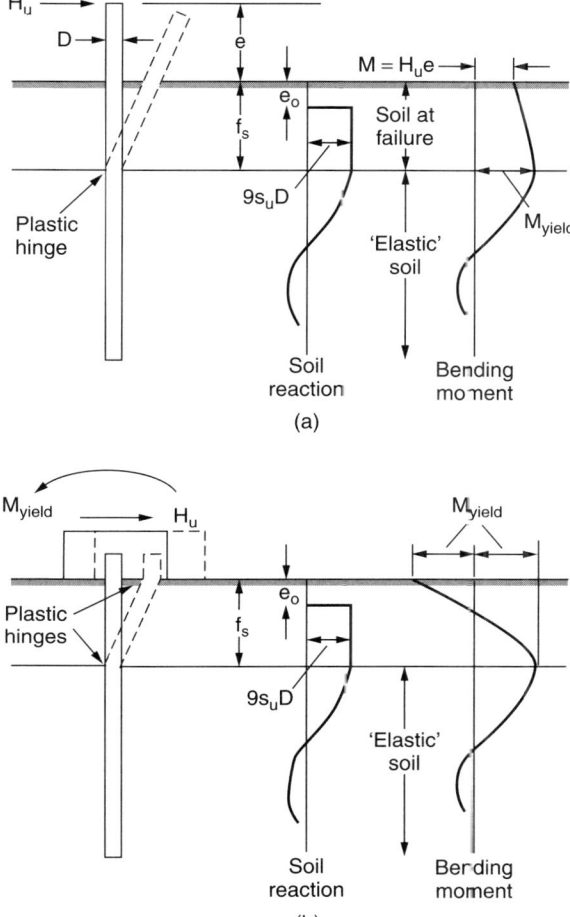

Figure 9.4 Ultimate pressure distribution against a laterally loaded long pile in cohesive soil (adapted from Broms, 1964a): (a) free-head and (b) fixed-head pile. (Reproduced by permission of the American Society of Civil Engineers)

For a fixed-head pile the ultimate lateral capacity will be given by

$$H_u = 2s_u D^2 n_c^{0.5}. \tag{9.18}$$

The above four equations are based on the same assumptions as Broms, with the exception that e_0 is 0.6 m for Budhu and Davies and $1.5D$ for Broms.

The minimum length of pile shaft required for this solution to be valid is

$$L_{\text{effc}} = 0.4 D n^{0.5}. \tag{9.19}$$

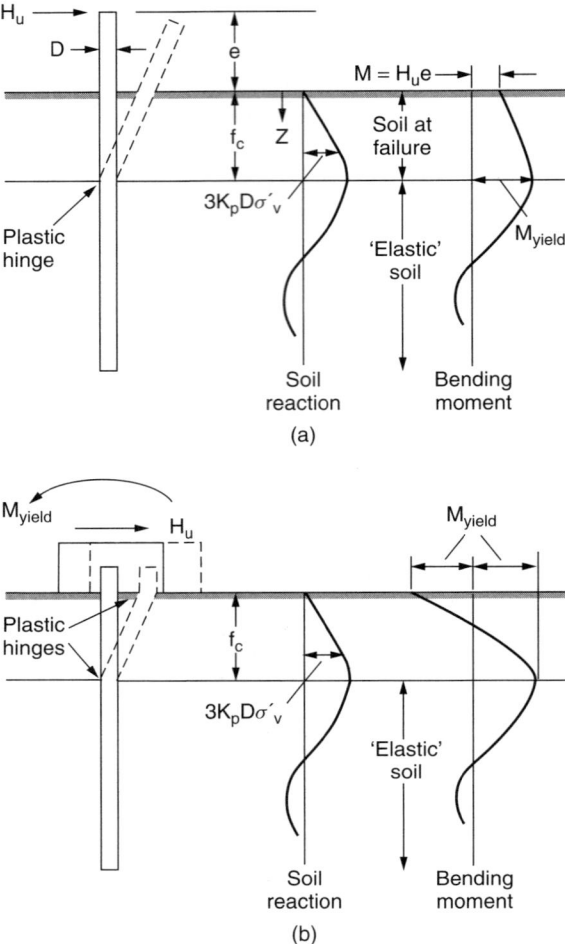

Figure 9.5 Ultimate pressure distribution against a laterally loaded long pile in cohesionless soil (adapted from Broms, 1964b): (a) free-head and (b) fixed-head pile. (Reproduced by permission of the American Society of Civil Engineers)

The above equations give essentially the same prediction for the ultimate lateral capacity as those of Broms (1964a).

The simplified pressure distributions adopted by Broms for *cohesionless soils* are given in Figure 9.5. It can be seen that Broms assumed that the soil pressure along the pile shaft is controlled by $3K_p$, where K_p is the coefficient of passive earth pressure.

Again, following the work of Budhu and Davies, their expression for the ultimate lateral capacity of a free-head pile in cohensionless soil is

$$H_u = 0.35 K_p \gamma D^3 n_s^{0.67} \exp(-1.6 f^{0.75} n_s^{-0.25}), \tag{9.20}$$

where $K_p = (1 + \sin\phi)/(1 - \sin\phi)$ and

$$n_s = \frac{10 M_y}{3 K_p \gamma D^4}. \tag{9.21}$$

The position of the yield moment (and the length of pile shaft over which failure occurs) is given by

$$f_s = \sqrt{\frac{2 H_u}{3 K_p \gamma D}}, \tag{9.22}$$

where γ is the unit weight of the sand chosen to give the effective vertical stresses (γ' for a saturated sand).

For the fixed-head case, the ultimate lateral capacity is

$$H_u = 0.56 K_p \gamma D^3 n_c^{0.67}. \tag{9.23}$$

The minimum length of pile shaft required for this long pile solution to be valid is

$$L_{\text{eff}s} = 0.8 D n^{0.33}. \tag{9.24}$$

As with the cohesive soil case, the above equations give essentially the same predictions of pile lateral capacity as those of Broms (1964b).

Other considerations for pile design analysis

In the foregoing discussion a number of aspects of pile behaviour have not been considered, such short piles, raking piles, dynamic stiffness of piles, vertical load capacity pile groups and other soil profiles. These all need specific consideration, for which the reader is referred to the literature, in particular Pender (1993) and Budhu and Davies (1987, 1988).

9.1.6 Foundations in liquefiable ground

When a structure is to be built on liquefiable ground, there are a number of procedures which may be used to ensure adequate foundation safety, which fall into two categories:

(1) pile foundations; and
(2) soil improvement.

Pile foundations

When using piles in liquefiable ground, the piles should be designed for the conditions induced by liquefaction, as the loss of soil support in the liquefied layer may cause large forces in the piles (Nishizawa *et al.*, 1984). Concrete piles should be detailed for strength and ductility as for columns (Section 10.3.3). However, sole reliance on piles should be practised with caution, because of the difficulty of determining the location and thickness of potential liquefaction layers. In many cases, it will be appropriate to combine piling with a degree of soil improvement to reduce the probability of liquefaction occurring. A range if issues related to the design of piles in liquefiable ground are discussed by Berrill and Yasuda (2002).

Soil improvement

In the latter part of the twentieth century, a number of techniques were developed to improve the properties of liquefiable ground. They can be divided into two main categories: densification techniques and reinforcement techniques (Kramer, 1996). These techniques are briefly discussed below.

Densification of soils
Densification increases the strength and stiffness of the soil, and reduces the tendency to generate excess porewater pressures under cyclic loading. It is thus one of the most effective and commonly used soil improvements methods. A number of methods are used to carry it out.

Vibro techniques use probes in the soil which vibrate horizontally (vibroflotation) or vertically (vibro rod). These techniques are most effective in granular deposits with fine contents less than 20% and clay contents less than 3%.

Ohsaki (1970) reported on the effectiveness of compaction using the vibroflotation technique in preventing liquefaction in the Tokachioki earthquake of 1968, which caused liquefaction at similar uncompacted nearby sites.

Dynamic compaction is effected by repeatedly dropping a heavy weight from a crane on to the ground surface in a grid pattern. It is effective to depths of 9–12 m. Because of its secondary effects of noise, transmission of vibration to nearby sites and dust problems on dry sites, dynamic compaction is sometimes unacceptable in urban areas. Another, more violent, method of using energy to cause compaction is *blasting*, which is effective in similar soil conditions to vibro techniques, but its use is clearly limited to remote sites.

Compaction grouting is carried out by injecting into the ground under pressure a low slump grout. A highly viscous grout forms intact bulbs or columns which densify the surrounding soil by displacement. It is at its most effective when the soil is softest and weakest. Compaction grouting is a very adaptable technique as it can be used to any depth and in all soil types.

In general, it is insufficient to densify the soils solely in the area vertically below the foundations. To minimize post-earthquake settlement of a structure on liquefiable soils, the area densified needs to be extended to a zone within an angle of 30–45° away from the base of the structure (Iai *et al.*, 1988), as shown in Figure 9.6.

Reinforcement of soils
The strength and stiffness of a soil deposit can sometimes be improved by installing in it discrete elements of other materials, such as concrete, steel, timber or dense gravel. For countering liquefaction these elements are generally in the form of piles.

Stone columns are made of either fine or coarse gravels. One way of installing them is the *Franki method*, in which a steel casing, initially closed at the bottom by a gravel

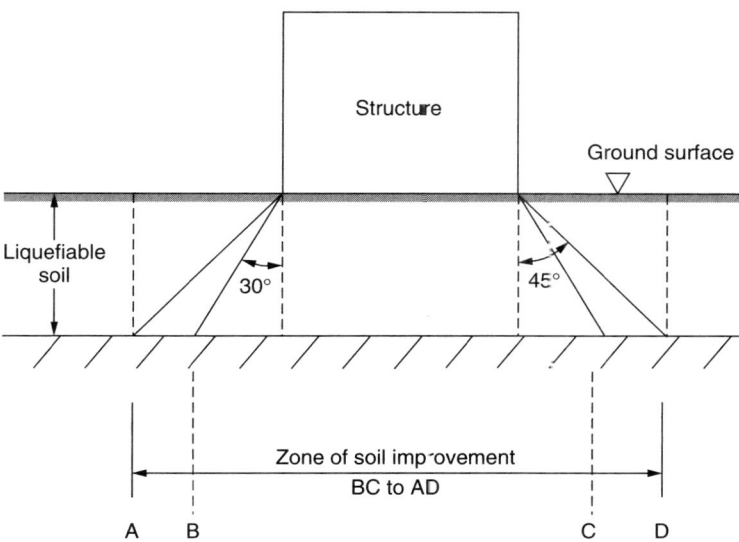

Figure 9.6 Zone of densification of potentially liquefiable soil required beneath and surrounding a structure

plug, is driven to the required depth, and gravel is placed in it as it is withdrawn. The resulting stone columns improve the soil deposit in four or more ways. First, the deposit is improved through the high density, stiffness and strength of the gravel and in this sense the soil is reinforced. Secondly, the stone columns provide drainage routes that inhibit the development of excess porewater pressures. Seed and Booker (1977) describe a design procedure for the use of gravel drains in liquefiable sand. The spacing of the drains is governed by the length of drainage path and the corresponding time required to permit safe dissipation of pore-pressure build-up. Thirdly, the processes used for installing the gravel densify the surrounding soil by both vibration and lateral displacement. Fourthly, as a result of the installation, the lateral stresses are increased in the soil surrounding the columns.

At the time of writing, the NEESR Grand Challenge project, 'Seismic Risk Management for Ports', was having promising results with testing the use of prefabricated vertical drains and colloidal silica grouting for mitigating liquefaction hazards. For more information, visit http://www.neesgc.gatech.edu.

Compaction piles

As compaction piles (usually of prestressed concrete or timber) are displacement piles, they improve the soil in the third and fourth ways discussed above for stone columns. In addition, the lateral strength of the piles helps suppress horizontal displacements of the soil. As compaction piles densify the soil within a distance from themselves of 7–12 pile diameters (Kishida, 1967), to be effective they need to be installed in a grid pattern. Between them relative densities of up to 75–80% are mostly achieved (Solymar and Reed, 1986).

Other considerations

While the above two main methods of designing foundations for liquefiable sites (i.e. piles and soil improvements) are sometimes used individually, they often need to be used at the same time. For example, dynamic compaction is sometimes part of site preparatory work prior to filing. Further, it is important to be aware that the effectiveness of soil improvement techniques may be difficult to accurately predict for a given site. Thus, it is advisable to construct *test sections* before commencing work, or even before finally selecting the techniques to be used.

Additional information on the above techniques soil improvement techniques should be sought in the literature (e.g. Kramer, 1996).

9.1.7 Further reading

Further advice on seismic foundation design is given in the specialist literature, such as Bhattacharya (2007), Kramer (1996), Pender (1993, 1996), Taylor and Williams (1979) and Zeevaert (1983).

9.2 Soil-Retaining Structures

9.2.1 Introduction

As the non-seismic design of soil-retaining structures is well discussed elsewhere, little other than seismic considerations are dealt with here. The principal types of structure covered in this section are retaining walls and basement walls.

The magnitude of the seismic soil pressures acting on a soil-retaining structure in part depends upon the relative stiffness of the structure and the associated soil mass. Two main categories of soil–structure interaction are usually defined:

(1) flexible structures which move away from the soil sufficiently to minimize the soil pressures, such as slender free-standing retaining walls;
(2) rigid structures, such as basement walls or tied-back retaining walls.

In case (1), active pressures will occur, and the amount of movement required to produce the active state is of the order indicated in Table 9.3. The amount of wall movement which will occur during earthquakes depends mainly upon the foundation fixity and the wall flexibility. Unless a more exact analysis is made, the following soil pressure states may be used:

Table 9.3 Movement of retaining wall required to produce the active state

Soil	Wall movement/height
Cohesionless, dense	0.001
Cohesionless, loose	0.001–0.002
Firm clay	0.01–0.02
Loose clay	0.02–0.05

Soil-Retaining Structures

(1) *Flexible:* walls founded on non-rock materials or cantilever walls higher than 5 m; assume active soil state.
(2) *Intermediate:* cantilever walls less than 5 m high founded on rock.
(3) *Rigid:* counterfort or gravity wall founded on rock or piles; at-rest soil state.

In general, wall stiffness will lie somewhere between the extremes of flexible and rigid, and interpolation between the forces generated in these states may be appropriate.

9.2.2 Seismic soil pressures

In the present state of knowledge, the recommended method of obtaining seismic soil forces is that using equivalent-static analysis. Only for exceptional structures would dynamic analyses using finite elements seem warranted.

In the equivalent-static method, a horizontal earthquake force equal to the weight of the soil wedge multiplied by a seismic coefficient is assumed to act at the centre of gravity of the soil mass. This earthquake force is additional to the static forces on the wall.

In general, the total soil pressure on a wall during an earthquake equals the sum of three possible components:

(1) static pressure due to gravity loads;
(2) dynamic pressure due to the earthquake;
(3) pressure due to the wall being displaced into the backfill by an external force, e.g. by the horizontal sway of a bridge deck at a monolithic abutment – design recommendations for this condition are given by Matthewson *et al.* (1980).

The soil pressures may be estimated by the following methods:

- elastic theory;
- approximate plasticity theory, e.g. Coulomb and Mononobe–Okabe; or
- numerical methods, modelling the soil as Winkler springs (Section 5.3.3) or as finite elements.

The results of dynamic analyses using finite elements are difficult to interpret because of the inability of such analyses to represent actual modes of failure. It should be noted that it is not appropriate to design all soil-retaining structures for earthquake soil pressures. For example, external retaining walls of modest height with no significant consequences of failure are generally not designed for earthquakes in many countries. Using the Mononobe–Okabe methods, it can be readily seen (Seed and Whitman, 1970) that it requires an effective ground acceleration of about $0.3g$ to produce an earthquake force increment equal to the static earth pressure for cohesionless soils. So, clearly a safety factor of 2.0 on a non-seismic design should permit walls to survive moderate earthquakes, with acceptable displacements. This rationale is applied to bridge design in New Zealand, where it has been recommended (Matthewson *et al.*, 1980) that higher risks should be accepted for lesser structures, by designing the walls (as wells as the decks) for earthquakes of lower return period.

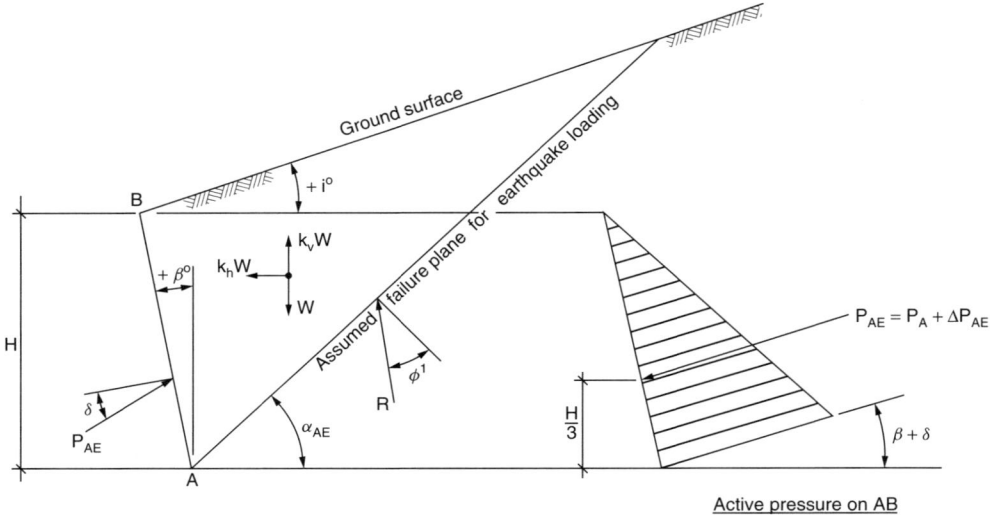

Figure 9.7 Active pressure due to unsaturated cohesionless soil on a flexible retaining wall during an earthquake, for use with Mononobe–Okabe equations (Coulomb conditions)

Active seismic pressures in unsaturated cohesionless soils

The most commonly used solution is that derived by Mononobe (1929) and Okabe (1926), based on Coulomb's theory. The effect of an earthquake is represented by a static horizontal force equal to the weight of the wedge of soil multiplied by the seismic coefficient. Referring to Figure 9.7, the Mononobe–Okabe equations are as follows.

The total force on a wall due to the static and earthquake active soil pressures due to unsaturated cohensionless soils is

$$P_{AE} = \frac{1}{2} K_{AE} \gamma_d H^2 (1 - \alpha_v), \tag{9.25}$$

where

$$K_{AE} = \frac{\cos^2(\varphi' - \beta - \theta)}{\cos\theta \cos^2\beta \cos(\delta + \beta + \theta) \left\{ 1 + \sqrt{\left[\frac{\sin(\varphi' + \delta) \sin(\varphi' - i - \theta)}{\cos(\delta + \beta + \theta) \cos(\beta - i)} \right]} \right\}}, \tag{9.26}$$

and

$$\cot(\alpha_{AE} - i) = -\tan(\varphi' + \delta + \beta - i) + \sec(\varphi' + \delta + \beta - i)$$
$$\times \sqrt{\left\{ \frac{\cos(\beta + \delta + \theta) \sin(\varphi' + \delta)}{\cos(\beta - i) \sin(\varphi' - \theta - i)} \right\}}, \tag{9.27}$$

where α_{AE} is the slope angle of the failure plane in an earthquake (Figure 9.7); β the angle of the back face of the wall to the vertical; γ_d the unit weight of the soil; δ the angle of wall friction; ϕ' the effective angle of shearing resistance; i the slope angle of the backfill; $\theta = \tan^{-1}[\alpha_h/(1-\alpha_v)]$; α_h is a seismic coefficient given by $1/g \times$ (horizontal ground acceleration); and α_v is a seismic coefficient given $1/g \times$ (vertical ground acceleration).

The ground accelerations $\alpha_h g$ and $\alpha_v g$ would normally correspond to those for the *design earthquake* ground motions (Chapter 4), except as modified to allow for wall inertia effects, as discussed below.

The effect of vertical acceleration on the wall pressures has been shown to be small (Seed and Whitman, 1970), except in the case of gravity walls (Richards and Elms, 1979), as discussed below. Therefore, for non-gravity walls, the term α_v disappears from equation (9.23), which thus reduces to

$$P_{AE} = \frac{1}{2} K_{AE} \gamma_d H^2 \qquad (9.28)$$

and the expression for θ becomes $\theta = \tan^{-1} \alpha_h$.

In the conditions assumed in Coulomb's theory, where the shearing resistance is mobilized between the back of the wall and the soil, the earthquake soil pressure is calculated directly (Figure 9.7). For concrete walls against formwork, the wall friction δ may be taken as $1/2\phi$. The static active force P_A (Figure 9.7) may be found from the Coulomb equation

$$P_A = \frac{1}{2} K_A \gamma_d H^2, \qquad (9.29)$$

where

$$K_A = \frac{\cos^2(\phi' - \beta)}{\cos^2 \beta \cos(\delta + \beta) \left[1 + \sqrt{\left\{\frac{\sin(\phi' + \delta) \sin(\phi' - i)}{\cos(\phi' + \beta) \cos(i - \beta)}\right\}}\right]^2}. \qquad (9.30)$$

Mononobe and Okabe apparently considered that the earthquake force $\Delta P_{AE} = P_{AE} - P_A$ (Figure 9.7), calculated by their analysis, would act on the wall at the same position as the initial static force P_A, that is, at a height of $H/3$ above the base. This assumption is reasonable for flexible walls (Matthewson *et al.*, 1980), which rotate as required in the active state. Suggestions that the earthquake force from the Mononobe–Okabe analysis acts at a higher level appear to have been based on tests of more rigid construction which, of course, are not applicable to the active state. Equation (9.26) describes the general case, and becomes considerably simplified in the case of a wall with a vertical back face and horizontal fill, so that $\beta = \omega = 0$. In this case, equation (9.26) reduces to

$$K_{AE} = \frac{\cos^2(\phi' - \theta)}{\cos^2 \theta \left[1 + \sqrt{\left\{\frac{\sin \phi' \sin(\phi' - \theta)}{\cos \theta}\right\}}\right]^2}. \qquad (9.31)$$

A simple way of obtaining K_{AE} from K_A (for which design charts are available) has been derived by Arango, and is described by Seed and Whitman (1970).

While the Mononobe–Okabe analysis is widely accepted as the basis for seismic design of retaining walls in Coulomb conditions, Richards and Elms (1979) have shown that for *gravity walls* it needs modification to allow for the effect of wall inertia, which causes pressures of the same size as the dynamic pressure derived from the Mononobe–Okabe analysis given above. Their design method, based on a deflection criterion rather than stresses or stability, involves the following steps:

(1) Select values of peak ground acceleration A_g and the ground velocity V.
(2) Select the maximum allowable permanent horizontal displacement, d_L.
(3) Find the resistance factor N (where N_g is the acceleration at which the wall begins to slide), such that the actual permanent displacement will just equal d_L. For finding N, Richards and Elms recommend using an expression for the dimensionless parameter N/A given by

$$\frac{N}{A} = \left[\frac{0.087 V^2}{d_L A_g}\right]^{1/4}. \tag{9.32}$$

(4) However, N is equivalent to the limiting value of the seismic coefficient acting on the wall, α_h. Hence, we may find the active pressure coefficient K_{AE} from equation (9.26). The horizontal force due to the wall (weight W_w) is $\alpha_h W_w$, and the effective design weight of the wall for sliding is $(1 - \alpha_v) W_w$. The resistance to sliding is

$$F = (1 - \alpha_v) W_w \tan \theta_b, \tag{9.33}$$

where θ_b is the friction angle for the base of the wall.

Equating F to $\alpha_h W_w$ plus the horizontal components of P_{AE}, it is found that

$$W_w = \frac{\frac{1}{2} \lambda_d H^2 [\cos(\delta + \beta) - \sin(\delta + \beta) \tan \theta_b]}{\tan \theta_b - \tan \theta} K_{AE}. \tag{9.34}$$

To allow for uncertainties in their design method, Richards and Elms (1979) suggest using a safety factor of 1.5 on the above wall weight, so that the weight of the wall as built should be $1.5 W_w$.

As an improvement on the above design procedure for gravity walls, in an enlightening study of the uncertainties involved, Whitman and Liao (1984) propose replacing equation (9.32) with

$$\frac{N}{A} = -\frac{1}{9.4} \ln\left[\left(\frac{d_L}{F_c}\right) \frac{A_g}{130 V^2}\right], \tag{9.35}$$

where F_c is a safety factor on the allowable displacement d_L. An appropriate value of F_c may be found from the probability distribution of $d_R (= d_L/F_c)$, which appears to be lognormal. Thus, if the 95th percentile value is required, a value of $F_c = 3.8$ should be used. Then, using the value of N obtained from equation (9.34) would lead directly to the value of W_w from equation (9.35), which would be used directly in the design without applying the factor 1.5, as in step 4 above.

Active seismic pressures in cohesionless soils containing water

For cohesionless soils containing water, the above solution using the Mononobe–Okabe equations is not realistic, and attempts to use them by applying factors to the densities and using the apparent angle of internal friction ϕ_u and the Mononobe–Okabe equations would lead to solving the wrong problem.

The undrained situation is not only undesirable physically, but also difficult to analyse, hence it is recommended that good drainage should be provided to obviate the problem. Such drainage should be effective to well below the potential failure zone behind the wall, and also in front of the wall if cohesionless soils exist there so that the required passive resistance is available.

Active seismic pressures in cohesive soils or with irregular ground surface

The trial wedge method (Figure 9.8) offers the easiest derivation of seismic soil pressure when the material is cohesive or the surface of the ground is irregular. Figure 9.8 is drawn for Rankine conditions, and where the ground surface is very irregular the direction of P_{AE} may be taken as approximately parallel to a line drawn between points A and C. For Coulomb conditions the principles of the trial wedge method are similar and the direction of P_{AE} will be at an angle δ to the surface on which the pressure is calculated, similar to Figure 9.7.

Note that in seismic conditions, tension cracks may be ignored on the assumption that this introduces relatively small errors compared with others involved in the analysis. For saturated soils the appropriate density will have to be taken in determining W in Figure 9.8.

Completely rigid walls

Where soil is retained by a rigid wall, pressures greater than active develop. In this situation, the static and earthquake soil pressures may be taken as

$$P_E = P_0 + P_{0E} = \tfrac{1}{2} K_0 \gamma H^2 + \alpha_h \gamma H^2, \tag{9.36}$$

where γ is the total unit weight of the soil and K_0 is the coefficient of at-rest soil pressure.

As with the active pressure case discussed above, this equation should not be applied to saturated sands. For a vertical wall and horizontal ground surface, and for all normally consolidated materials, K_0 may be taken as

$$K_0 = 1 - \sin \phi' \tag{9.37}$$

where ϕ' is the effective angle of shearing resistance. For other wall angles and ground slopes, K_0 may be assumed to vary proportionately to K_A. The at-rest soil pressure force $P_0 = 1/2 K_0 \gamma H^2$ may be assumed to act at a height $H/3$ above the base of the wall, while the dynamic pressure $\Delta P_{0E} = \alpha_h \gamma H^2$ may be assumed (Matthewson et al., 1980) to act at a height of $0.58H$.

For gravity retaining walls, the at-rest force should be taken as acting normal to the back of the wall, while for cantilever and counterfort walls it should be calculated on the vertical plane through the rear of the heel, and taken as acting parallel to the ground surface.

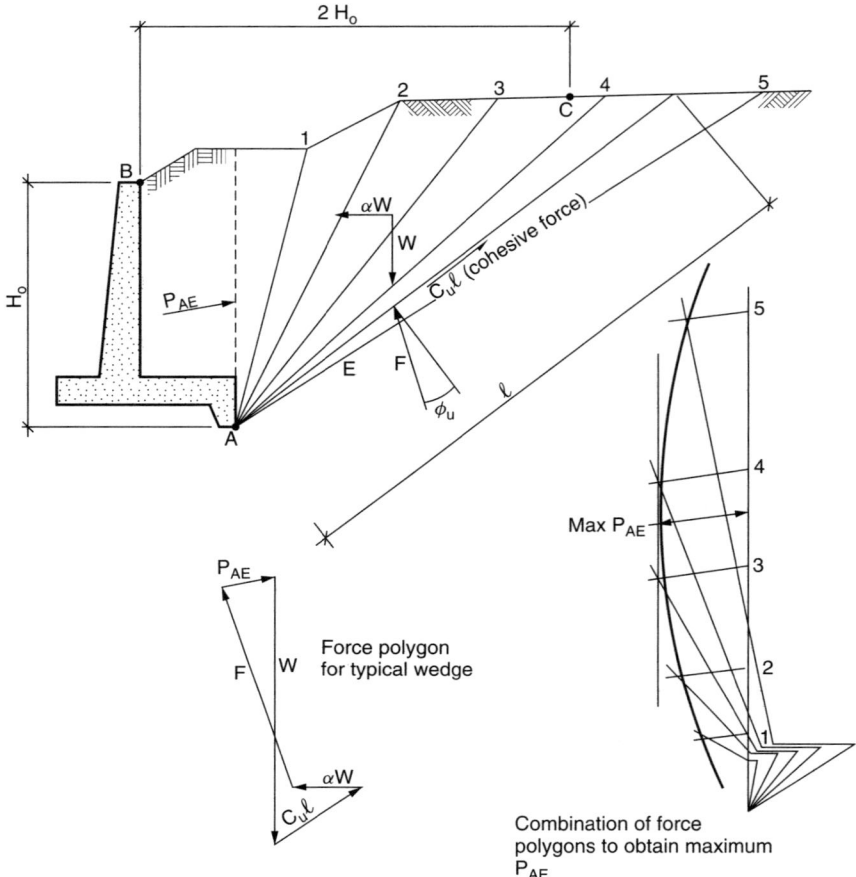

Figure 9.8 Trial wedge method for earthquake loading in Rankine conditions for cohesive soil or irregular ground surface

Seismic displacements of retaining walls

The serviceability of retaining walls after earthquakes is often dictated by whether their displacements are acceptable, rather than the forces that they can withstand. The design methods of Richards and Elms (1979) and Whitman and Liao (1984), referred to above, include methods of estimating displacements. In addition, such estimates may be made by finite-element analysis. However, the prediction of displacements is much less reliable than is desirable.

References

Berrill J and Yasuda S (2002) Liquefaction and piled foundations: Some issues. *J Earthq Eng* **6** (Special Issue): 1–41.

References

Bhattacharya S (ed.) (2007) *Design of Foundations in Seismic Areas: Principles and Applications*. National Centre of Earthquake Engineering, Indian Institute of Technology, Kanpur.

Blaney GW, Kausel E and Roesset JM (1976) Dynamic stiffness of piles. *2nd Int Conf on Numerical Methods in Geomech*, Blacksburg, USA, ASCE.

Broms BB (1964a) Lateral resistance of piles in cohesive soils. *Proc. ASCE, J Soil Mech and Found Div* **90**(SM2): 27–63.

Broms BB (1964b) Lateral resistance of piles in cohesionless soils. *Proc. ASCE, J Soil Mech and Found Div* **90**(SM3): 123–156.

Budhu M and Davies TG (1987) Nonlinear analysis of laterally loaded piles in cohesionless soils. *Canadian Geotech J* **24**: 286–296.

Budhu M and Davies TG (1988) Analysis of laterally loaded piles in soft clays. *Proc. ASCE, J Geotech Eng* **114**(1): 21–39.

Davies TG and Budhu M (1986) Nonlinear analysis of laterally loaded piles in heavily overconsolidated clays. *Geotechnique* **36**(4): 527–538.

Gazetas G and Dobry R (1984) Simple radiation damping model for piles and footings. *J Eng Mech* **110**(6): 937–956.

Iai S, Noda S and Tsuchida H (1988) Basic considerations for designing the area of the ground compaction as a remedial measure against liquefaction. *Proc. US–Japan Joint Workshop on Remedial Measures for Liquefiable Soils*.

Ishihara K, Koyamachi N and Kasuda K (1974) Strength of cohesive soil in irregular loading. *Proc. 8th World Conf on Earthq Eng*, San Francisco **III**: 7–14.

Kishida H (1967) Ultimate bearing capacity of piles driven in loose sand. *Soils and Found* **7**(3): 20–29.

Kramer SL (1996) *Geotechnical Earthquake Engineering*. Prentice Hall, Upper Saddle River, NJ.

Mathewson MB, Wood JH and Berrill JB (1980) Earth retaining structures. *Bull NZ Nat Soc Earthq Eng* **13**(3): 280–293.

Mononobe N (1929) Earthquake-proof construction of masonry dams. *Proc. World Eng Conf* **9**: 275–293.

NAVFAC (1982) *Foundations and Earth Structures*. Design Manual 7.2. Naval Facilities Engineering Command, Department of the Navy, Alexandra, VA, USA.

Nishizawa T, Tajiri S and Kawamura S (1984) Excavation and response analysis of damaged rc piles by liquefaction. *Proc. 8th World Conf on Earthq Eng*, San Francisco **III**: 593–600.

New Zealand Ministry of Works (1970) *Design of Public Buildings*, Code of Practice PW 81/10/1. NZMOW.

Ohsaki Y (1970) Effects of sand compaction on liquefaction during the Tokachioki earthquake. *Soils and Found* **10**(2): 112–128.

Okabe S (1926) General theory of earth pressures. *J Jap Soc Civil Engrs* **12**(1).

Pender MJ (1993) Aseismic pile foundation design analysis. *Bull NZ Nat Soc Earthq Eng* **26**(1): 49–160.

Pender MJ (1996) Earthquake resistant design of foundations. *Bull NZ Nat Soc Earthq Eng* **29**(3): 155–171.

Penzien J (1970) Soil-pile foundation interaction. In *Earthquake Engineering*. (ed. RL Wiegel). Prentice Hall, Englewood Cliffs, NJ, pp. 349–381.

Poulos HG and Davis EH (1980) *Pile Foundation Analysis and Design*. John Wiley & Sons, Inc., New York.

Richards R and Elms DG (1979) Seismic behaviour of gravity retaining walls. *J Geotech Eng Div* **105**(GT4): 449–464.

Seed HB (1960) Soil strength during earthquakes. *Proc. 2nd World Conf on Earthq Eng*, Tokyo, **1**: 183–194.

Seed HB and Booker JR (1977) Stabilization of potentially liquefiable sand deposits using gravel drains. *J Geotech Eng Div* **103**(GT7): 757–768.

Seed HB and Whitman RV (1970) Design of earth retaining structures for dynamic loads. *ASCE Specialty Conf on Lateral Stresses and the Design of Earth Ret Struc*, New York 103–118.

Solymar ZV and Reed DJ (1986) A comparison of foundation compaction techniques. *Can Geotech J* **23**(3): 271–280.

Taylor PW and Williams RL (1979) Foundations for capacity designed structures. *Bull NZ Nat Soc Earthq Eng* **12**(2): 101–113.

Whitman RV and Liao S (1984) Seismic design of gravity retaining walls. *Proc 8th World Conf on Earthq Eng*, San Francisco **III**: 533–540.

Wyllie LA, McClure FE and Degenkolb HJ (1973) Performance of underground structures at the Joseph Jensen filtration plant. *Proc. 5th World Conf on Earthq Eng*, Rome **1**: 66–73.

Zeevaert L (1983) *Foundation Engineering for Difficult Sub-soil Conditions*, 2nd edn. Van Nostrand Reinhold, New York.

10

Design and Detailing of New Structures for Earthquake Ground Shaking

10.1 Introduction

The design of any new item of the built environment (e.g. structures or lifelines) provides both an opportunity and a challenge to minimize earthquake risk to people and property within the resources available. To minimize risk, designers must minimize the seismic vulnerability of whatever is being designed. Any given structure may be subject to one or more of the earthquake induced hazards listed in Section 3.1, but this chapter is restricted to design for the basic phenomenon of ground shaking. Except for liquefaction, the other phenomena are largely matters for site selection (Chapter 3), site response (Chapter 5), regional planning (Chapter 7) or foundation design (Chapter 9), rather than the design of superstructure.

As discussed in Chapter 8, the designer will have a design brief, which leads to a preferred structural form and construction materials. In addition, the desired performance of the structure will have been agreed, based on accepting that implied by a code, or performance requirements for selected hazard levels or limit states such as may be selected from Table 8.1, or those of performance-based design (Section 8.2.2). In cases where it is desired to have a structure with especially low damage in the design earthquake, design approaches such as that discussed in Section 8.6 should be considered.

A key part of the design of any structure relates to its ductility: how much ductility it should have, how to design for it, and how to achieve the desired level of ductility. The structural ductility ratio μ, defined in Section 5.4.7, is widely used as the basis for determining the loading. Its effect on loading is considerable, as shown for example in Figure 10.1, which is the basic seismic hazard acceleration coefficient as a function of period of vibration and μ, as adopted by the New Zealand loadings standard. In general

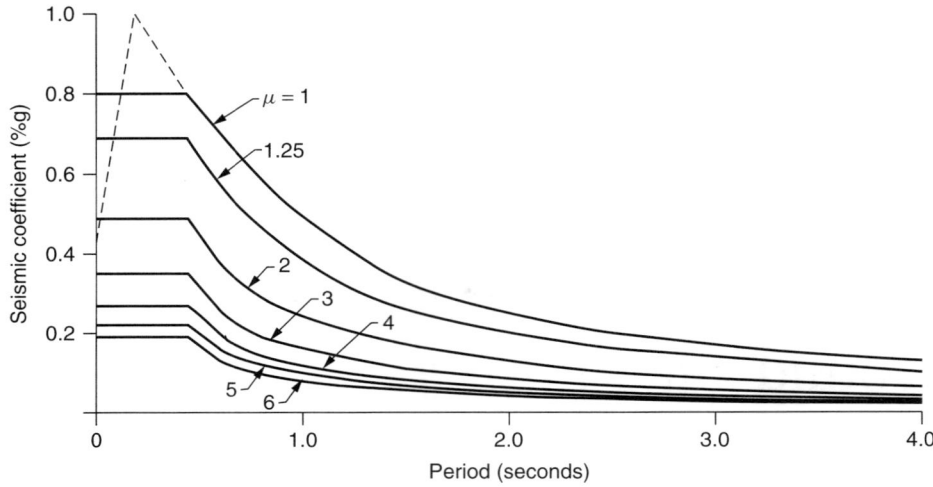

Figure 10.1 Basic seismic hazard acceleration coefficients for site ground class B from NZS 4203 (1992). (Reprinted by permission of Standards New Zealand)

it is difficult to achieve high ductilities of (say) $\mu \geq 6$. Structures are said to be fully ductile where $\mu \geq 4$ and of limited ductility if μ is 2 or 3.

The ductility factor μ that is appropriate for a given structure depends not only on the construction materials, but also on the structural configuration. This is illustrated in Table 10.1, which comes from a representative loadings code. It is noted that some materials (i.e. steel, reinforced concrete, and reinforced masonry) may be deemed to be slightly ductile ($\mu = 1.25$), allowing a small reduction in design loads, even when they are designed as elastically responding structures. Except where research shows otherwise, the values of μ in Table 10.1 are recommended in the code to be maximum permitted values. Also such ductility factors may not be appropriate for all structural elements of the construction material under consideration. For example, a squat reinforced concrete structural wall is not as ductile as a slender one and consequently the structural ductility factor shown in Table 10.1 should not be applied to such a wall without modification. (Note that more recent New Zealand earthquake codes have moved the ductility values from the loadings to the materials standards, but Figure 10.1 and Table 10.1 are retained as they illustrate the principles well.)

10.1.1 Strength-based vs. displacement-based design

Up to the time of writing this book, earthquake design codes worldwide have based their design methods on providing strength to resist earthquake forces. However, the advantages and rationality of using displacement criteria for ensuring adequate seismic performance have been widely recognized for over two decades, and procedures now known as *direct displacement-based design* (DDBD) have been in development for possible eventual adoption by design codes (e.g. He *et al*., 1984; and notably the book of Priestley *et al*., 2007). The latter demonstrate the application of inelastic design spectra to

Introduction

Table 10.1 Typical structural ductility factors, μ, for different categories of structure

Category of structure	Structural steel	Reinforced concrete	Prestressed concrete	Reinforced masonry	Timber
1. Elastically responding structures	1.0	1.0	1.0	1.0	1.0
2. Nominally ductile structures	1.25	1.25	1.25	1.25	1.25
3. Structures of limited ductility					
(a) Braced frames:					
(i) Tension and compression yielding	≤ 3	–	–	–	≤ 3
(ii) Tension yielding only (two storeys maximum)	≤ 3	–	–	–	≤ 3
(b) Moment resisting frame	≤ 3	≤ 3	≤ 2	≤ 2	≤ 3
(c) Walls	≤ 3	≤ 3	–	≤ 2	≤ 3
(d) Cantilevered face loaded walls (single storey only)	–	≤ 2	–	≤ 2	–
4. Fully ductile structures					
(a) Braced frames (tension and compression yielding)	4	≤ 6	–	–	–
(b) Moment resisting frames	≤ 6	≤ 6	≤ 5	≤ 5	≤ 6
(c) Walls	3	≤ 5	–	≤ 5	≤ 6
(d) Eccentrically braced frames	≤ 6	–	–	–	–

the displacement-based design of structures, and show that the procedure that uses elastic design spectra and equivalent linear systems does not necessarily satisfy the design criteria. In particular, it can erroneously imply that the allowable plastic rotation requirement has been satisfied. On the other hand, many engineers are not yet convinced that the displacement-based design procedure is superior to the traditional force-based approach. Problems that are cited include the difficulty in choosing the damping level and the fact that forces are in any case needed for the design of cladding on buildings.

The following discussion of the DDBD method is drawn from Priestley *et al.* (2007), who point out that the fundamental difference between it and force-based design is that DDBD characterizes the structure to be designed by a single-degree-of-freedom (SDOF) representation of the performance at peak displacement response, rather than its initial elastic characteristics. This is based on the *substitute structure* approach (see Shibata and Sozen, 1976). The basic philosophy is to design a structure which would achieve, rather than be bounded by, a given performance limit state under a given seismic intensity. The design procedure determines the strength required at designated plastic hinge locations to achieve the design aims through defined displacements.

In Figure 10.2(a) a frame building is represented by a SDOF model, while Figure 10.2(b) shows the bilinear envelope of the lateral force–displacement response of the SDOF representation. The initial elastic stiffness is K_i and the post-yield stiffness is rK_i. DDBD characterizes the structure by the secant stiffness K_e at maximum displacement Δ_d (Figure 10.2(b)), and a level of equivalent viscous damping ξ, representing the combined elastic damping and the hysteretic energy dissipated during inelastic response. Thus (Figure 10.2(c)), for a given level of ductility demand, a structural steel frame building will be assigned a higher level of equivalent viscous damping than a reinforced concrete

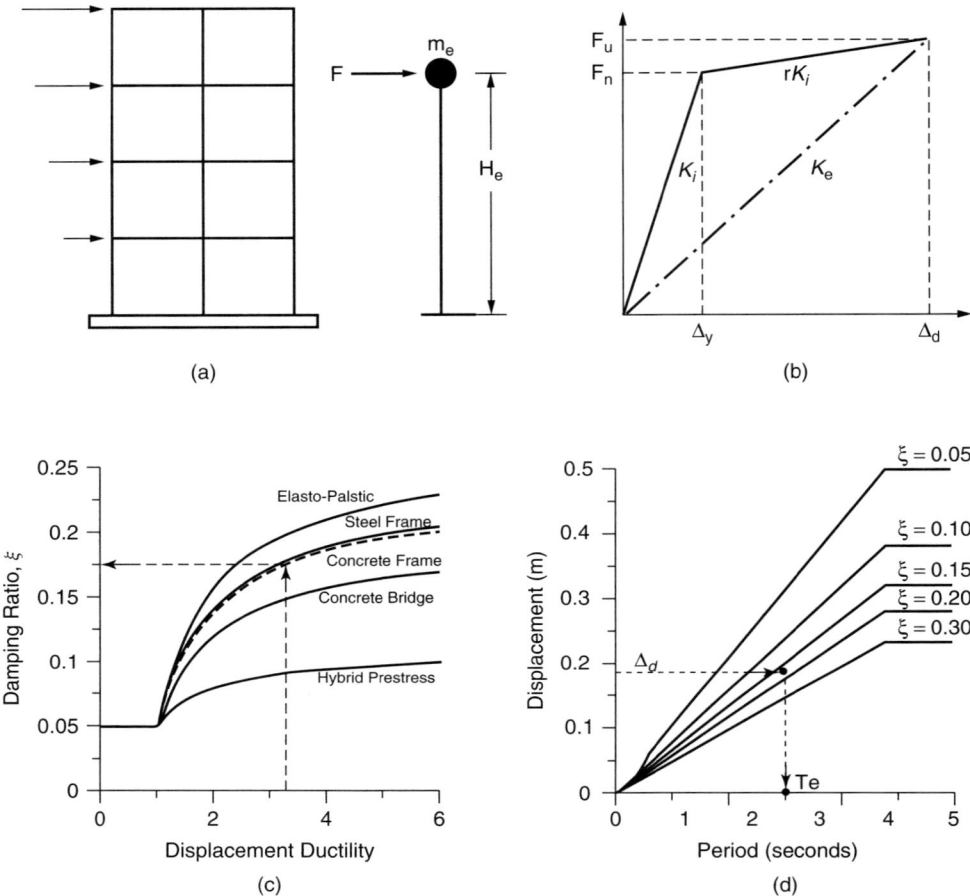

Figure 10.2 Fundamentals of direct displacement-based design (after Priestley *et al.*, 2007): (a) SDOF simulation; (b) effective stiffness K_e; (c) equivalent damping vs. ductility; (d) design displacement spectra. (Reproduced by permission of IUSS Press)

bridge column designed for the same level of ductility demand, as a consequence of having wider hysteresis loops, as seen in Figure 10.3.

Having determined the design displacement at maximum response (see Priestley *et al.*, 2007, Section 3.4.1) and the corresponding damping estimated from the expected ductility demand (see Priestley *et al.*, 2007, Section 3.4.3) measured at the effective height H_e (Figure 10.2(a)), the effective period T_e can be read from a set of displacement spectra for different levels of damping, as illustrated by the example of Figure 10.2(d). The effective stiffness K_e of the equivalent SDOF system at maximum displacement is found from the normal equation for the period of the SDOF system, rewritten as

$$K_e = 4\pi^2 m_e / T_e^2, \tag{10.1}$$

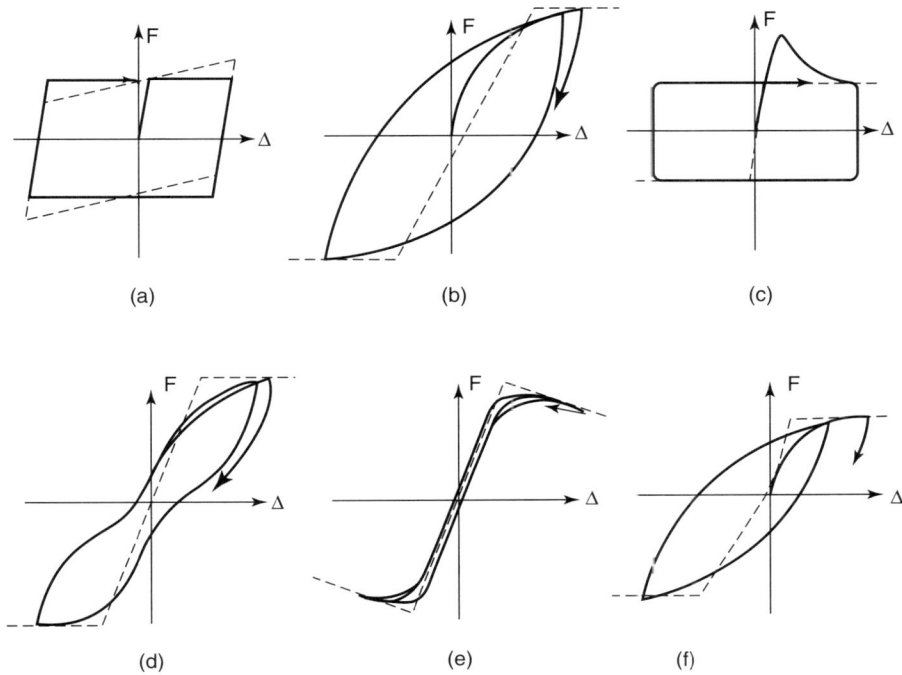

Figure 10.3 Common structural force–displacement hysteresis response shapes (after Priestley et al., 2007): (a) idealized steel frame response; (b) reinforce concrete frame response; (c) friction slider response; (d) bridge column with high axial load; (e) non-linear elastic with P-Δ; (f) asymmetrical strength. (Reproduced by permission of IUSS Press)

where m_e is the effective mass of the structure participating in the fundamental mode vibration. From Figure 10.2(b), the design lateral force, which is also the design base shear force, is

$$F = V_{\text{Base}} = K_e \Delta_d. \tag{10.2}$$

Thus the design concept is very simple. Any complexities arise in determining the character of the *substitute structure*, the design displacement, and the design displacement spectra. It is also necessary to give careful consideration to the distribution of the design base shear force V_{Base} to the discretized mass locations, and to the analysis of the structure under the distributed seismic force.

Priestley et al. also offer a slightly simplified design procedure which removes one step in the design process, by combining the damping–ductility relationship for a specific hysteresis rule with the seismic displacement spectral demand in a single inelastic displacement spectra set, where the different curves relate directly to displacement ductility demand. This is illustrated in the example of Figure 10.4. The inelastic displacement spectra set is entered with the design displacement and the design effective period is

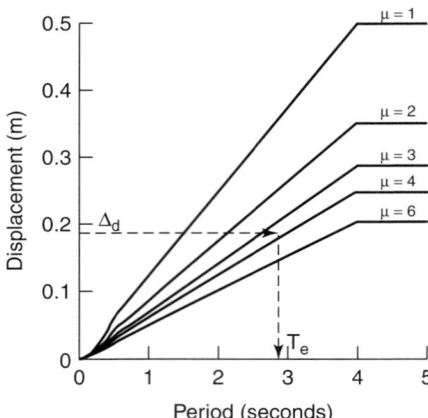

Figure 10.4 Example of an inelastic displacement spectra set related to effective period for a specific hysteresis rule (after Priestley et al., 2007). (Reproduced by permission of IUSS Press)

read off for the level of design displacement ductility. Although this is a slightly simplified procedure, it requires the inelastic displacement spectra to be generated for different hysteresis rules for each new seismic intensity considered.

The above design procedure is illustrated by Priestley et al. by the following example. A five-storey reinforced concrete frame building is to be designed to achieve a design displacement of 0.185 m, corresponding to a displacement ductility demand of $\mu = 3.25$. The seismic weight contributing to first mode response is 4500 kN. Using the design information of Figures 10.2(c) and (d), the required base shear strength is to be determined.

Entering Figure 10.2(c) at a ductility of 3.25, and moving up to the curve for concrete frames (follow the dashed lines and arrows and arrows), the damping ratio is found to be $\xi = 0.175$. Entering Figure 10.2(d) with a design displacement of $\Delta = 0.185$ m and moving horizontally to the line corresponding to a damping ratio of $\xi = 0.175$, the effective period is found to be $T = 2.5$ s. The effective mass is $m_e = 4500/g$ where $g = 9.805$ m/s^2. Hence, from equation (10.1), the effective stiffness is

$$K_e = 4\pi^2 m_e / T_e^2 = 4\pi^2 \times 4500/(9.805 \times 2.5^2) = 2900 \text{ kN/m}.$$

Finally, from equation (10.2), the required base shear force is

$$V_{\text{Base}} = K_e \Delta_d = 2900 \times 0.185 = 536 \text{ kN}.$$

The above design procedure must then extend to capacity design procedures to ensure that plastic hinges occur only where intended, and that brittle modes of deformation do not develop. The capacity design procedures must be calibrated to the displacement-based design approach, as discussed by Priestley et al., who show that the capacity design requirements are generally less onerous than those for force-based designs, resulting in more economical structures.

10.2 Steel Structures

10.2.1 *Introduction*

Because of the inherent ductility available in appropriately manufactured and fabricated steel, structures made from this material have been less liable to collapse in earthquakes than traditionally designed concrete or masonry ones. Nevertheless, care is needed in the design and fabrication of steel structures to ensure that local or even global failure is not induced by strong earthquake shaking.

10.2.2 *Seismic response of steel structures*

The seismic response of steel structures depends mainly upon:

(1) the onset of instability (local or global);
(2) the nature of the steel members;
(3) the nature of the connections; and
(4) the nature of other components interacting with the frame.

Under ideal conditions of lateral restraint, repeatable high ductility of a very stable nature can be obtained, as shown by the hysteresis loops for bending illustrated in Figure 5.23(b), but with less restraint to webs or flanges marked loss of strength and stiffness may occur as shown in Figures 10.8 and 10.9. Similar deterioration in strength occurs as a result of damage from *low-cycle fatigue*, a phenomenon which increases with increasing strain. For example, Bertero and Popov (1985) found that a 100×100 mm WF beam failed after 607 cycles when strained to $\pm 1.0\%$, while strain amplitudes of $\pm 2.5\%$ produced failure after only 16 cycles.

Commonly, members fail at plastic hinges after local buckling has increased the strains at the surface sufficiently to initiate cracking which rapidly reaches the critical size for fast fracture. In members which are well detailed for local buckling (Section 10.2.4), with workmanship that keeps notches or notch-like defects to a minimum, good welding details (Heavy Engineering Research Association (HERA), 1995) and the use of normally available weldable steels (Section 10.2.3), low-cycle fatigue should be controlled.

Different types of connection affect response, depending upon the damping that they produce. For example, elastically responding steel structures typically have up to 2% of critical damping if fully welded, compared with up to 7% if fully bolted.

The response of steel structures, particularly in terms of damping, is greatly influenced by *components interacting* with the steel structure itself (e.g. cladding, partitions, floors and chimney linings). The combined effects of stress level, connections, and interacting components (clad or unclad) are indicated by the typical design values of equivalent viscous damping given in Table 10.2 (damping is further discussed in Section 5.4.4).

A further feature of the seismic response of steel is the increase in yield strength exhibited with increase in *rate of loading*. Normal quasi-static tests of yield stress f_y are conducted at low strain rates of about 10^{-3}/s. Under seismic loading conditions in short-period structures local strain rates may be in excess of 1.0/s, causing increases in f_y of 30% or so over the quasi-static value. While no increase in strength exists at fracture, the increase in yield strength is reflected in the stress–strain relationship throughout the usable inelastic strain range.

Table 10.2 Typical damping ratios for steel structures, ξ, percentage of critical from NZS 3404 (1997). (Reproduced by permission of Standards New Zealand)

Type of structure	Stress state	Welded connections	Bolted connections
Clad[1]	Elastic	4–6	5–10
Clad	Inelastic	5–7	10–15
Unclad[1]	Elastic	2–3	5–7
Unclad	Inelastic	5–7	10–5

[1] *Unclad* refers to open industrial frameworks (perhaps with web grating steel flooring or platforms). *Clad* refers to most other structures such as offices.

Wakabayashi *et al.* (1984) found from experimental work that the dynamic yield stress f_{yd} is given by

$$\frac{f_{yd}}{f_y} = 1 + 0.473 \log\left(\frac{\dot{\varepsilon}}{\dot{\varepsilon}_0}\right), \tag{10.3}$$

where $\dot{\varepsilon}$ is the strain rate, and $\dot{\varepsilon}_0 = 50 \times 10^{-6}$/s.

Separate studies by Udagawa *et al.* (1984) and Wakabayashi *et al.* (1984) found that the monotonic and cyclic loading strength of steel members was up to 20% greater at their rate of loading than the strength under quasi-static conditions.

10.2.3 Reliable seismic behaviour of steel structures

Introduction

For obtaining reliable seismic response behaviour the principles concerning choice of form, materials, and failure mode control discussed in Section 8.3 apply to steel structures, while further factors specific to steel are discussed below.

Designing for failure mode control requires consideration of the structural forms used, with all the forms discussed in Section 8.4 being appropriate for steel:

(1) moment resisting frames;
(2) frames tube structures;
(3) structural walls;
(4) concentrically braced frames;
(5) eccentrically braced frames; and
(6) hybrid (or composite) structures.

In designing these structural forms with failure mode control in mind, in addition to the discussion in Section 8.3.8, it should be noted that the essential objectives are as follows:

(a) beams should fail before columns (unless extra column strength is provided);
(b) premature instability failure modes should be suppressed; and
(c) an appropriate degree of ductility should be provided.

If a structure is designed to have 'limited ductility' (i.e. $\mu \leq 3$), and hence designed to resist higher code loadings as discussed in Section 5.4.7 the element design criteria of most non-seismic steel codes are likely to be adequate. For structural elements of fully ductile structures (i.e. where $\mu > 3$), the design details must ensure that full plastic deformations can be obtained. The rotation of plastic hinges under strong seismic shaking is considerably in excess of those envisaged by non-seismic steel codes, and more stringent stability requirements are needed to maintain section capacity.

While hot rolled steel sections are generally preferable to *cold formed sections*, because the latter have limited ductility, cold formed sections may be used in earthquake resistant structures provided that the appropriate measures are taken. In the same way that timber (which is brittle) may be protected by ductile connections so as to form ductile structures (Section 10.5), so may relatively brittle cold formed steel sections be joined by ductile steel plates at appropriate locations to ensure that ductile failure modes occur.

The collapse mechanisms of the so-called plastic design method may not always be consistent with the objectives of seismic failure mode control, but it may be used as long as the objectives of the latter are met and excessive lateral sway is avoided.

Material quality of structural steel

For the construction of reliable earthquake resistant ductile steel frames, the basic steel material must, of course, be of good quality. While steels suitable for seismic resistance are found amongst those produced for general structural purposes, not all normal structural grades are sufficiently ductile. The main properties required are as follows:

(1) adequate ductility;
(2) consistency of mechanical properties;
(3) adequate *notch* ductility;
(4) freedom from lamination;
(5) resistance to lamellar tearing;
(6) good weldability.

Ductility
Ductility may be described generally as the post-elastic behaviour of a material (Section 5.4.2). For steel it may be expressed simply from the results of elongation tests on small samples, or more significantly in terms of moment–curvature of hysteresis relationships, as discussed later in this chapter. Steels manufactured in various countries may have sufficient ductility, and earthquake codes of practice often recommend suitable steels. There are many different steel products on the market, such that the New Zealand steel structures standard NZS 3404 (1997) lists four Australian, ten British and seven Japanese standards that suitably control the performance of steel for ductile behaviour as required for earthquake resistance. These standards are listed here in Appendix B.

Consistency of mechanical properties
In economically designing so that beams fail before columns, it is desirable that the maximum and minimum strengths of members are as nearly equal in magnitude as possible. This means that the standard deviation of strengths should be as small as possible. While

it is satisfactory for non-seismic design, it is unfortunate for earthquake resistant design that steel manufacturers have been more concerned with simply achieving their minimum guaranteed yield strengths, than in producing consistent ultimate strengths.

Notch ductility

Notch ductility is a measure of the resistance of a steel to brittle fracture and is a separate property from that of general ductility discussed earlier in this section. Adequate notch ductility is required in *all* structural steelwork, not only in seismic areas of the world. It is generally expressed as the energy required to fracture a test piece of particular geometry. Three widely used tests are the Charpy V-notch test, the Izod test, and Charpy keyhole test. The results, although quantitative, are generally empirical and are not comparable between tests and between materials.

Laminations

Laminations are large areas of unbonded steel found in the body of a steel plate or section. This implies a layering of the steel with little structural connection between the layers. The laminated areas originate in the casting and cropping procedures for the steel ingots, and may be several square metres in extent. Steel may be screened ultrasonically for lamination before fabrication.

Lamellar tearing

It should first be pointed out that lamellar tearing should not be confused with laminations, the two being different phenomena. Lamellar tearing is a tear or stepped crack which occurs under a weld where sufficiently large shrinkage stresses have been imposed in the through thickness direction of susceptible material (Figure 10.5). It commonly occurs in T-butt welds and in corner welds and is caused by inclusions which act as 'perforations' in the steel. Lamellar tearing has been discussed in some detail by Farrar and Dolby (1972). Unfortunately, no non-destructive method of screening for susceptibility to lamellar tearing is as yet available. The usual method of checking is by measuring the ductility in a through plate tensile test.

Electric arc steelmaking incorporating vacuum degassing can produce steels with reduced (not eliminated) susceptibility to lamellar tearing – although at some extra cost. The risk of lamellar tearing can also be reduced by the use of suitable welding techniques and details (Farrar and Dolby, 1972).

Weldability

Weldability may be considered simply as the capacity of the parent metal to be joined by sound welds. The weld metal should be able to closely match the properties of the parent plates, and few material defects should arise. To some extent, weldability will be assured by the use of steels produced according to major national standards, such as those referred to above, or British Standard 7668, 'Weldable structural steels'. The weldability of a steel is often assessed by means of a formula based on the chemical analysis of the steel. Such methods determine the preheating temperature necessary to avoid hydrogen cracking. In general, the higher the tensile strength of the steel, the lower is its weldability (Coe, 1973; Wade, 1972).

Steel Structures

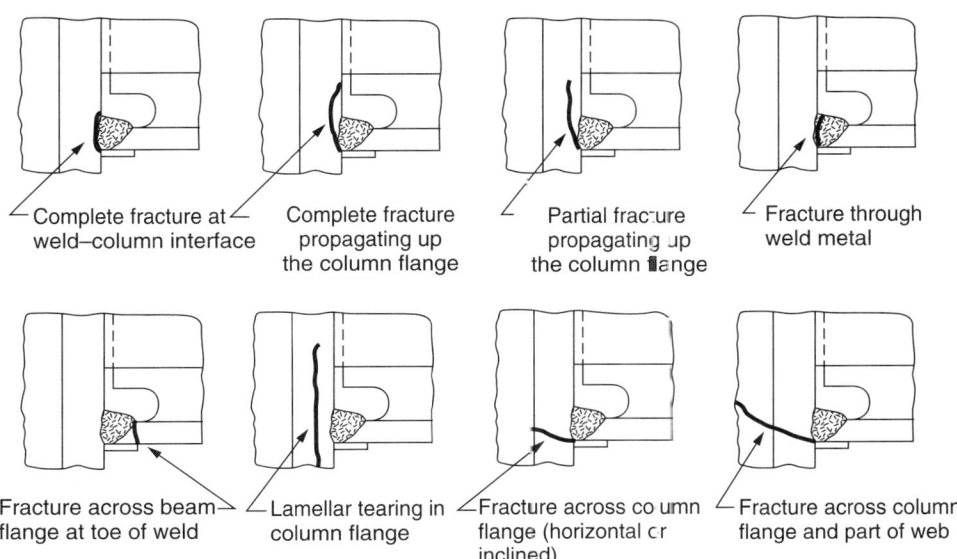

Figure 10.5 Fracture modes observed at welded beam bottom flange to column flange connection in the 1994 Northridge, California, earthquake (from Krawinkler, 1996)

It is sobering to consider the fact that even after several decades of constructing thousands of welded structures, the 1994 Northridge earthquake caused many local failures associated with welds in moment resisting frames in California (Bertero *et al.*, 1994; Krawinkler, 1996). Examples of such failures are shown in Figure 10.5. Those failures caused extensive experimental investigations aimed at finding more reliable welding details (e.g. Uang and Latham, 1995). It is noteworthy that, in contrast to the Northridge experience, post-1981 steel buildings in Kobe performed very well in the powerful earthquake experienced there in 1995 (Park *et al.*, 1995).

10.2.4 Steel beams

In this section, the behaviour and design of beams acting primarily in bending will be considered. In most beams axial forces are small enough to be neglected, but where large axial forces may occur column design procedures should be employed.

Moment–curvature relationships for steel beams under monotonic loading

For the adequate seismic design of the steel beams, and the associated connections and columns, the moment–curvature or moment–rotation relationship should be known. A long stable plastic plateau is required which is not terminated too abruptly by lateral or local buckling effects, such as indicated by terminating at points A, B and C in Figure 10.6. The curves terminating at D and E are typical of the desired behaviour achieved by well-designed beams under moment gradient and uniform moment, respectively. The

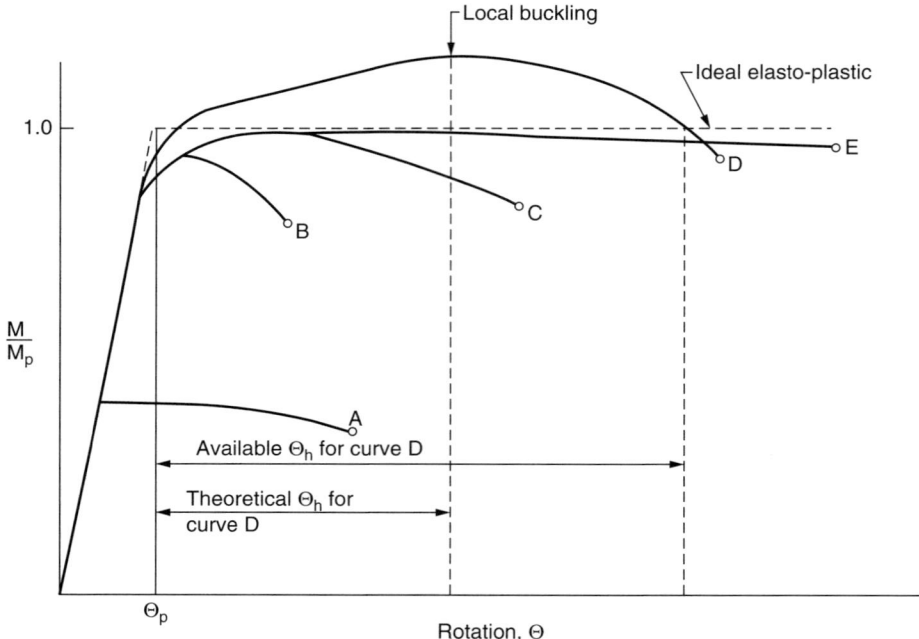

Figure 10.6 Behaviour of steel beams in bending

moments in Figure 10.6 have been normalized in terms of the plastic moment capacity,

$$M_p = S f_y, \qquad (10.4)$$

where S is the plastic modulus of the section, and f_y the characteristic yield stress of the steel.

Moment gradient is the usual loading condition to be considered with plastic hinges forming at the ends of beams in laterally loaded frames. The localization of high stresses produced by the moment gradient causes strain hardening to occur during plastic rotation, resulting in an increase in moment capacity above the ideal plastic moment M (curve D in Figure 10.6). Strain hardening may increase the plastic moment by as much as 40% (Erasmus, 1984). Local buckling and lateral buckling arising from plastic deformation of the compression flanges generally produce a reduction of moment capacity in the later stages of rotation, as illustrated by curve D in Figure 10.6.

To predict the rotation capacity of a plastic hinge the following expression presented by Lay and Galambos (1967) for the monotonic inelastic hinge rotation θ_h (Figure 10.7) of a beam under *moment gradient* may be used:

$$\theta_h = 2.84 \varepsilon_y (\beta - 1) \frac{b \, t_f}{d \, t_w} \left(\frac{A_w}{A_f} \right)^{1/4} \left(1 + \frac{V_1}{V_2} \right), \qquad (10.5)$$

where b is flange width, d overall depth of section, t_f flange thickness, t_w web thickness, A_f flange area, A_w web area, V_1 and V_2 are absolute values of shears acting either

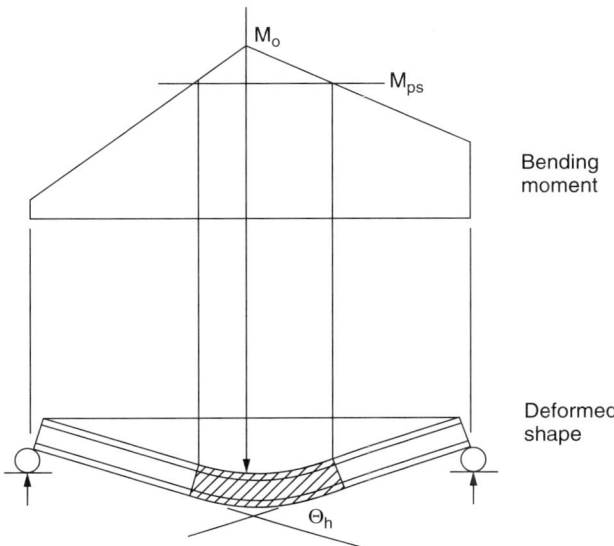

Figure 10.7 Beam under moment gradient with plastic hinge deformations and the hinge rotation θ_h of equation (10.5) as defined by Lay and Galambos (1967). (Reproduced by permission of the American Society of Civil Engineers)

side of the hinge, arranged so that $V_1 \leq V_2$, β is the ratio of strain at onset of strain hardening to strain at first yield, and ε_y is strain at first yield. θ_h represents a substantial proportion of the total rotation capacity of the beam (Figure 10.6). For the American section 10 WF25 (A36 steel), equation (10.5) predicts that $\theta_h = 0.07$ radians. It should be noted that equation (10.4) incorporates simplifications which lead to underestimations of θ_h of 20% or more.

The degree to which the plasticity of a section is utilized in rotation may be expressed by the rotation capacity R, which is a ratio of the plastic hinge rotation to the rotation at or near first yield. Under monotonic loading, the rotation capacity is a function only of the beam section properties and its lateral supports, and decreases as some inverse function of the slenderness ratio l/r_y. Using the definition

$$R_1 = \frac{\theta_h}{\theta_p} - 1, \qquad (10.6)$$

where $\theta_p = M_p l / EI$, Takanashi *et al.* (1973) have shown for typical Japanese beam sections that R_1 exceeds 10 for l/r_y less than about 40, and $R = 2$ for l/r_y of about 100.

A similar alternative to equation (10.6) calculates the rotation capacity from

$$R = \frac{\theta_h}{\theta_y}, \qquad (10.7)$$

where θ_y is the elastic rotation between the far ends of the beam segment up to the formation of the hinge.

The rotation capacity of plastic hinges may be subject to reduction under cyclic loading, as discussed below.

Behaviour of steel beams under cyclic loading

In steel frames designed to make good use of inelastic resistance in earthquakes, several reversals of strain of 1.5% or more may have to be withstood. As discussed in Section 10.2.2, stable repetition of the monotonic ductile capacity of beams, as measured by θ_h or R in equations (10.5) and (10.7), may not be possible under cyclic loading to higher strains. The hysteretic degradation of strength observed by Vann *et al.* (1973) was mainly due to web buckling, but flange buckling and lateral torsional buckling, plus low-cycle fatigue, can also have similar effects.

The rate at which strength degradation occurs is, of course, significant to design. It is of interest that Vann *et al.* (1973) found that the strength of an American W8 × 13 I-section (Figure 10.8) had degraded to 72% of its plastic moment after 11 load cycles. This behaviour would be acceptable for full ductile design as strength degradation of up to 20% after four load cycles represents robust cyclic behaviour.

Design of steel beams

As may be concluded from the above discussion, the key factor in maintaining beam strength under seismic loading is the provision of stiffness or restraints to control local and lateral buckling. The requirements of the New Zealand code set out in Tables 10.3

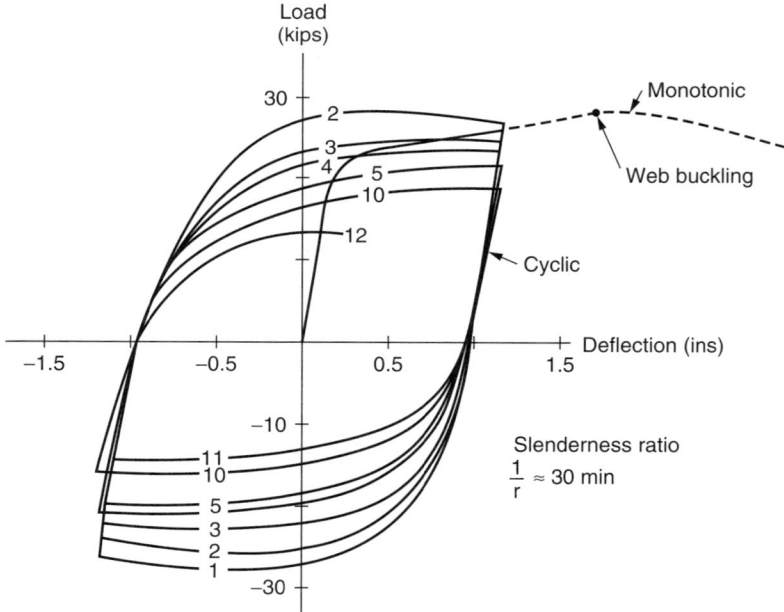

Figure 10.8 Hysteresis loops for a steel beam under moment gradient (after Vann *et al.*, 1973)

Table 10.3 Values of plate element slenderness limits for hot rolled steel, simplified from NZS 3404 (1997). (Reproduced by permission of Standards New Zealand)

Case number	Plate element type	Longitudinal edges supported	Category 1 members[1] (λ_{e1})	Category 2 members[2] (λ_{e2})	Category 2 members[3] (λ_{e3})	Category 4 members[4] (λ_{e4})
1	Flat (Uniform compression)	One	9	9	10	25
2	Flat (Maximum compression at unsupported edge, zero stress or tension at support edge)	One	9	9	10	25
3	Flat (Uniform compression)	Both	25	30	40	60[3]
4	Flat (Either non-uniform compression or compression at one edge, tension at the other)	Both	30	40	55	75
5	Circular hollow section		35	50	65	170

[1] Fully ductile structure $\mu > 3$.
[2] Limited ductility, $3.0 \geq \mu > 1.25$.
[3] $\mu = 1.25$.
[4] $\mu = 1.0$.

Table 10.4 General limit on $N^*/\phi N_s$ as a function of member category (ductility demand) from NZS 3404 (1997). (Reproduced by permission of Standards New Zealand)

Category member (see Table 10.2)	$N^*/\phi N_s$ not to exceed:	
	in a column member	in a brace in an eccentrically braced frame
1	0.5	–[1]
2	0.7	–[1]
3	0.8	0.8
4	1.0	1.0

[1] This category of member is not appropriate for this application.

and 10.4, show the increasing stability needs with increasing ductility demands. In Table 10.3,

$$\lambda_e = \frac{b}{t}\sqrt{\frac{f_y}{250}}, \tag{10.8}$$

where b is the clear width of the element, and t is plate thickness.

10.2.5 Steel columns

Monotonic and hysteretic behaviour of steel columns

Columns are often required to resist appreciable bending moments as well as axial forces. The moment–curvature relationships for the so-called 'beam column' are similar to those for beams under uniform moment, except that the capacity is reduced below the beam plastic moment M_p by the presence of axial load, as shown in Figure 10.9.

As indicated in Figure 10.9, the full moment M_{pc} of columns may not be developed because of local buckling or lateral torsional buckling as for beams. Although columns should generally be protected against inelastic cyclic deformation by prior hinging of the beams, some column hysteretic behaviour is likely in strong earthquakes in most structures, and even with beam hinging mechanisms column hinges (or pins) are required at the lowest point in columns, as shown in Figure 8.9(b).

The behaviour of steel columns under cyclic bending is similar to that of beams without axial load, except that the axial force added to the bending moment concentrates the yielding in the regions of larger compressive stress. This leads to a more rapid decay of load capacity owing to more extensive buckling, as may apparently be inferred by comparing Figure 10.10 with Figure 10.8. Second-order bending ($P \times \Delta$ effect) may also be important in the inelastic range.

Design recommendations which allow for the effects to axial load and restraint are discussed below.

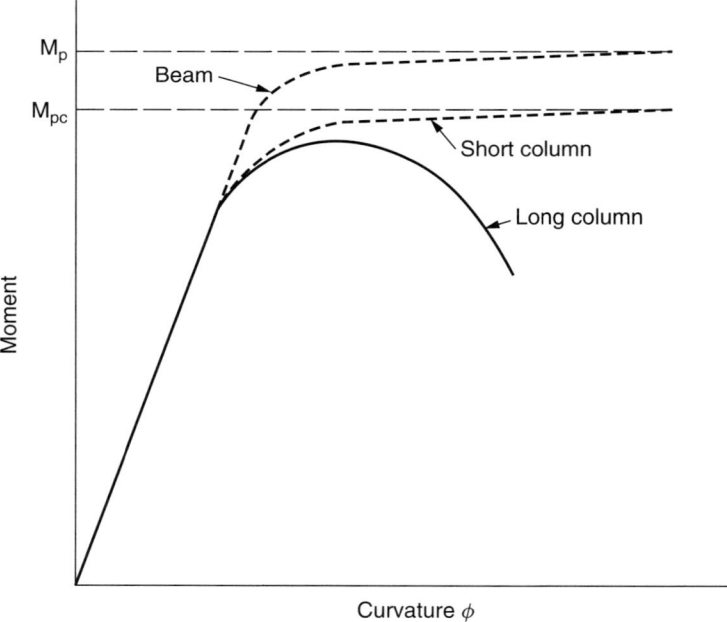

Figure 10.9 Typical moment–curvature relationships for short and long columns compared with pure beam behaviour

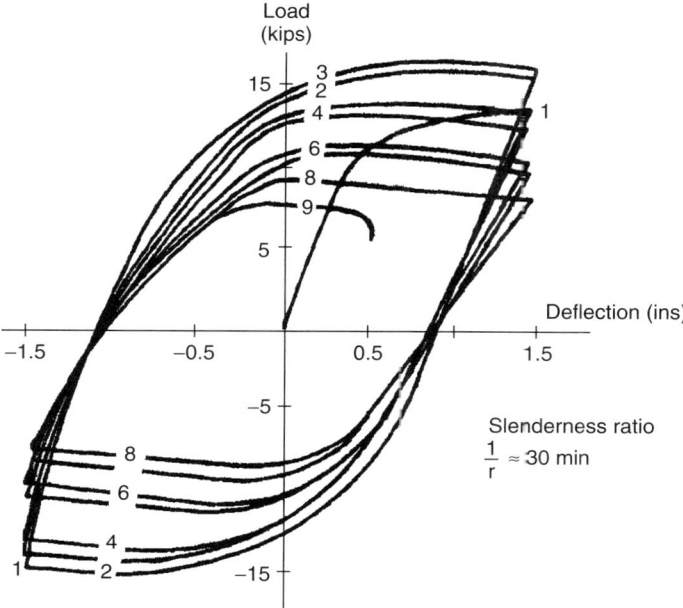

Figure 10.10 Hysteresis loops for a steel member under cyclic bending with a constant axial force (after Vann *et al.*, 1973)

Design of steel columns

Column ductility

According to NZS 3404 (1997), the degree of ductility which a column should be designed to supply is a function of the level of axial compression N^* expressed as a fraction of its yield compression capacity ϕN_s, where ϕ is strength reduction factor (typically 0.8 or 0.9), and $N_s = A f_y$ is nominal section capacity, as given in Table 10.4.

In addition, for category 1, 2 and 3 column members, excluding brace members of concentrically and eccentrically braced frames, the following limitation on design axial compression applies. When capacity design (see Section 8.3.3) is not undertaken,

$$\frac{N^*}{\varphi N_s} \leq \left[\frac{0.263(\beta_m + 1)^{0.88}}{e^{0.19/(\beta_m+1)}} \right]^{\lambda_{\text{EYC}}}, \qquad (10.9a)$$

where N^* is the design axial compression force, β_m is 0 for columns forming part of a seismic resisting system and 0.5 for columns forming part of an associated structural system, and

$$\lambda_{\text{EYC}} = \sqrt{\frac{N_s}{N_{OL}}}$$

in which

$$N_{OL} = \frac{\pi^2 EI}{L^2},$$

with I being the second moment of area for the axis about which the design moment acts, i.e. the axis perpendicular to the plane of the structural system, and L the actual length of the member.

When capacity design is undertaken and the column is a secondary element,

$$\frac{N^*_{OC}}{\varphi N_s} \leq \left[\frac{0.263(\beta_m + 1)^{0.88}}{e^{0.19/(\beta_m+1)}}\right]^{\lambda_{EYC}} \quad (10.9b)$$

where N^*_{OC} is the capacity design derived design axial compression force on the column when the column is a secondary element.

Effective length

In earthquake resistant design, special considerations regarding effective length of columns arise through the effects of inelasticity and drift limitations. Alternatively, if the effective lengths of columns are to be calculated it should be noted that effective length factors for inelastic columns will be the same only when the column acts as an independent member. In other cases (i.e. when an inelastic column is part of a continuous frame) its effective length should be calculated appropriately. Estimating the effective length of a steel member can be quite complex, and the reader is recommended to read modern earthquake code requirements, such as in NZS 3404 (1997).

Lateral buckling

Columns designed to respond elastically in earthquakes may be designed to the normal non-seismic rules for lateral restraint against buckling, but extra precautions are required for the development of limited or full ductility as set out in Table 10.5. As shown in the

Table 10.5 Spacing of lateral restraints for steelwork (adapted from Walpole and Butcher, 1985)

	Parts of members requiring full ductility $R = 24$, $\alpha = 0.75$		Parts of members requiring limited ductility $R = 10$, $\alpha = 1.0$	
Flange length L_y where the compression flange is fully yielded	$L_y \geq 480a$	$L_y \geq 480a$	$L_y \geq 640a$	$L_y \geq 640a$
Spacing of braces with length L_y	$\leq 480a$	one brace required	$\leq 640a$	one brace required
Spacing to brace adjacent to length L_y	$\leq 720a$	$\leq 720a$	$\leq 960a$	$\leq 960a$

[1] Parts of members responding elastically should be braced according to allowable stress.
[2] Symbols are defined in Section 10.2.5.

table, appropriate levels of rotation capacity R, as defined by equation (10.7), are $R = 10$ for structures of limited ductility ($R = 10$ is normally used for non-seismic plastic design), and $R = 24$ for fully ductility structures. The spacings in Table 10.5 are based upon the length $640\alpha a$, where

$$\alpha = \frac{1.5}{(1 - R/8)^{1/2}}, \tag{10.10}$$

$$a = \frac{r_y}{f_y^{1/2}} \tag{10.11}$$

and L_y is the length of column (or beam) over which the compression flange is fully yielded, which may be taken as occurring where

$$L_y = \text{length of member for } M_{res}^* > C_1 \phi M_s \tag{10.12}$$

where M_{res}^* is the design bending moment at the point under consideration for calculation of the length of yielding region, $C_1 = 0.85$ for $N^*/\phi N_s \leq 0.15$, $C_1 = 0.75$ for $N^*/\phi N_s \leq 0.15$, ϕ is a strength reduction factor, typically 0.8 or 0.9, $M_s = M_s$ or M_r, and M_r is the section moment capacity reduced by axial load.

Local buckling
The section geometry limits for controlling local buckling of steel columns are the same as for beams as given in Table 10.3.

Forces in struts
The maximum compressive load capacity of struts not subjected to bending may be taken as

$$N_{ac} = \frac{A_s F_{ac}}{0.6}, \tag{10.13}$$

where F_{ac} is the maximum compressive stress as a function of the slenderness ratio, calculated on a permissible stress basis, and A_s is the sectional area of the member.

Combined axial load and moment
Where uniaxial bending occurs about the major principal axis, the design bending moment M_x^* should satisfy

$$M_x^* \leq \phi M_{rx}, \tag{10.14}$$

where ϕ is the strength reduction factor and

$$M_{rx} = M_{sx}(1 - (N^*/\phi N_s))$$

is the nominal section moment capacity, reduced by axial force (tension or compression), with N^* being the design axial force, N_s the nominal section capacity for axial force, and M_{sx} the nominal section capacity for bending moment.

The equivalent y-axis expression is used for uniaxial being about the minor axis, and where biaxial bending occurs the following condition should be satisfied:

$$\frac{N^*}{\phi N_s} + \frac{M_x^*}{\phi M_{sx}} + \frac{M_y^*}{\phi M_{sy}} \leq 1.0. \qquad (10.15)$$

Shear in columns
In the unusual circumstances where the shear stress is high in a column (i.e. $V/V_p \geq 2/3$), the interaction formula for moment, axial load and shear given by Neal (1961) is appropriate:

$$\frac{M}{M_p} + \left(\frac{N}{N_y}\right)^2 + \frac{(V/V_p)^4}{1 - (N/N_y)^2} \leq 1.0, \qquad (10.16)$$

where $V_p = 0.55\ dt f_y$.

10.2.6 Steel frames with diagonal braces

Diagonally braced frames are discussed under two classifications, depending on whether the braces create perfect triangulation or not, namely concentrically braced frames (CBFs) and eccentrically braced frames (EBFs). Reliable ductility under cyclic loading is much more readily obtained from EBFs than CBFs, as will be evident from the following discussion.

Concentrically braced steel frames

Introduction
The general characteristics of CBFs have been described in Section 8.4.4, to which some further points are added here. As illustrated in Figure 10.11, the braces may be arranged in either of two ways: (1) the braces may lie along the lines between beam–column joints (Z- or X-braces); or (2) pairs of braces may meet at a point along a beam (V-braces, also called K-braces). In case (1) the diagonals of Z-bracing should be arranged in opposing pairs (as in Figure 10.11), rather than all sloping in the same direction, so as to avoid the larger residual sway deflections that occur in asymmetrically braced frames. V-bracing is likely to be inferior to X-bracing because the resistance of V-bracing to storey shears will be governed by the compression resistance of the braces.

The effects of instability in reducing the strength of braces under cyclic loading include those described for columns (Section 10.2.5), but are more serious because of additional effects of tensile yielding and because braces are generally more slender than columns. Various workers (e.g. Asteneh-Asl et al., 1984) have described the rapid loss in strength, stiffness, and energy dissipation capacity that occurs under inelastic cyclic loading of V- and X-braced frames.

Design of concentrically braced frames
The ultimate limit-state design seismic loads for CBFs with bracing effective in tension and compression may be obtained using a coefficient C_h appropriate to the chosen structural ductility factor (Figure 10.1), multiplied by the factor C_s which varies according to

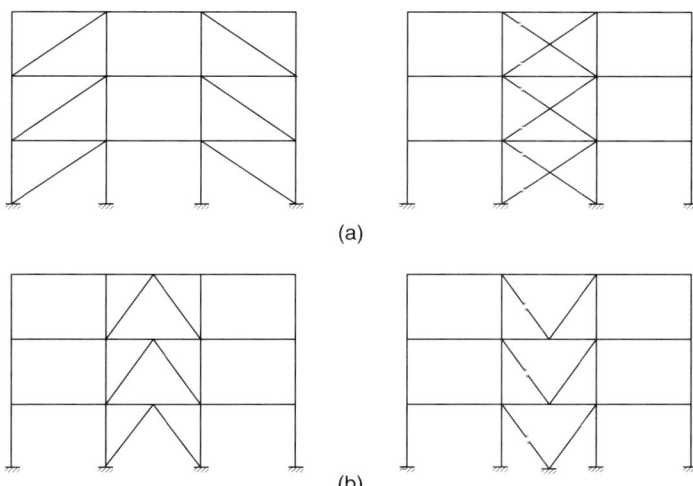

Figure 10.11 Typical concentrically braced frames with recommended symmetry of bracing for (a) Z- or X-braces and (b) V- or K-braces

Table 10.6 C_s factors for determination of ultimate limit-state design seismic loads for CBFs with bracing effective in tension and compression (extract from NZS 3404, 1997). C_s factors for (a) Category 1 and (b) Category 2 CBFs. (Reproduced by permission of Standards New Zealand)

(a) Category 1 systems	Compression brace slenderness ratio $\dfrac{k_e L}{r}\sqrt{\dfrac{f_y}{250}}$		
Number of storeys	≤30	≤80	≤120
1	1.0	1.3	1.6
2–3	1.1	1.45	1.75
4–5	1.2	1.55	1.9
6–8	1.3	1.7	2.1
(b)			
1	1.0	1.2	1.45
2–4	1.1	1.3	1.55
5–8	1.15	1.4	1.7
9–12	1.25	1.5	1.8

the slenderness ratio $k_e L/r$ and the ductility category of the structure. Values of C_s for two categories (1 and 2) are given in Table 10.6. The values of C_s decrease further for the reduced ductility demands implied by structure categories 3 and 4; see NZS 3404 (1997).

More severe restrictions (i.e. higher loads) are placed on CBFs in which the bracing is effective in tension only.

High-ductility CBFs are penalized in comparison to low-ductility ones in terms of the maximum numbers of storeys that are considered safe, as illustrated in Table 10.7. If the

Table 10.7 Height restrictions for category 1 and 4 CBFs by number of storeys (extract from NZS 3404, 1997). (Reproduced by permission of Standards New Zealand)

Brace type		Compression brace slenderness ratio $\dfrac{k_e L}{r}\sqrt{\dfrac{f_y}{250}}$		
		≤30	≤80	≤120
X-brace	Cat. 1	8	4	2
	Cat. 4	32	24	16
V-brace	Cat. 1	4	2	1
(Chevron)	Cat. 4	16	12	8

bracing of CBFs is made from threaded rods, it is recommended (Walpole, 1985b) that the structure be treated as elastically responding, i.e. it should be designed using $\mu = 1$.

CBFs are well suited to protection from earthquakes using base isolation techniques (e.g. Figure 8.17), partly because they are relatively low-period structures (Section 8.5.2). Also, a suitable location for energy-dissipating devices is in the diagonals either as discussed in Section 8.5.7, or by simply deliberately bending the diagonals so that they must form moment hinges. The use of such devices would greatly increase the number of mass levels which could be built with acceptable reliability compared with the restrictions noted above.

Eccentrically braced steel frames

Introduction
The general characteristics of EBFs have been described in Section 8.4.5, to which some further points are added here. As illustrated in Figure 10.12, EBFs are formed by deliberately creating eccentricities, e, with Z- and K-braces (V-braces) such that moments and shears exist in the short length of beams known by terms such as the *link beam* or *shear link*. The ductility of this link beam may be utilized to obtain reliable seismic ductile response of the frame as a whole.

Referring to Figure 10.12, the following comparative merits are noted:

- Figure 10.12(a) is well suited to short column spacings.
- Figure 10.12(b) is the best for column safety, as the link (with its heavy welds) is located away from the columns.
- Figure 10.12(c) minimizes the rotation angle θ for a given lateral displacement.

As noted earlier, one of the advantages of braced frames is the reduction in lateral drift compared with moment resisting frames (MRFs). This is illustrated by the graph of lateral stiffness (Figure 10.3) for a rectangular frame plotted as a function of the bracing eccentricity e, which varies from zero (a CBF) to unity (an MRF).

Considerable research has been done in California (e.g. Roeder and Popov, 1978; Kasai and Popov, 1984; Hjelmstad and Popov, 1983) on the response of I-section shear links and eccentric Z-bracing subject to cyclic loading, and the design recommendations given below are based mainly on this work.

Steel Structures

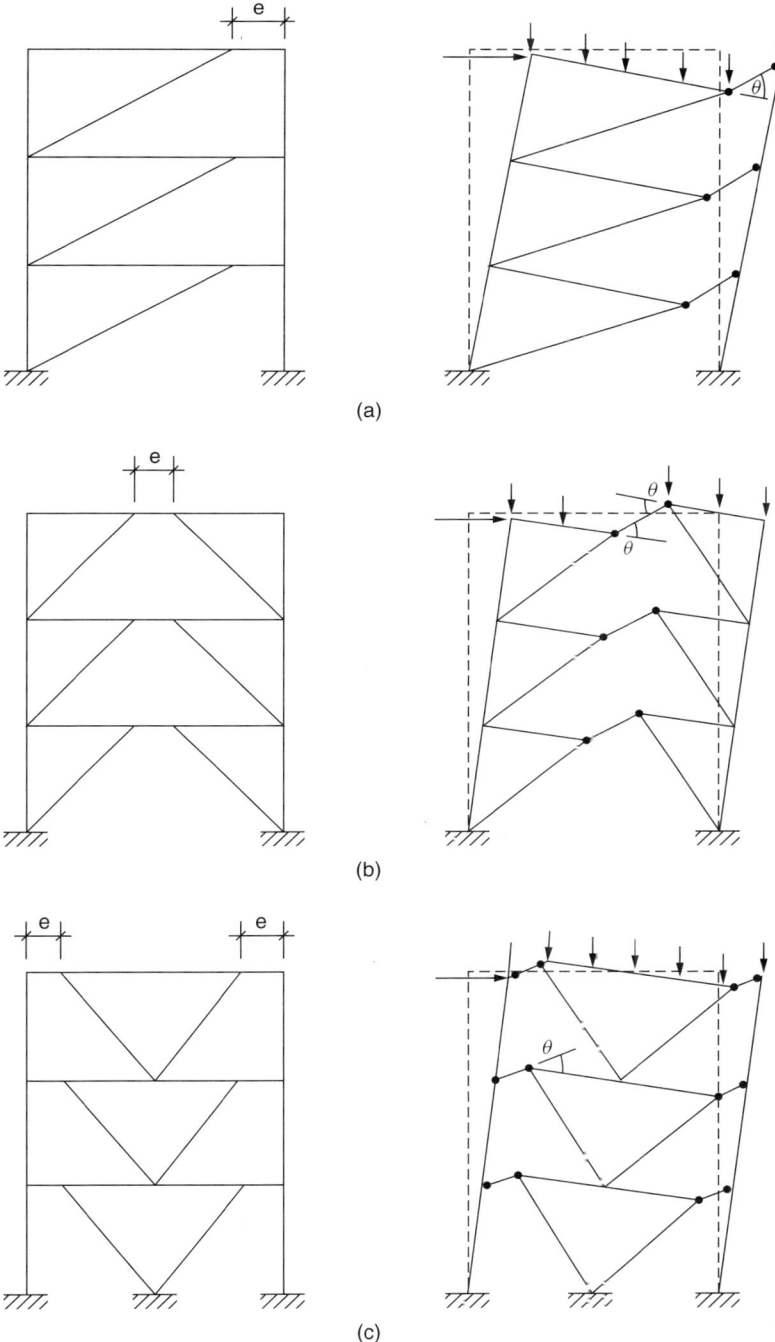

Figure 10.12 The three preferred eccentrically braced frames, showing kinematics of deformation

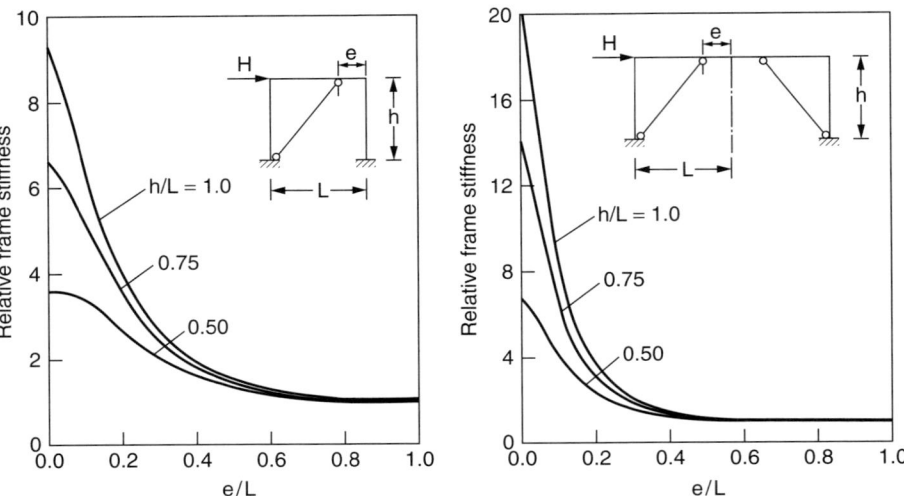

Figure 10.13 Stiffness of braced frames varying with eccentricity (after Hjelmstad and Popov, 1983)

Design of ductile link beams

For EBFs in which the active links are intended to dissipate inelastic energy mainly in shear, the clear length of the active link, e, is

$$e \leq 1.6 M_s/V_w, \tag{10.17}$$

where M_s and V_w are the nominal capacities in moment and shear, respectively.

The geometry of the inelastically deformed EBF bay shall be such that the rotation angle, θ_p, between a beam and active link, for active links not attached to columns, shall not exceed the following:

$$\theta_p = \pm 0.09 \text{ rad}, \quad \text{for } e \leq 1.6 M_{sp}/V_w,$$

$$\theta_p = \pm 0.045 \text{ rad}, \quad \text{for } e \leq 3 M_{sp}/V_w,$$

θ_p is determined by interpolations for $1.6 M_{sp}/V_w < e < 3 M_{sp}/V_w$, where M_{sp} is the nominal plastic moment capacity. For active links attached to column flanges or column webs, the limits on θ_p should be 0.09 and 0.045 radians, respectively.

Design of other components of EBFs

For failure mode control, it is necessary to ensure that the remainder of the frame is sufficiently stronger than the link beams. This requires consideration of the potential overstrength of the link beam based on an upper limit on its yield strength, rather than the characteristic strength, and the cracking strength of the floor at the link zone should be minimized and allowed for in the frame design.

10.2.7 Steel connections

Introduction

Connections as well as members should be designed to conform to the failure mode controls for the structure concerned. Thus, unless a connection is required to yield prior to the adjacent members as part of an energy-absorbing scheme, as is sometimes done with holding-down bolts, it is usual to design each connection to carry greater loads than the members entering it. In addition, the panel zones of beam–column joints should have stiffness appropriate to the assumptions made in the analysis of frame response.

Connections should also be designed to make fabrication and erection of the framework as simple and quick as possible. They should not be too sensitive to factory or field tolerances, and should minimize the use of highly skilled crafts. Connections should also permit adequate inspections to be made at the time of construction as proper quality control of fabrication processes, particularly welding, is, of course, essential. Important aspects of workmanship are discussed elsewhere (McKay, 1985).

Butt welding, fillet welding, bolting, and riveting may be employed for seismic connections, either individually or in combination. As fully bolted or riveted connections tend to be very large and expensive, fully welded connections or a combination of welding and bolting are most frequently used. Bolts have the advantage of contributing more damping to frames than welding (Section 10.2.2).

Behaviour of steel connections under cyclic loading

Compared with beam and column elements, relatively few cyclic load tests have been carried out on steel connections. Popov and Pinkney (1969) tested five types of joint, two involving minor-axis bending of the column and three involving major-axis bending of the column. The latter three joints (Figure 10.14) were of the following types: a butt-welded joint, a fillet-welded joint using flange plates, and a joint using high-strength bolts and flanges plates. In the tests it was found that the butt-welded joints were superior to the other two types in terms of total energy dissipation. In the bolted joints, the hysteresis loops were reduced in area considerably by slippage, although the use of smaller than normal oversize holes reduced this effect. All the joints sustained loads in excess of their design limit values until the onset of cracking.

In tests on connections using fully welded and flange welded-web bolted joints by Popov and Stephen (1972) very large increases in bending strength (up to 69%) due to strain hardening were observed.

The comparative cyclic load behaviour of bolted connections in the snug tightened and fully tightened conditions has not been well established. In tests by Popov and Pinkey (1969), although the degree of tightness is not clear, some pinching of the hysteresis hoops indicates the effect of slip on the faying surfaces. The extent of pinching was reduced for holes drilled only 0.4 mm oversize instead of the 1.6 mm oversize, which was standard in the USA.

Deformation behaviour of steel panel zones

The panel zone of a connection between two members is the intersection zone common to the two members. This zone is assumed to deform in shear, as indicated in

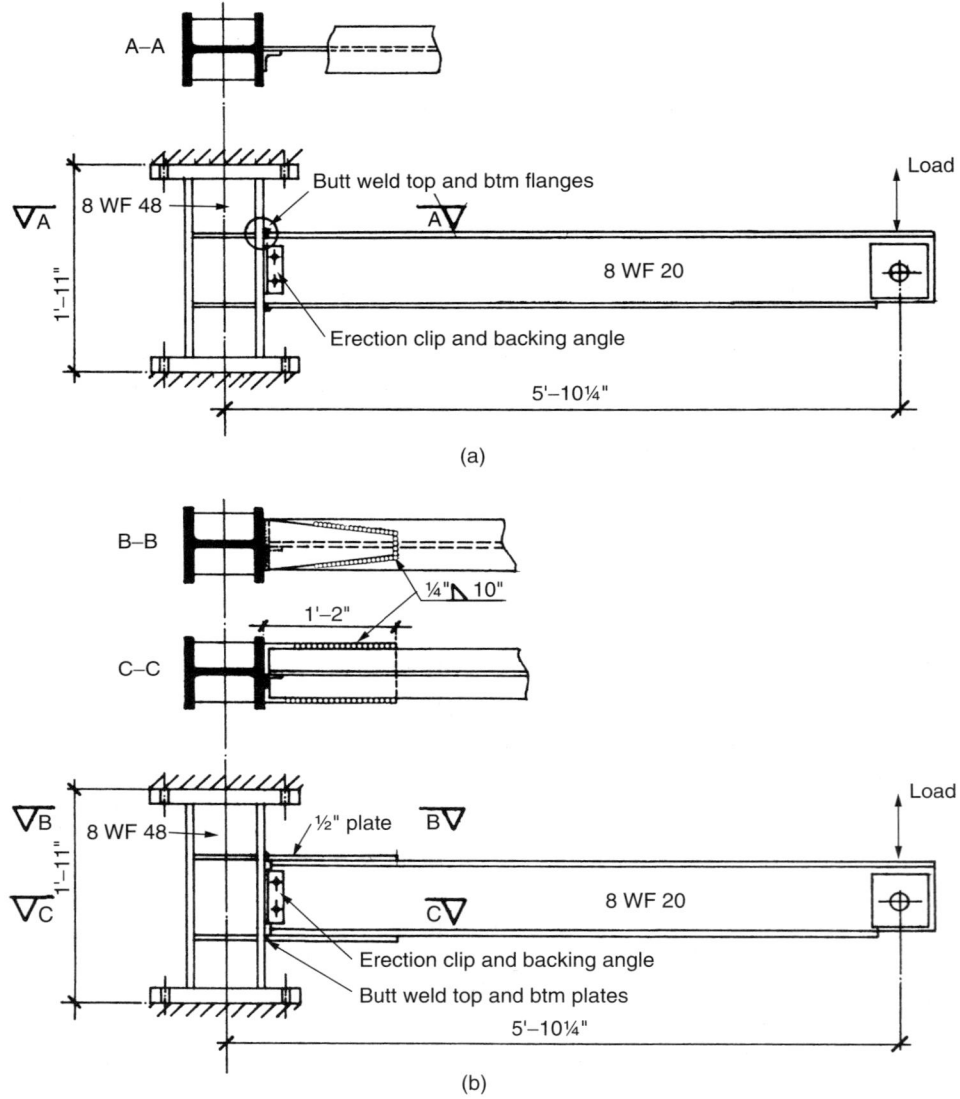

Figure 10.14 Beam–column connections with major axis column bending tested by Popov and Pinkney (1969): (a) butt-welded beam–column joint; (b) fillet welded beam–column joint; (c) bolted beam–column joint. (Reproduced by permission of the American Society of Civil Engineers)

Figure 10.15(a). Kato and Nakao (1973) have suggested a trilinear relationship between the shear stress and shear strain as a good approximation to the results of their monotonic tests on Japanese **H**-section connections (Figure 10.15(b)). Although little testing has been done on the deformation characteristics of panel zones, specially under cyclic loading, it has been demonstrated (Teal, 1968) that the deformation of beam–column connections

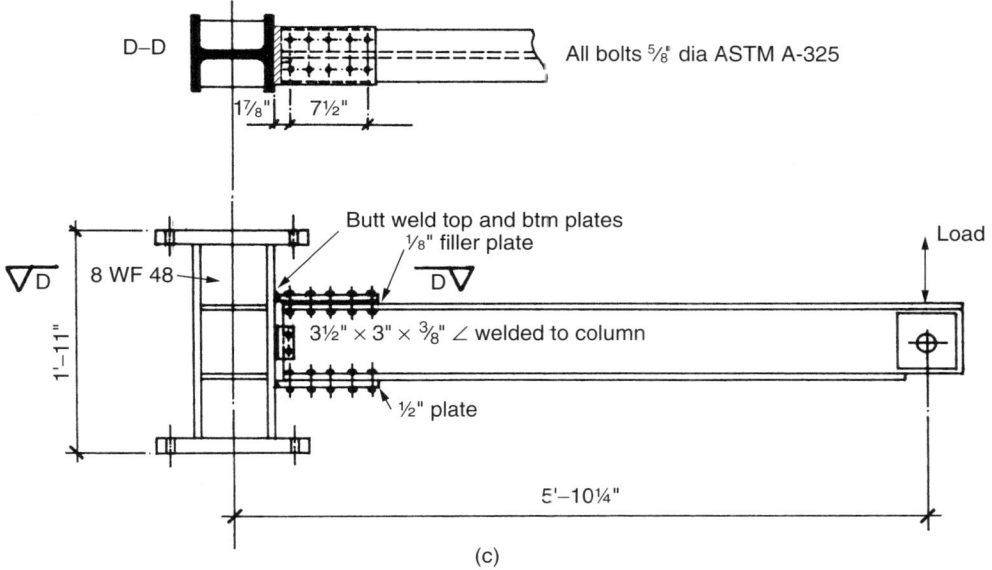

Figure 10.14 (*continued*)

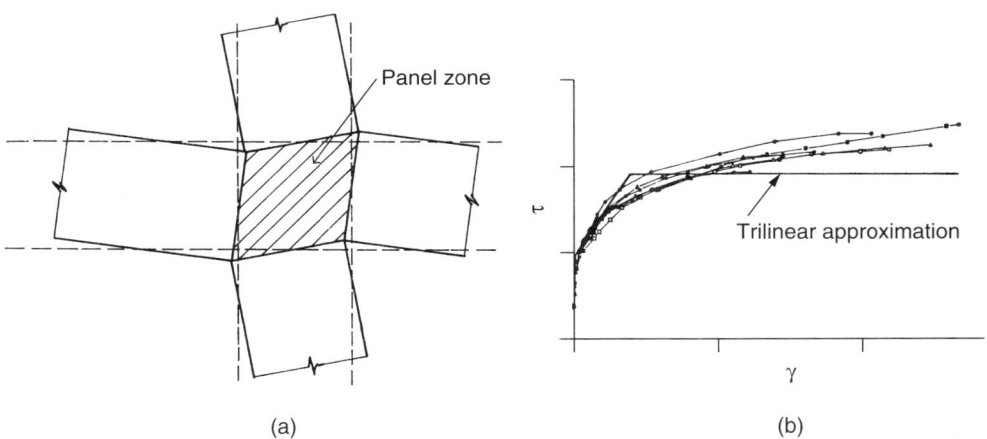

Figure 10.15 Idealized shear deformation of beam–column panel zones

may contribute up to about one-third of the inter-storey deflection in multi-storey buildings, and of this deformation about half may arise from the shear deformation of the panel zone itself.

The large influence of panel zone behaviour on overall frame strength and stiffness has also been indicated by Kato (1974). If bilinear hinges were assumed in the panel zones, the ultimate shear resistance of the frame was developed only in the frame members.

However, in fully ductile frames it appears to be economic to permit some yielding in the panel zones ($\mu \approx 3$), as well as in the beam plastic hinges.

Design of steel connections for seismic loading

The previous sections give the background to the following introduction to the exacting task of providing well-detailed connections for seismic steelwork. In addition, it is noted that all the components of connections should be arranged to give a smooth stress flow between members, so that stress-raising notches and sharp re-entrant angles should be avoided. Fuller recommendations on connection design have been given elsewhere (HERA, 1995; Nicholas, 1985; Walpole, 1985a).

Design forces for connections

As discussed earlier, connections are usually designed to be stronger than the adjacent members, the strength of which should be based on some probability that the actual strength will exceed the guaranteed minimum strength. Typical increases are indicated below:

Guaranted minimum f_y	Average f_y
250 MPa (36 psi)	$1.15 f_y$ min.
350 MPa (51 psi)	$1.10 f_y$ min.

Allowance should also be made for increase in strength beyond yield point due to strain hardening. Combining both those effects, the design forces for connections in *fully ductile* structures should be derived using $1.5 f_y$, and in structures of *limited ductility* $1.35 f_y$ should be used. Such design forces need not exceed those applicable if the structure were designed to be elastically responding.

There is also concern (Nicholas, 1985) that connections may be subjected to larger forces than those given by the analysis due to unpredictable movements of the structure, and it was therefore recommended that connections be able to withstand the following minimum forces:

- 50% of the member strength in compression or tension ($0.5 A_s f_y$) – this requirement is severe and need only be applied when the design axial forces are significant;
- 30% of the member strength in flexure ($0.3 Z f_y$);
- 10% of the member strength in shear ($0.5 A_v f_y$).

Welding

Full penetration *butt welds* are the best means of load transfer, while partial penetration butt welds should not be used in areas of stress reversal. *Fillet welds* are also acceptable for load transfer provided that a variety of design controls are practised. For example, the throat thickness should not be less than half the thickness of the plate being welded. Nicholas (1985) and McKay (1985) describe seismic design rules for welds which should be used in conjunction with normal practice for welding.

Bolting

The design and performance of bolted connections are affected by the following factors:

(1) The size of the hole and method of protection, i.e. the hole should be snug on the bolt and/or the bearing stress should not exceed f_y under ultimate load conditions.
(2) The conditions of the faying (mating) surfaces affect the frictional load transfer, and the effect of different paint systems should be considered.
(3) The threaded portion of the bolt may lie in the shear zone. On smaller contracts it may be impractical to avoid this.
(4) The bolt-tightening procedure affects the design method because fully tightened bolts are more reliable under seismic loading than snug tightened bolts.

More detail on bolting is given by Nicholas (1985) and Walpole (1985a).

Cleats

Cleats should be treated as sub-members in their own right and should be designed for effects such as eccentricity of the applied forces, buckling, bearing, punching and splitting.

Beam–column joints

Beam–column joints are obviously one of the most common types of connection in steelwork, and the principles given above apply to them. Allowance needs to be made for reduction in section due to bolts holes, and for the stiffness that may be required for local stress effects in webs and flanges both within the panel zone and adjacent to it. Within the panel zone (Figure 10.15) the shear strength may be found using the von Mises criterion for yielding,

$$\left[\frac{N}{N_y}\right]^2 + \left[\frac{V}{V_p}\right]^2 \leq 1.0, \tag{10.18}$$

where $N_y = A_s f_y$ is the axial load yield capacity of the beam, $V_p = 0.55 d t_w f_y$, and

$$V = \frac{M_{b1} + M_{b2}}{d_b} - V_{c1} + V_{c2} \tag{10.19}$$

in which d_b is the depth of the beam.

It will often be necessary to use doubler plates to increase the web area to comply with equation (10.18). Alternatively, diagonal stiffeners may be used (Figure 10.16), in conjunction with horizontal stiffeners, to reduce the shear force V acting on the web by

$$V = f_y A_{st} \cos \beta, \tag{10.20}$$

where A_{st} and f_y are the area and specified yield stress of the diagonal stiffener, respectively, and β is the angle of the stiffener to the horizontal.

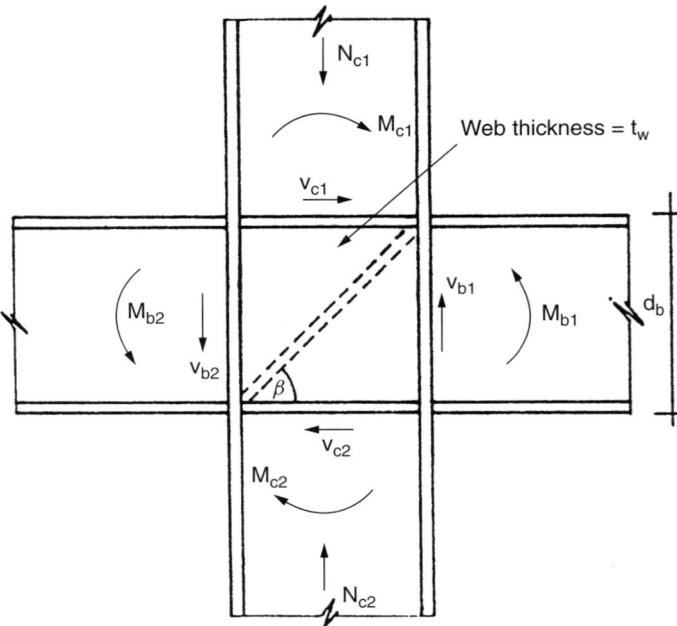

Figure 10.16 Forces acting on a typical panel zone

The seismic design of beam–column joints has been discussed in more detail by Walpole (1985a).

Connections in diagonally braced frames

In addition to the above principles, special considerations relating to failure mode control arise in the design of connection in diagonally braced frames, as discussed elsewhere (Sidwell, 1985; Walpole, 1985b; HERA, 1999).

10.2.8 Composite construction

In many steel structures, particularly multi-storey buildings, the steel acts compositely with concrete which is used for floors or fire protection of columns. Obviously, the concrete may add strength and stiffness to the steel frame, but for failure mode control it may also imply additional forces to be dealt with in any given member, increasing the overstrength demands from adjacent members.

The principles outlined in the previous sections of this chapter apply to composite construction, but special seismic requirements exist, as discussed by Clifton (1985).

10.2.9 Further reading

For further information on the design and detailing of steel structures, various special design guides should be referred to, such as those for New Zealand conditions by HERA,

Reports R4-76 (HERA, 1995) including Amendment No.1 (2000), or the equivalent document of the American Institute of Steel Construction.

10.3 Concrete Structures

10.3.1 Introduction

There is more information available about the seismic performance of reinforced concrete than any other material. No doubt this is because of its widespread use, and because of the difficulties involved in ensuring its adequate ductility (robustness). Well-designed and well-constructed reinforced concrete is suitable for most structures in earthquake areas, but achieving both these prerequisites can be problematical even in countries of advanced technology.

Reinforced concrete is generally desirable because of its wide availability and economy, and its stiffness can be used to advantage to minimize seismic deformations and hence reduce damage to non-structure. Difficulties arise due to reinforcement congestion when trying to achieve high ductilities in framed structures, and the problem of detailing beam–column joints to withstand strong cyclic loading remains a difficult and contentious problem. It should be recalled that no amount of good detailing will enable an ill-conceived structural form to survive a strong earthquake.

10.3.2 Seismic response of reinforced concrete

The seismic response of structural materials has been discussed generally in Section 5.4, where some stress–strain diagrams were presented. The hysteresis loops of Figure 5.23(d) indicate that considerable ductility without strength loss can be achieved in doubly reinforced beams having adequate confinement reinforcing. This is in distinct contrast to the loss of strength and stiffness degradation exhibited by plain unconfined concrete under repeated loading as shown in Figure 5.23(c). Because the hysteretic behaviour of reinforced concrete is so dependent on the amount and distribution of the longitudinal and transverse steel, mathematical models of hysteresis curves need to be chosen with care to reflect the details of the actual construction, using methods such as those outlined by Paulay and Priestley (1992).

Reinforcement controls and delays failure in concrete members, the degradation process generally being initiated by cracking of the concrete. Inelastic elongation of reinforcement within a crack prevents the latter from closing when the load direction is reversed and cyclic loading leads to progressive crack widening and steel yielding (Figure 10.17). Fenwick (1983) argued that shear in plastic hinge regions of beams is resisted by truss action until the phase of rapid strength degradation in which large shear displacements occur.

10.3.3 Reliable seismic behaviour of concrete structures

Introduction

For obtaining reliable seismic response behaviour the principles concerning choice of form, materials, and failure mode control discussed in Section 8.3 should be applied

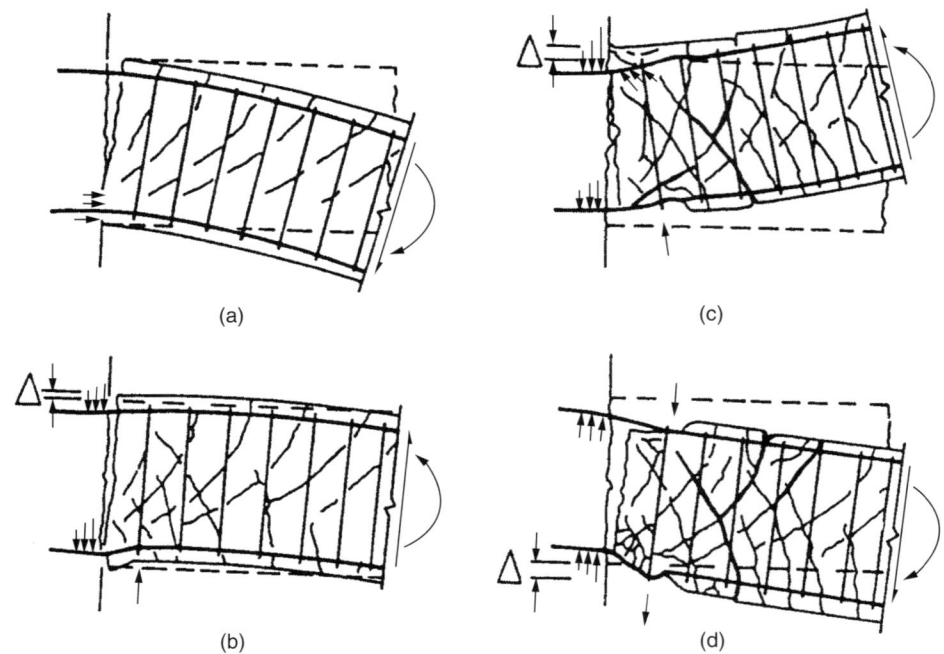

Figure 10.17 Significant stages of development of a plastic hinge in reinforced concrete during cyclic flexural and shear loading (after Paulay, 1981)

to concrete structures. Designing for failure mode control requires consideration of the structural form used, with most of the forms discussed in Section 8.4 being appropriate for concrete, i.e.:

(1) moment resisting frames;
(2) structural walls (i.e. shear walls);
(3) concentrically braced frames;
(4) hybrid structures.

For concrete structures, in addition to the discussion in Section 8.3.8, it should be noted that the essential objectives of failure mode control are as follows:

(a) Beams should fail before columns (unless extra column strength is provided).
(b) Brittle failure modes should be suppressed.
(c) An appropriate degree of ductility should be provided.

To help fulfil these three objectives, some concrete codes have specific strength factors for enhancing column strength in relation to beams and for enhancing shear strength in relation to flexural strength. Also for highly ductile structures, some concrete codes (led by New Zealand) seek to attain objectives (a)–(c) by requiring a *capacity design* procedure to be followed, wherein greater strength capacity has to be supplied in the brittle modes than in the ductile ones.

Unfortunately, the full rigour of this capacity design approach outlined in the code commentary is difficult to apply in all but the simplest of structures. Because the capacity design procedure appears to lead to very reliable failure mode control it is to be hoped that a simpler and more usable version of it becomes available, such as by increasing the column design forces by a single easily determined factor.

Returning specifically to objective (b) above, the best-known brittle failure mode in concrete which should be suppressed is shear failure. To prevent shear failure occurring before bending failure it is good practice to design so that the flexural steel in a member yields while the shear reinforcement is working at a stress less than yield (say, 90%). In beams a conservative approach to safety in shear is to make the shear strength equal to the maximum shear demands which can be made on the beam in terms of its bending capacity.

Referring to Figure 10.18(a), the shear strength of the beam should correspond to

$$V_{\max} = \frac{M_{u1} - M_{u2}}{l} + V_g, \qquad (10.21)$$

where V_g is the dead load shear force and

$$M_u = A_s f_{su} z,$$

in which A_s is all steel in the tension zone (Figure 10.18(b)), f_{su} is the maximum steel strength after strain hardening, say the 95th percentile for the steel samples (Figure 10.18(c)), and z is the lever arm.

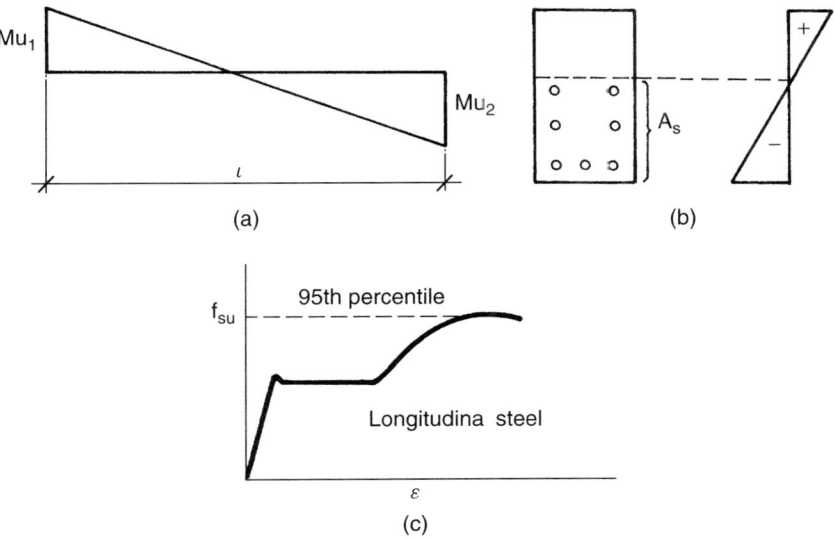

Figure 10.18 Shear strength considerations for reinforced concrete beams

Required ductility (robustness) of concrete structures

With regard to to objective (c) above, the degree to which ductility should be enhanced is debatable. Until the 1990s, research and codes had rightly been preoccupied with overcoming the excessive brittleness and unreliability of ill-reinforced concrete. However, there may have been too much emphasis on creating ductility for ductility's sake. The high cost of design and the complexity of some of the reinforcement of highly ductile concrete has raised the valid question of how we design less ductile structures which are sufficiently reliable in earthquakes. This question has long been raised regarding structures in regions of lesser seismicity. In any seismic region, the question applies not only to whole structures but also to parts of structures, e.g. beams and columns in buildings where the primary earthquake resisting elements are structural walls.

Methods of adjusting the design loading for different degrees of ductility have been discussed earlier, such that the value of the ductility factor may be chosen in the range from $\mu = 1$ (non-ductile) to about $\mu = 6$ (highly ductile). Some concrete codes gives recommendations for the design of structures of *limited ductility* implying a value of $\mu \leq 3$.

Ductility and robustness have been discussed in the general terms of inelastic behaviour in Section 5.4.2, and the problems of analysing inelastic behaviour and hence assessing the *ductility demand* in a structure have been considered in Section 5.4.7. While most concrete structures are designed by equivalent-static analysis and codified reinforcing rules aimed at providing ductility, it is important for designers to understand how the ductility demand arises. This is now discussed using a simplified method of determining hinge rotations in reinforced concrete frames, which involves the assumption of a hinge mechanism (Figure 8.9) and the imposition of an arbitrary lateral deflection ductility factor μ on the frame.

As mentioned above, it is preferable that beams should fail before columns (for safety reasons). Considering ten storeys above the column hinges of a column sidesway mechanism, Park (1980) found that for an overall frame deflection ductility factor $\mu = 4$, the required section ductility ratio was $\phi_u/\phi_y = 122$, which is impossibly high as shown by Figure 10.24. ϕ_u and ϕ_y are the hinge curvatures at ultimate and first yield, respectively. On the other hand, for a beam sidesway mechanism the required section ductility was found to be less than 20.

Having made an estimate of the ductility demands in the structure, the members should be detailed to have the appropriate section ductility, the theory for which is discussed below.

Available ductility for reinforced concrete members

The available section ductility of a concrete member is most conveniently expressed as the ratio of its curvature at ultimate moment ϕ_u to its curvature at first yield ϕ_y. The expression ϕ_u/ϕ_y may be evaluated from first principles, the answers varying with the geometry of the section, the reinforcement arrangement, the loading, and the stress–strain relationships of the steel and the concrete. Various idealizations of the stress–strain relationships give similar values for ductility, and the following methods of determining the available ductility should be satisfactory for most design purposes. It should be noted that the ductility of walls is discussed elsewhere (Section 10.3.2).

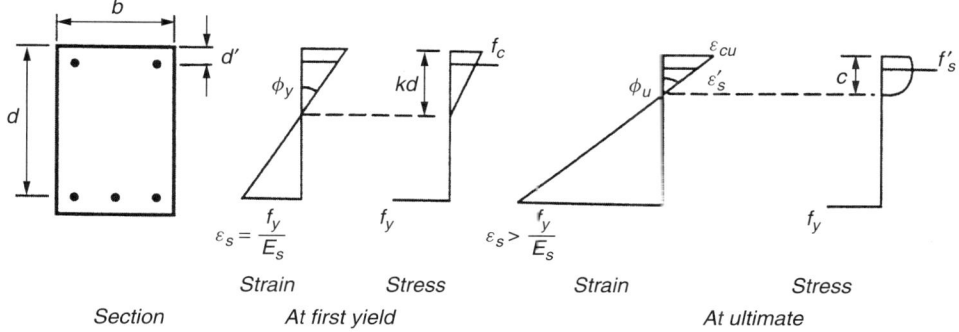

Figure 10.19 Doubly reinforced beam at first yield and ultimate curvatures

Singly reinforced sections

Consider conditions at first yield and ultimate moment as shown in Figure 10.19. Assuming an under-reinforced section, first yield will occur in the steel and the curvature

$$\phi_y = \frac{\varepsilon_{sy}}{(1-k)d} = \frac{f_y}{E_s(1-k)d}, \tag{10.22}$$

where

$$k = \sqrt{(\rho n)^2 + \rho n} - \rho n, \tag{10.23}$$

in which $\rho = A_s/bd$ is the tensile reinforcement ratio and $n = E_s/E_c$ is the modular ratio, E_s and E_c being the modulus of elasticity of the steel and the concrete, respectively.

Strictly, this formula for k is true for linear elastic concrete behaviour only, i.e. for

$$f_{cu} = \frac{2\rho f_y}{k} \leq 0.7 f_c',$$

where f_y is the steel yield stress and f_c' is the concrete cylinder compressive strength. For higher concrete stresses the true non-linear concrete stress block should be used. Referring again to Figure 10.19, it can be shown that the ultimate curvature is

$$\phi_u = \frac{\varepsilon_{cu}}{c} = \frac{\beta_1 \varepsilon_{cu}}{a}, \tag{10.24}$$

where

$$a = \frac{A_s f_y}{0.85 f_c' b} \tag{10.25}$$

and β_1, which describes the depth of the equivalent rectangular stress block, may be taken as $\beta_1 = 0.85$ for $f_c' = 27.6$ MPa, otherwise

$$\beta_1 = 0.0308(f_c' - 27.6), \tag{10.26}$$

From the above derivation, the available section ductility may be written as

$$\frac{\phi_u}{\phi_y} = \frac{\varepsilon_{cu}d(1-k)E_s}{cf_y}. \qquad (10.27)$$

The ultimate concrete strain ε_{cu} is given various values in different codes for different purposes. For estimating the ductility available from reinforced concrete in a strong earthquake a value of 0.004 may be taken as representing the limit of useful concrete strain, although some codes conservatively recommend a value of 0.003.

Doubly reinforced sections

The ductility of doubly reinforced sections (Figure 10.20) may be determined from the curvature in the same way as for singly reinforced sections above. Once again, the expression for available section ductility is as given by equation (10.27), but to allow for the effect of compression steel ratio ρ' the expressions for c and k become

$$c = \frac{a}{\beta_1} = \frac{(\rho - \rho')f_y d}{0.85 f'_c \beta_1} \qquad (10.28)$$

and

$$k = \sqrt{(\rho + \rho')^2 n^2 + 2[\rho + (\rho' d'/d)]n} - (\rho + \rho')n. \qquad (10.29)$$

The above equations assume that the compression steel is yielding, but if this is not so, the *actual* value of the steel stress should be substituted for f_y. As k has been found assuming linear elastic concrete behaviour, the qualifications mentioned for singly reinforced members also apply.

The results of a set of calculations for member ductility ϕ_u/ϕ_y, using equations (10.27)–(10.29), are presented graphically in Figure 10.21. In this figure, the ultimate compressive

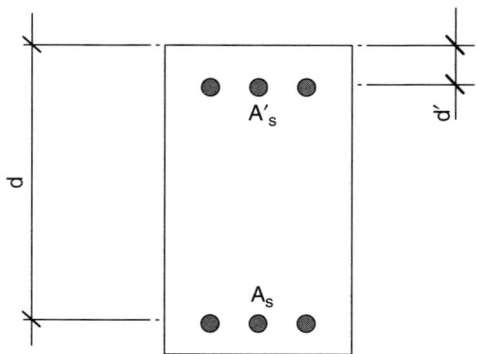

Figure 10.20 Doubly reinforced section

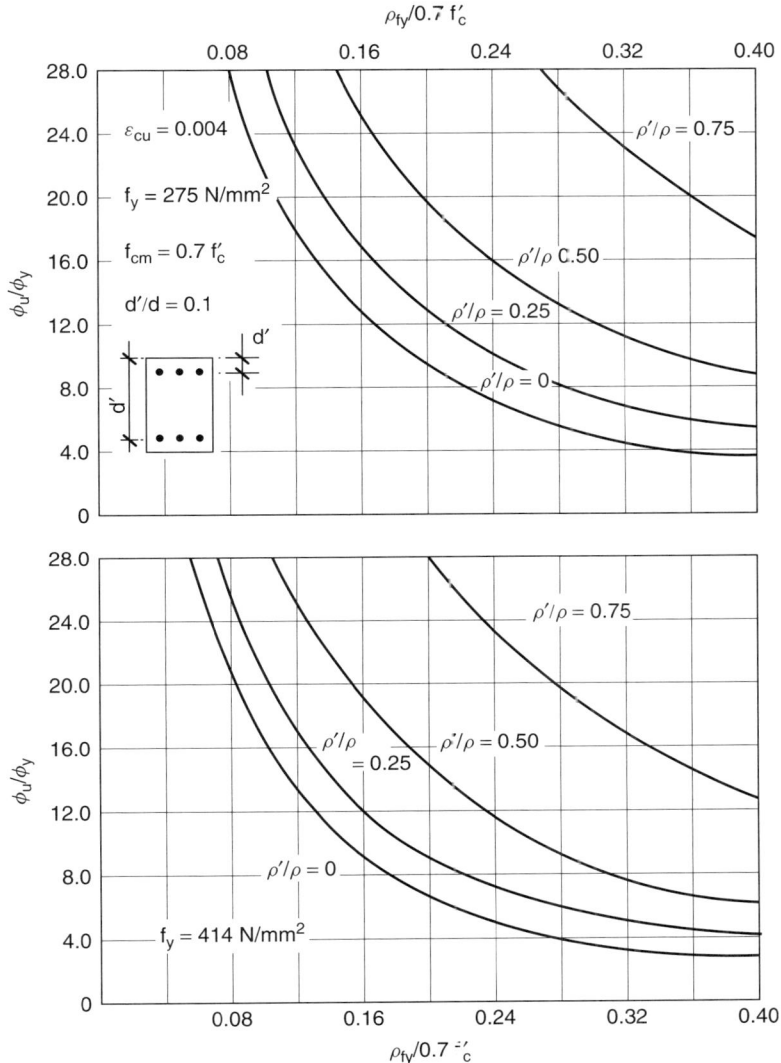

Figure 10.21 Variation of ϕ_u/ϕ_y for singly and doubly reinforced unconfined concrete (after Blume *et al.*, 1961). (Reproduced by permission of the Portland Cement Association)

force in the concrete was taken as $0.7 f'_c bc$ acting at a distance of $0.4c$ from the extreme compressive fibre. It can be seen from Figure 10.21 that ductility:

(1) reduces with increasing tension steel ρ;
(2) increases with increasing compression steel ρ';
(3) reduces with increasing yield stress f_y.

The effect of confinement of ductility

That the ductility and strength of concrete is greatly enhanced by confining the compression zone with closely spaced lateral steel has been demonstrated by various workers. In order to quantify the ductility of confined concrete, a number of stress–strain curves for monotonic loading of confined concrete have been derived from research going back to the 1960s (e.g. Kent and Park, 1971).

A good model for our purposes is the modified Kent and Park model shown in Figure 10.22, the relationships for which are given by Park *et al.* (1982). Figure 10.22 illustrates the beneficial effect of ductility of confinement, with curve (d) being for unconfined concrete and curves (a)–(c) being for $\rho_s = 2.55\%$, 1.7% and 0.85% of confining reinforcement content, respectively. Using the modified Kent and Park stress–strain model for concrete and an appropriate stress–strain model for the longitudinal reinforcement, flexural strengths and moment–curvature diagrams of the type shown in Figure 10.31 can be reliably predicted for a wide variety of member properties (e.g. Park *et al.*, 1982).

In addition, much research has been done on the response to cyclic loading of various shapes of beams and columns with different arrangements and details of confining steel in an endeavour to find construction methods that ensure strength retention under inelastic cycling, as discussed in Section 10.3.8.

The procedure for calculating the section ductility ϕ_u/ϕ_y is the same as that for unconfined concrete described above, the only difference being in determining an appropriate value of ultimate concrete strain ε_{cu} for use in equation (10.27). Corley (1966) recommended that a lower bound for the maximum concrete strain for concrete confined with

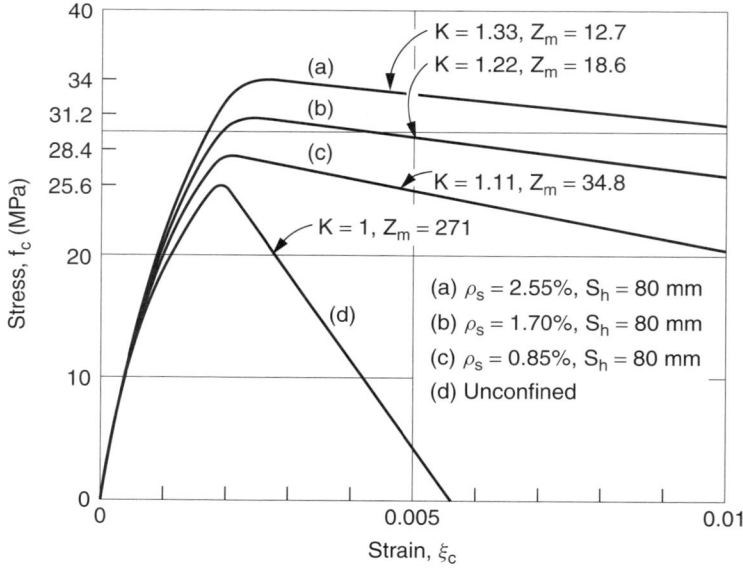

Figure 10.22 Modified Kent and Park stress–strain model for concrete in compression: curves (a)–(c) confined; curve (d) unconfined (Park *et al.*, 1982)

rectangular links is

$$\varepsilon_{cu} = 0.003 + 0.002\frac{b}{l_c} + \left(\frac{\rho_v f_{yv}}{138}\right)^2, \qquad (10.30)$$

where b/l_c is the ratio of the beam width to the distance from the critical section to the point of contraflexure, ρ_v is the ratio of volume of confining steel (including the compression steel) to volume of concrete confined, and f_{yv} is the yield stress of the confining steel (MPa).

Because of the high strains at ultimate curvature, the increased tensile force due to strain hardening should be taken into account, or the calculated ultimate curvature may be too large and the estimated ductility will be unconservative. Spalling of the concrete in compression is ignored in Corley's method.

As will be apparent from the following example, it is easy to increase the ultimate concrete strain to 0.01 or higher. As confinement and shear reinforcement are generally provided by the same bars, and as it is necessary for controlling the width of diagonal shear cracks to limit the strength of shear reinforcement to $f_y \approx 500$ MPa, only modest advantage may be taken by increasing f_{yv} in equation (10.30).

Example 10.1: Section ductility of reinforced concrete beam
Consider the beam shown in Figure 10.23. The confining steel consists of 12 mm diameter mild steel bars ($f_y = f_{yv} = 275$ MPa) at 75% mm centres, and the concrete strength is $f'_c = 21$ MPa). Estimate the section ductility ϕ_u/ϕ_y.

To find the curvature at first yield, first estimate the depth of the neutral axis using equation (10.23), the section being effectively singly reinforced. As the modular ratio $n = 9$, and $\rho = 0.0193$, we have $\rho n = 0.174$ and

$$k = \sqrt{(\rho n)^2 + 2\rho n} - \rho n = 0.441$$

Although this implies a computed maximum concrete stress greater than f'_c, the triangular stress block gives a reasonable approximation. Using equation (10.22), the yield curvature

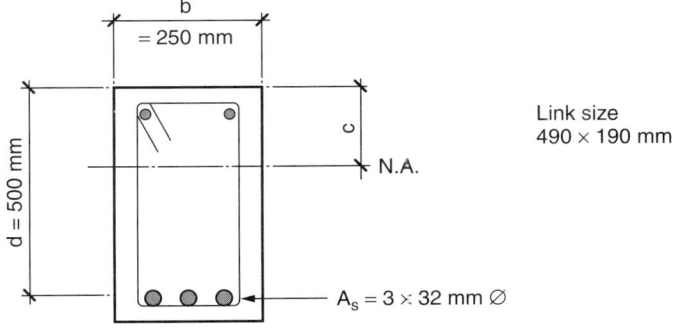

Figure 10.23 Reinforced concrete beam for ductility calculation example

is found to be

$$\phi_y = \frac{f_y}{E_s(1-k)d} = \frac{275}{2 \times 10^5 \times (1-0.441)500} = 4.92 \times 10^{-6} \text{ rad/mm}.$$

To find the ultimate curvature for the confined section, first determine the ultimate concrete strain from equation (10.30). Assume that for this beam

$$\frac{b}{l_c} = \frac{1}{8},$$

$$\rho_v = \frac{113 \times 2(490+190)}{490 \times 190 \times 75}$$

and

$$\varepsilon_{cu} = 0.003 + 0.02\frac{b}{l_c} + \left(\frac{\rho_v f_{yv}}{138}\right)^2$$

$$= 0.003 + \frac{0.02}{8} + \left(\frac{0.022 \times 275}{138}\right)^2 = 0.00742.$$

(10.31)

Next, find the depth of the neutral axis at ultimate from

$$c = \frac{a}{\beta_1} = \frac{A_s f_y}{\beta_1 \times 0.85 f'_c b}$$

$$= \frac{2412 \times 275}{0.85 \times 0.85 \times 21 \times 250} = 175 \text{ mm}$$

Hence, the ultimate curvature is

$$\phi_u = \frac{\varepsilon_{cu}}{c} = \frac{0.00742}{175} = 4.24 \times 10^{-5} \text{ rad/mm}.$$

The available curvature ductility for the confined section can now be found:

$$\frac{\phi_u}{\phi_y} = \frac{4.24 \times 10^{-5}}{4.92 \times 10^{-6}} = 8.6$$

It is of interest to observe that the ultimate concrete strain $\varepsilon_{cu} = 0.00742$, computed in the above example, is about twice the value of 0.004 noted earlier for unconfined concrete. Hence, the available section ductility has been roughly doubled by the use of confinement steel. This can be checked by reference to the curves of Figure 10.21, which gives ϕ_u/ϕ_y for unconfined flexural members. Now for the example beam and as the beam is singly reinforced, $\rho'/\rho = 0$. Hence, from Figure 10.21 it can be seen that for the unconfined section

$$\frac{\rho f_y}{0.7 f'_c} = \frac{0.193 \times 275}{0.7 \times 21} = 0.36,$$

$$\frac{\phi_u}{\phi_y} \approx 4.25,$$

which is about half the value of $\phi_u/\phi_y = 8.6$ determined above.

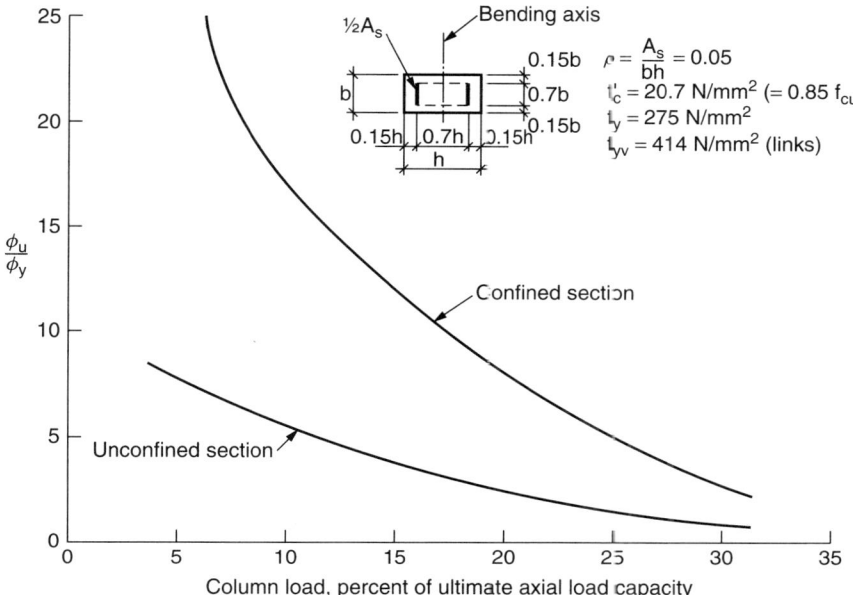

Figure 10.24 ϕ_u/ϕ_y for columns with confined or unconfined concrete (after Blume *et al.*, 1961). (Reprinted by permission of the Portland Cement Association)

Ductility of reinforced concrete members with flexure and axial load

Axial load unfavourably affects the ductility of flexural members, as can be seen from Figure 10.24. Indeed, it has been shown (Pfrang *et al.*, 1964) that only with axial compression less than the balanced load does ductile failure occur. It is evident from Figure 10.24 that, for practical levels of axial load, columns must be provided with confining reinforcement. For rectangular columns with closely spaced hoops, and in which the longitudinal steel is mainly concentrated in two opposite faces, the ratio ϕ_u/ϕ_y may be estimated from Figure 10.25. In that figure A_s is the area of tension reinforcement and

$$\beta_h = \frac{A_h f_{yh}}{s h_h f'_c}, \qquad (10.32)$$

where A_h is the cross-sectional area of the hoops, f_{yh} is the yield stress of the hoop reinforcement, s is the spacing of the hoop reinforcement, and h_h is the longer dimension of the rectangle of concrete enclosed by the hoops. The value ϕ_u/ϕ_y for a particular section is obtained by following a path parallel to the arrowed zigzag on the diagram.

10.3.4 Reinforced concrete structural walls

Introduction

Great structural advantage may be gained from reinforced concrete structural walls (shear walls) in seismic construction, provided they are properly designed and detailed for strength and ductility. Favourably positioned structural walls can be very efficient in

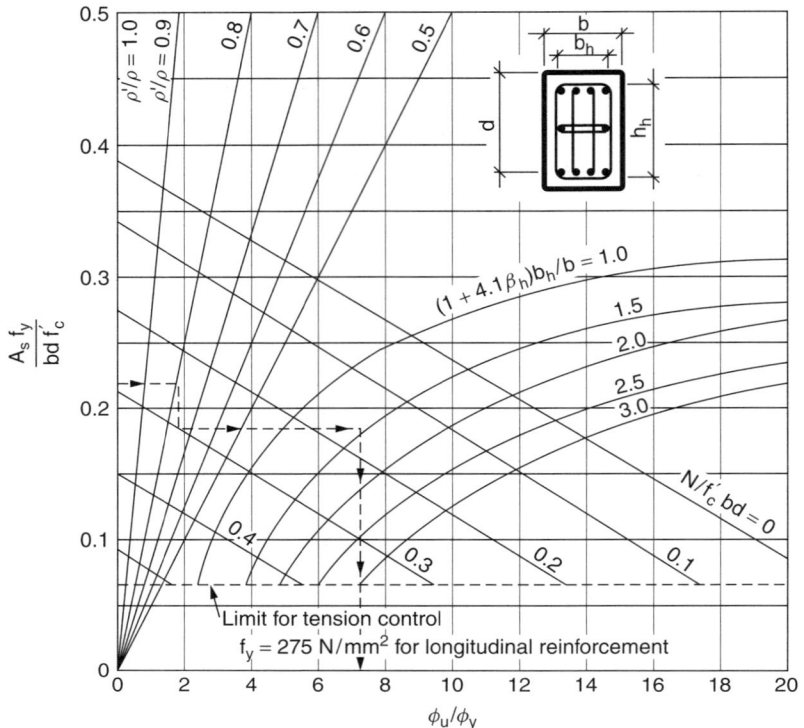

Figure 10.25 ϕ_u/ϕ_y for columns of confined concrete (after Blume *et al.*, 1961). (Reproduced by permission of the Portland Cement Association)

resisting horizontal wind and earthquake loads. The considerable stiffness of walls not only reduces the deflection demands on other parts of the structure, such as beam–column joints, but may also help to ensure development of all available plastic hinge positions throughout the structure prior to failure. A valuable bonus of structural wall stiffness is the protection afforded to non-structural components in earthquakes due to the small storey drift compared with that of beam and column frames. Further discussion of stiff and flexible construction can be found in Section 8.3.6.

A notable early example of the confidence accorded concrete structural wall construction is the 44-storey Parque Central apartment buildings in Caracas (Paparoni *et al.*, 1973) built after the 1967 Caracas earthquake.

It should be noted that simpler methods of analysis, particularly equivalent-static seismic analysis, may give markedly inaccurate force distributions, especially in upper storeys due to the interaction of structural walls with rigid-jointed frames. This interaction may have undesirable effects, resulting in greater than expected ductility demands such as in captive spandrel columns (e.g. Selna *et al.*, 1980).

Most walls are fairly lightly loaded vertically and behave essentially as cantilevers, i.e. as vertical beams fixed at the base. A discussion of the basic design criteria for structural walls follows under the headings of tall and squat cantilevers. Coupled walls are then discussed as a special case of cantilever walls. Irregular arrangements of openings in structural walls require individual consideration and may require analysis by finite-element techniques which, however, may not lead to adequate prediction of ductility requirements. Such structures may invite disaster by concentrating energy dissipation in a few zones which are unable to develop the strength or ductility necessary for survival.

Cantilever walls

A single cantilever wall can be expected to behave as an ordinary flexural member if its height to depth ratio H/h is greater than about 2.0. Some distinctions between the two types of wall are made in the following sections.

Having obtained the design ultimate axial force N_u, moment M_u and shear force V_u for a given wall, it will usually be appropriate first to check the wall size and reinforcement for bending strength. This should be followed by a check that its ductility is adequate, and then that the wall's shear strength is somewhat greater than its bending strength (these two procedures may be implicit in detailing rules of advanced concrete codes). While considering the shears it should be ensured that the safe maximum applied shear stress is never exceeded and that the construction joints are adequately reinforced. These considerations are discussed more fully below.

Bending strength of cantilever walls

When *wall or rectangular sections* are designed for small bending moments, the designer may be tempted to use a uniform distribution of vertical steel as for walls in non-seismic areas, but it may be shown from first principles that with this steel arrangement the ductility reduces as the total steel content increases.

When the flexural steel demand is larger, it will be better to place much of the flexural steel near the extreme fibres, while retaining a minimum of 0.25% vertical steel in the remainder of the wall. Apart from efficient bending resistance, this steel arrangement will considerably enhance the rotational ductility.

In rectangular wall sections in which the reinforcement is concentrated at the extremities, the bending strength may be calculated from first principles following accepted codes of practice, or use may be made of column design charts which are frequently available. As design charts for *uniformly* reinforced members are not so readily available, their bending strength is discussed below.

If the contribution of the reinforcement in the elastic core of a uniformly reinforced wall with $H/h > 1.0$ is neglected, the following simple, conservative expression found by Cardenas *et al.* (1973) for ultimate bending strength arises:

$$M_u = 0.5 A_s f_y h \left(1 + \frac{N_u}{A_c f_y}\right)\left(1 - \frac{c}{h}\right) \tag{10.33}$$

where

$$\frac{c}{h} = \frac{\alpha + \beta}{2\beta + 0.85\beta_1};$$

$$\alpha = \frac{A_s f_y}{bh f'_c};$$

$$\beta = \frac{N_u}{bh f'_c};$$

M_u = design resisting moment (ultimate (MNm));
A_s = total area of vertical reinforcement (mm²);
f_y = yield strength of vertical reinforcement (MPa);
h = horizontal length of shear wall (mm);
c = distance from extreme compression fibre to neutral axis (mm);
b = thickness of shear wall (mm);
N_u = design axial load (ultimate), positive if compressive (N);
f'_c = characteristic cylinder compression strength of concrete (MPa);
β_1 = 0.85 for strength up to 27.6 MPa and reduced continuously at a rate of 0.05 for each 6.9 MPa (1.0 ksi) of strength in excess of 27.6 MPa (4 ksi).

Alternatively, the bending strength of uniformly reinforced rectangular walls can be predicted from non-linear beam theory as discussed by Salse and Fintel (1973), who derived the axial load–moment interaction curves shown in Figure 10.26.

Flanged walls are desirable for their high bending resistance and ductility, and arise in the form of I-sections or as channel sections which may be coupled together as lift shafts. Figure 10.31 shows the interaction curves for a channel section with bending about the minor axis.

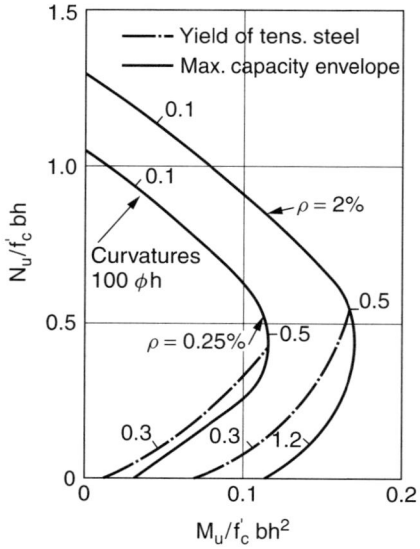
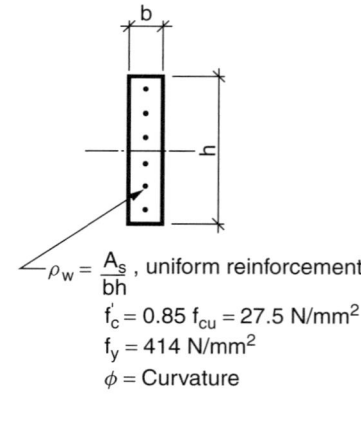

Figure 10.26 Axial load–moment interaction curves for rectangular uniformly reinforced concrete walls (after Salse and Fintel, 1973)

Behaviour effects on different reinforcement arrangements can be seen in Figure 10.27, which shows axial load–moment interaction curves for I-sections or channel sections derived from non-linear beam theory. The curves are general for all values of b and h, and the web reinforcement is 0.25% in all cases except curve (1), being largely due to the assumptions of *high concrete confinement* in the flanges in this case.

Park and Paulay (1980) summarized the design procedure for *squat walls* as follows:

> In low-rise buildings the height of a structural wall may be less than its length. Such walls cannot be designed with the customary techniques of reinforced concrete theory. However, because the earthquake load for squat walls is seldom critical, approximate design that ensures at least limited ductility will often suffice. The strength of many low-rise walls will be limited by the capacity of the foundations to resist the overturning moments. In such cases a rocking structure results and thus ductility becomes irrelevant.

As Figure 10.28 indicates, after diagonal cracking the horizontal shear introduced at the top of a low-rise cantilever will need to be resolved into diagonal compression and vertical tensile forces. Thus, distributed vertical flexural reinforcement will also enable the shear to be transmitted to the foundations. The equilibrium condition of the free body marked ② shows this in Figure 10.28. Where the diagonal compression field does not find a support at foundation level, as is the case with the triangular free body marked ①, an equal amount of horizontal shear reinforcement will be required. Figure 10.28 thus shows that for a squat shear wall a steel mesh with equal area in both directions will be required if a compression field acting at 45° is conservatively assumed. The flexural strength at the base must be carefully evaluated, taking the contribution of all vertical bars into account, to ensure that the required shear strength can be provided. This way most squat shear walls can be made ductile and a brittle failure will be avoided.

Ductility of cantilever walls

The general problem of ductility in concrete structures is discussed elsewhere (Section 10.3.3); suffice it here to say that adequate ductility under seismic loadings implies inelastic cyclic deformations without appreciable loss of strength. As mentioned above, walls will exhibit greater ductility in bending if much of the reinforcement is concentrated near the extreme fibres, and consequently flanged sections are more ductile than rectangular walls. A comparison of the ductility of rectangular and I- (or channel) sections is given in Figure 10.29 where it was taken that

$$\text{Available section ductility} = \frac{\phi^*}{\phi_y}. \tag{10.34}$$

where ϕ^* is curvature at maximum moment, and ϕ_y is curvature at initiation of tension steel yield.

The ductility calculation was based on monotonic loading only, and hence Figure 10.29 serves better for qualitative comparisons than for quantitative purposes; the true ductility under reversible loading may be less than that shown, depending on the reinforcement quantities and disposition.

From Figure 10.29, it can be seen that both increasing steel percentages and increasing axial loads will decrease ductility. By comparing curve A with B, and curve C with D,

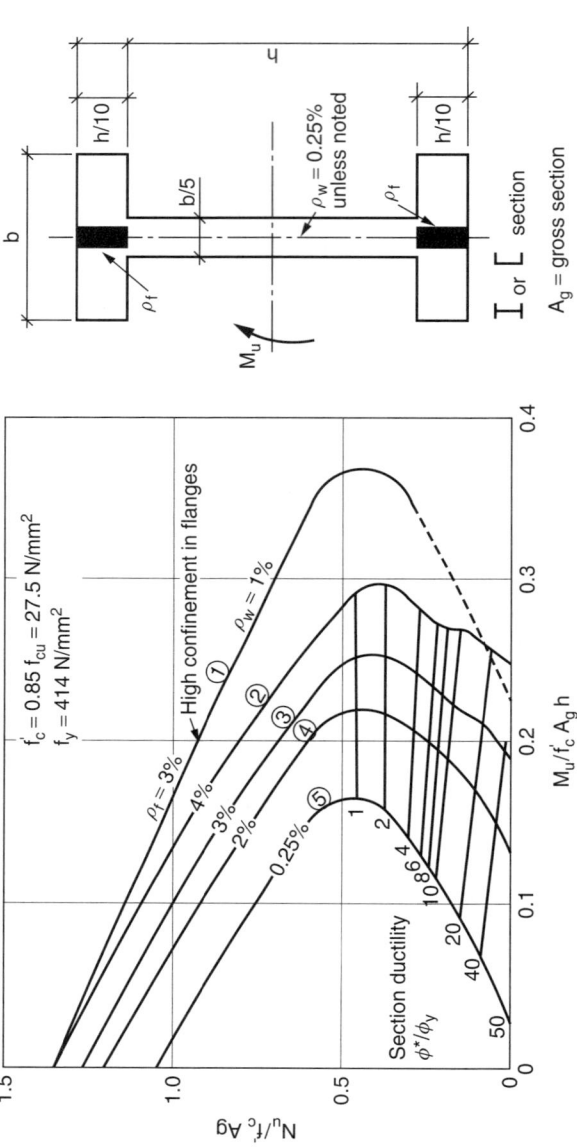

Figure 10.27 Axial load-moment interaction curves for I and [- section reinforced concrete walls (after Salse and Fintel, 1973)

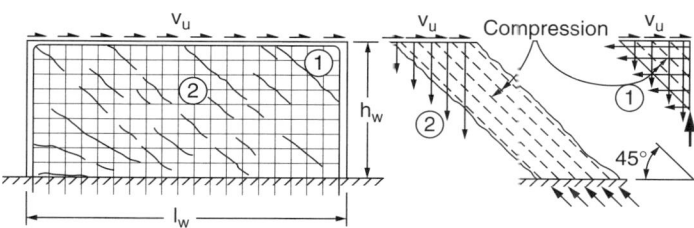

Figure 10.28 The shear resistance of squat shear walls (after Paulay, 1972)

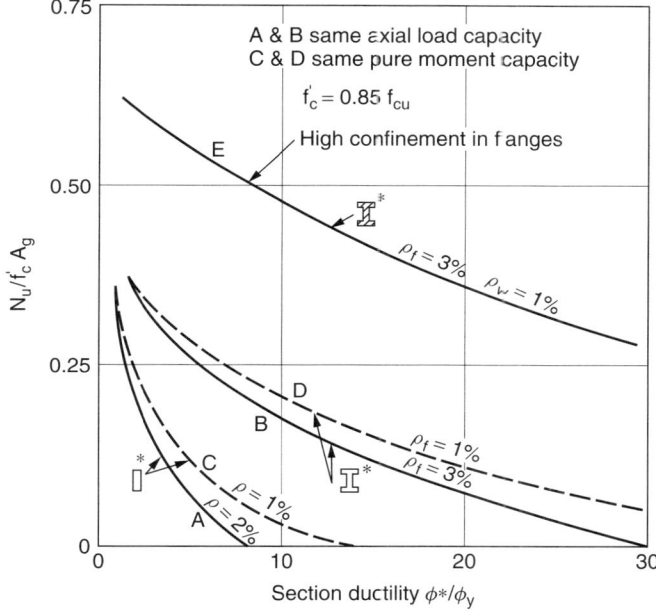

Figure 10.29 Ductility of walls as affected by cross-sectional shape, steel distribution, and concrete confinement (after Salse and Fintel, 1973)

it can be seen that the section ductility for I shapes is three to four times greater than that for uniformly reinforced rectangular sections. By comparing curve E and the remainder in Figure 10.29 the great effect on ductility of concrete confinement in the flanges can be seen.

In design situations it may be convenient to refer to an interaction diagram as shown on Figure 10.27, which incorporates ductility factors, thus allowing suitable strength and ductility to be chosen simultaneously.

Squat walls (i.e. those with height to depth ratio $H/h \leq 1.0$), are not amenable to the above ductility calculations as discussed in page 381.

Shear strength of structural walls

Some special considerations may apply in the design of potential plastic hinge zones for shear. Tests conducted in the USA (Oesterle *et al.*, 1980; Bertero *et al.*, 1977) showed that web crushing in the plastic hinge zone may occur after a few cycles of reversed loading involving displacement ductility ratios of 4 or more. When the imposed ductilities were only 3 or less, a shear stress equal to or in excess of $0.16 f'_c$ could be attained. Web crushing then occurs, which eventually spreads over the entire length of the wall. When boundary elements with a well-confined group of vertical bars were provided, significant shear after failure of the panel (web) could be carried because the boundary elements acted as short columns or dowels. However, according to Paulay and Priestley (1992), it is advisable to rely more on shear resistance of the panel, by preventing diagonal compression failure, rather than on the second line of defence of the boundary elements. To ensure this, either the ductility demand on a wall with high shear stresses must be reduced, or, if this is not done, the shear stress, used as a measure of diagonal compression, should be limited as follows:

$$v_{i,\max} \leq \left(\frac{0.22 \phi_{0,w}}{\mu} + 0.03 \right) f'_c < 0.16 f'_c \leq 6 \text{ MPa}. \tag{10.35}$$

Paulay and Priestley add that,

> for example, in coupled walls with typical values of the overstrength factor $\phi_{0,w} = 1.4$ and $\mu_\Delta = 5$, $v_{i,\max} = 0.092 f'_c$. In a wall with restricted ductility, corresponding values of $\phi_{0,w} = 1.4$ and $\mu_\Delta = 2.5$ would give $v_{i,\max} = 0.153 f'_c$, close to the maximum suggested. The expression also recognizes that when the designer provides excess flexural strength, giving a larger value of $\phi_{0,w}$, a reduction in ductility demand is expected, and hence [equation (10.35)] will indicate an increased value for the maximum admissible shear stress.

Coupled walls

Design approach

It is common practice to utilize the inherent lateral resistance of adjacent walls by coupling them together with beams at successive floor levels. Vertical access shafts punctured by door openings, as shown in Figure 10.30, form the classical example of this type member. The analysis of coupled shear walls requires consideration of axial deformations of the walls and shear distortions of the coupling beams.

Ideally, the designer would like the coupled walls to act as a box or I-unit as if the openings did not exist, such a structure would be much stronger than the two constituent channel units acting independently. In an efficiently couple pair of walls, the beam stiffness will be such thant between one-third and two-thirds of the total overturning moment,

$$M_0 = M_1 + M_2 + Nl, \tag{10.36}$$

is resisted by the fixity moments of the walls $(M_1 + M_2)$, the remainder being resisted by the overturning moment being taken by the push-pull couple Nl due to the vertical reactions N at the base of the walls and their lever arm l. This implies the development of high shears in the coupling beams acting as a web, and the existence of large longitudinal

Concrete Structures

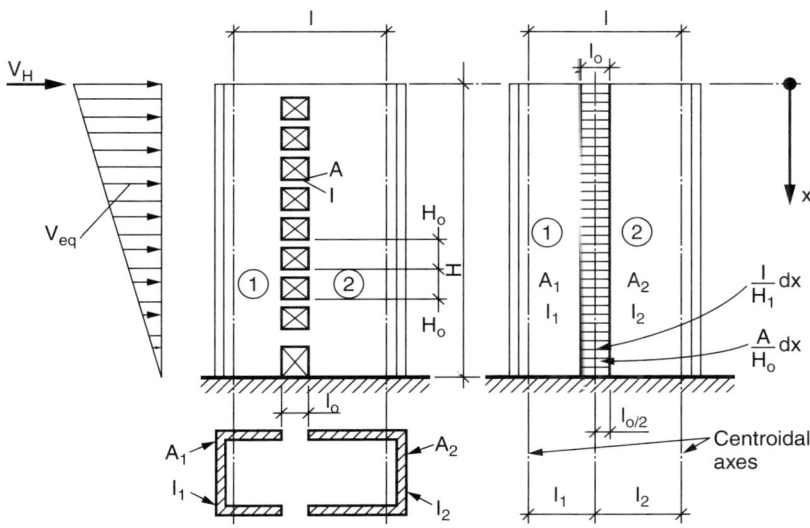

Figure 10.30 A typical coupled wall structure and its mathematical laminar model (after Paulay, 1972)

forces in each wall unit. The failure of the coupling beams in coupled walls exposed to strong earthquakes indicates insufficient ductility of the beams. This has been due partly to inadequate detailing of the beams and partly to the use of elastic analysis which has not been adjusted to model the behaviour adequately. Standard frame analysis may suffice as long as the extra stiffness of the beam ends (within the walls) is taken into account, and redistribution of beam moments due to inelastic effects is properly done (Paulay and Williams, 1980).

Strength of coupled walls

Having derived the bending moments and forces acting on the wall elements of the coupled systems, it will be necessary to design the walls to withstand those forces. The bending moment pattern will be similar to that of simple cantilever walls. In addition, because of the coupling system there will be considerable axial forces which may produce net tensions in the walls.

It is evident that the design considerations are as for cantilever walls discussed above. In the design of a high-rise structure with many similar horizontal sections to consider, it may be worth producing a family of axial load–moment interaction diagrams (Figure 10.31). Note that similar diagrams for different ratios of biaxial bending may be necessary for the same section.

Strength and ductility of coupling beams

The classical failure mode of the coupling beams in earthquakes is that of diagonal tension. To avoid this brittle type of failure two alternative methods of reinforcing the coupling beams are available. It has been recommended that when the earthquake shear stress in

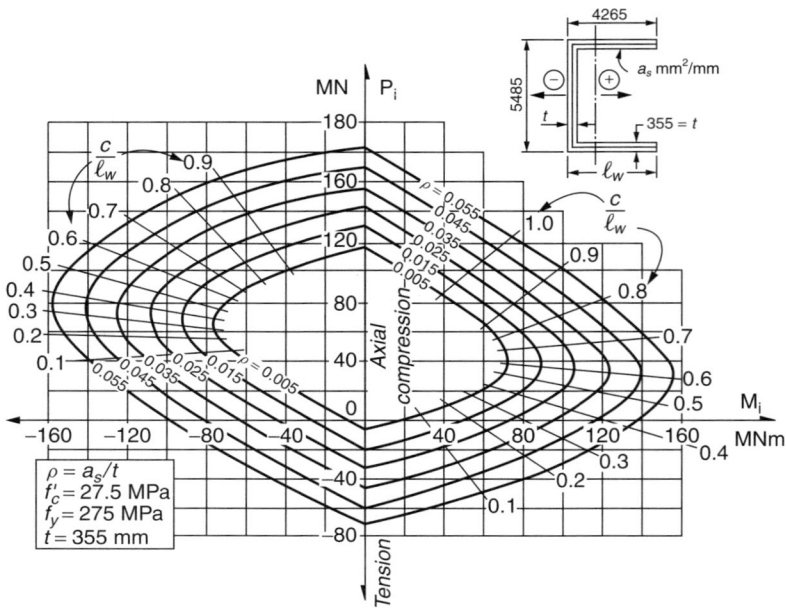

Figure 10.31 Axial load–moment interaction relationships for a channel-shaped wall section. (Reproduced from Paulay and Priestley, MJN (1992) *Seismic Design of Reinforced Concrete and Masonry Buildings*. Copyright © (1992). Reprinted by permission of John Wiley & Sons, Inc)

the beam is less than

$$v = 0.1 \frac{l_n}{h} \sqrt{f'_c}, \qquad (10.37)$$

then the beam may be detailed in the normal manner, otherwise all of the shear force should be resisted by diagonal reinforcement. The danger of shear failure and the inhibition of ductility increases with increasing depth to span ratio h/l_n (where l_n is clear span of the beams), as reflected in equation (10.37). This severe limitation is recommended because coupling beams of shear walls can be subjected to very large rotational ductility demands, as noted below.

Where coupling beams may experience high seismic stresses, diagonal reinforcement of the type shown in Figure 10.32 provides far greater seismic resistance than conventional steel arrangements, as the comparison of ductilities in Figure 10.33 shows.

Conventionally reinforced deep coupling beams having a ductility ratio of 4–5 (Figure 10.33) would often be unsatisfactory, whereas the diagonal reinforcement arrangement easily provides the commonly required ductility ratio of about 12. Thus in areas of moderate or strong ground motion, diagonal reinforcement of deep coupling beams is seen to be required. The importance of restraining the diagonals against buckling in compression must, however, be realized, and careful detailing to suit this and still allow the proper placement of the diagonal bars will be necessary.

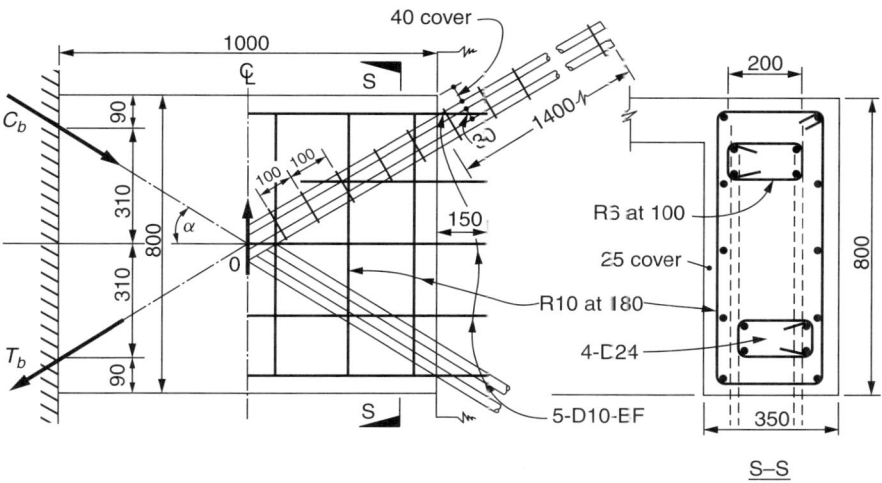

Figure 10.32 Typical details of a typical coupling beam. (Reprinted from Paulay and Priestley, 1992) *Seismic Design of Reinforced Concrete and Masonry Buildings*. Copyright © (1992). Reprinted by permission of John Wiley & Sons, Inc)

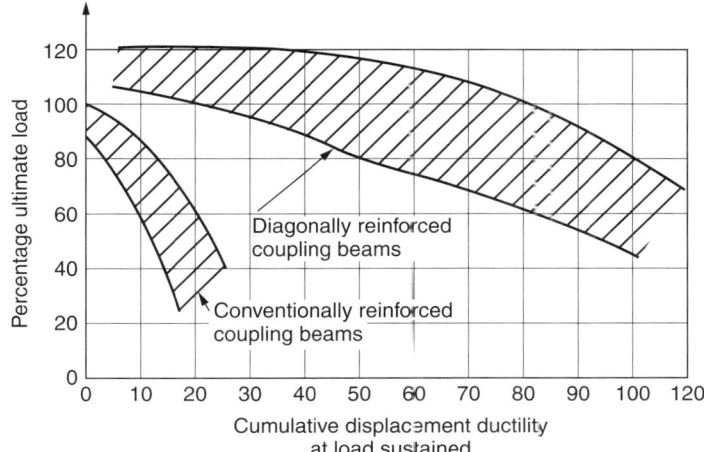

Figure 10.33 Comparison of ductilities of diagonally and conventionally reinforced deep coupling beam (after Paulay, 1972)

10.3.5 In situ *concrete design and detailing: general requirements*

The following notes and the associated detail drawings have been compiled to enable the elements of reinforced concrete structures to be detailed in a consistent and satisfactory manner for earthquake resistance. These details should be satisfactory in regions of medium and higher seismic risk in so far as they reflect the present state of the art. Supplementary information on the seismic design of reinforced concrete is available in manuals such as those produced for New Zealand (Cement and Concrete Association of New Zealand, 1995; New Zealand Concrete Society (NZCS), 2008). In regions of lower seismic hazard, relaxations may be made to the following requirements, such as recommended in codes in the USA, but the principles of splicing, containment and continuity must be retained if adequate ductility is to be obtained.

For discussions of the all-important capacity design procedure, see elsewhere (e.g. Paulay and Priestley, 1992; NZS 3101, 2006).

Splices

Splices in earthquakes resisting frames must continue to function while the members of joints undergo large deformations. As the stress transfer is accomplished through the concrete surrounding the bars, it is essential that there be adequate space in a member to place and compact good-quality concrete.

Splice laps should not be made in joints or in plastic hinge zones (see the bottom-storey column in Figure 10.34). Tensile reinforcement in beams or columns should not be spliced in regions of tension or reversing stress unless the spliced region is confined by hoops or

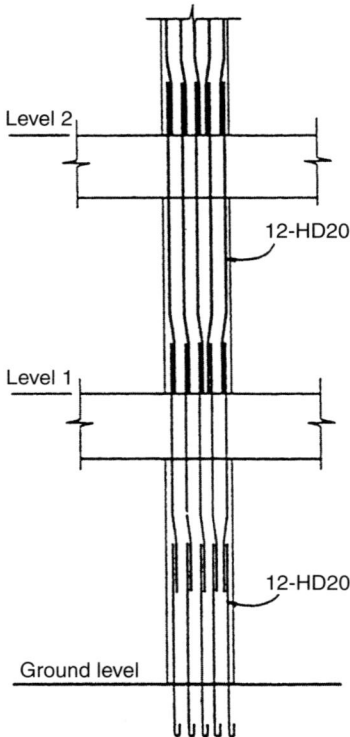

Figure 10.34 Elevation of a column in a ductile moment resisting frame, showing splice located out of plastic hinge zone at column base

stirrups so that the area of the confinement steel is not less than

$$A_t = d_b s f_y / 50 f_{yt},$$

where d_b is the diameter of the bars being spliced, s is the spacing on the confining steel, and f_{yt} is the specified yield stress of the transverse reinforcement.

Tests have shown that contact laps perform just as well as spaced laps, because the stress transfer is primarily through the surrounding concrete. Contact laps (as with welded splices) reduce the congestion and give better opportunity to obtain well-compacted concrete over and around the bars.

Laps should preferably be staggered but where this is impracticable and large numbers of bars are spliced at one location (e.g. in columns) adequate links or ties should be provided to minimize the possibility of splitting the concrete. In columns and beams, even when laps are made in regions of low stress, at least two confining links should be provided.

Development (anchorage)

Satisfactory development may be achieved by extending bars as straight lengths, or by using 90° and 180° bends, but development efficiency will be governed largely by the state of stress of the concrete in the anchorage length. Tensile reinforcement should not be anchored in zones of high tension. If this cannot be achieved, additional reinforcement in the form of links should be added, especially where high shears exist, to help to confine the concrete in the development length. It is especially desirable to avoid anchorage bars in the 'panel' zone of beam–column connections. Large amounts of the reinforcement should not be curtailed at any one section.

Bar bending

The minimum bend radius depends upon the ductility of the steel being used and upon the stress in the bar, so that earthquake related codes have a range of values on this subject. As an example, NZS 3101 (2006) requires that bends in longitudinal reinforcement have the minimum diameters measured to the inside of the bars given in Table 10.8, and that stirrups and ties conform to Table 10.9.

Cover

Minimum cover to reinforcement should comply with local codes of practice.

Table 10.8 Minimum diameters of bends in longitudinal reinforcing bars, from NZS 3101 (2006). (Reproduced from NZS 3101: 2006 by permission of Standards New Zealand under Licence 000710)

f_y (MPa)	Bar diameter, d_b (mm)	Minimum diameter of bend, d_i (mm)
300 or 500	6–20	$5d_b$
	24–40	$6d_b$

Table 10.9 Minimum diameters of bends in stirrups and ties, from NZS 3101 (2006). (Reproduced from NZS 3101:2006 by permission of Standards New Zealand under Licence 000710)

f_y (MPa)	Bar diameter, d_b (mm)	Minimum diameter of bend d_i (mm)	
		Plain bars	Deformed bars
300 or 500	6–20	$2d_b$	$4d_b$
	24–40	$3d_b$	$6d_b$

Concrete quality

For earthquake resistance, the minimum recommended characteristic cylinder crushing strength for structural concrete is 25 MPa.

The use of lightweight aggregates for structural purposes in seismic zones should be very cautiously proceeded with, as many lightweight concretes are very brittle. Appropriate advice should be sought in selecting the type of aggregate and mix proportions and strengths in order to obtain a suitably ductile concrete. It cannot be overemphasized that quality control, workmanship and supervision are of the utmost importance in obtaining earthquake resistant concrete.

Reinforcement quality

For adequate earthquake resistance, suitable quality of reinforcement must be ensured by both specification and testing. As the properties of reinforcement vary greatly between countries and manufacturers, much depends upon knowing the source of the bars, and on applying the appropriate tests. Particularly in developing countries the role of the resident engineer may be decisive and indeed onerous.

The following points should be observed:

(a) An adequate minimum yield stress may be ensured by specifying steel to an appropriate standard.
(b) The variability of the strength of reinforcing steels as currently manufactured is so great as to inhibit design control of failure modes, because of overstrength effects on capacity design. As noted by Paulay and Priestley (1992), overstrength results primarily from variability of the actual yield strength above the specified nominal value, and from strain hardening of reinforcement at high ductility levels. Thus, the overstrength factor λ_o can be expressed as

$$\lambda_o = \lambda_1 + \lambda_2, \tag{10.38}$$

where λ_1 represents the ratio of actual to specified yield strength and λ_2 represents the potential increase resulting from strain hardening.

λ_1 will depend upon where the local supply of reinforcing steel comes from, and considerable variability is common, as noted earlier in this section. With tight control of steel manufacture, values of $\lambda_1 = 1.15$ are appropriate. It is recommended that designers make the effort to establish the local variation in yield strength, and where this is excessive, to specify in construction specifications the acceptable limits to yield strength. Since steel suppliers keep records of yield strength of all steel in stock, this should does not cause any difficulties with supply.

λ_2 depends primarily upon yield strength and steel composition, and again should be locally verified. If the steel exhibits trends as shown in Figure 5.23(a) the appropriate values may be taken as

$$\lambda_2 = 1.10, \quad \text{for } f_y = 275 \text{ MPa},$$
$$\lambda_2 = 1.25, \quad \text{for } f_y = 400 \text{ MPa}.$$

For $\lambda_1 = 1.15$ these result in $\lambda_o = 1.25$, and 1.40 for $f_y = 275$ and 400 MPa, respectively.

(c) Grades of steel with characteristic strength in excess of 500 MPa are not permitted in some earthquake areas, e.g. California and New Zealand, but slightly greater strengths may be used if adequate ductility is proven by tests.

(d) Cold worked steels are not recommended in California or New Zealand, but cold worked steel to the appropriate British standard is sufficiently ductile.

(e) Steel of higher characteristic strength than that specified should not be substituted on site.

(f) The elongation test is particularly important for ensuring adequate steel ductility. Appropriate requirements are set out for steels conforming to American and British standards. Steels to other standards require specific consideration.

(g) Bending tests are most important for ensuring sufficient ductility of reinforcement in the bent condition. Steels to different standards require specific consideration.

(h) Resistance to brittle fracture should be checked by a notch toughness test conducted at the minimum service temperature, where this is less than about $3-5\,°C$.

(i) Strain-age embrittlement should be checked by rebend tests, similar to those for British steels.

(j) Welding of reinforcing bars may cause embrittlement and hence should only be allowed for steel of suitable chemical analysis and when using an approved welding process.

(k) Galvanizing may cause embrittlement and needs special consideration.

(l) Welded steel fabric (mesh) is unsuitable for ductile earthquake resistance because of its potential brittleness. However, mesh may be used in slabs or walls where little ductility is required.

Codes and standards

The reinforcing details recommended in this book are derived from a wide range of experience. Greatest reliance has been placed on American and New Zealand opinion, and their codes and leading research results have been applied.

In some earthquake countries, local codes may overrule some of the recommendations given in this book, but generally the requirements herein reflect the mainstream of current good seismic detailing. As such they are imperfect and generalized, and will need updating from time to time and at the discretion of earthquake experienced engineers.

10.3.6 Foundations

Column bases and pile caps

The following rules apply:

(1) Minimum percentage of steel: 0.15% each way.
(2) Bars should be anchored at the free end.
(3) Piles and caps should be carefully tied together to ensure integral action in earthquakes, and sufficient reinforcement should be provided in non-tension piles to prevent separation of pile and cap due to ground movements.

Foundation tie-beams

In the absence of a thorough dynamic analysis of the substructure, tie-beams may be designed for arbitrary longitudinal forces of up to 10% of the maximum vertical column load into which the particular beam connects (Section 9.1.2). As the axial loads may be either tension or compression, the following rules are appropriate:

(1) Minimum percentage of longitudinal steel: 0.8%.
(2) Maximum percentage of longitudinal steel: see Section 10.3.9.
(3) Maximum and minimum spacing as for columns.
(4) The design check for the compressive case should be carried out as for design of columns with regard to such items as permissible compressive stresses, slenderness effects, and confining reinforcement.

Tie-beams taking bending

In some cases, it may be required to transmit part of the bending moment at the column base into the tie-beams. Such tie-beams must therefore be designed for bending combined with axial compression or tension. The design should be carried out using the rules for beams or columns depending on the level of compressive stress. Requirements for column bases and pile caps above are applicable.

10.3.7 Walls

For determining the reinforcement in structural walls or coupling beams, refer to Section 10.3.4. More general requirements are as follows:

(1) The minimum content of vertical and horizontal steel should be 0.25% and 0.15%, respectively.
(2) The detailing around openings is important, and the details applicable to holes through suspended slabs may also be appropriate for smaller holes in walls.
(3) Horizontal construction joints should be cleaned and roughened to match the design assumptions.

10.3.8 Columns

General

The design notes given in this section are aimed primarily at columns which form part of ductile moment resisting frames. Columns in other situations, such as

(1) trapped spandrel columns in wall/frame systems,
(2) columns in flat slab structures, and
(3) pilaster columns

require specific consideration, as outlined by Selna *et al.* (1980). Other general design requirements for columns are as follows:

(4) The minimum width of the compression face of a member should be 200 mm.
(5) The minimum content of longitudinal steel should be 0.8% of the gross sectional area. The maximum content should be 6% for grade 300 steel (8% at lap splices) and 4.5% for grade 400 steel (6% at lap splices).

Column confinement reinforcing

Confinement steel in columns is provided to confine the concrete core and prevent buckling of the longitudinal reinforcement, as illustrated in Figure 10.35. Transverse reinforcement required for confinement should be provided unless a large amount is required for shear. In potential plastic hinge zones of columns, when spirals or circular hoops are used the volumetric ratio of the transverse reinforcement should not be less than that given by either

$$\rho_s = \frac{(1.3 - p_t m)}{2.4} \frac{A_g}{A_c} \frac{f'_c}{f_{yt}} \frac{N^*}{\phi f'_c A_g} - 0.0084, \qquad (10.39)$$

where A_g/A_c shall not be taken less than 1.2 and $\rho_t m$ shall not be taken greater than 0.4, or

$$\rho_s = \frac{A_{st}}{110 d''} \frac{f_y}{f_{yt}} \frac{1}{d_b}, \qquad (10.40)$$

where
A_g = gross cross-sectional area of the concrete;
A_c = area of core of column measured to outside of the hoops;
f_{yt} = yield strength of confining steel;
ρ_s = (volume of confining steel)/(volume of core);
ρ_t = ratio of longitudinal steel A_{st}/A_g;
N^* = design ultimate load;
$m = f_y/0.85 f'_c$;
d'' = diameter of concrete core measured to outside of hoop or spiral.

In equations (10.39) and (10.40), f_{yt} should be no greater than 800 MPa.

Alternatively, in potential plastic hinge regions, when rectangular hoops with or without supplementary cross-ties are used. The total effective area of hoop bars and supplementary

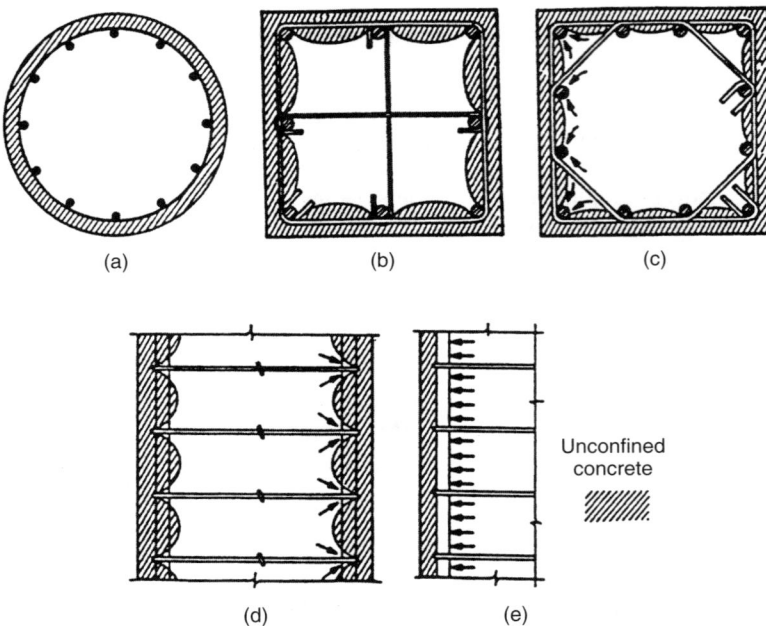

Figure 10.35 Arrangements of reinforcement which confine the concrete and prevent premature buckling of longitudinal reinforcement in columns: (a) circular hoops or spiral; (b) rectangular hoops with cross ties; (c) overlapping rectangular hoops; (d) confinement by transverse bars; (e) confinement by longitudinal bars

cross-ties in each of the principal directions of the cross section within spacing s_h shall not be less than that given by:

$$A_{sh} = \frac{(1.3 - \rho_t m)}{3.3} \frac{A_g}{A_c} \frac{f'_c}{f_{yt}} \frac{N^*}{\phi f'_c A_g} - 0.006 s_h h'' \qquad (10.41)$$

where A_g/A_c shall not be taken less than 1.2 $\rho_t m$ shall not be taken greater than 0.4 and f_{yt} shall not be taken larger than 800 MPa (115 ksi). The symbol h'' is the dimension of the concrete core measured perpendicular to the direction of the hoop bars to the centre of the peripheral hoop.

Outside of potential plastic hinge zones, of course, less confinement reinforcement is usually needed, the requirements being according to equations (10.39) and (10.41).

Care is necessary in selecting the details for anchorage of the transverse reinforcement, as some commonly used details have ineffective anchorage in reversed cyclic loading, i.e. bars with only 90% hooks are unsatisfactory in most situations. This has been demonstrated by Tanaka *et al.* (1985), who studied four rectangular sections incorporating a range of confinement details used in California and New Zealand, as shown in Figure 10.36. Three conclusions were drawn:

(1) The hoop and cross-tie arrangements in units 1 and 2 were satisfactory.
(2) Perimeter hoop U bars (unit 3) were unsatisfactory.

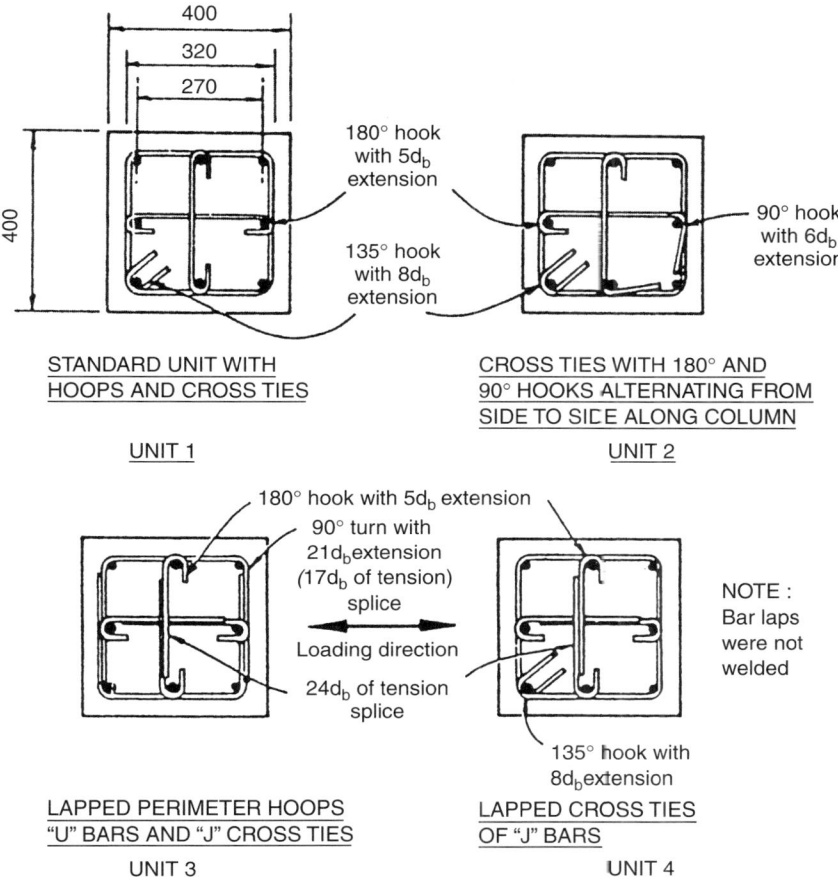

Figure 10.36 Alternative column confinement details of different effectiveness – see text (after Tanaka *et al.*, 1985)

(3) J bar interior cross-ties with a $24d_b$ tension splice were satisfactory in the columns tested, with a measured stress in the ties of $0.6f_y$. Their effectiveness at higher stresses was not known.

Shear strength of columns, beams and walls

To avoid premature diagonal compression failure before the onset of yielding of shear reinforcement the nominal shear stress needs to be limited as follows:

(a) Generally,

$$v_i \leq 0.2 f'_c \leq 6 \text{ MPa}. \tag{10.42}$$

(b) In potential plastic hinge zones,

$$v_i \leq 0.16 f'_c \leq 6 \text{ MPa}. \tag{10.43}$$

According to Paulay and Priestley (1992), the contribution of the concrete to shear strength is limited, in all regions except potential plastic hinges, as follows. In cases of flexure only,

$$v_c = v_b = (0.07 + 10\rho_w)\sqrt{f'_c} \leq 0.2\sqrt{f'_c} \text{ (MPa)} \tag{10.44}$$

where the ratio of the flexural tension reinforcement ρ_w is expressed in terms of the web width b_w. In flexure with axial compression,

$$v_c = (1 + 3N_u/A_g f'_c)v_b. \tag{10.45}$$

In flexure with axial tension,

$$v_c = (1 + 12N_u/A_g f'_c)v_b. \tag{10.46}$$

In structural walls,

$$v_c = 0.27\sqrt{f'_c} + N_u/4A_g. \tag{10.47}$$

When the force load N_u produces tension, its value in equations (10.46) and (10.47) must be taken as negative.

In regions of plastic hinges, the contribution of the concrete to shear strength is limited as follows. In beams,

$$v_c = 0. \tag{10.48}$$

In columns,

$$v_c = 4v_b\sqrt{N_u/A_g f'_c}. \tag{10.49}$$

In walls,

$$v_c = 0.6\sqrt{N_u/A_g}. \tag{10.50}$$

Equations (10.49) and (10.50) apply when the axial load N_u results in compression. When N_u represents tension, $v_c = 0$.

To prevent a shear failure resulting from diagonal tension, shear reinforcement, generally in the form of stirrups, placed at right angles to the axis of a member, is to be provided to resist the difference between the total shear force V_i and the contribution of the concrete V_c. The required area of a set of stirrups A_v with spacing s along a member is

$$A_v \geq \frac{(v_i - v_c)b_w s}{f_y}. \tag{10.51}$$

Reinforcement provided for confinement purposes (see previous section) may be assumed also to act in shear.

10.3.9 Beams

Beam longitudinal steel

(1) In potential plastic hinge regions, the maximum longitudinal tensile reinforcement should not be so large as to impair ductility, and thus should conform to

$$\rho_s \leq \frac{f'_c + 10}{6 f_y} \leq 0.025. \tag{10.52}$$

(2) The minimum longitudinal steel content as a fraction of the gross cross-sectional area of the web ($h \times b$) should be

$$\frac{\sqrt{f'_c}}{4 f_y} \quad (f_y \text{ in MPa}),$$

where h is the overall depth of the beam and b is the width of the web for both the top and bottom reinforcement.

(3) Curtailment of longitudinal steel should allow for the most adverse loading conditions. Large numbers of bars should not be cut off at the same section.
(4) The distance between bars should be according to the code adopted, but not less than 25 mm.
(5) In beams forming part of a moment resisting framework, the positive moment capacity at columns should not be less than half the negative moment capacity provided. At least two bars of 16 mm diameter should be provided both top and bottom throughout the length of the member, and the bending strength (of either sign) at any section along the beam should not be less than a quarter of the maximum bending strength at either end.

Shear strength of beams

See shear strength of columns, beams and walls in Section 10.3.8.

10.3.10 Beam–column joints in moment resisting frames

The strength of a beam–column core should be at least as great as, and preferably greater than, the strengths of the members it joins. This is because the joint area is subject to failure under cyclic loads and is obviously difficult to repair. Also, as it is part of the vertical load-carrying system it should comply with the principle that beams fail before columns.

A model of the forces involved in a beam–column joint is shown in Figure 10.37, from which it can be shown (NZS 3101, 2006) for conventionally reinforced members that the nominal horizontal shear force across a typical interior joint is

$$V_{jh} = 1.25 f_y (A_{s1} + A_{s2}) - V_{col}, \tag{10.53}$$

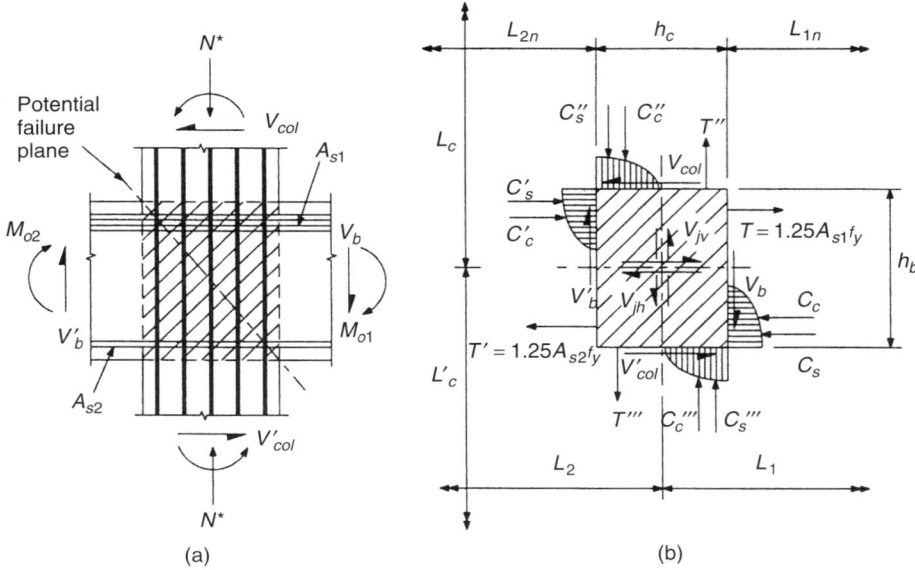

Figure 10.37 External actions and internal forces of a typical interior beam–column joint (after Part 2 of NZS 3101, 2006). (Reproduced from NZS 3101:2006 by permission of Standards New Zealand under Licence 000710)

where

$$V_{\text{col}} = 2\left(\frac{L_1}{L_{1n}}M_{o1} + \frac{L_2}{L_{2n}}M_{o2}\right)/(L_c + L'_c) \tag{10.54}$$

in which the symbols involving L are beam and column lengths as defined in Figure 10.37.

As shown in Figure 10.38, beam–column joints resist shear through two mechanisms, a single diagonal concrete strut (force V_{ch}) and a truss mechanism (force V_{sh}) comprising a diagonal compression field created by horizontal and vertical shear reinforcement.

The area of total effective horizontal joint shear reinforcement corresponding to each direction of horizontal joint shear force in interior joints is

$$A_{jh} = \frac{6v_{ih}}{f'_c}\alpha_i\frac{f_y}{f_{yh}}A_s^* \tag{10.55}$$

where $\alpha_i = 1.4$, or

$$\alpha_i = 1.4 - 1.6\frac{C_j N^*}{f'_c A_g},$$

whereby the beneficial effects of the axial compression load acting on the column above the joint may be included. A_s^* is the greater of the area of the top or bottom beam reinforcement passing through the joint. It excludes bars in effective tension flanges.

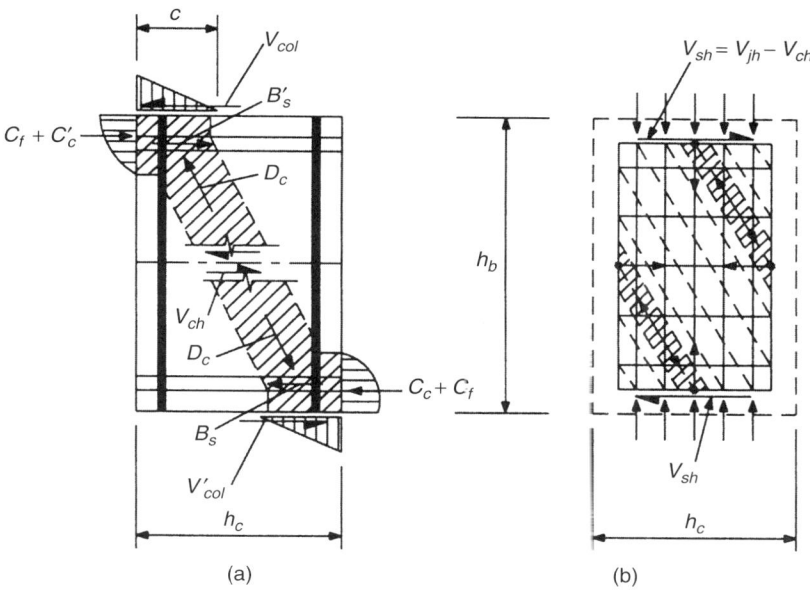

Figure 10.38 Models of the two modes of transfer of horizontal shear in a beam–column joint: (a) diagonal concrete strut; (b) truss mechanism (after Part 2 of NZS 3101, 2006). (Reproduced from NZS 3101:2006 by permission of Standards New Zealand under Licence 000710)

An example of the reinforcement arrangement for a beam–column joint in a precast concrete moment resisting frame designed to NZS 3101 (2006) is illustrated in Figure 10.39. This figure comes from NZCS (2008). The ducts are grouted after the joint members are all assembled.

As well as horizontal shear, beam–column joints need to be designed for vertical shear forces, as discussed elsewhere (e.g. Paulay and Priestley, 1992; NZS 3101, 2006).

10.3.11 Structural precast concrete detail

Introduction

Precast concrete structures have given mixed performance in earthquakes, difficulties mainly being experienced at connections between members. Prior to the mid-1980s, a negligible number of properly documented test results had been published on the behaviour of connections under cyclic loading. Much of the testing which has been done has been related to specific proprietary precast systems, and such results as have been published are usually either lacking in essential detail or are not readily applicable to other precasting assemblies.

Nevertheless, precast concrete, either reinforced or prestressed, has been used to some extent in most forms of structure in earthquake areas, often in conjunction with a cautiously large amount of unifying *in situ* concrete. The nature of the seismic response of a precast structure must be inferred from the response of the reinforced or prestressed members involved. Allowance for the effect of the connection on the stress flow must also

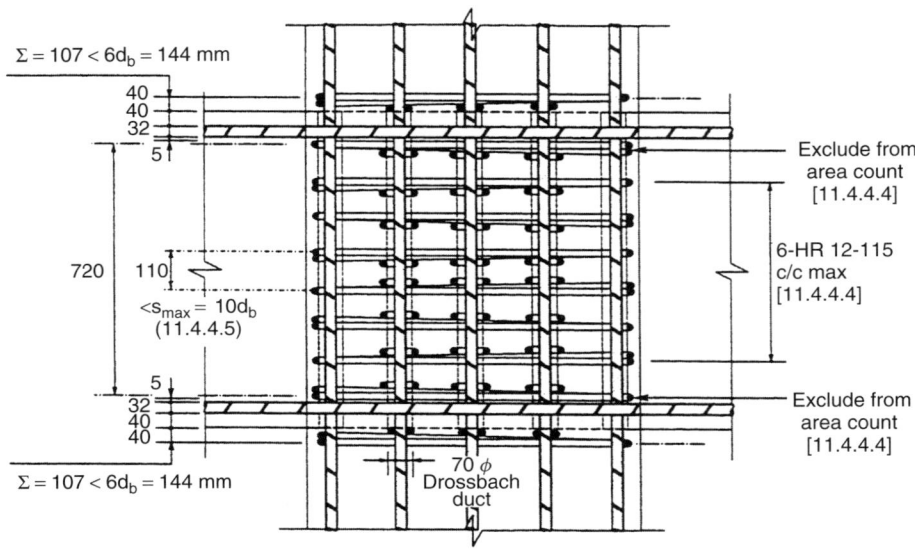

Figure 10.39 Detail of a beam–column joint in a precast concrete moment resisting frame (after NZCS, 2008)

be made. This is particularly important when adapting proprietary precast products made for general purposes, especially when intended originally for use in non-seismic areas. Without appropriate dynamic test results the effect of the connections may be difficult or impossible to assess, especially if they depart substantially from providing full continuity and homogeneity between adjacent members.

Dealing with building tolerances is a major problem in the design of connections. Constructional eccentricities may result in large secondary stresses in earthquakes, and should be either designed for or minimized by the manner of connection. It may be advantageous to design structural joints which permit generous constructional tolerances and restrict the amount of expensive fine tolerance work to the cladding for visual or drainage purposes.

To overcome the connection problem, partial precasting is often done. For example, precast beams may be used with *in situ* columns, or precast walls may be used with in situ floors, or vice versa.

As the basic detailing of reinforced and prestressed concrete has been discussed elsewhere in this book, only the essential problem of precast construction, that of *connection*, is considered in this section.

For further reading, see Fédération Internationale du Béton (2003) and NZSG (1999), and papers such as those of Pampanin (2005), Clough (1984) and Sauter (1984).

Connections between bases and precast columns

The following typical details (Details 10.1–10.4) must be individually designed for the forces acting on the joint. The base considered may be at foundation level or on suspended members higher up the structure. Member reinforcement is not shown.

Concrete Structures

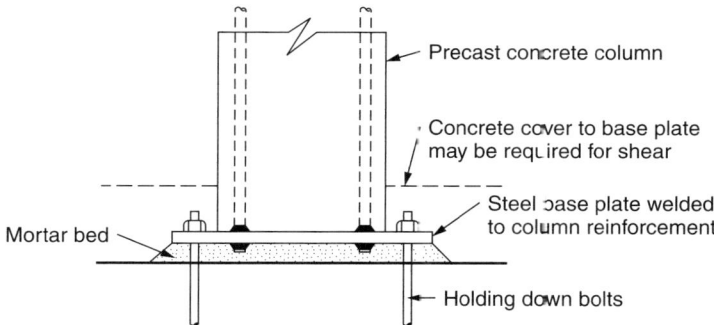

Detail 10.1 Site bolted. Moment transfer controlled by base plate

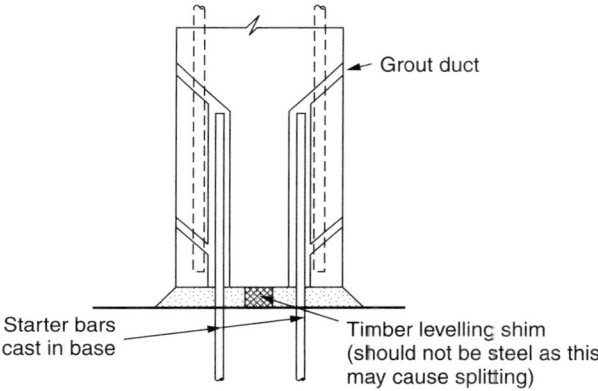

Detail 10.2 Site grouted. Effectiveness depends on grouting

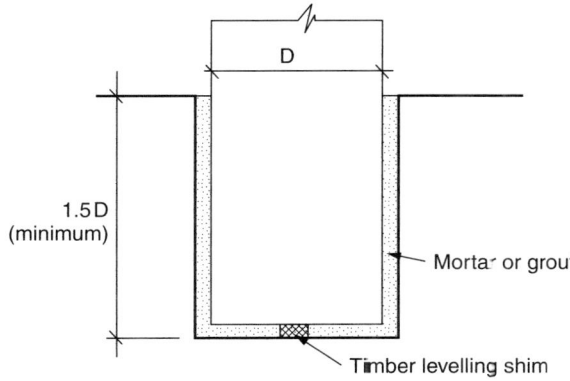

Detail 10.3 Site grouted. Best all-round joint of this type. Method of transfer of vertical load to base must be checked

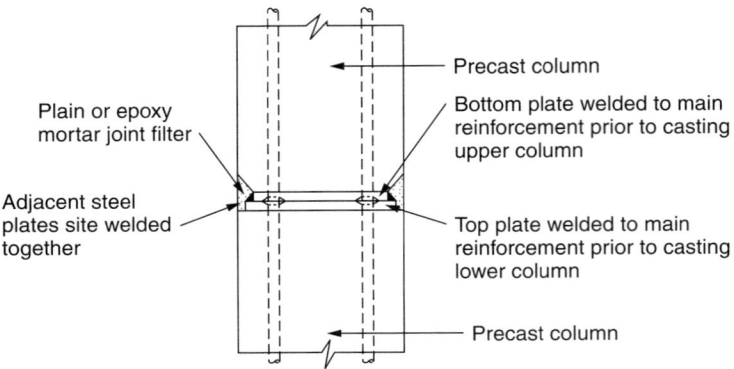

Detail 10.4 Site welded

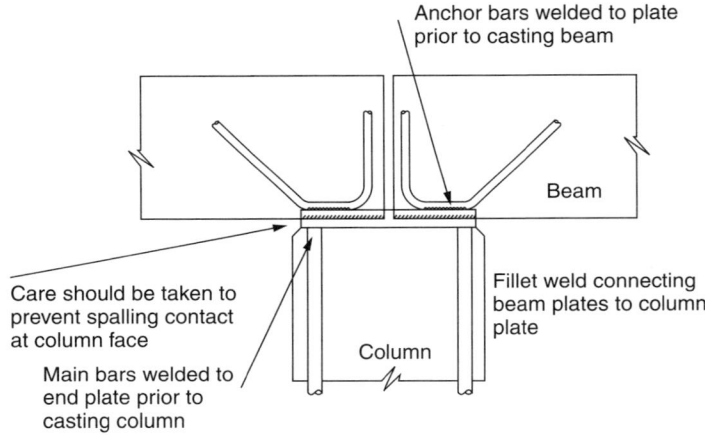

Detail 10.5 Site welded. Low moment capacity

Connections between precast columns and beams

The following typical details (Details 10.5–10.8) must be designed for the forces acting on the joint under construction. Variations on these connections may be made to suit the circumstances. Member reinforcement is not shown. Note that welding of bent bars should only be done with suitable steels. High carbon steels are prone to brittle fracture in this situation.

Connections between precast floors and walls

The following typical details (Details 10.9–10.12) must be designed for the forces acting on the joint under consideration. Member reinforcement is not shown.

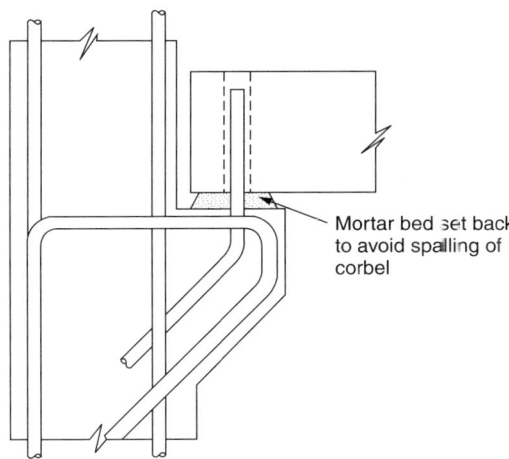

Detail 10.6 Site grouted. Low moment capacity, poor in horizontal shear

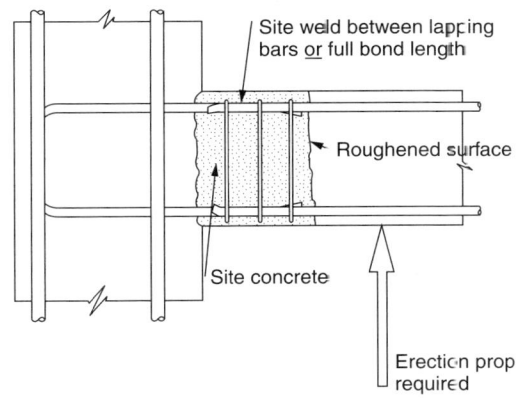

Detail 10.7 Site concreted and welded and stirrups fixed

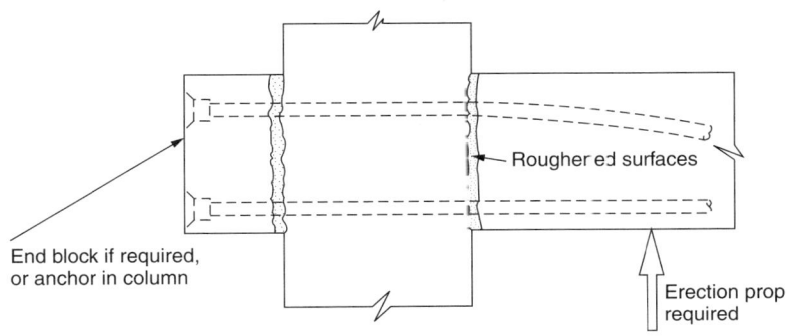

Detail 10.8 Site mortared and post-tensioned

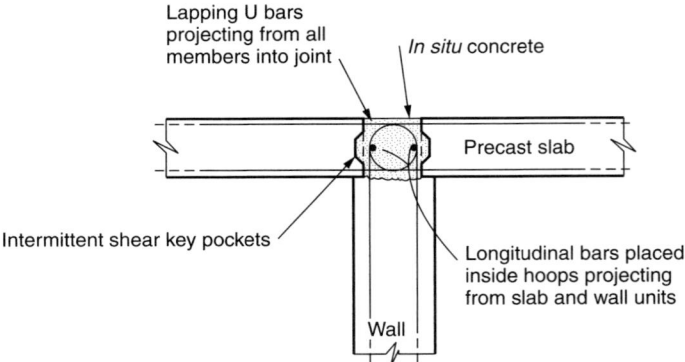

Detail 10.9 Site concrete and reinforcement

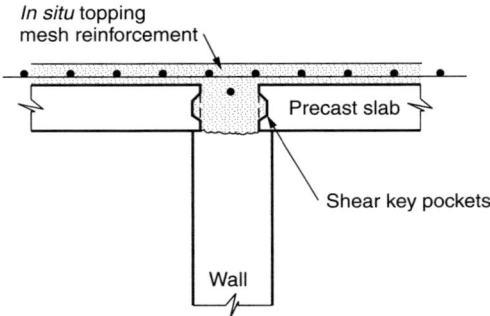

Detail 10.10 Site concrete and reinforcement

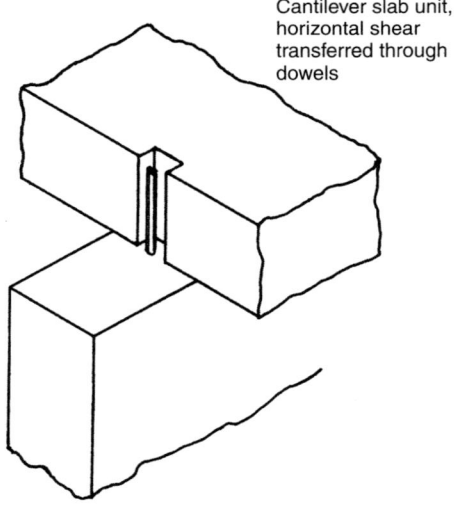

Detail 10.11 Site grouting

Concrete Structures

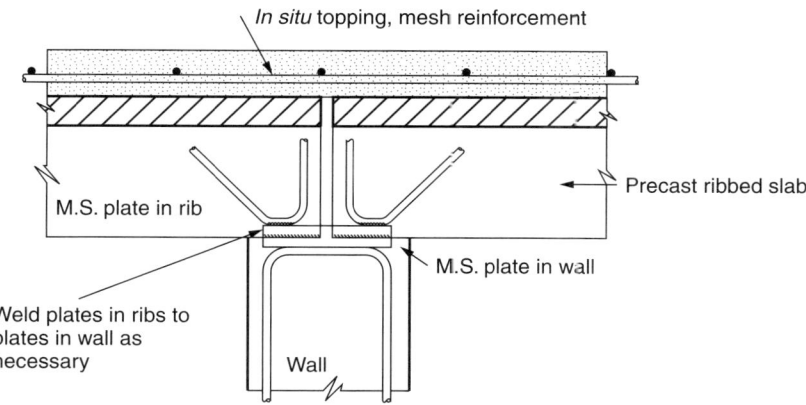

Detail 10.12 Site concrete, reinforcement and welding

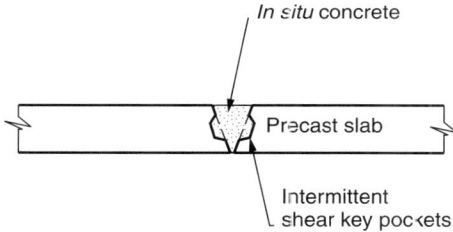

Detail 10.13 Site concreting. This joint depends on perimeter reinforcement to complete shear transfer system

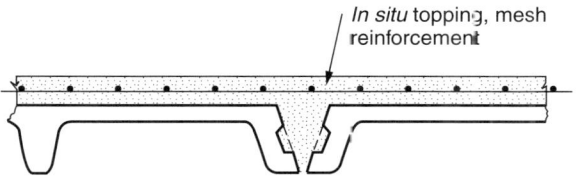

Detail 10.14 Site concreting and reinforcement

Connections between adjacent precast floor and roof units

The following typical details (Details 10.13–10.16) must be designed for the forces acting on the joint under consideration. Floor slabs should be designed as a whole, to act as diaphragms distributing the shear between the vertical members of the structure. The unifying effect of perimeter beams should be taken into account. Member reinforcement is not shown.

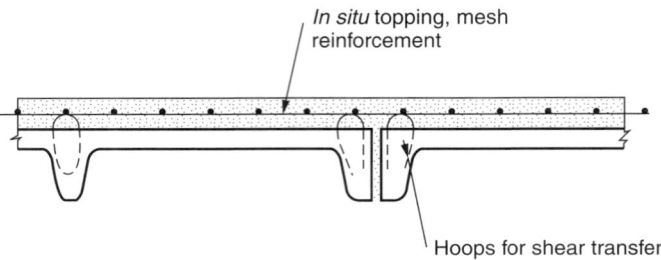

Detail 10.15   Site concreting and reinforcement

Connections adjacent precast wall units

The following typical details (Details 10.17–10.22) should be designed for the forces acting on the joint under consideration. Great problems occur in producing a ductile and easily erected precast shear wall, and no universal solution has as yet been evolved. The details below may be adapted for use with internal or external walls, i.e. cladding. Member reinforcement is not shown.

10.3.12   *Precast concrete cladding detail*

Precast concrete cladding varies in its relationship to the building structure, from being fully integrated to being fully separated from frame action. Ideally, the cladding should be either fully integrated or fully separated, with no intermediate conditions. Fully integrated *structural* precast concrete cladding should be treated like any other precast structural element, as discussed in Section 10.3.11. Cladding which is not treated as part of the structure is considered below.

In flexible beam and column buildings it is desirable to effectively separate the cladding from the frame action, both to protect the cladding from seismic deformations and also to ensure that the structure behaves as assumed in the analysis. For very flexible buildings in strong earthquakes the storey drift may be so large as to make full separation difficult to achieve, and some interaction of frame and cladding through bending of the connections may have to be accepted. Ductile behaviour of the cladding and of its connections to the structure is most important in such cases, to ensure that the cladding does not fall from the building during an earthquake.

In stiff (shear wall) buildings the storey drift will generally be small enough to significantly reduce the problem of detailing of connections which give full separation. On the other hand, protection of the cladding from seismic motion is less necessary in stiff buildings, and connections permitting movement through bending may be satisfactory as long as the interaction between cladding and frame can be allowed for in the frame analysis.

It is common for precast cladding to be fully separated from the frame in strong motion areas such as California, Japan and New Zealand. This has been done on major buildings, such as the 47-storey Keio Plaza Hotel in Tokyo. Unfortunately, little has been published regarding the connection details for separated cladding, although some reference is made to this problem by Uchida *et al*. (1973). In Uchida's structure, a 25-storey steel-framed

Concrete Structures

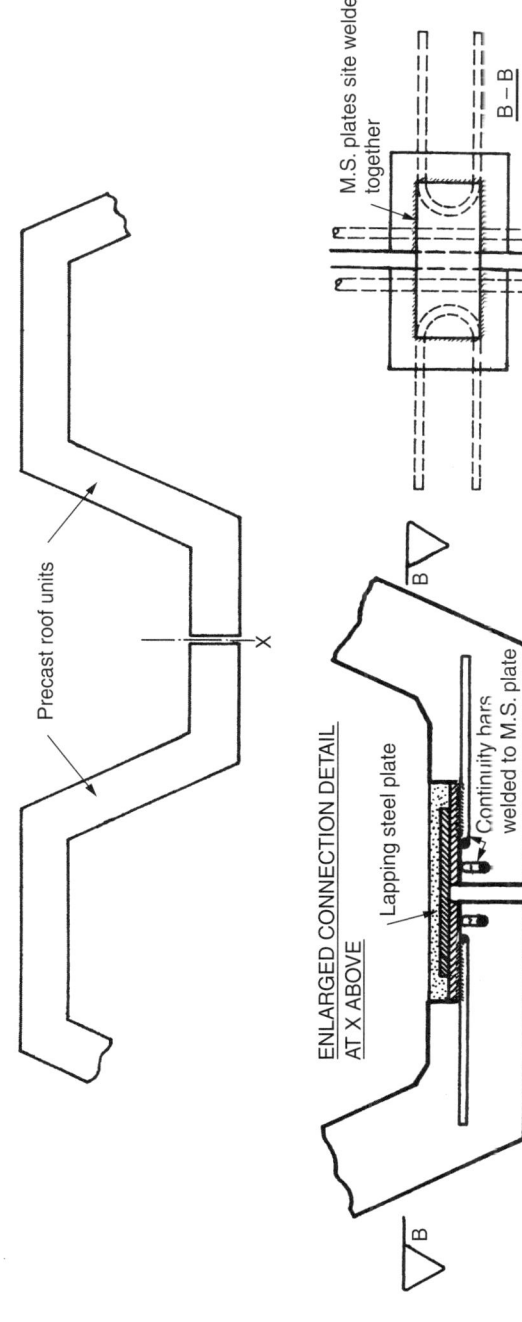

Detail 10.16 Site welding and mortaring. Lapping steel plate bent on site to suit differential camber of adjacent precast units

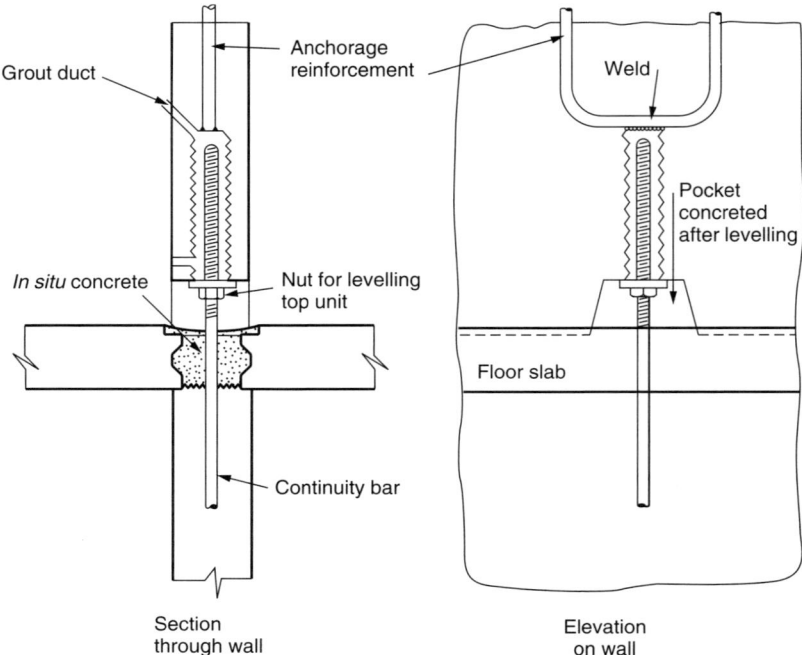

Detail 10.17 Site concreting and grouting

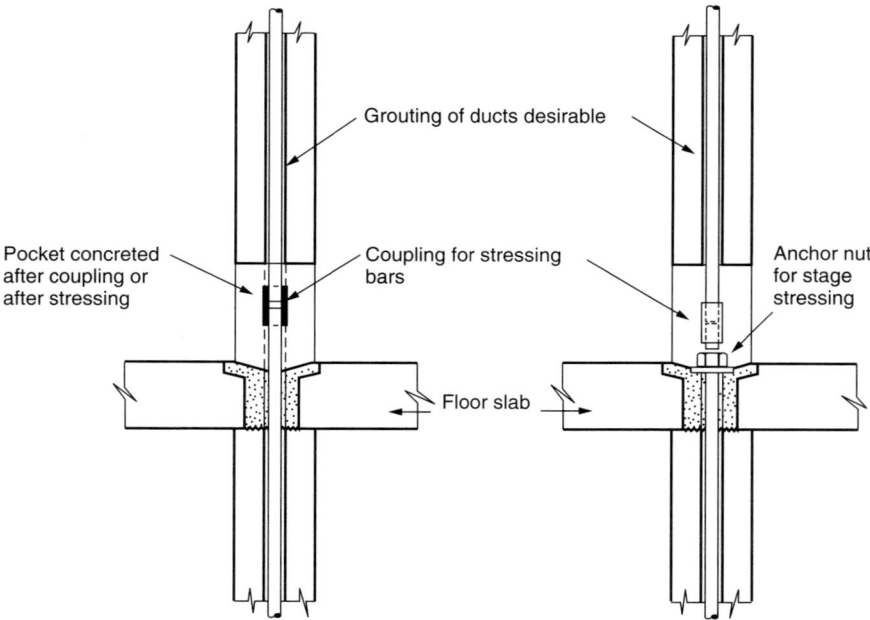

Detail 10.18 Site concreting and post-tensioning and grouting of ducts

Concrete Structures

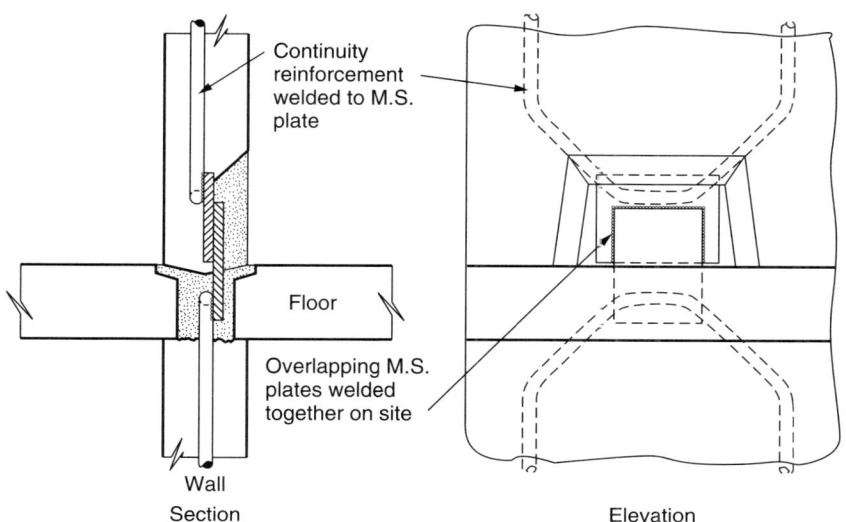

Detail 10.19 Site welded and concreted

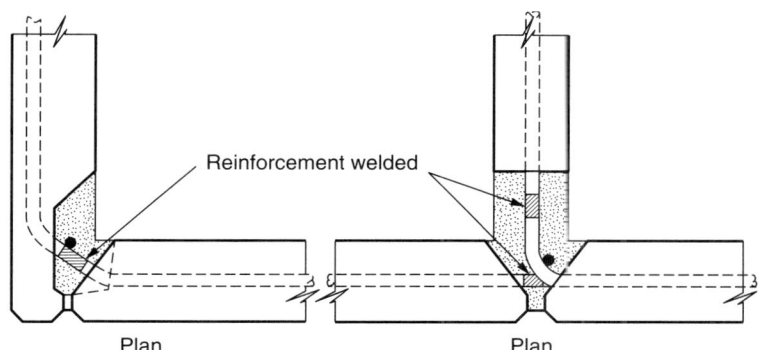

Detail 10.20 Site welded and concreted

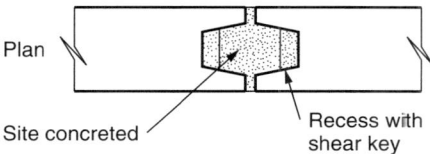

Detail 10.21 Site concreted

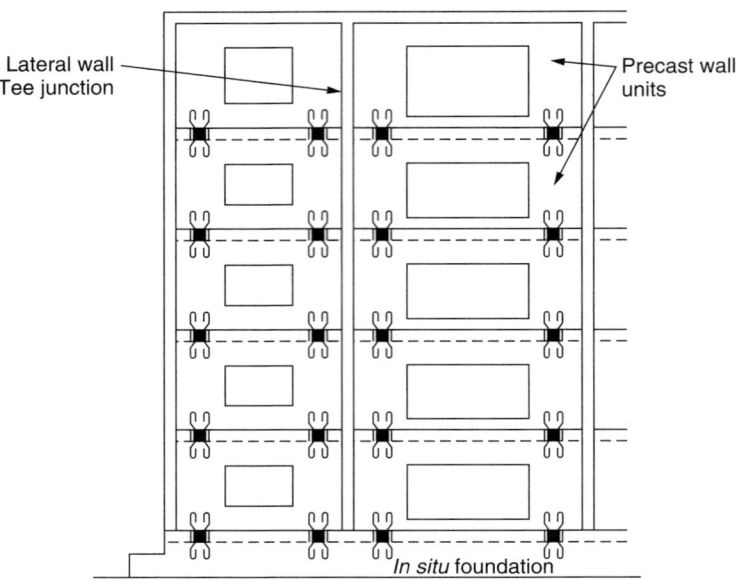

Detail 10.22 Layout of joints in wall elevation

building, separation of the cladding was only partial, and the connections were designed so that the panels would not fall off if the storey drift was 50 mm.

Gaps between adjacent precast units are often specified to be 20 mm to allow for seismic movements and construction to tolerances, but smaller or larger gaps may be determined from drift calculations. Waterproofing of gaps may be effected by baffled drain joints or mastic, but the performance of mastic-filled joints in earthquakes is not known to the author.

The principles of support for fully separated precast cladding are illustrated diagrammatically in Figure 10.40. Such connections should be made of corrosion resistant materials, and must be designed to carry the gravity and wind loads of the cladding back into the structure as well as to allow the free movement of the frame to take place.

10.3.13 Prestressed concrete design and detail

Introduction

Prestressed concrete elements in structures which have been subjected to earthquakes have mostly performed well. Failures have been mainly due to inadequate connection details or supporting structure (Blakeley, 1973; Pond, 1972).

Although prestressed concrete is well established in bridge construction and various civil engineering applications, it is less widely used in building structures, and relatively few structures have been fully framed in prestressed concrete. This is true in both seismic and non-seismic areas. The comparative neglect of prestressed concrete for building structures has occurred partly for constructional and economic reasons, and in earthquake

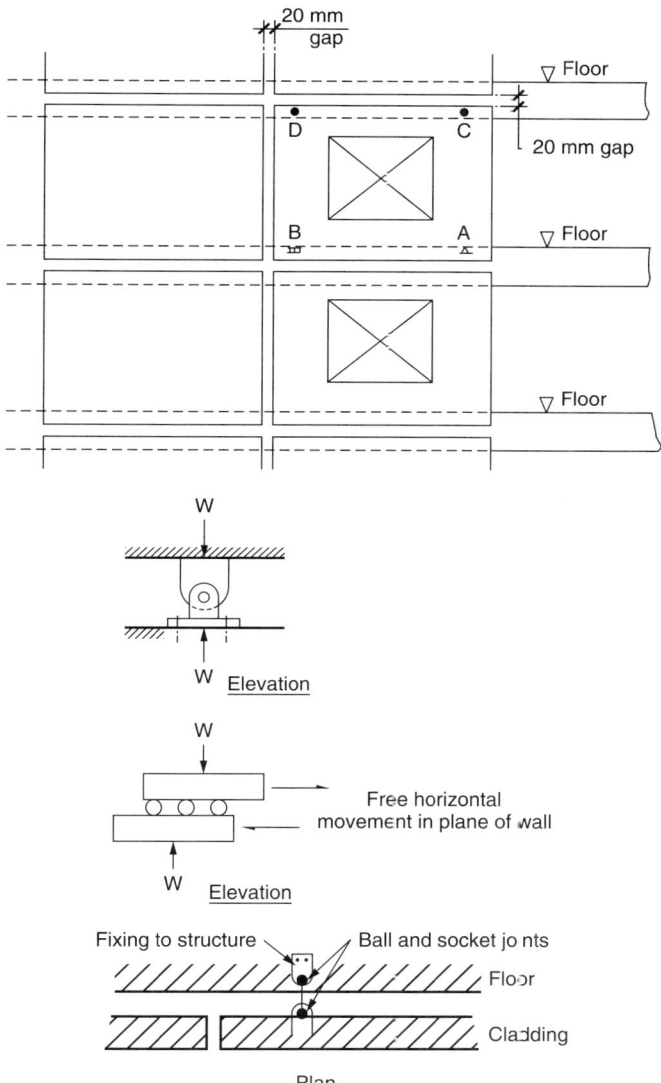

Figure 10.40 Schematic illustration of supports for precast concrete cladding fully separate from frame action

areas it has also occurred because of divergent opinions on the effectiveness of prestressed concrete in resisting earthquakes.

Recommendations for seismic design of prestressed concrete

A good overview of recent trends in the use of prestressed concrete has been given by Pampanin (2005). Seismic design recommendations include Fédération Internationale du

Béton (2003) and NZS 3101 (2006). In contrast, the major US concrete codes only discuss prestressed concrete in non-seismic terms, which unfortunately implies a reluctance to assess the seismic design requirements of this valuable material.

Seismic response of prestressed concrete

The seismic response of structural materials has been discussed generally in Section 6.6, where some stress–strain diagrams were presented. The main characteristics of prestressed concrete under cyclic loading may be inferred from Figure 5.23(e), from which Blakeley (1973) proposed the idealized hysteresis loop shown in Figure 10.41.

It is evident from the narrowness of the hysteresis loops that the amount of hysteretic energy dissipation of prestressed concrete will be relatively small compared to steel or reinforced concrete. On the other hand, the capacity of prestressed concrete to store elastic energy is higher than for a comparable reinforced concrete member.

Prestressed concrete suffers in comparison to reinforced concrete because of its lack of compression steel, so that its performance is poorer once concrete crushing begins. Compared to reinforced concrete, prestressed concrete undergoes relatively more uncracked deformation and relatively less deformation in the cracked state. This means that prestressed concrete structures should exhibit less structural damage in moderate earthquakes. In the event of structural repairs being necessary after an earthquake, there are obvious difficulties in restoring the prestress to sections of replaced concrete, and conversion of the failure zones to reinforced concrete may be necessary.

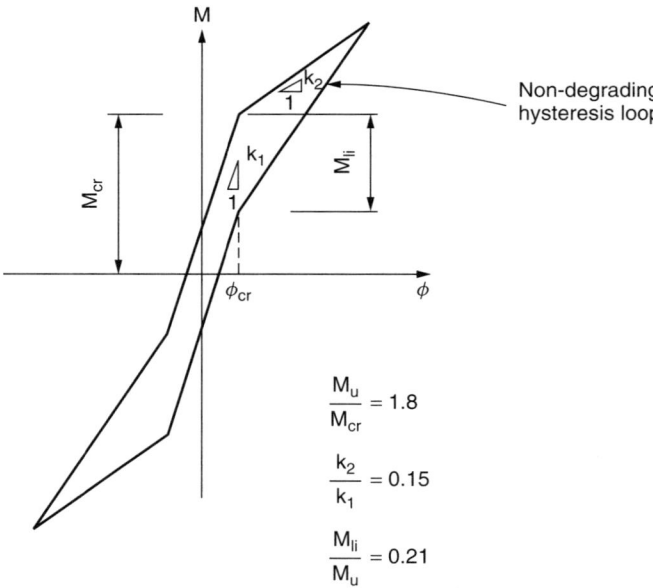

Figure 10.41 Moment–curvature idealization for plastic hinge regions in prestressed concrete (after Blakeley, 1973)

It has been suggested that prestressed concrete buildings may be more flexible than comparable reinforced concrete structures, and that more non-structural damage may occur. However, differences in flexibility will be small in practical design terms, and structures in either material will generally be less flexible than steelwork. In any case, proper detailing of the non-structure will be necessary regardless of the materials used in the structure.

For notes on the damping of prestressed concrete structures, see Table 5.11.

Factors affecting ductility of prestressed concrete members

For the satisfactory seismic resistance of prestressed concrete members, brittle failure must be avoided by the creation of sufficient useful ductility, as discussed in Section 5.4.2. In the case of prestressed concrete, the useful available section ductility may be defined as

$$\frac{\phi_{0.004}}{\phi_{cr}}$$

where $\phi_{0.004}$ is the curvature at a nominal maximum concrete strain of 0.004, and ϕ_{cr} is the curvature at first cracking.

The ductility or rotation capacity of prestressed concrete is affected by:

(1) the longitudinal steel content;
(2) the transverse steel content;
(3) the distribution of longitudinal steel;
(4) the axial load.

Each of these variables is discussed below.

Longitudinal and transverse steel content
Figure 10.42 shows that ductility decreases markedly with increasing prestressing steel content. As seen in Section 10.3.3, *unstressed* longitudinal reinforcement also reduces the ductility. Thus, to ensure that reasonable ductility is obtained in potential plastic hinge zones, the content of prestressed plus non-prestressed flexural steel should be such that at the flexural capacity of the section, the depth of the equivalent rectangular stress block is

$$a \leq 0.2h, \tag{10.56}$$

where h is the overall depth of the section, unless confinement reinforcing is provided, in which case

$$a \leq 0.3h. \tag{10.57}$$

The confinement should be provided as for reinforced concrete columns, Park *et al.* (1984) reporting good cyclic behaviour using half the amount of confinement steel required for columns by the New Zealand code. As with reinforced concrete, the use of confinement steel greatly increases the ductility of prestressed members compared with unconfined ones.

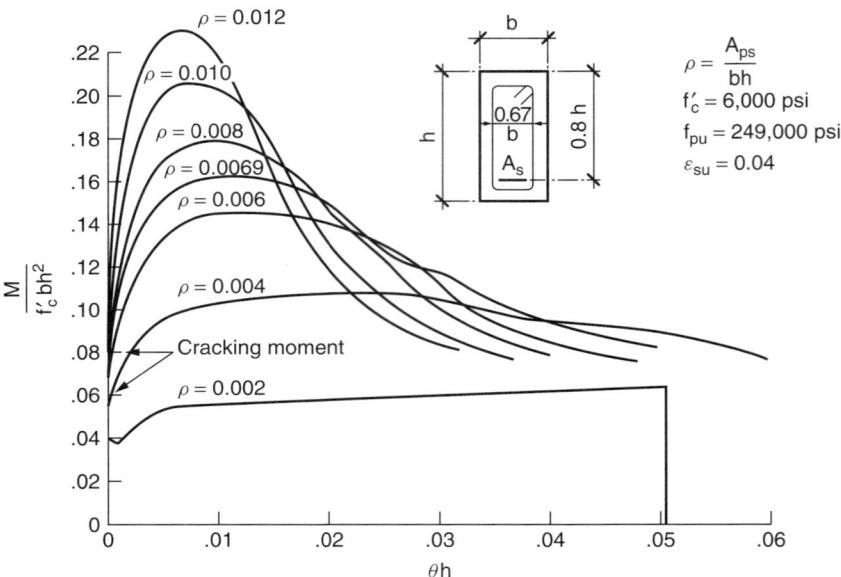

Figure 10.42 Moment–curvature relationship for rectangular prestressed concrete beams showing the effect of prestressing steel content on ductility (after Blakeley and Park, 1971)

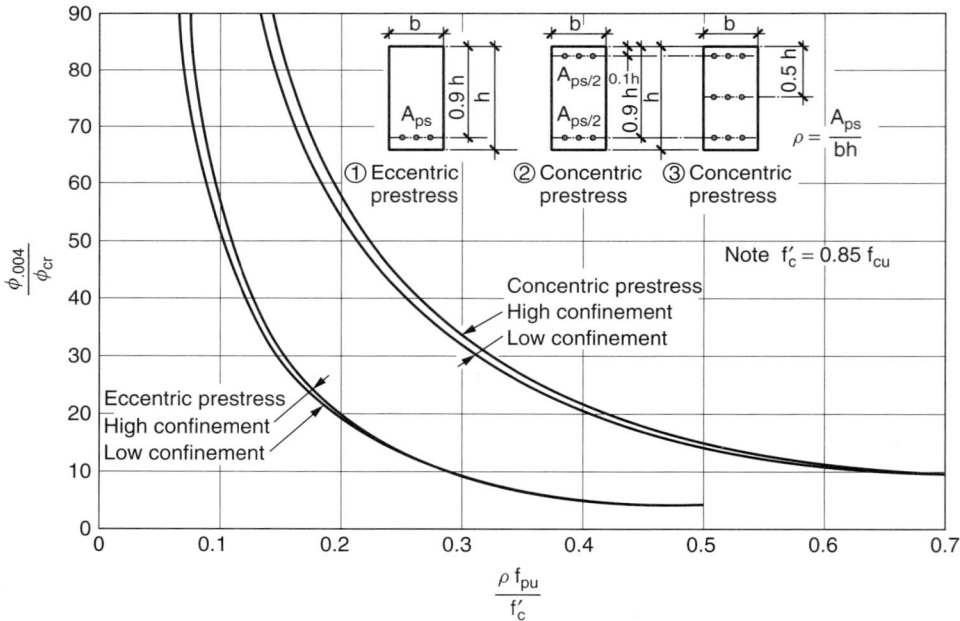

Figure 10.43 Variation of curvature ratio at crushing (section ductility) for prestressed concrete beams (after Blakeley, 1973)

Distribution of longitudinal steel

At positions of moment reversal where the greatest ductility requirements exist, the required distribution of prestress will usually be nearly axial. Blakeley (1973) demonstrated that a single axial tendon produced a less ductile member than that achieved by multiple tendons placed nearer the extreme fibres. At points in structures where stress reversals do not occur, eccentric prestress may be used. Where no unstressed reinforcement exists, an eccentrically prestressed beam is notably less ductile than a concentrically stressed one with equal prestressing steel content (Figure 10.43).

The tendon distribution ③ in Figure 10.43 is not only as ductile as ②, but has the advantage that the axial tendon will be practically unharmed by large rotations, and should hold the structure together after the tendons near the extreme fibres have failed.

Effect of axial load on ductility of prestressed concrete columns

The section ductility $\phi_{0.004}/\phi_{cr}$ decreases rapidly with increasing column axial load N. This effect is seen in Figure 10.44, where ductility is plotted against the level of prestress for columns carrying varying axial loads.

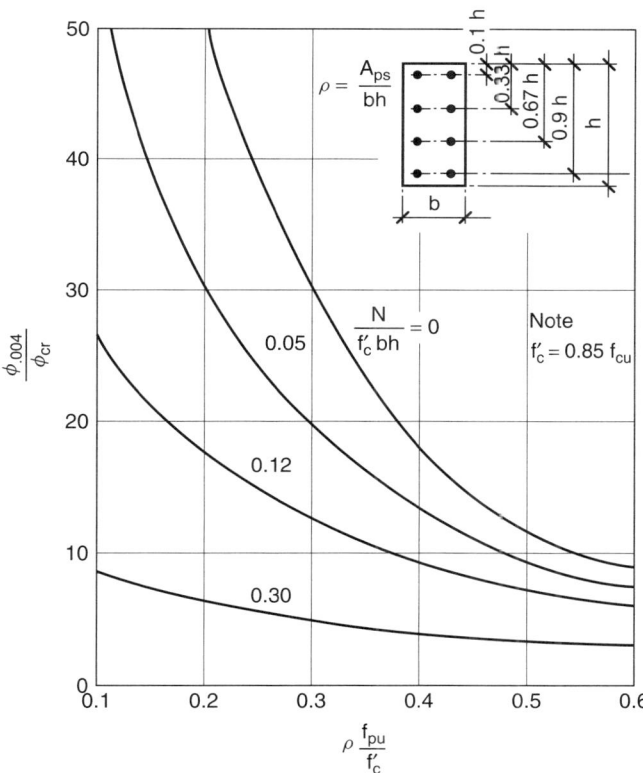

Figure 10.44 Variation of curvature ratio at crushing (section ductility) for columns with varying prestress and varying axial load (after Blakeley, 1973)

Detailing summary for prestressed concrete

For the adequate performance of prestressed concrete in earthquakes, its ductility and continuity should be maximized by careful consideration of the following items:

(1) Transverse and longitudinal steel content (page 391).
(2) Longitudinal steel distribution (page 391).
(3) Continuity, ensured by adequate lapping of prestressing tendons or reinforcing bars.
(4) Anchorages in post-tensioned construction, carefully positioned to avoid congestion and stress-raising in highly stressed zones. They should be situated as far from potential plastic hinge positions as possible.
(5) Joints between prestressed members involving ordinary reinforced concrete, properly designed as outlined in Sections 10.3.3–10.3.11.
(6) Joints using mechanical details, as suggested in Section 10.3.11.
(7) Unbonded tendons are acceptable for use in primary earthquake resistant members if a substantial quantity of bonded steel is also present (Goodsir *et al.*, 1984). In general, they should be used with anchorages of proven reliability under cyclic loading, and proper crack control measures should be taken.

10.4 Masonry Structures

10.4.1 Introduction

The term *masonry* covers a very wide range of materials such as adobe, brick, stone, and concrete blocks, and each of these materials in turn varies widely in form and mechanical properties. Also, masonry may be used with or without reinforcement, or in conjunction with other materials. As well as its use for primary structure, masonry is used for infill panels creating partitions or cladding walls.

The variety available in form, colour and texture makes masonry a popular construction material, as does its widespread geographic availability and, in some cases, its comparative cheapness. Properly used, it also has reasonable resistance to horizontal forces. However, masonry has a number of serious drawbacks for earthquake resistance: it is naturally brittle; it has high mass and hence has high inertial response to earthquakes; its construction quality is difficult to control; and relatively little research has been done into its seismic response characteristics compared with steel and reinforced concrete.

Because of the poor performance of some forms of masonry in earthquakes, official attitudes towards masonry are generally cautious in most moderate- or strong-motion seismic areas. For example, in 2008 Japan masonry had not yet been permitted for buildings more than 13 m in height and the use of masonry for new buildings had almost died out.

In contrast, carefully designed apartment buildings of 15 storeys or more have been built in California. No doubt there is considerable inherent strength in the 'egg-crate' form of apartment buildings, but there has been little dynamic testing or seismic field experience of tall masonry structures to date.

The other types of construction in which masonry is most popular are low-rise housing and industrial buildings.

10.4.2 Seismic response of masonry

This discussion is supplementary to the general introduction to seismic response given in Section 5.4.

The tendency to fail in a brittle fashion is the central problem with masonry. While unreinforced masonry may be categorically labelled as brittle, uncertainty exists as to the degree of ductility which should be sought in reinforced masonry.

Based on static load-reversal tests, Meli (1973) contended that

> for walls with interior reinforcement, where failure is governed by bending, behaviour is nearly elasto-plastic with remarkable ductility and small deterioration under alternating load except for high deformation.... If failure is governed by diagonal cracking, ductility is smaller and, when high vertical loads are applied, behaviour is frankly brittle. Furthermore ... important deterioration [occurs] after diagonal cracking.

Meli concluded that bending failure was the most favourable design condition for walls. In dynamic tests, Williams and Scrivener (1971) confirmed this conclusion. Their tests on brick walls showed fairly stable hysteretic behaviour at drifts up to about 1%, despite the absence of horizontal reinforcement.

In later cyclic load tests on concrete masonry walls, Priestley and Elder (1982) showed that ductility could be increase even in masonry by using confinement steel, and repeatable ductility factors of $\mu = 3$ or a little more can be obtained (Figure 10.45). The confinement consisted of 600 mm long confining plates in the bottom seven mortar courses at each end of the wall.

As in concrete walls, it has been found that the ductility of rectangular walls decreases as axial load, reinforcement ratio, reinforcement yield stress, or aspect ratio increase, but increases as the masonry compression strength increases.

Research in various countries has examined various masonry products and wall-reinforcing layouts, sometimes under slow cyclic reversed loading and shake-table dynamic tests. The value of having vertical horizontal reinforcement distributed throughout walls is apparent, but the use of reinforcement only at the perimeter of wall panels is surprisingly effective for both in-plane and out-of-plane loading (Gulkan *et al*., 1979). The latter is more true for masonry of higher tensile strength (i.e. concrete blocks), and also is probably more true for stiff structures with low lateral displacements. Perimeter-only reinforcement is very cost-effective as a minimum provision for low-cost construction and for strengthening of existing buildings.

The adequacy of very light reinforcement in low-rise construction has been remarkably demonstrated in shake-table tests at the Earthquake Engineering Research Center (EERC) of hollow concrete block houses, where three simultaneous components of shaking with peak accelerations up to about $0.9g$ were applied (Gulkan *et al*., 1979; Manos *et al*., 1984). The walls had no intermediate height horizontal reinforcement, but were well connected at top and bottom to continuous horizontal perimeter members of the foundation and roof construction. The vertical reinforcement content varied from zero to a maximum of four bars (of 10 and 12 mm diameter) in walls about 4 m long.

The good performance of lightly reinforced masonry is shown by the mean damage ratios at intensity MM9 in the 1987 Edgecumbe, New Zealand, earthquake. In

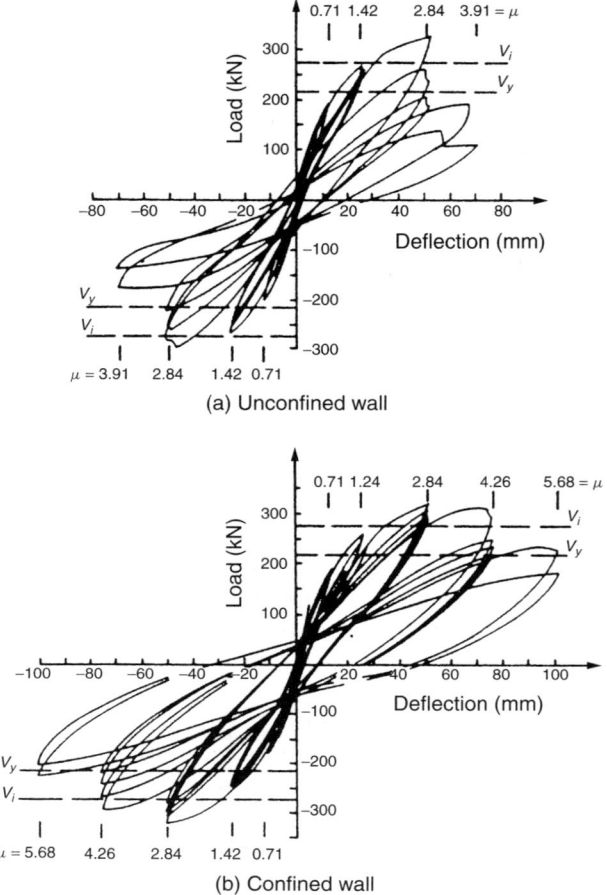

Figure 10.45 Load–deflection behaviour of concrete masonry test walls with high aspect ratio. (Reprinted from Paulay, T. and Priestley, MJN (1992), *Seismic Design of Reinforced Concrete and Masonry Buildings*. Copyright © (1992). Reprinted by permission of John Wiley & Sons, Inc)

Figure 10.53(a), it can be seen that the mean damage ratio D_{rm} was only 0.04 for single-storey buildings of the 1935–1979 era, when little reinforcement was used in concrete block masonry construction. (D_{rm} is defined in Section 6.3.1.)

In California in the 1994 Northridge earthquake, also with intensities up to MM9, lightly reinforced masonry again did well. According to Holmes and Somers (1995):

> In the greater Los Angeles area, and particularly in the epicentral region, very little distress was shown by modern single- or multi-story reinforced masonry bearing-wall buildings. In general, masonry structures built since the 1950s that were engineered, reinforced, grouted, and inspected in accordance with the then-current building codes experienced little damage in the earthquake.

10.4.3 Reliable seismic behaviour of masonry structures

Introduction

For obtaining reliable seismic response behaviour, the principles concerning choice of form, materials and failure mode control discussed in Section 8.3 apply to masonry structures, while further factors specific to masonry are discussed below.

The wide range of masonry products, of clay and concrete types, means a wide range of material behaviour and hence of seismic reliability. Reinforced hollow concrete blocks, which have been more studied than other masonry materials, are probably the most reliable type. However, with the growth of research interest in reinforced clay bricks (e.g. Wakabayashi and Nakamura, 1984) and other masonry products, the full reliability potential and relative merits of the various masonry materials are becoming better understood. Where a choice between relatively unresearched masonry materials has to be made, those which are weaker in compression and tension will obviously tend to be less reliable in earthquakes.

In considering the reliability of seismic behaviour of masonry structures through structural forms and failure mode control, fewer alternatives need be considered than for other structural materials. Masonry is best suited to forming walls and less suited to columns and lintel beams, and is constructionally and seismically ill suited for forming other structural members. Thus, this discussion mainly relates to the reliable seismic behaviour of walls.

While quite high repeatable ductilities can be achieved in masonry walls and columns by using thin steel plates between block courses (Figure 10.45), the constructional complications and costs of such measures suggest that seeking high ductilities for masonry structures is an example of seeking high ductility for its own sake. The more pragmatic traditional approach of seeking limited ductility, so well demonstrated as successful (at least for single-storey buildings) by the EERC tests (Section 10.4.2), seems likely to remain appropriate for most masonry structures, namely:

(1) Suppress the more brittle failure modes (e.g. shear).
(2) Design for limited ductility and adequate strength.
(3) Use sound structural forms (as discussed below).

Structural form for unreinforced or nominally reinforced masonry buildings

The general principles of earthquake resistant structural form have been given in Section 8.3, but additional guidance peculiar to masonry is given here. Five interrelated criteria for consideration in masonry construction are that:

(1) the aspect ratio H/B for the structure should be a minimum;
(2) the aspect ratio H'/B' for vertical members should be a minimum;
(3) the ratio of aperture area to wall area, $\sum A_a/HB$, should be a minimum;
(4) the distribution of apertures should be as uniform as possible;
(5) stress-raising apertures should be located away from highly stressed zones.

Considering these criteria in relation to a single-storey structure with zero or minimum reinforcement, Figure 10.46(a) represents a good structure and Figure 10.46(b) represents

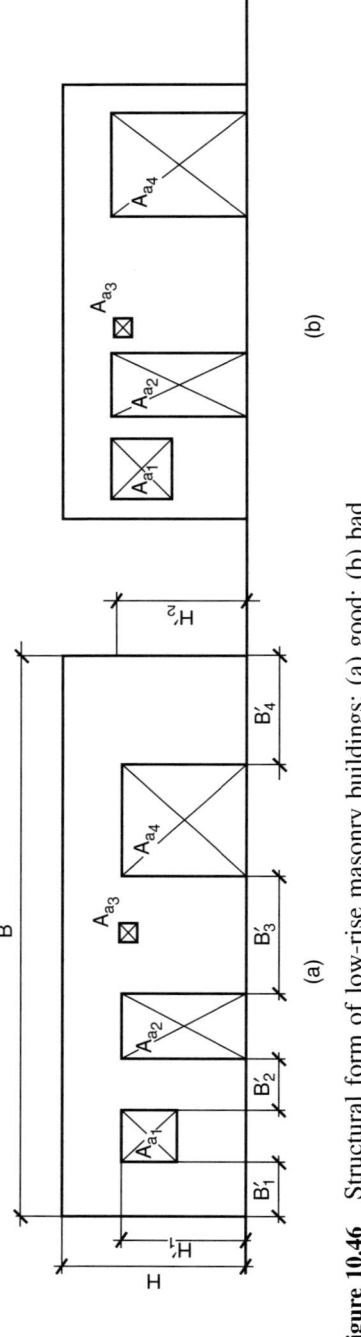

Figure 10.46 Structural form of low-rise masonry buildings: (a) good; (b) bad

a bad one. Neither case has a bad overall aspect ratio, as is typical for single-storey buildings. For unreinforced buildings with a maximum aperture area, a value of $H/B \not> 2/3$ might be taken.

The aspect ratio of vertical members, particularly those at the ends of walls (H'_1/B'_1 and H'_2/B'_4 in Figure 10.46(a)), should be little greater than unity in buildings of minimum reinforcement. This is clearly not so in Figure 10.46(b).

It is commonly recommended that the total area of holes should not exceed one-third of the wall area, i.e. $H/B \not> 1/3$. If criteria (2) and (3) above have been satisfied, it is likely that the distribution of apertures will be reasonably uniform. Small holes, such as those used for the passage of services pipes and ducts, should be kept away from corners of load-bearing members; A_{a_3} in Figure 10.46(b) is badly placed compared with that in Figure 10.46(a).

The main objective of the above criteria is to distribute the strength as uniformly as possible; in brittle structures the early failure of one member causes the remaining members to share the total load, which can lead to incremental collapse.

Structural form for reinforced masonry

The criteria set out above are also applicable to reinforced masonry, although some relaxation of the suggested limits may be made. The degree of relaxation will depend upon the degree of protection against early brittle failure afforded by the reinforcement. In high-quality construction, building aspect ratios, H/B, of 2 or 3 are reasonable, and in apartment block construction with small aperture ratios even higher values of H/B have been used.

The aspect ratio H'/B' for vertical members when reinforced may also be increased over those given in the previous subsection, perhaps to values of 2 or 3 for members of low strength, but much higher values may be taken for column members of higher-quality design and construction.

10.4.4 Design and construction details for reinforced masonry

For the reasons given in Section 10.4.3, most masonry structures should be designed primarily for strength, with limited, rather than high, ductility being sought. Design and construction procedures should conform to well-established codes of practice, such as the International Building Code (International Code Council, 2006 or later), NZS 4230 (2004) or the *Masonry Designers' Guide* (Masonry Society, 2007), as followed in the design handbook by Amrhein (1994), or the advice of Paulay and Priestley (1992) should be followed. Seismic loadings set out in local regulations should be used with larger load factors than those used for steel or concrete, unless a specific factor for materials is used in finding the horizontal loads.

Some of the more important details of seismic reinforced masonry construction are commented on below.

Minimum reinforcement

Although code recommendations on this subject vary somewhat, the following practice is fairly representative. At least one vertical bar, not less than 12 mm in diameter, should be placed at all corners, wall ends and wall junctions. Such bars should be anchored into the upper and lower walls beams or foundations, and adequately lapped at splices.

Walls should be reinforced with a minimum steel content of 0.07% both horizontally and vertically, and the total of the two directions should not be less than 0.2%. A minimum spacing of 1.2 m between bars in both directions is also recommended. Some codes require closer minimum spacing, but the above nominal requirements are well supported both field experience and the EERC test series discussed in Section 10.4.2.

At least one bar, not less than 12 mm in diameter, should be placed on all sides of apertures exceeding about 600 mm in any direction. Such framing bars should extend not less than 600 mm beyond each corner of the aperture, or be equivalently anchored.

Horizontal continuity

Horizontal continuity around the perimeter of the building should be ensured at least at the levels of the base, the floors and the roof. The walls should be tied into an effective ring beam at these levels. Connections to floors or roofs other than of *in situ* concrete have proved especially vulnerable in earthquakes.

Grouting

The reinforcement of masonry depends for its effectiveness upon the transfer of stress through the grout from steel to masonry. Every effort should be made to ensure that compacted grout completely fills the cavities. A low-shrinkage grout is essential to minimize separation of the grout from the masonry. Grout for cavities of up to 60 mm in width should contain aggregates up to 5 mm in size, while for larger cavities coarser aggregate may be suitable. Grout should have a characteristic cube strength of 20 MPa at 28 days.

Hollow concrete blocks

Although the shape of hollow concrete blocks varies in detail in different countries, those shown in Figure 10.47 illustrate the principal types used in walls 150–200 mm thick.

Supervision of construction

To ensure adequate standards of construction, more supervision is required for reinforced masonry than on equivalent projects in other materials. The following points in particular need watching:

(1) Cavities should be clean and free from mortar dropping.
(2) Reinforcement should be placed centrally or properly spaced from the masonry.
(3) Reinforcement should be properly lapped.
(4) The grouting procedure should be properly carried out.
(5) The grout mix should conform to the specification.

In multi-storey hollow concrete block construction, inspection holes at the bottom of walls on the line of vertical reinforcement are advisable to facilitate the checking of items (1)–(4) above (Figure 10.48).

Masonry Structures

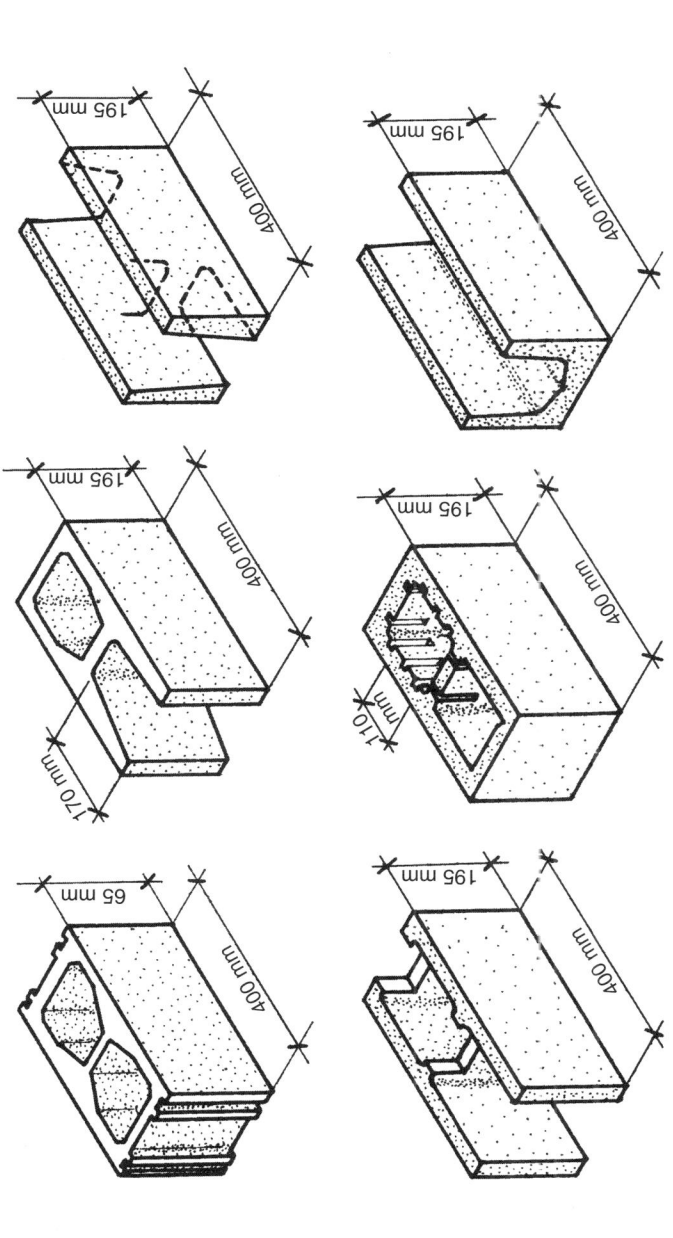

Figure 10.47 Principal structural concrete blocks used in New Zealand for reinforced masonry construction, 150–200 mm nominal wall thickness

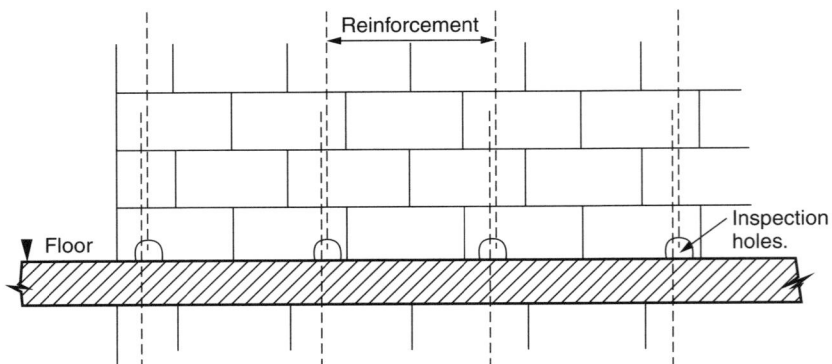

Figure 10.48 Inspection holes in reinforced concrete block construction for checking reinforcement and grouting (see text)

Typical reinforcement details

Typical details of reinforcement for selected masonry elements derived from New Zealand practice, are shown for walls (Figures 10.49 and 10.50), for columns (Figure 10.51) and for beam–column joints (Figure 10.52).

10.4.5 Construction details for structural infill walls

Masonry is often used as structural infill, either as cladding or as interior partitions. It should be either effectively separated from frame action or fully integrated with it, as discussed in Section 12.2.2. The analysis of the interaction between frames and integrated infill panels has been discussed in Section 5.4.6. Few data on the detailing of masonry in this situation exist, but the following points may be made:

(1) No gap should be left between the infill and the frame, so as to prevent accidental pounding damage during earthquakes.
(2) The top of the panel should be structurally connected to the structure above to ensure lateral stability of the infill in earthquakes.
(3) Ideally the form of the structure and the strength of the infill panel would be such that shear failure of neither the infill nor the frame would occur. For this reason infill panels should be the full height of the aperture in which they are built.

Placing of full-height vertical reinforcement is obviously difficult in hollow block or brick construction of the form shown in Figure 10.47, when the infill is erected after the upper frame member has been constructed, as is usually the case in this form of construction. This difficulty is obviated by the use of external reinforcement in the form of expanded metal sheets bonded to the side of the block wall by means of a layer of mortar. Tso *et al*. (1973) have reported favourable behaviour of this type of construction in cyclic loading tests, using washers and bolts through the wall to improve the bonding of the expanded metal to the wall.

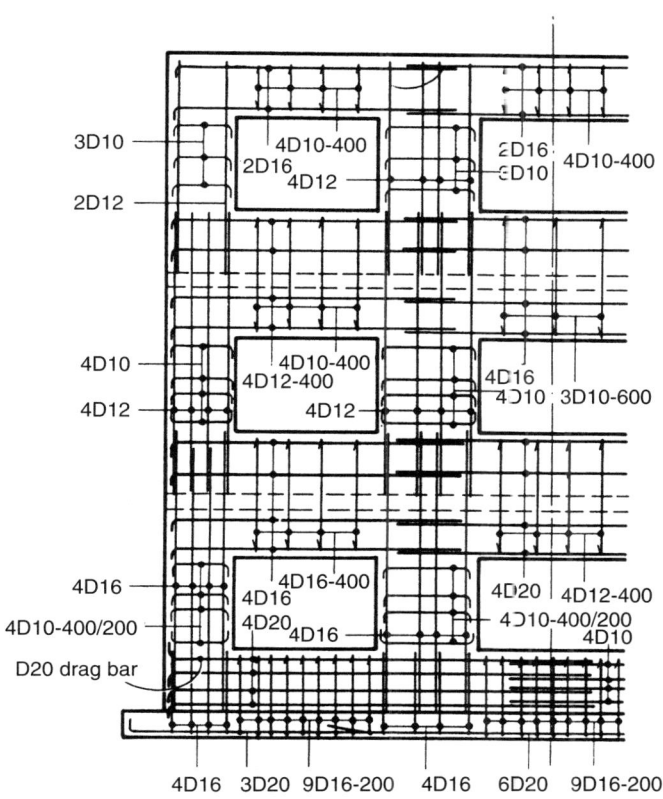

Figure 10.49 Details of reinforcement for a three-storey masonry structure. (Reprinted from Paulay T and Priestley MJN (1992) *Seismic Design of Reinforced Concrete and Masonry Buildings*. Copyright © 1992. Reprinted by permission of John Wiley & Sons, Inc)

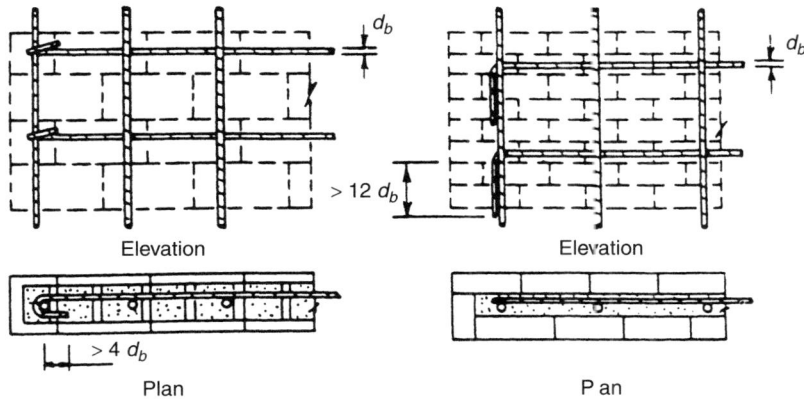

Figure 10.50 Anchorage of shear reinforcement in masonry walls. (Reproduced from NZS 4230:2004 by permission of Standards New Zealand under Licence 000710)

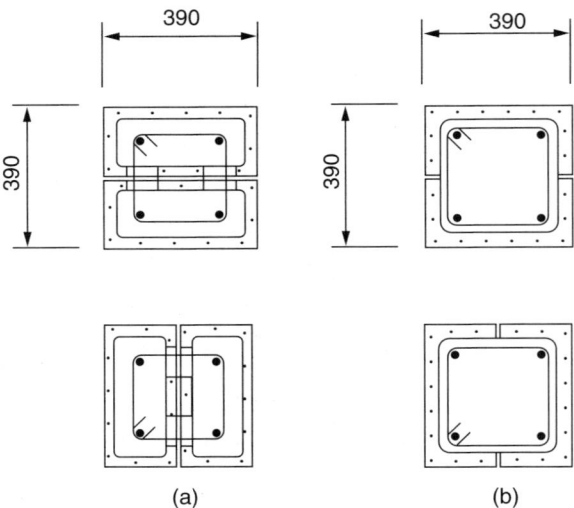

Figure 10.51 Typical construction of masonry columns showing alternating courses of two types of concrete masonry units (a) and (b) (from NZS 4230: 2004). (Reproduced from NZS 4230:2004 by permission of Standards New Zealand under Licence 000710)

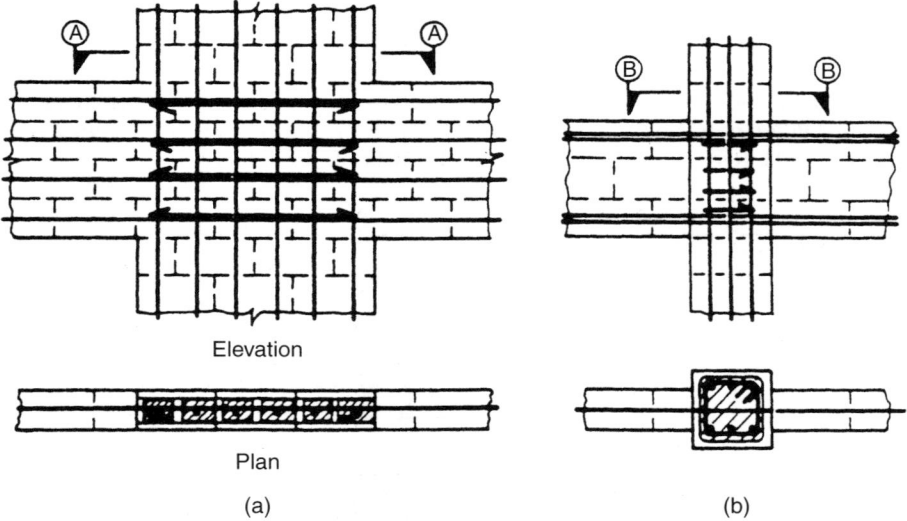

Figure 10.52 Reinforcement arrangements for masonry interior beam–column joints: (a) planar beam–wall joint (unconfined) using open-end bond beam units; (b) one-way beam–column joint using pilaster units for column joint (confined) (from NZS 4230, 2004). (Reproduced from NZS 4230:2004 by permission of Standards New Zealand under Licence 000710)

10.4.6 Masonry structures in regions of low and moderate seismic hazard

Page (1996) reviewed the use of *unreinforced masonry* in Australia a region of mostly *low seismic hazard*, concluding with the following useful and perceptive remarks:

> Unreinforced masonry is a commonly used building material in Australia as it is economical, attractive and durable, with good thermal and sound insulation and excellent fire resistance. Masonry is thus widely used for loadbearing elements as well as for infill and cladding in domestic and framed construction. Unreinforced masonry does not have good seismic performance as it is a heavy, brittle material with low tensile strength and exhibits little ductility when subjected to seismic effects. It is therefore unsuited for areas of high seismicity. However, in regions of lower seismic activity such as Australia unreinforced masonry can be used in most instances provided it is designed and detailed correctly and built to the required standard.
>
> This paper has given an overview of the use of unreinforced masonry in Australia and in particular the impact of the new seismic loading provisions of AS1170.4. Despite the restrictions imposed by the provisions, correctly designed and constructed unreinforced masonry can still be used in most applications. There is, however, a need for research into the performance of unreinforced masonry systems under dynamic loading, particularly with regard to wall-floor connections, membranes and flashings, and tying of veneer walls. It is also important that the structural engineer be involved in both the design and supervision of all aspects of the masonry construction, even if the masonry is considered to be non-structural. This is not the current practice, with the masonry usually being considered as an architectural rather than a structural material, and the responsibility for detailing and supervision resting with the architect and/or builder.

Regarding reinforced masonry in regions of low and moderate hazard, the use of confined masonry construction has some advantages over traditional reinforced concrete frames with masonry infill, as discussed in Section 10.6.2.

10.5 Timber Structures

10.5.1 Introduction

In some countries timber structures have a well-deserved reputation for high resistance to earthquakes. For instance, pre-code timber buildings performed very well at intensity MM10 in the $M_W = 7.8$ Hawke's Bay, New Zealand, earthquake of 1931, as shown in Figure 6.2. This high performance is due to a number of factors, particularly the high strength-to-weight ratio of timber, its enhanced strength under short-term loading and the ductility of its steel fastenings such as nails and bolts. However, despite these qualities timber has traditionally been used mainly in domestic construction, and has been the least used of the main structural materials for engineered structures even in those earthquake areas where timber is a plentiful resource. This apparent reluctance to use timber for engineered structures has probably been due mainly to the difficulty of making structural connections and to reservations about fire resistance. However, a change in attitudes was marked by the success of the 1984 Pacific Timber Engineering Conference held in Auckland, New Zealand, and by the surge of research and development in seismic design of timber structures which started in New Zealand in the late 1970s.

The fruitfulness of the above research is demonstrated by the marked improvement of the performance of 1980–1987 non-domestic timber buildings in the Edgecumbe, New Zealand, earthquake of 1987, compared with that of buildings of the earlier code era of 1935–1979. As seen in Figure 10.53, the mean damage ratio D_{rm} for one-storey timber buildings at intensity MM10 was 0.07 for the earlier era and only 0.002 for the post-1979 era. This difference is statistically significant at the 0.01 level, despite the post-1979 subset being small in number.

However, in some countries the performance of timber buildings has been less satisfactory than in New Zealand, e.g. Managua (Falconer, 1968) and the USA (National Bureau of Standards (NBS), 1971; Earthquake Engineering Research Laboratory (EERL), 1973; Holmes and Somers, 1995). According to the latter report:

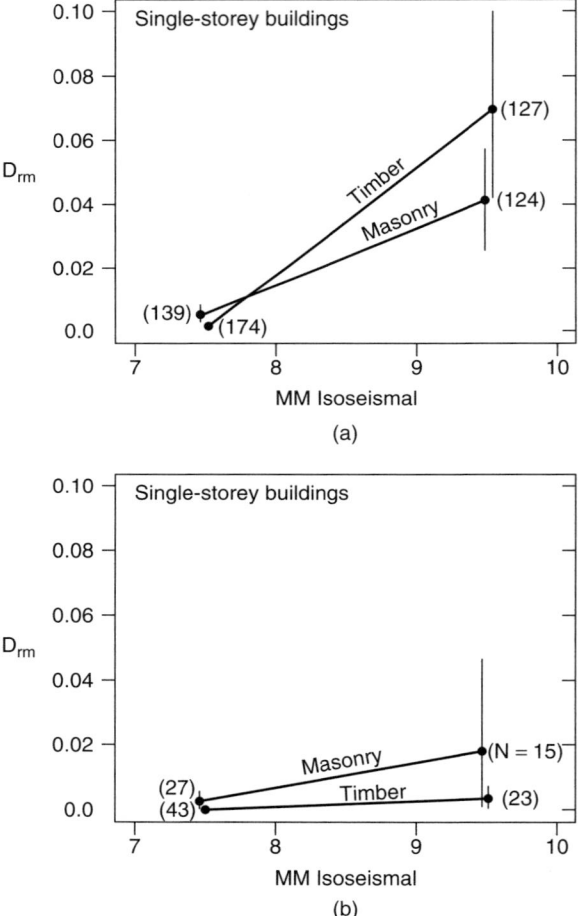

Figure 10.53 D_{rm} and its 95% confidence limits for one-storey non-domestic buildings of timber and masonry from design eras (a) 1935–1979 and (b) 1980–1987, in the 1987 Edgecumbe earthquake (from Dowrick and Rhoades, 1997)

The Northridge earthquake should dispel the myth that wood construction is largely immune to earthquake shaking. Although the 1971 San Fernando and 1989 Loma Prieta earthquakes provided good evidence, the inferred $10 billion of damage to wood buildings by an earthquake of the moderate magnitude of 6.7 is rather convincing. Though it is true that many wood structures shook hard and behaved satisfactorily, problems emerged that need to be addressed. These relate to both design and construction.

Much of the poor performance can be traced to poor construction that would not have escaped inspections had they been thorough. Inspection needs to be increased at job sites in terms of both time spent and the inspector's familiarity with the structural concept. Active involvement of the design engineer or architect to see that the plans are implemented is essential. It is hoped that the changes in this regard adopted by the City of Los Angeles will prove effective and be adopted elsewhere.

Considering the effects of various earthquakes, the main causes of inadequate performance of timber construction have been as follows:

(1) large response on soft ground;
(2) lack of integrity of substructures;
(3) asymmetry of the structural form;
(4) insufficient strength of chimneys;
(5) inadequate structural connections;
(6) use of heavy roofs without appropriate strength of supporting frame;
(7) deterioration of timber strength through decay or pest attack;
(8) inadequate resistance to earthquake induced fires; and
(9) inadequate supervision of construction.

Within certain limitations, means are available for dealing with all these aspects of earthquake resistance of timber construction, as discussed below.

10.5.2 *Seismic response of timber structures*

The response of timber structures to earthquakes depends on the combined response properties of the components, i.e. the timber and the connections. As shown by the typical monotonic stress–strain curves in Figure 10.54, timber is ductile in compression and brittle in tension, so that members failing in axial tension or in bending of the parent timber do so in a brittle manner. Column behaviour will be ductile or brittle, depending on the ratio of compression to bending stress, as indicated by the monotonic moment–curvature relationships found by Buchanan (1984a) for a species group of average strength (Figure 10.55). For strong green timber the zero axial load line in Figure 10.55 would be more curved, as more compression yielding occurs before tension failure finally takes place.

The effect of connections on the response of timber structures depends upon the nature of the fastening, and whether the timber or the fastening governs the behaviour. For example, steel nails may be detailed to yield so that nailed timber structures can have ductile response, as shown by the hysteresis loops for particle board sheathed walls subjected to slow cyclic loading as reported by Thurston and Flack (1980) (Figure 10.56).

It is well established that the *duration of loading* has an important effect on the ultimate strength of timber, the latter increasing as the duration decreases. As shown in Figure 10.57

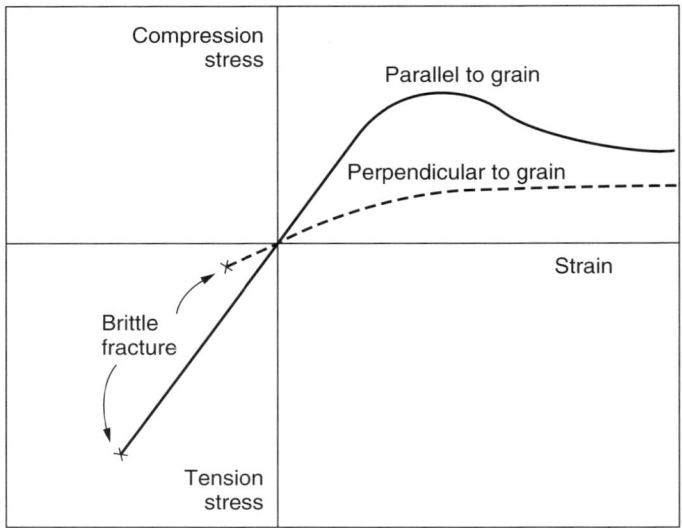

Figure 10.54 Stress–strain relationships for timber (after Buchanan, 1984b)

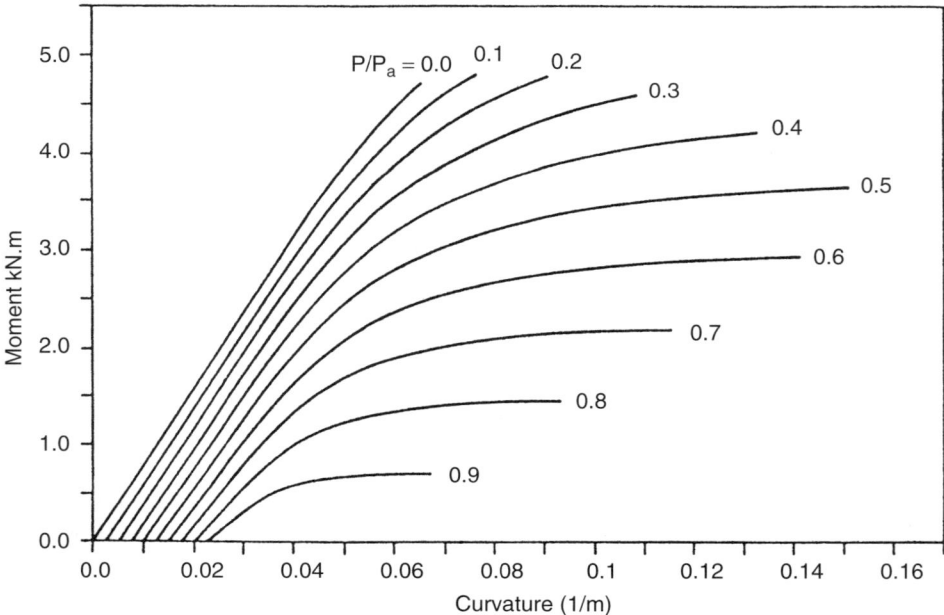

Figure 10.55 Typical moment–curvature relationships for commercial-quality sawn timber (Canadian spruce-pine-fir) (after Buchanan, 1984a)

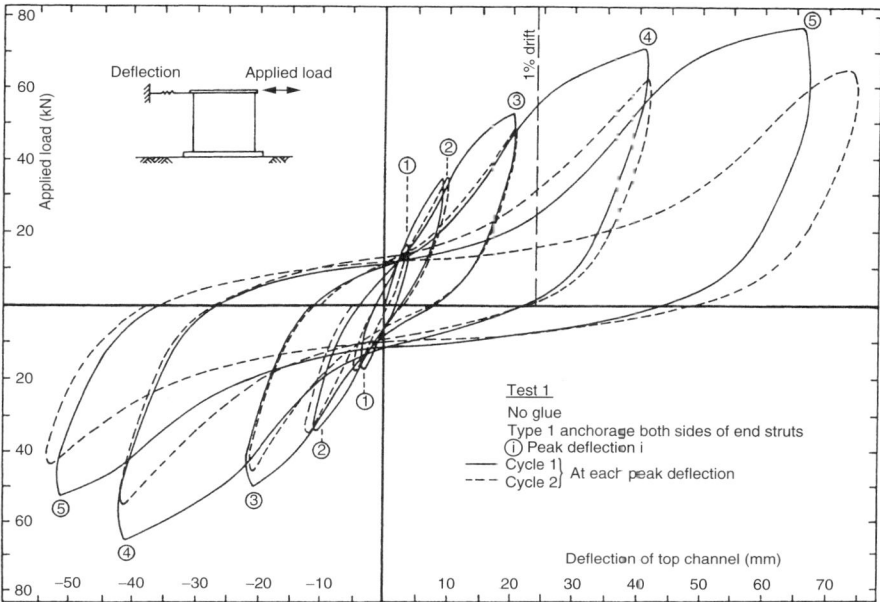

Figure 10.56 Hysteretic behaviour in a cyclic load test on a 2.4 m square particle board sheathed wall (after Thurston and Flack, 1980)

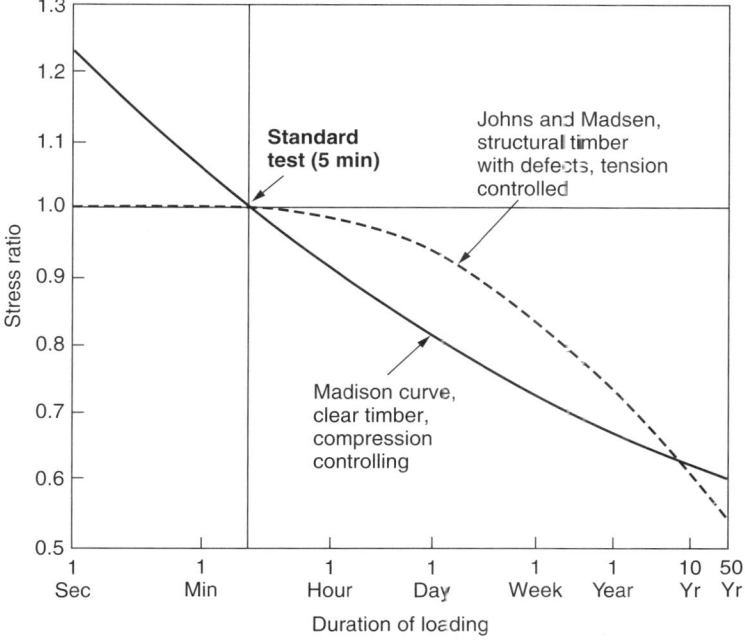

Figure 10.57 Relative strength of timber as a function of load duration (after Buchanan, 1984b)

by the 'Madison' curve (traditionally used in US and Canadian codes), it was originally thought that this strength enhancement continued into the dynamic load range, but subsequent research on shorter durations of loading (down to 0.05 s) by Spencer (1978) indicates that enhancement ceases for load durations less than about 5 min, as shown by the Johns and Madsen (1982) curve in Figure 10.57. This would imply that slow cyclic load tests reflect the actual dynamic response behaviour correctly, rather than underestimating it, as suggested by the Madison curve. However, the strength of timber under earthquake loads is, of course, substantially higher than for long-term dead loads, and an enhancement factor of 1.75 times the strength under permanent load is used in the Australian and New Zealand timber codes (NZS 3603, 1993, incl. amendments).

Hysteretic behaviour with fat loops of the type shown in Figure 10.56 implies that the equivalent viscous *damping* available in such structures may be considerable. For example, from tests on sheathed diaphragms Medearis (1966) found the equivalent viscous damping to be 8–10% regardless of amplitude. This figure bears comparison with the range of damping of 3–10% found for timber houses with plywood sheathed walls in Japan, as reported by Sugiyama (1984). The latter also found that traditional timber housing in Japan had much higher damping at 7–25%. Thus the figure of 15% of critical damping in Table 5.11 for timber shear wall construction obviously includes additional damping, which is often available from other parts of a timber building, but which was apparently not available in the Japanese shear walls houses noted above.

The hysteretic behaviour shown in Figure 10.56 exhibits the phenomenon often referred to as *pinching* or *pinched loops*, where there is reverse curvature on rising and falling arms of the loop, and sometimes the loop is thinner in the middle than near its ends. Pinching is typical of hysteresis of many forms of timber construction where nail yield occurs and also occurs in bolted construction, as reviewed by Dowrick (1986).

For a discussion of the influence of microzoning on the seismic response of timber houses, see Section 6.3.5.

10.5.3 Reliable seismic behaviour of timber structures

By the beginning of the twenty-first century, technology developments meant that buildings of up to about six storeys could be built in timber without introducing steel or concrete frames for lateral resistance.

For obtaining reliable seismic response behaviour, the principles concerning choice of form, materials and failure mode control discussed in Section 8.3 apply to timber structures. Regarding the form (configuration) or timber structures, the building's resistance against horizontal forces should be derived from walls or frames providing reasonably symmetrical resistance in two orthogonal directions in plan (Figure 10.58(a)). If one façade only consists mainly of window and door apertures, horizontal diaphragm action at eaves level should be capable of transferring the resulting earthquake torque to the end walls at right-angles to that façade (Figure 10.58(b)). It should be noted that because of the inherent high torsional flexibility of buildings with essentially only three resisting walls or frames, some short elements resisting horizontal shear should be introduced into the window or door façade. The resulting reduced torsions will nevertheless need to be distributed through a horizontal diaphragm. Damage arising from excessive asymmetry

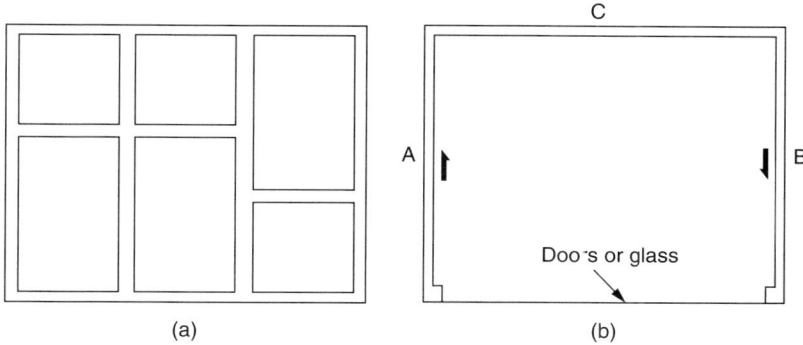

Figure 10.58 Schematic plans showing layout of shear walls in low-rise houses: (a) good; (b) unsatisfactory

occurred in the San Fernando earthquake, as shown in Figures 27 and 28 of the paper by the California Institute of Technology (EERL, 1973).

Elements which are stiffer and heavier than the rest of the building cause a great deal of damage in earthquakes. Concrete and masonry chimneys in basically timber houses are particularly vulnerable (NBS, 1971; Cooney, 1979). In many cases, the ideal solution would be to make the stiff elements structurally independent of the rest of the building, but difficulties arise in detailing the movement gaps. Otherwise, the stiff and the flexible elements should be much more strongly tied to a stiff element, the latter becomes a major horizontal shear resisting element for the whole building, and the building should be designed accordingly.

Designing for failure mode control requires consideration of the structural form used, and of the forms discussed in Section 8.4 the main options for earthquake load resistance in timber construction are:

(1) sheathed walls (shear walls);
(2) moment resisting frames;
(3) concentrically braced frames;
(4) hybrid moment resisting/braced frames (e.g. see Figure 10.67).

The failure mode of each of these forms can be made ductile, generally by using the ductility of the steel connections of holding-down bolts (Figure 10.66). As this requires that the timber is designed to be stronger than the connections, the lateral load design may be governed by earthquakes, even in cases when the wind base shear exceeds the earthquake base shear (as is often the case for timber structures).

Further discussion of failure mode control is given in the following sections on different member types.

10.5.4 Foundations of timber structures

The principles for design of timber structures are, of course, the same as those for other materials, as discussed in Section 9.1, but some details specific to timber construction need

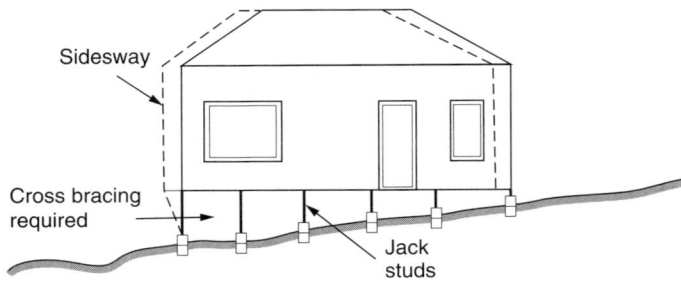

Figure 10.59 Substructure in timber stud construction requiring extra horizontal shear strength

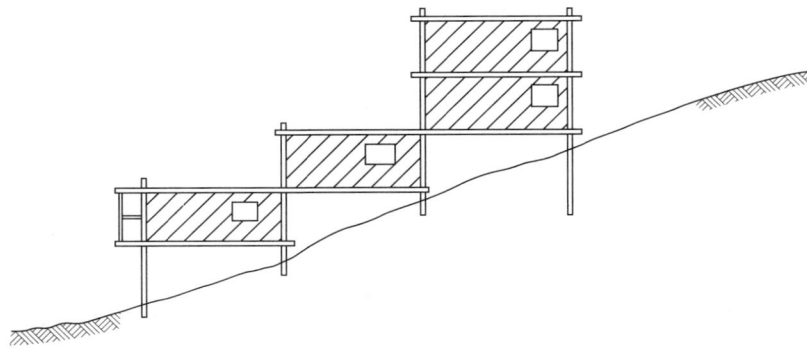

Figure 10.60 Pole frame apartments as built at Lugunda Beach, California

to be given. For example, in timber housing the substructure between the footings and the first occupied floor tends to have inadequate horizontal shear resistance, and sidesway damage (Figure 10.59) occurs in earthquakes (EERL, 1973). Pole frame construction as illustrated in Figure 10.60 readily overcomes this problem.

Another common failing has been that the timber structure is inadequately connected to the concrete foundation blocks or strips. The detail shown in Figure 10.66(b), for example, should be provided with adequate bolts.

The provisions of suitable foundations for earthquake resistance of low-rise construction *on soft ground* is a basic engineering problem for commercial-industrial buildings. Because the foundation requirements for gravity and wind loading are minimal in such buildings, the extra cost for providing protection at source against differential ground movements is large compared with that for taller structures.

The holding down of timber structures to concrete foundation has been prone to problems which occur wherever disparate materials are connected and holding-down fastenings of the type shown in Figure 10.67 should be designed for strength and ductility to ensure that the desired ductile failure mode occurs.

The type of foundation provided by pole construction (Figure 10.60) overcomes some of the weaknesses of orthodox substructures to timber building, as the poles themselves provide vertical continuity and the pole frameworks develop the necessary resistance to horizontal forces (Buchanan, 1999). For some species of timber, preservative treatment such as tanalizing will, of course, be essential for durability below ground.

10.5.5 Timber-sheathed walls (shear walls)

Most timber buildings derive their strength and stiffness from shear panels or diaphragms which may constitute walls, floors, ceilings, or roof slopes. Individual shear elements are built up from planks, plywood, metal, plaster or other sheeting which is fixed to the basic timber framework by nails, screws or glue. The effectiveness of different types of wall or diaphragm for resisting in-plane shears depends upon:

(1) its overall size and shape;
(2) the size, shape and position of any apertures;
(3) the nature of the timber framework;
(4) the nature and disposition of the diagonal or sheeting members;
(5) the connections between elements (3) and (4).

Regarding walls in particular, a useful study of some of the above factors was carried out by the US Forest Products Laboratory in 1946, the results of which are shown in Figure 10.61. The superiority of plywood for the panelling compared with diagonal or strip boarding is obvious in Figure 10.61. More recent tests (Thurston and Flack, 1980) show that modern composite timber panels (particle board) are similar to plywood panels in this respect. While Figure 10.61 shows that glued construction has higher strength and rigidity than nailed construction, more recent cyclic load tests in New Zealand (Yap, 1985) did not confirm that an increase in stiffness occurs, but, more important, they showed this form of construction to be very brittle at failure load.

Extensive research in New Zealand into plywood sheathed walls using cyclic and dynamic testing and non-linear dynamic analyses has provided a basis for design procedures (Dean *et al.*, 1986; Dowrick and Smith, 1986), which ensure ductile behaviour. This is done by providing extra strength in the framing chords and holding-down fixings so that yielding occurs in the sheathing nails only.

Further, it is noted that openings in sheathed walls require special attention, especially in highly stressed walls (Dowrick and Smith, 1986), and a simplified analysis method has been proposed by Dean *et al.* (1984).

From Figure 10.61 it is also clear that diagonals are much more effective when continuous between opposite framing members of a panel, rather than when broken by apertures. In domestic building it is, of course, common for only one or two diagonals to be used within an individual wall unit, and such diagonals should clearly be inclined between 30° and 60° to the horizontal for greatest effectiveness. In timber which is likely to split, nail holes near the ends should be predrilled slightly smaller than the nail diameter. In framing up shear panels, care should be taken that the perimeter members, and diagonals if used, are made from sound timbers. The framing members for door and window apertures should similarly be good-quality timber.

External timber framed walls are often clad with plaster, and the earthquake performance of such walls has been greatly improved using expanded metal lath. In Japan and California expanded metal is now commonly used in conjunction with cement, lime,

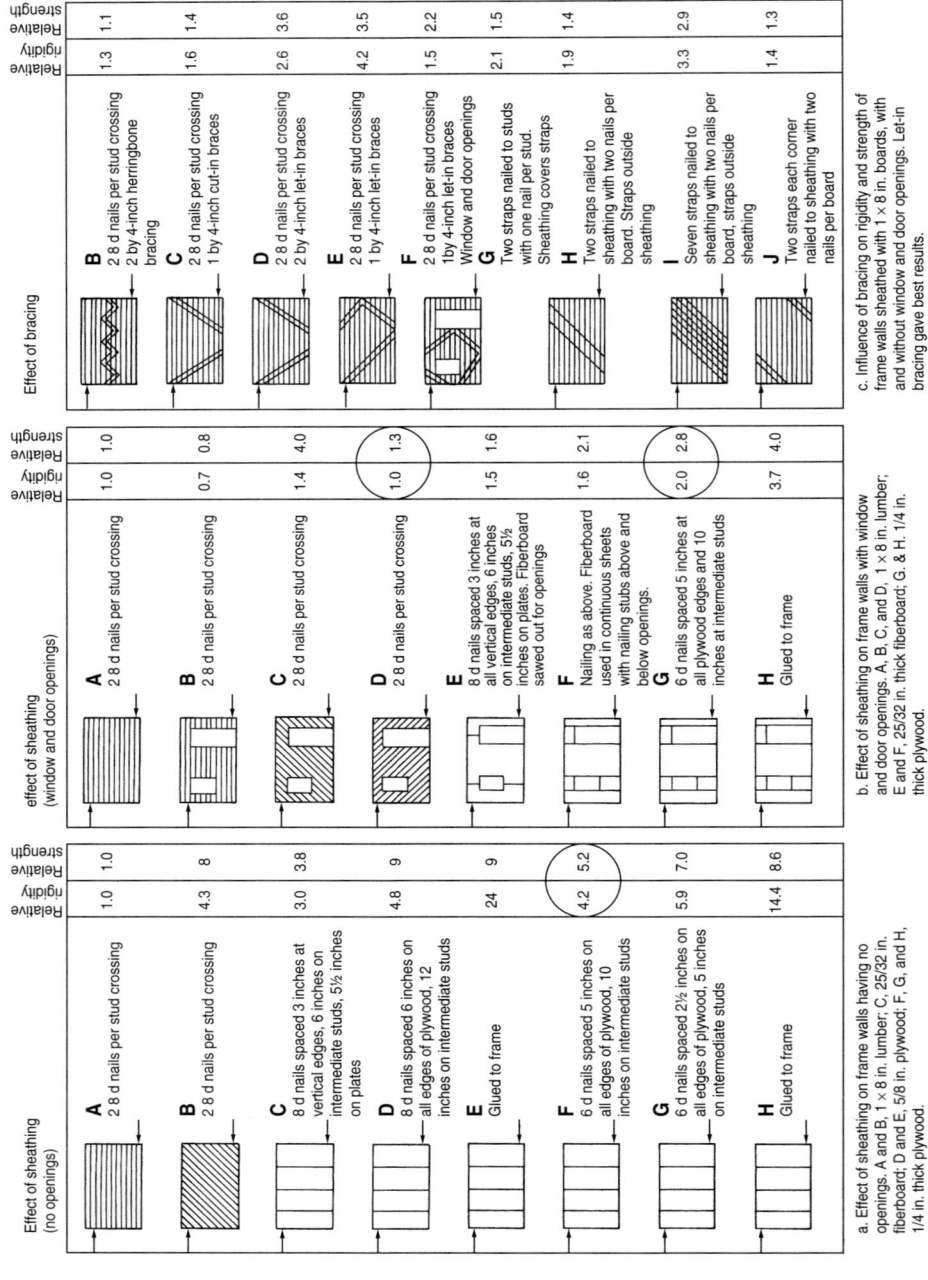

Figure 10.61 Tests on timber-framed walls with various forms of sheeting and fixing carried out by the US Forest Products Laboratory

and sand plaster (mixes of between 1:1:3 and $1:1:4\frac{1}{2}$ are found to be durable), the lime reducing the brittleness of the plaster. The application of two or three costs of plaster, giving a total thickness of about 20 mm, is normal practice.

10.5.6 Timber horizontal diaphragms

Horizontal diaphragms such as floors or roofs in timber construction require more design consideration than corresponding concrete or steel diaphragms. This occurs largely because of the greater flexibility of timber diaphragms, which may render invalid the usual simplifying assumption made in analysis that horizontal diaphragms are rigid, and which may lead to troubles with excessive deflections, as shown by Figure 10.62. Thus stiffness rather than strength may be the controlling design criterion.

There are two alternative (and contrasting) design approaches that may be adopted (Dean *et al.*, 1984) for diaphragms, both of which relate to the high degree of non-linearity which readily occurs in diaphragms under design earthquake loading:

(1) *Suppression of non-linearity*. The distribution of loading through the diaphragm to the vertical structure and the control of deflections can be more easily and more reliably predicted if the non-linearity in the nail deformations is suppressed sufficiently so that the diaphragm may be considered effectively linear elastic for design purposes. In the majority of buildings, this will be the preferable and economical design procedure.

(2) *Recognition of inelasticity*. In this approach the inelastic (non-linear) behaviour of the diaphragm at design loads is allowed for explicitly in calculating the loads, load distribution and deflections. The energy absorption and ductility of the diaphragm implied by the inelasticity may be utilized to reduce the loadings generated within the diaphragm (e.g. by the use of reduced code structural factors permitted by higher ductility). This results in a reduction in the estimated forced transferred from the diaphragm to the supporting structure.

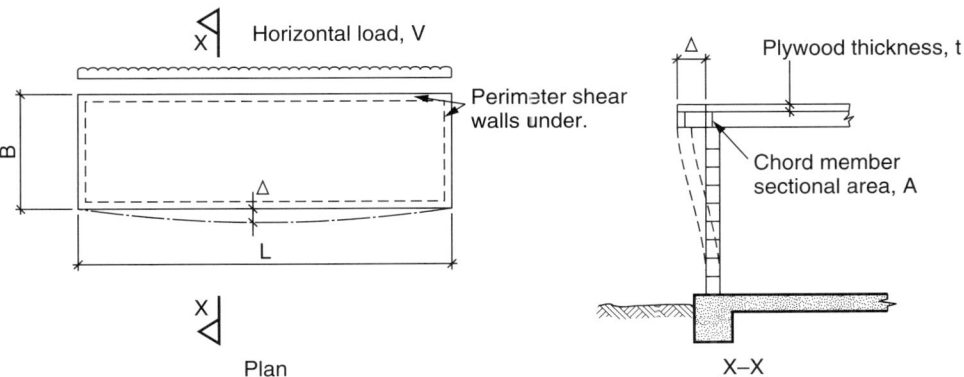

Figure 10.62 A typical horizontal timber diaphragm showing the effect on supporting walls of deflections under horizontal loading

For the strength design of horizontal diaphragms, the forces are usually found using the girder analogy (Smith *et al.*, 1986), where the sheathing is the 'web' of the girder and the top plate of perimeter wall or a continuous perimeter joist is the flange. However, because of the great depth of typical diaphragms, strength is not generally problematical. Brittle failure is avoided by providing extra strength in the chords (flanges), so that yielding (and failure) occurs in the sheathing nails only.

Excessive deflection of plywood diaphragms is to some extent controlled by limiting the aspect ratio of diaphragms.

As horizontal diaphragms may deflect sufficiently to endanger supporting or attached wall components (Figure 10.62), a means of calculating their deflections under in plane loading is desirable. This deflection involves contributions from three main sources, i.e. bending, shear and nail slip, which for a single-span diaphragm sheathed with plywood or particle board panels is given by NZS 3603 (1993, incl. amendments) by

$$\Delta = \frac{5WL^3}{192EAB^2} + \frac{WL}{8GBt} + 0.5(1+a)me_n, \qquad (10.58)$$

where
Δ = horizontal deflection (mm);
W = total horizontal load on diaphragm (N);
L = length of diaphragm (mm);
B = width of diaphragm (mm);
A = cross-sectional area of chord (mm^2);
E = modulus of elasticity of chords (MPa);
G = shear modulus of plywood (MPa);
t = thickness of sheathing (mm);
e_n = nail slip (mm);
a = aspect ratio of each sheathing panel, which is 0 when relative movement along sheet edges is prevented, 1 when square sheathing panels are used, and 2 when 2.4 m × 1.2 m panels are orientated with the 2.4 m length parallel with diaphragm chords (= 0.5 alternative orientation).

The nail deformation, e_n, is a non-linear function of load level, nail type, and sheathing thickness, as shown by Figure 10.63, and is also dependent on other factors such as the nature of the sheathing and framing and the surface of the nail. Thus, it is clearly desirable to obtain e_n from the results of tests matching the components to be used. For example, in New Zealand for local plywood and medium-density particle board sheathing, the recommended nail slip at design load level is 0.8 mm, regardless of sheathing thickness.

Finally, it is noted that design problems arise from large openings in diaphragms, as discussed elsewhere (Dean *et al.*, 1986; Applied Technology Council, 1981).

10.5.7 *Timber moment resisting frames and braced frames*

The most common form of *moment resisting frame* in timber is the portal frame. Because of their high flexibility, and hence long period of vibration, in many cases the design of timber portals may be governed by wind loads and by deflections rather than by seismic strength. Multi-storey moment resisting frames in timber are unlikely to reach more than

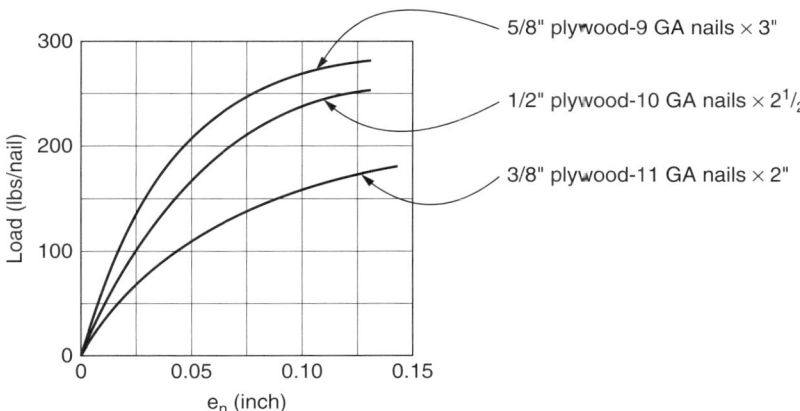

Figure 10.63 Nail deformation in plywood diaphragms framed on to Douglas fir as derived in the USA (Timber Engineering Company, 1956)

a few storeys in height without excessive beam and column sizes, or without adding other means of providing horizontal stiffness, because of the high inter-storey drifts that would occur and the difficulties of making such structures cost-competitive.

As reviewed by Buchanan and Fairweather (1993), ductility is obtained in moment resisting frames by yielding in the connections, which may be of five types:

(1) steel side plates, with yielding in the nails;
(2) steel side plates, necked so that yield occurs in the plate.
(3) plywood side plates, with yielding in the nails;
(4) steel dowels;
(5) epoxied steel rods with or without steel brackets.

Four different ways of using type (5) joints are shown in Figure 10.64. Buchanan *et al.* (2001) comment that:

> If *moment resisting glulam frame structures* are to be designed for ductile response to seismic forces, it is essential that the steel components in the connections be able to provide sufficient ductility without premature failure elsewhere, including the following failure modes:
>
> - tensile yielding of the steel connecting bracket;
> - tensile yielding of steel rods;
> - pullout failure of the epoxy;
> - tensile failure of the threaded couplers;
> - compression punching of the couplers into the steel connecting bracket;
> - tensile failure of the timber near the end of the epoxied rods; and
> - shear failure of the timber between the two groups of rods.

Fairweather (1992) carried out cyclic tests on moment resisting connections obtaining good ductility and energy dissipation. Best results were obtained with the glulam members bolted to ductile steel connecting brackets. Ductile damage to the brackets allows inspection and replacement after an earthquake.

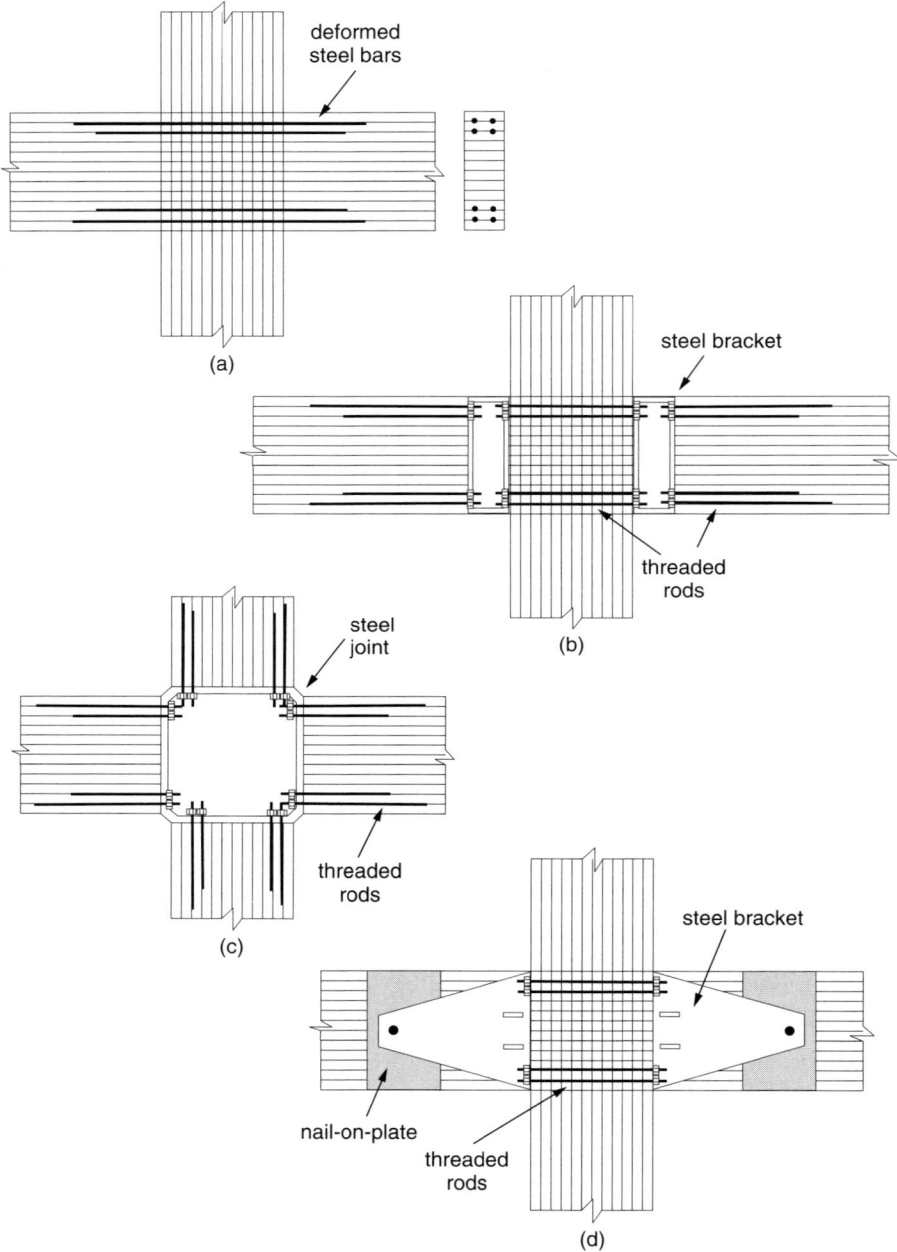

Figure 10.64 Multi-storey beam column connection using epoxied steel joints (from Buchanan and Fairweather, 1993)

The most critical part of moment resisting glulam frames are the connections, and only some of the connections available are suitable for achieving adequate ductility. To obtain the latter, Buchanan and Fairweather note that

it is necessary to:

(a) Provide steel components that are capable of sufficient ductile yielding.
(b) Use a capacity design procedure to ensure that the chosen mechanism can occur with no failure of the wood, and adhesives, or other non-ductile component.
(c) Provide careful detailing so that the connection performs as intended.

Large structural ductility factors are not necessary in the design of glulam frame structures because low building masses result in low seismic forces, and inter-storey deflections often govern the design process.

Braced frames in timber construction may be created by traditional timber braces or by steel (e.g. Figure 10.67). It appears that double bracing is better than single bracing, because of progressive deflection in one direction only in the latter asymmetrical cases, as shown by the lopsided hysteretic behaviour found by Sakamoto *et al.* (1984).

10.5.8 Connections in timber construction

Connections between timber members may be formed in the timber itself, or may involve glue, nails, screws, bolts, metal straps, metal plates or toothed metal connectors. Under earthquake loading, joints formed in the timber are inferior to most other forms of joint. In light timber construction such as smaller dwellings, the use of metal nail plates (hurricane braces) or toothed steel connectors is now widespread.

The nailed joints, the nail load, size and spacing require careful attention. A nail driven parallel to the timber grain should be designed for not more than two-thirds of the lateral load which would be allowed for the same size of nail driven normal to the grain. Nails driven parallel to the grain should not be expected to resist withdrawal forces. Edge or end distance of nails should not be less than half the required nail penetration.

The effectiveness of nails in enhancing seismic response behaviour improves with increasing thickness of (1) side plates in moment resisting connections, and (2) sheathing of walls (Dean *et al.*, 1986). More research, however, is required before the effect of thickness can be fully described in simple design rules.

In diaphragms, perimeter framing may need jointing capable of carrying the longitudinal forces arising from wind or seismic loading. A simple method of connection is shown in Figure 10.65.

Connections between shear walls and foundation or between successive storeys of shear walls must be capable of transmitting the horizontal shear forces and the overturning moments applied to them. Details which are considered good practice in California for these connections are illustrated in Figure 10.66, but obviously both details on the right-hand side of the figure are capable of resisting only small overturning tensions.

For some comments on the connections of timber roof diaphragms to walls of other materials, see Section 10.5.6.

Pole frame buildings are usually jointed using bolts, steel straps and clouts (Figure 10.67), as described in detail elsewhere. An effective means of obtaining resistance

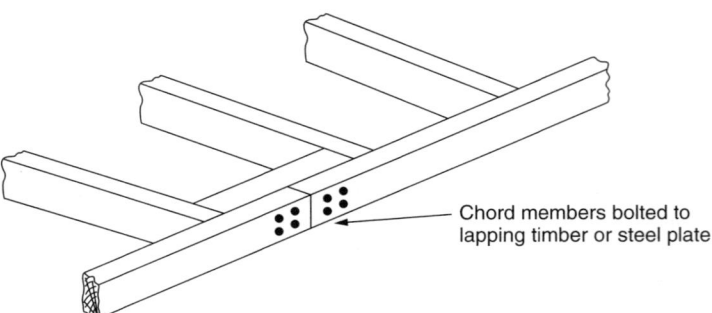

Figure 10.65 Method of jointing chord members in timber diaphragms (Nail-on metal straps may also be used)

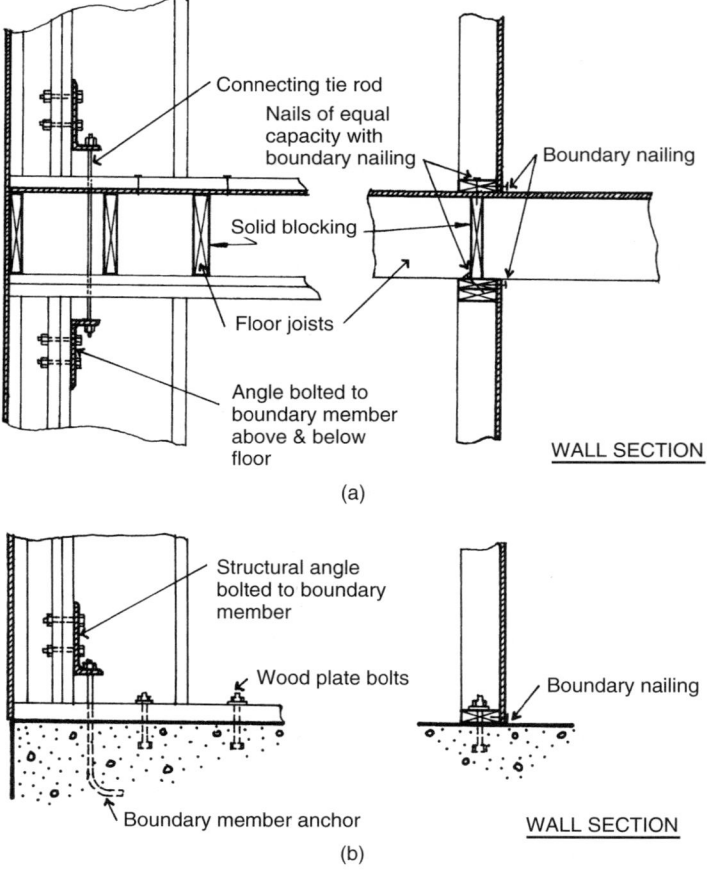

Figure 10.66 Connection details for plywood shear walls: (a) inter-storey connections in timber buildings; (b) connection of timber members to concrete foundations. An alternative to bolting is to use nail-on metal straps on the outside of the wall

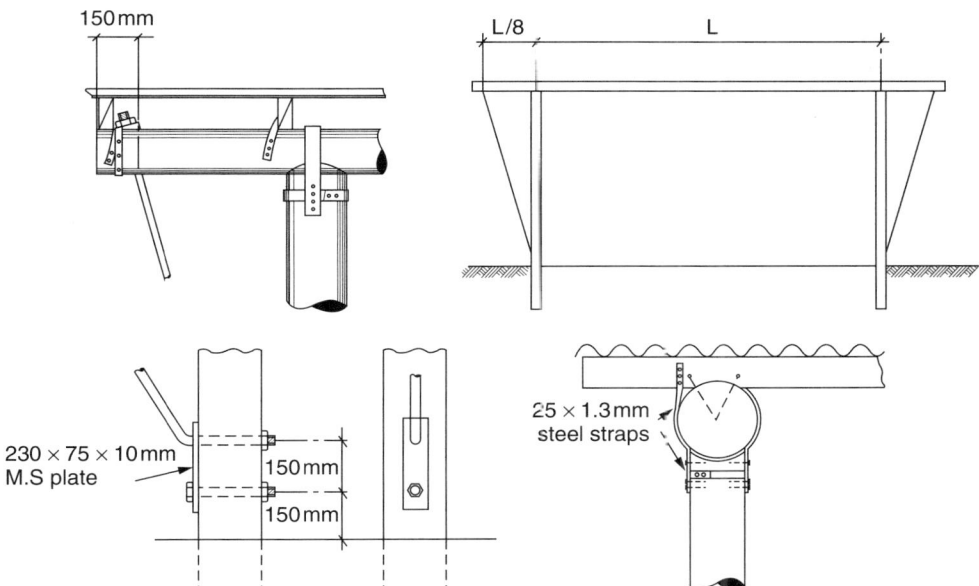

Figure 10.67 Tied rafter pole building showing typical connection details developed in Australia

to lateral shear forces is to create moment resisting triangles at the knees of portals (Figure 10.67) using steel rods as the diagonal member.

10.5.9 Fire resistance of timber construction

Although fire resistance is a problem common to all materials, and is not solely related to seismic areas, it is mentioned here because of the occurrence of earthquake induced fires and the general flammability of timber.

The fire resistance of timber construction varies widely, depending on the thickness of the timbers used. Pole frames and other thick timbers such as used in moment resisting frames have relatively low fire risk because of their large volume to surface area ratio. Surface charring is relatively shallow and protects the interior of the member from the flame. Apparently, badly burned structures of heavy timber have been found to be strong enough to continue in service. Such behaviour in fire is likely to be superior to equivalent unprotected steelwork, and its fire risk merits are becoming recognized for insurance purposes.

Because of its obvious flammability, light timber construction should only be used where low fire rating is acceptable. Various chemical fire retardants have been marketed in some parts of the world in an attempt to improve the resistance of flammable construction, but their value for timber construction seems limited as they appear to increase the time to ignition by only a few minutes.

For further information on fire protection of timber structure, specialist literature should be consulted, such as the book by Buchanan (2001).

10.6 Design of New Structures in Developing Countries

10.6.1 Introduction

Earthquakes continue to cause unacceptably high casualties and damage to the built environment in developing countries (e.g. 75,000 deaths in the 2005 Kashmir earthquake), despite the very effective mitigating measures affordable by wealthier societies, and the much cheaper but still substantially mitigating measures devised over the past several decades for developing countries. In particular, 'adobe and other forms of unreinforced masonry construction are proven "serial killers" in many earthquakes worldwide'. As so rightly stated by Comartin *et al.* (2004), this situation 'is the most important challenge facing the global earthquake engineering community'. An introduction to efforts to manage earthquake risk in developing countries is given in Section 7.8, and the crucial design issue of structual form is discussed in Section 8.3. The rest of this section gives sources of design and detailing advice with some specific discussion of masonry construction.

10.6.2 Some aspects of design and detailing of buildings

A recent cost-effective and simple method for improving the performance of masonry construction called *confined masonry* is currently practised in Latin America, the Mediterranean and the Middle East. Reinforced concrete frames with infill masonry and confined masonry construction both use the same materials, the basic difference being the construction sequence. In confined masonry construction the walls are constructed first, and the confining elements are poured around the walls. Confining elements are not designed as moment resisting frames, which means the reinforcing details are simple. In general the confining elements have smaller cross-sections than the corresponding elements of a reinforced concrete frame, but the important difference is that confined masonry walls are load-bearing, while infill walls are not. For more information, see Brzev (2008).

A very useful, simple and wide-ranging book is that of the International Association for Earthquake Engineering (IAEE, 1986) which gives guidelines for earthquake resistance of non-engineered construction. Among other things, it gives design guidance on new construction in masonry, timber, earth, reinforced concrete, and retrofitting of existing buildings.

For a discussion of factors helping to achieve reliable seismic behavour of masonry structures see Section 10.4.3, including Figure 10.46.

10.6.3 Further sources of design information

A major source of information regarding seismic design, earthquake performance and retrofitting of *houses* is the World Housing Encyclopedia (WHE), developed by the EERI and IAEE. Primarily a volunteer effort, by more than 180 engineers and architects from 50 countries, it comprises a comprehensive, global, searchable database of housing contruction types, to be found at http://www.world-housing.net. A forthcoming publication (expected in 2009), prepared by the WHE in conjuncion with the EERI, is a design guide for stone masonry buildings by Bothara *et al.* (in preparation). This book will effectively supersede part of IAEE (1986).

The National Information Center of Earthquake Engineering (NICEE) in Kanpur, India, produces many publications, some of them free to individuals; see http://www.nicee.org. Examples are their journal *Earthquake Engineering Practice* and CVR Murty's book *Earthquake Tips: Learning Earthquake Design and Construction*.

References

Amrhein JE (1994) *Reinforced Masonry Engineering Handbook*, 5th edn. Masonry Institute of America, Los Angeles.

Applied Technology Council (1981) *Guidelines for the Design of Horizontal Wood Diaphragms*, ATC-7. Applied Technology Council, Berkeley, CA.

Asteneh-Asl A, Goel SC and Hanson RD (1984) Cyclic behaviour of double angle bracing members with bolted connections. *Proc. 8th World Conf on Earthq Eng*, San Francisco **VI**: 249–256.

Bertero VV and Popov EP (1985) Effect of large alternating strains on steel beams. *J Struc Div* **91**(ST1): 1–12.

Bertero VV, Popov EP, Wang TY and Vallenas J (1977) Seismic design implications of hysteretic behaviour of reinforced concrete structural walls. *Proc. 6th World Conf on Earthq Eng*, New Delhi **5**: 159–165.

Bertero VV, Anderson JC and Krawinkler H (1994) Performance of steel structures during the Northridge earthquake. Report No. UCB/EERC-94/09, Earthquake Engineering Research Center, University of California, Berkeley.

Blakeley RWG (1973) Prestressed concrete seismic design. *Bull Nat Soc Earthq Eng* **6**(2): 2–21.

Blakeley RWG and Park R (1971) Ductility of prestressed concrete members. *Bull NZ Nat Soc Earthq Eng* **4**(1): 145–170.

Blume JA, Newmark NM and Corning LH (1961) *Design of Multi-story Reinforced Concrete Buildings for Earthquake Motions*. Portland Cement Association, Skokie, IL.

Bothara J *et al.* (in preparation) *Tutorial on Stone Masonry Buildings*. Earthquake Engineering Institute, Oakland, CA.

Brzev SN (2008) *Earthquake Resistant Confined Masonry Construction*. National Information Center of Earthquake Engineering, Indian Institute of Technology, Kanpur

Buchanan AH (1984a) Strength model and design methods for bending and axial board interaction in timber members. PhD thesis, Dept of Civil Engineering, University of British Columbia.

Buchanan AH (1984b) Wood properties and seismic design of timber structures. *Proc. Pacific Timber Eng Conf*, Auckland **II**: 462–469.

Buchanan AH (ed.) (1999) *Timber Design Guide*. NZ Timber Industry Federation.

Buchanan AH (2001) *Structural Design for Fire Safety*. John Wiley & Sons, Ltd, Chichester.

Buchanan AH and Fairweather RH (1993) Seismic design of glulam structures. *Bull NZ Nat Soc Earthq Eng* **26**(4): 415–436.

Buchanan A, Moss P and Wong N (2001) Ductile moment-resisting connections in glulam beams. Technical Conf NZ Soc Earthq Eng, Wairakei, Paper No. 6.02.01.

Cardenas AE, Hanson JM, Corley WG and Hognestad E (1973) Design provisions for structural walls. *AcI J* **70**(3): 221–230.

Cement and Concrete Association of New Zealand (1995) *New Zealand Reinforced Concrete Design Handbook*. CCANZ, Porirua.

Clifton GC (1985) Composite design. *Bull NZ Nat Soc Earthq Eng* **18**(4): 381–936.

Clough DP (1984) Development of seismic design criteria for connections in jointed precast prestressed structures. *Proc. 8th World Conf on Earthq Eng*, San Francisco **V**: 645–652.

Coe FR (1973) *Welding Steels without Hydrogen Cracking*. Welding Institute, Cambridge.

Comartin C, Brzev S, Naeim F *et al.* (2004) A challenge to earthquake engineering professionals. *Earthq Spectra* **20**(4): 1049–1056.

Cooney RC (1979) The structural performance of houses in earthquakes. *Bull NZ Nat Soc Earthq Eng* **12**(3): 223–237.

Corley WG (1966) Rotational capacity of reinforced concrete beams. *J Struct Divn* **92**(ST5): 121–146.

Dean JA, Moss PJ and Stewart W (1984) A design procedure for rectangular openings in shear walls and diaphragms. *Proc. Pacific Timber Eng Conf*, Auckland **II**: 513–518.

Dean JA, Stewart WG and Carr AJ (1986) The seismic behaviour of plywood sheathed shearwalls. *Bull NZ Nat Soc Earthq Eng* **19**(1): 48–63.

Dowrick DJ (1986) Hysteresis loops for timber structures. *Bull NZ Nat Soc Earthq Eng* **19**(2): 143–152.

Dowrick DJ and Rhoades DA (1997) Vulnerability of different classes of low-rise buildings in the 1987 Edgecumbe, New Zealand, earthquake. *Bull NZ Nat Soc Earthq Eng* **30**(3): 227–241.

Dowrick DJ and Smith PC (1986) Timber sheathed walls for wind and earthquake resistance. *Bull NZ Nat Soc Earthq Eng* **19**(2): 123–134.

Earthquake Engineering Research Laboratory (1973) Engineering features of the San Fernando earthquake. *Bull NZ Nat Soc Earthq Eng* **6**(1): 22–45.

Erasmus L (1984) The mechanical properties of structural steel sections and the relevance of those properties to the capacity design of structures. *Trans Inst Prof Engineers, New Zealand* **11**(3/CE): 105–111.

Fairweather RH (1992) Beam column connections for multi-storey timber buildings. Research Report 92/5, Dept of Civil Engineering, University of Canterbury, Christchurch, New Zealand.

Falconer BH (1968) Preliminary comments on damage to buildings in the Managua earthquake. *Bull NZ Soc Earthq Eng* **1**(2): 61–71.

Farrar JCM and Dolby RE (1972) *Lamellar Tearing in Welded Steel Fabrication*. Welding Institute, Cambridge.

Fédération Internationale du Béton (2003) *Seismic Design of Precast Concrete Building Structures*. FIB, Lausanne.

Fenwick RC (1983) Strength degradation of concrete beams under cyclic loading. *Bull NZ Nat Soc Earthq Eng* **16**(1): 25–38.

Goodsir WJ, Paulay T and Carr AJ (1984) A design procedure for interacting wall-frame structures under cyclic actions. *Proc. 8th World Conf on Earthq Eng*, San Francisco **V**: 621–628.

Gulkan P, Mayes RL and Clough RW (1979) Shaking table study of single-storey masonry houses, Volume 1: Test structures 1 and 2. Earthquake Engineering Research Center Report No. UCB/EERC-79/23.

He G, Wei L, Zhong L and Dai G (1984) New procedure for aseismic design of multistorey shear type building based upon deformation checking. *Proc. 8th World Conf Earthq Eng*, San Francisco **V**: 483–490.

Heavy Engineering Research Association (1995) Seismic design procedures for steel structures. Report R4-76, HERA, Manukau City, New Zealand.

Heavy Engineering Research Association (1999) Structural steelwork connections guide. Report R4-100, HERA, Manukau City, New Zealand.

Hjelmstad KD and Popov EP (1983) Seismic behaviour of active beam links in eccentrically braced frames. Report No UBC/EERC-83/15, Earthquake Engineering Research Center, University of California, Berkeley.

Holmes WT and Somers P (eds) (1995) Northridge earthquake of January 17, 1994. Reconnaissance report – Volume 2. *Earthq Spectra* **11**(Supplement C).

International Association for Earthquake Engineering (1986) *Guidelines for Earthquake Resistant Non-engineered Construction*. IAEE, Tokyo.

International Code Council (2006 or later) *International Building Code*. ICC, Washington, DC.

Johns KC and Madsen B (1982) Duration of load effects in lumber, Parts I, II, III. *Canadian J Civil Eng* **9**(3): 502–536.

Kasai K and Popov EP (1984) On seismic design of eccentrically braced steel frames. *Proc. 8th World Conf on Earth Eng*, San Francisco **V**: 387–394.

Kato B (1974) A criteria of beam-to-column joint panels. *Bull NZ Nat Soc Earthq Eng* **7**(1): 14–26.

Kato B and Nako M (1973) The influence of elastic plastic deformation of beam-to-column connections on the stiffness, ductility and strength of open frames. *Proc. 5th World Conf on Earthq Eng*, Rome **1**: 825–828.

Kent DC and Park R (1971) Flexural members with confined concrete. *Proc. ASCE* **97**(ST7): 1969–1990.

Krawinkler H (1996) Earthquake design and performance of steel structures. *Bull NZ Nat Soc Earthq Eng* **29**(4): 229–241.

Lay MG and Galambos TV (1967) Inelastic beams under moment gradient. *J Struct Div* **93**(ST1): 389–399.

Manos GC, Clough RW and Mayes RL (1984) A three component shaking table study of the dynamic response of a single storey masonry house. *Proc. 8th World Conf on Earthq Eng*, San Francisco **VI**: 855–862.

Masonry Society (2007) *Masonry Designers' Guide*, 5th edn. Masonry Society, Boulder, CO.

McKay GR (1985) Materials and workmanship. *Bull NZ Nat Soc Earthq Eng* **18**(4): 400–405.

Medearis K (1966) Static and dynamic properties of shear structures. Proc Intl Symp Effects of repeated loading on Materials and Structures, RILEM – Inst Eng, Mexico VI.

Meli R (1973) Behaviour of masonry walls under lateral loads. *Proc. 5th World Conf on Earthq Eng*, Rome **1**: 853–862.

Murty CVR (2002) *IITK:BMTPC Earthquake Tips: Learning Earthquake Design and Construction*. National Information Centre of Earthquake Engineering, IIT Kanpur, India.

National Bureau of Standards (1971) The San Fernando California earthquake of February 9, 1971. NBS Report 10 556, US Dept of Commerce, National Bureau of Standards.

Neal BG (1961) The effect of shear and normal forces on the fully plastic moment of an I beam. *J Mech Eng Sci* **3**.

New Zealand Concrete Society (2008) *Examples of Concrete Structural Design to New Zealand Standard 3101*. Cement and Concrete Association of New Zealand, Porirua.

Nicholas CJA (1985) Connection design. *Bull NZ Nat Soc Earthq Eng* **18**(4): 360–368.

NZS 3101 (2006) *Concrete structures standard*, Parts 1 and 2. Standards New Zealand, Wellington.

NZS 3603 (1993) *Timber structures standard*. (incl. amendments) Standards New Zealand, Wellington.

NZS 4203 (1992) *Loadings Standard*. Standards New Zealand, Wellington.

NZS 4230 (2004) *The Design of Reinforced Concrete Masonry Structures*. Standards New Zealand, Wellington.

NZS 3404 (1997) *Steel structures standard*, Parts 1 and 2. Standards New Zealand, Wellington.

NZSG (1999) *Guidelines for the Use of Structural Precast Concrete in Buildings*, 2nd edn. Report of a Study Group of NZ Concrete Society and the NZ Society for Earthquake Engineering.

Oesterle RG, Fiorato AE and Corley WG (1980) Reinforcement details for earthquake resistant structural walls. *Concrete International Design and Construction* **2**(12): 55–66.

Page AW (1996) Unreinforced masonry structures – an Australian overview. *Bull NZ Nat Soc Earthq Eng* **29**(4): 242–255.

Pampanin S (2005) Emerging solutions for high seismic performance of precast/prestressed concrete buildings, *J Advanced Conc Technology* **3**(2): 207–223.

Paparoni M, Ferry Borges J and Whitman RW (1973) Seismic studies of Parque Central buildings. *Proc. 5th World Conf on Earthq Eng*, Rome **2**: 1991–2000.

Park R (1980) Ductility of reinforced concrete frames under seismic loading. *NZ Engineering* **23**(11): 427–435.

Park R and Paulay T (1980) Concrete structures. In *Design of Earthquake Resistant Structures* (ed. E Rosenblueth). Pentech Press, London.

Park R, Priestley MJN and Gill WD (1982) Ductility of square confined concrete columns. *J Struct Divn* **108**(ST4): 929–950.

Park R, Priestley MJN, Falconer TJ and Joen PH (1984) Detailing of prestressed concrete piles for ductility. *Bull NZ Nat Soc Earthq Eng* **17**(4): 251–271.

Park R, Billings IJ, Clifton GC et al. (1995) The Hyogo-ken Nanbu earthquake of 17 January 1995. *Bull NZ Nat Soc Earthq Eng* **28**(1): 1–98.

Paulay T (1972) Some aspects of shear wall design. *Bull NZ Nat Soc Earthq Eng* **5**(3): 89–105.

Paulay T (1981) Developments in the seismic design of reinforced concrete frames in New Zealand. *Canadian J. Civil Eng* **8**(2): 91–113.

Paulay T and Priestley MJN (1992) *Seismic Design of Reinforced Concrete and Masonry Buildings*. John Wiley & Sons, Inc., New York.

Paulay T and Williams RL (1980) The analysis and design of and the evaluation of design actions for reinforced concrete ductile shear wall structures. *Bull NZ Nat Soc Earthq Eng* **13**(2): 108–143.

Pfrang EO, Seiss CP and Sozen MA (1964) Load-moment-curvature characteristics of r.c. cross sections. *ACI J* **61**: 763–778.

Pond WF (1972) Performance of bridges during the San Fernando earthquake. *J Prestressed Conc Institute* **17**(4): 65–75.

Popov EP and Pinkney RB (1969) Cyclic yield reversal in steel building connections. *J Struct Div* **95**(ST3): 327–353.

Popov EP and Stephen RM (1972) *Cyclic loading of full size steel connections*. American Iron and Steel Institute, Steel Research for Constr Bull No. 21.

Priestley MJN and Elder DM (1982) Cyclic loading tests of slender concrete masonry shear walls. *Bull NZ Nat Soc Earth Eng* **15**(1): 3–21.

Priestley MJN, Calvi, GM and Kowalsky MJ (2007) *Displacement-Based Seismic Design of Structures*. IUSS Press, Pavia, Italy.

Roeder CW and Popov EP (1978) Eccentrically braced steel frames for earthquakes. *J Struct Div* **104**(ST3): 391–412.

Sakamoto I, Ohashi Y and Shibata M (1984) Some problems and considerations on aseismic design of wooden dwelling houses in Japan. *Proc. 8th World Conf on Earthq Eng*, San Francisco **V**: 669–776.

Salse EAB and Fintel M (1973) Strength, stiffness and ductility properties of slender shear walls. *Proc. 5th World Conf on Earthq Eng*, Rome **1**: 919–928.

Sauter F (1984) Earthquake resistant criteria for precast concrete structures. *Proc. 8th World Conf on Earthq Eng*, San Francisco **V**: 629–636.

Selna L, Martin I, Park R and Wyllie L (1980) Strong and tough concrete columns for seismic forces. *J Struct Divn* **106**(ST8): 1717–1734.

Sidwell GK (1985) Eccentrically braced frames. *Bull NZ Nat Soc Earthq Eng* **18**(4): 355–359.

Shibata A and Sozen M (1976) Substitute structure method for seismic design in reinforced concrete. *ASCE J Struc Eng* **102**(1): 1–18.

Smith PC, Dowrick DJ and Dean JA (1986) Horizontal timber diaphragms for wind and earthquake resistance. *Bull NZ Nat Soc Earthq Eng* **19**(2): 135–142.

Spencer RA (1978) Rate of loading effects for Douglas-fir lumber. First Intl Conf on Wood Fracture, Banff, Canada.

Sugiyama H (1984) Japanese experience and research on timber buildings in earthquakes. *Proc Pacific Timber Eng Conf*, Auckland **III**: 431–438.

Takanashi K, Udugawa K and Tanaka H (1973) Failure of steel beams due to lateral buckling under repeated loads. *Symp on Resistance and Ultimate Deformability of Structures Acted on by Well Defined Loads*, IABSE, Lisbon, Preliminary Report: 163–169.

Tanaka H, Park R and MacNamee B (1985) Anchorage of transverse reinforcement in rectangular reinforced concrete columns in seismic design. *Bull NZ Nat Soc Earthq Eng* **18**(2): 165–190.

Teal EJ (1968) Structural steel seismic frames – draft ductility requirements. *Proc. 37th Annual Conv, Struct Engrs Assn of Calif*.

Thurston FJ and Flack PF (1980) Cyclic load performance of timber sheathed bracing walls. Central Laboratories Report No 5-80/10, NZ Ministry of Works and Development.

Timber Engineering Company (1956) *Timber Design and Construction Handbook*. FW Dodge Corporation, New York.

Tso WK, Pollner E and Heidebrecht AC (1973) Cyclic loading on externally reinforced masonry walls. *Proc. 5th World Conf on Earthq Eng*, Rome **1**: 1177–1186.

Uang C-M and Lathan CT (1995) *Cyclic testing of full-scale MNH-SMRF moment connections*. Structural Systems Research, University of California, San Diego.

Uchida N, Aoyagi T, Kawamura M and Nakagawa K (1973) Vibration test of steel frame having precast concrete panels. *Proc. 5th World Conf on Earthq Eng*, Rome **1**: 1167–1176.

Udagawa K, Takanashi and Kato B (1984) Effects of displacement rates on the behaviour of steel beams and composite beams. *Proc. 8th World Conf on Earthq Eng*, San Francisco **VI**: 177–184.

Vann WP, Thompson LE, Walley LE and Ozier LD (1973) Cyclic behaviour of rolled steel members. *Proc. 5th World Conf on Earthq Eng*, Rome **1**: 1187–1193.

Wade J (1972) The weldability of modern structural steels. *Symposium on Modern Applications of Welding Technology in Steel Structures*, University of New South Wales: 233–238.

Wakabayashi M and Nakamura T (1984) Reinforcing principle and seismic resistance of brick masonry walls. *Proc. 8th World Conf on Earthq Eng*, San Francisco **V**: 661–668.

Wakabayashi M, Nakamura T, Iwai S and Hayashi Y (1984) Effects of strain rate on the behaviour of structural members subjected to earthquake force. *Proc. 8th World Conf on Earthq Eng*, San Francisco **IV**: 491–498.

Walpole WR (1985a) Beam-column joints. *Bull NZ Nat Soc Earthq Eng* **18**(4): 369–380.

Walpole WR (1985b) Concentrically braced frames. *Bull NZ Nat Soc Earthq Eng* **18**(4): 351–5.

Walpole WR and Butcher GW (1985) Beam design. *Bull NZ Nat Soc Earthq Eng* **18**(4): 337–343.

Williams D and Scrivener JC (1971) Behaviour of reinforced masonry shear walls under cyclic loading. *Bull NZ Nat Soc Earthq Eng* **4**(2): 316–332.

Yap KK (1985) Slow cyclic shear testing of shear panels bonded with adhesives. Central Laboratories Report No 5-85/14, NZ Ministry of Works and Development.

11

Earthquake Resistance of Services, Equipment and Plant

11.1 Seismic Response and Design Criteria

11.1.1 Introduction

This chapter sets out to advise engineers on the earthquake resistant design of services, components and other equipment and plant. Much of the background information on the earthquake problem is contained in other chapters of this book or in the literature on structural engineering and seismology. Until about 1970, only a comparatively small effort had been made in this field by services engineers on their own account. Since the 1971 San Fernando, California, earthquake (Housner and Jennings, 1972; US Geological Survey, 1971), in which about 10% of the total cost of damage was attributed to damage to mechanical and electrical equipment, there has been increasing awareness that equipment needs its own specialist seismic design and detailing. The following points are worthy of attention:

(1) Seismic design of equipment is a problem of dynamics, which cannot be treated adequately with equivalent-static methods alone.
(2) Earthquake accelerations applied in the design of equipment generally should be much larger than the corresponding values used in the design of the buildings housing the equipment.
(3) The response spectrum method (see Section 11.1.4) provides ready-worked solutions of the equations of motion, and is a powerful aid to understanding the true dynamic nature of the earthquake problem.
(4) In many cases, a high level of earthquake resistance can be provided at relatively small extra cost (Hitchcock, 1969).

Equipment ranges in earthquake vulnerability from inherently *robust* to inherently *fragile*, as defined and discussed in Section 6.3.3. In this chapter, the means of providing earthquake protection to equipment of a range of inherent vulnerabilities is discussed.

Earthquake Resistant Design and Risk Reduction D. Dowrick
© 2009, John Wiley & Sons, Ltd

11.1.2 Earthquake motion – accelerograms

Strong-motion earthquakes are most commonly recorded by accelerographs, which produce accelerograms which are a plot of the variations with time of acceleration in a given direction. A formerly widely used accelerogram is that obtained at El Centro, California, during the Imperial Valley earthquake of 18 May 1940. Figure 5.19 shows the north–south accelerogram of this earthquake, with a peak acceleration of $0.33g$. By integration of the acceleration record, the ground velocity was deduced, showing a maximum value of 34 cm/s. Similarly, by integrating the velocity, the displacement of the ground was inferred, showing a maximum of 21 cm. The record of acceleration in the east–west direction was similar, with a maximum value of $0.22g$. The vertical component showed considerably more rapid variations, reaching a maximum of $0.2g$.

More detailed discussions of earthquake motion are given in Chapter 4.

11.1.3 Design earthquakes

It is common for a design earthquake to be adopted in a given region for certain types of construction. The design earthquake is defined as the worst earthquake likely to occur in that region with a given average return period (say, 500 years), and may be specified in terms of peak ground accelerations, a response spectrum, or an earthquake magnitude. Although there is a risk of a worse real earthquake occurring, the standard for officially acceptable minimum risk is set by the design earthquake. Individual structures or equipment items may be designed to some authorized or discretionary fraction (greater or less than unity) of the design earthquake, depending on the design levels of risk and ductility.

Design earthquakes are discussed more fully in Chapter 4.

11.1.4 The response spectrum design method

A direct analytical approach to the problem of earthquake strength is to make a mathematical model of the structure and subject it to accelerations as recorded in actual earthquakes. Many structures, including items of equipment, approximate to the single-degree-of-freedom model shown in Figure 11.1, where a mass is supported by a spring and is connected to a damping device. If linear material behaviour is assumed, the ratio of spring stiffness to horizontal shear is constant. For mathematical convenience the damping force is usually taken as proportional to velocity, which is generally a satisfactory approximation.

Figure 5.20 is a typical response spectrum diagram. It shows the maximum acceleration response to a given earthquake motion of a linear single-degree-of-freedom structure, with any fundamental period in the range $T = 0$ to $T = 4.0$ s. Note the considerable reduction in response resulting from an increase in damping.

As individual earthquakes give different irregular responses dependent on local ground conditions, a design criterion is sought by averaging the response curves for a number of earthquakes (Figure 11.2). The curves in Figure 11.2 clearly show that for earthquakes recorded on firm ground:

- accelerations not much larger than the maximum applied acceleration occur;
- structures with small flexibility (periods of 0.2–0.6 s) act as mechanical amplifiers and experience accelerations up to four times the peak applied (ground) acceleration;

Seismic Response and Design Criteria

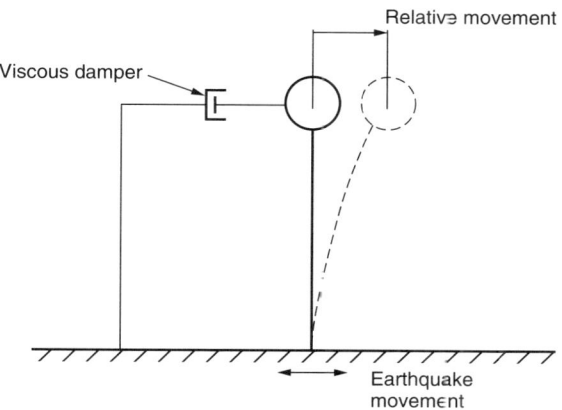

Figure 11.1 Single-degree-of-freedom model for studying earthquake response of equipment

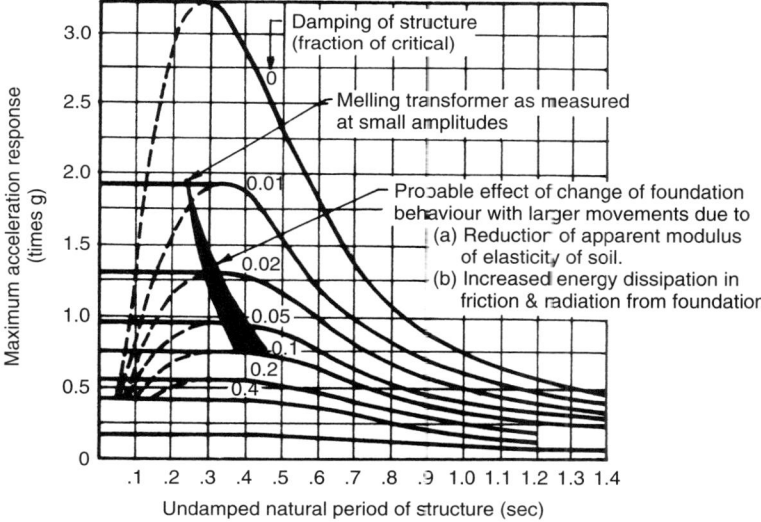

Figure 11.2 Response of 30-tonne transformer on concrete pad on soft subsoil to design earthquake (after Hitchcock, 1969). (Reproduced by permission of the Institution of Professional Engineers NZ)

- structures with large flexibilities experience accelerations less than the peak applied acceleration;
- structures with strong damping, whatever their natural period, experience greatly reduced response to ground motion.

11.1.5 Comparison of design requirements for buildings and equipment

The principles underlying the choice of earthquake design loads and level of material response are, of course, the same for buildings and equipment, depending on the consequences of any given response and their acceptability (Section 1.3). Thus, where equipment and buildings have the same risk characteristics, they should be designed as follows:

(1) They should be designed to the same design earthquake, except as modified by the building (Section 11.1.6).
(2) They should be designed to the same response (stress) levels, so that they will generally be damaged to a comparable degree in a given event (NZS 4219, 1983).

However, in some cases it is acceptable for a building to be deformed into the post-elastic range, using its ductility, while some of its contents may have to respond elastically to remain operational or safe. Thus vital services may be designed to more stringent earthquake criteria than the buildings housing them, e.g. safety equipment used for firefighting or emergency ventilation. Also dangerous substances should not be released, such as gas, steam or toxic chemicals.

The typical difference between traditional design loads for structures designed to be fully ductile ($\mu = 4$ to 6) and those designed to remain elastic ($\mu = 1$) may be seen by comparing the bottom curve in Figure 11.2 with the upper family of curves. Obviously there will be a large difference in design loads for equipment required to remain elastic and the building housing the equipment, if that building is designed to respond in a fully ductile manner. The question of the effect of ductility μ on the design loads is discussed in Section 5.4.7.

The continuation of electricity supply is a major factor in the success of emergency plans after earthquakes. A notable failure of electricity supply equipment occurred in the San Fernando earthquake, where there was $30 million damage to the then newly built $110 million Pacific Intertie Electric Converter Station (Housner and Jennings, 1972).

The importance of good earthquake resistant design of equipment in a building was highlighted at a hospital in the 1987 Edgecumbe, New Zealand, earthquake. The main building of the hospital was of six storeys, with its main water supply tanks on the top floor. Located in a zone of low damage of only intensity MM7, the main block suffered only slight structural damage. However, the unfastened water tanks moved, rupturing the piping connections, and water flooded down through all floors of the building (Pender and Robertson, 1987). The non-structural damage was severe enough to cause the hospital to be closed for several weeks for refurbishment.

11.1.6 *Equipment mounted in buildings*

Equipment mounted in a building should be designed to withstand the earthquake motions to which it will be subjected by virtue of its dynamic relationship to the building. The design of the equipment and its mountings should take into account the dynamic characteristics of the building, both as a whole and in part.

A building tends to act as a vibration filter, and transmits to the upper floors mainly those frequencies close to its own natural frequencies. Thus, on the upper floors there will be a reduction in width of the frequency band of the vibrations affecting the equipment. As a rough guide, the fundamental period of a flexible building may be taken as $0.1N$ seconds, where N is the number of storeys, but individual parts of the structure such as floors (on which the equipment is mounted) may have lower fundamental periods. Also the magnitude of the horizontal accelerations will generally increase with height up the building; and hence amplification of the ground accelerations usually takes place.

The accurate prediction of the vibrational forces occurring in equipment mounted in a building is a complex dynamical problem, which at present is only attempted on major installations. In ordinary construction, a simpler approach has to suffice, such as the response spectrum technique. With such methods, however, it is difficult to make realistic allowance for the filtering and amplification characteristics of the building. (See also Sections 11.2.2. and 11.3.3.)

11.1.7 Material behaviour

Failure modes

As noted in Section 11.1.5, whether a material behaves elastically or inelastically has a great influence on the seismic response (or loading). It is thus of great importance to understand whether a material is brittle or what degree of ductility may be obtained, and for maximum reliability the failure modes need to be controlled, as discussed for general structures in Section 8.3.8. Some further points relating to equipment are noted below.

Brittle materials

Whereas structural engineers generally try to avoid the use of brittle materials, electrical engineers have no choice but to use one of the most brittle of all materials, namely porcelain, in many of their structures. Because there is no ductility, any failure of such a material is total. Therefore seismic design accelerations for these structures should be 10–20 times those used for ordinary buildings. Diagonally braced structures carrying heavy loads, even though made of steel, may also have to be designed for large accelerations to avoid sudden failure by buckling of struts.

Ductility

For massive rigid bodies such as transformers all the energy imparted by an earthquake has to be absorbed in holding-down bolts or clamps, which are very small compared with the mass of the whole structure. Thus, if reliance is to be placed on the ductility of these fastenings to justify using reduced seismic accelerations for elastic design or for protecting the transformer, considerable knowledge of the post-yield behaviour and energy-absorbing capacity of the fastenings is essential. Reference to the behaviour of steel (Section 10.2), particularly to steel connections in Section 10.2.7, is relevant to the design of holding-down fastenings.

Damping

In the design of buildings, as only a fairly small amount of damping is readily available, survival in a large earthquake depends largely upon post-yield energy dissipation and ductility. On the other hand, with much electrical equipment the provision of high damping becomes a practical possibility because of the smaller masses involved. Such damping may be in the form of rubber pads, stacks of Belleville washers (Figure 11.10), or true viscous damping units.

The beneficial effect of damping on seismic response is apparent in Figures 11.2 and 11.6.

11.1.8 Cost of providing earthquake resistance of equipment

Many smaller items of electrical equipment can withstand horizontal accelerations of $1.0g$ as currently designed and installed. Even a 10-tonne transformer with the height of its centre of gravity equal to the width of its base could be secured against a horizontal acceleration of $1.0g$ with four 20 mm diameter holding-down bolts without exceeding the yield stress – an inexpensive protection for such a valuable piece of equipment.

Provided the nature of earthquake loading is understood and taken into account from the beginning of a design, earthquake resistance can often be obtained at little cost especially for inherently *robust* equipment (Section 6.3.3). The introduction of additional earthquake strength into an existing design is bound to be more expensive.

11.2 Seismic Analysis and Design Procedures for Equipment

Seismic analysis techniques and design procedures are the same in principle for equipment as those described for general structures elsewhere in this book, particularly Section 5.4. The following discussion, which considers the main points specific to equipment, separately describes procedures using dynamic and equivalent-static analyses.

In either case the basic considerations are, of course, the same. For example, vertical seismic accelerations of a similar size to the horizontal accelerations occur, and may exceed gravity in a severe earthquake. Thus, as the motions reverse in direction, equipment may need to survive net vertical accelerations ranging from about zero to $2.0g$, prior to considering the dynamic response of the equipment itself. Such accelerations would obviously greatly affect the stability of equipment. For example, friction between the base and the floor could not be relied on to locate the equipment horizontally.

Further background on the aseismic design of equipment may be found in a paper by Schiff (1984) regarding electric power substations, and in the New Zealand code for equipment of various types in buildings (NZS 4219, 1983).

11.2.1 Design procedures using dynamic analysis

Single-degree-of-freedom structures

For structures which can be thought of as having effectively only one mode of vibration in a given direction, the following simple *response spectrum* design procedure will usually prove to be both easy to carry out and seismically realistic:

(1) Ascertain the natural period of vibration in the direction being studied (by calculation or by measurement of similar structures). This should be done for its condition as installed, including the effects of supports and foundations.
(2) Determine an appropriate value of equivalent viscous damping by measurement of similar structures, by inference from experience, or by calculation if special dampers are provided.

(3) Read the acceleration response to the standard earthquake from the appropriate spectrum, (e.g. Figure 11.2). For natural periods less than 0.3 s, the maximum value for the damping concerned should generally be used, unless there is convincing proof that the structure concerned will remain very rigid throughout strong shaking, i.e. that it will always retain a natural period less than 0.1 s. This latter condition is often very difficult to prove, and should not normally be used.
(4) Combine the stresses from this earthquake loading with other stresses such as those from dead loads, and working pressures, including short-circuit loads, but not with stresses due to wind loads.
(5) For ductile structures, design to meet total loadings with stresses not exceeding normal working stresses, or such higher stresses as may be shown to meet the specification for survival in a major earthquake.
(6) For brittle structures, design to meet total loading with stresses that allow a factor of safety of a least 2.0 on the guaranteed breaking load of brittle components, or at least 2.5 if the breaking load is not based on statistically adequate information.

Multi-degree-of-freedom structures

For structures that have more than one mode of vibration two main methods of dynamic analysis exist as described below:

(1) A *response spectrum* technique similar to that described above can be used, but it is more complex in that the responses due to a number of modes must be combined.
(2) A more powerful dynamic analysis involves the application of *time-dependent forcing functions* directly to structures, rather than using response spectra. The equations of motion for the structure are solved using *full modal analysis* or *direct integration*, as described in Section 5.4.7. This approach is used in the dynamic analysis of a wide range of engineering structures, and it has been used by some manufacturers of high-voltage circuit breakers.

11.2.2 Design procedures using equivalent-static analysis

Where dynamic analysis is not used, it is desirable to establish suitable equivalent-static forces expressed as coefficients of gravity, C_p, such that the horizontal base shear, F, acting on the equipment is

$$F = C_p W_p R_p, \qquad (11.1)$$

where W_p is the weight of the item of equipment concerned and R_p is its importance factor (Section 5.4.7). Such coefficients should preferably be determined only for structures that fall into well-defined groups within which dynamic characteristics do not vary greatly. This has been done for a wide range of types of equipment in some codes as discussed in relation to Table 11.2. Preferably, each equipment group should have its coefficients derived from fundamental principles in such a way as to cover reasonable variations from the chosen dynamic characteristics. Hitchcock (1969) suggested three such groups: (1) base-mounted free-standing equipment; (2) equipment mounted on suspended floors;

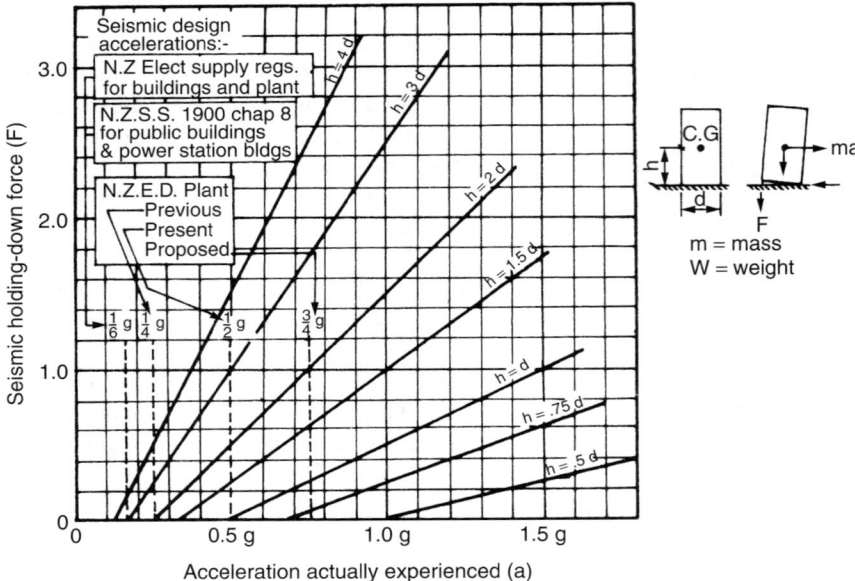

Figure 11.3 Relationship between seismic holding-down bolt loads, applied accelerations, and geometry of base-mounted free-standing equipment (after Hitchcock, 1969). (Reproduced by permission of the Institution of Professional Engineers NZ)

and (3) equipment that would fail in a brittle manner. The following discussion of these groups is largely based on Hitchcock's seminal paper.

Base-mounted free-standing equipment

Transformers are the chief members of this group of equipment. Figure 11.3 shows the forces acting on such equipment; the graph shows how the calculated holding-down force, expressed as a fraction of the weight, varies with the maximum acceleration experienced and with the ratio of height of centre of gravity to effective width of base.

As an example of equipment in this group, consider the power transformer mounted on a concrete pad shown in Figure 11.4. Some field measurements with small-amplitude vibrations in the transverse direction gave the damping as 0.9% of critical, and the fundamental period as 0.24 s. Plotting these values of damping and period on Figure 11.2 shows that the acceleration response of this transformer to the proposed New Zealand standard earthquake would be nearly $2.0g$ if the period and the damping remain unchanged. It is known, however, that when foundations rock in this manner, the subsoil properties may be modified; its modulus of elasticity (and hence natural frequency) decreases while the energy dissipated per cycle (and hence equivalent damping) increases.

The equivalent damping of rocking foundations can reach about 10% of critical as compared with about 20% for foundations moving vertically without rocking. As the overall equivalent damping factor for this example will probably lie in the range 2–10%, it can be seen from Figure 11.2 that this particular transformer would be subjected to a peak acceleration of $1.3-0.75g$ in an earthquake corresponding to this response spectrum.

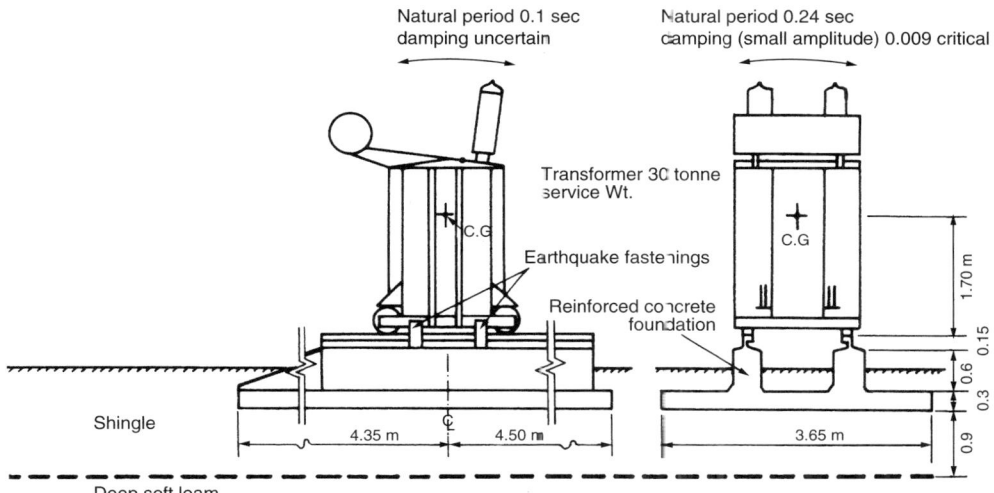

Figure 11.4 110/33 kV transformer on concrete pad (after Hitchcock, 1969). (Reproduced by permission of the Institution of Professional Engineers NZ)

Therefore, the equivalent-static design method for base-mounted free-standing equipment (including transformers and fastenings) should be as follows:

(1) If the natural period of vibration of the equipment as finally installed on its foundations is not known or is known to be larger than 0.1 s, then a design acceleration of $0.7g$ should be used, in conjunction with normal working stresses and with properly designed ductile material behaviour in the weakest part of the fixings.
(2) If the natural period of vibration of the equipment as finally installed on its foundations can be shown to be less than 0.1 s (and to remain so for accelerations up to $0.4g$) then a design acceleration of $0.4g$ should be used, in conjunction with normal working stresses and properly designed ductile material behaviour. As mentioned previously, it is very difficult to be sure that the lower portion of the response spectrum for very small values of the period (T) can be safely used, and this provision should seldom be applied in practice.

In a study (Chandrasekeran and Singhal, 1984) of the dynamic response of transformers, it was found that the equivalent-static force representative of the dynamic loads was best made by a single force applied at the top, this force varying depending on the type of mountings and whether shear or moment was being considered.

Since then more insight has been gained into the dynamics of rocking bodies. In particular, Makris and Roussos (1998) and Makris *et al.* (1999) have investigated the dynamics of rocking items of rigid equipment such as transformers, with geometry, including slenderness parameter α and radius R, as shown in Figure 11.5. They proposed a simple procedure of seven steps requiring only hand calculations for evaluating the overturning potential of an electrical equipment item in a near-source ground motion containing a long-duration pulse. Makris and Roussos found that transformers with approximate

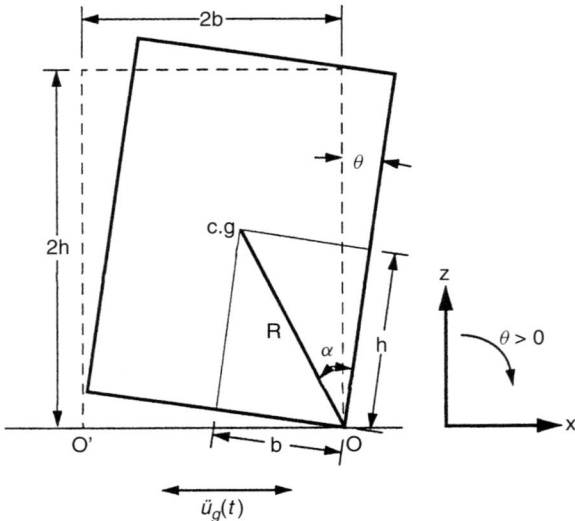

Figure 11.5 Geometry of a rocking rigid block (from Makris *et al.*, 1999)

Table 11.1 Response of equipment mounted in buildings. Equipment resonant with fundamental periods in range 0.2–0.4 s (from Shibata *et al.*, 1964)

Fraction of critical damping for building	0.07	0.07	0.07	0.07
Fraction of critical damping for plant item	0.007	0.02	0.1	0.2
Peak response of plant item to 1940 El Centro shaking	8–10g	5–7g	3g	2g

values of slenderness $\alpha \approx 20°$ and frequency parameter $p = \sqrt{3g/4R} = 2$ rad/s are likely to overturn due to damaging near-source short-period pulses of the type first observed in the 1979 Imperial Valley earthquake.

Equipment mounted on suspended floors of buildings or other structures

Equipment mounted in buildings is generally subjected to modified earthquake effects; this can mean amplification especially in the upper floors of buildings. This amplification of the ground and building motions is worse when the building and the equipment resonate, i.e. when they have equal periods of vibration. Fortunately, damping between the building and the equipment can be used to drastically reduce amplification, as it is not always possible to avoid the resonance effect. Table 11.1 illustrates the effect of resonance and damping as obtained in a simple analysis by Shibata *et al.* (1964).

Any rule of thumb for seismic design should require all equipment items in a building above the ground floor to be designed and fastened for higher excitations than on the ground, as reflected in the difference in the provisions for equipment in single- and multi-storey buildings in New Zealand (NZS 4219, 1983).

When closer design modelling is required, such as when resonance may occur, the excitation at any given floor level may be found from the dynamic characteristics of the

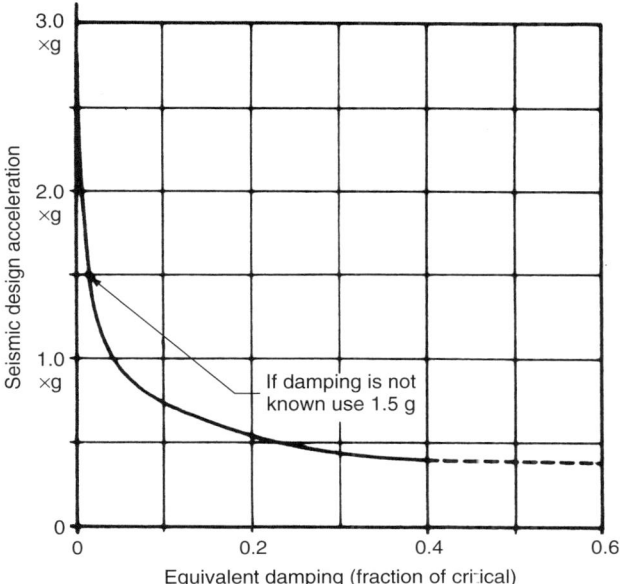

Figure 11.6 Seismic design accelerations for brittle structures with fundamental period less than 0.4 s (after Hitchcock, 1969) (Reproduced by permission of the Institution of Professional Engineers NZ)

building. This is conveniently done in the form of floor response spectra (Igusa and Der Kiureghian, 1985).

Equipment that would fail in a brittle manner in earthquakes

Hitchcock (1969) reports on some tests on the dynamic characteristics of porcelain-supported equipment, some ground-mounted, some supported on concrete posts. The natural periods of vibration were found to be in the range 0.2–0.4 s, corresponding to the peak of the response spectra in Figure 11.2. As the damping ranged from 0.018 to 0.006 of critical, the expected response varied from about $1.5g$ to greater than $2.0g$. Most of the items of equipment involved had strengths appreciably less than those required to withstand such accelerations. To deal with this situation, Hitchock (1969) suggested three alternative procedures as follows:

(1) Provide the required strength with factors of safety of the order of 2–3 to cover uncertainties in the assessment of the strength of brittle materials.
(2) Provide ductile components that yield early enough to prevent the brittle components reaching breaking load (Gilmour and Hitchcock, 1971).
(3) Provide additional damping. This solution is quite practicable when dealing with small masses of electrical equipment (Winthrop and Hitchcock, 1971).

Equipment used in an electrical installation must in general be suitably rigid to avoid variations in clearance between live parts, and to limit the amount of flexibility to be provided in electrical connections. In fact, any acceptable structure of equipment is unlikely to have a period longer than about 0.4 s. From Figure 11.2 it can be seen that for periods less than about 0.4 s acceleration is taken as constant for any given damping. Hence, the relationship between acceleration response and damping can be plotted as in Figure 11.6.

Assuming that the New Zealand design earthquake is to be used, the following design rules for this type of brittle equipment may be adopted:

(1) If the amount of damping in the equipment is not accurately known, the equivalent-static acceleration for equipment that fails in brittle components under horizontal loading should be $1.5g$. This should be used as a factor of safety of 2.0 on the guaranteed breaking strength of the brittle portions, and with ordinary working stresses in the ductile parts of the structure.
(2) Alternatively, if satisfactory evidence is available of the amount of damping inherent in the equipment, the seismic coefficient may be that read from Figure 11.5 for that amount of damping.

These rules would be suitable for the design of standard items of equipment installed in any part of a seismic country, because any type of foundation, from extremely rigid to highly flexible, could be used without invalidating the underlying assumptions.

Code seismic coefficients for equipment

The horizontal design coefficient C_p in equation (11.1) needs to account for resonance with the supporting structure. As a qualitative guide to the sorts of coefficient to expect (without resonant amplification) for different types of equipment, the values given in Table 11.2 may be helpful if used with caution. The severity of the requirements compared with those for buildings is clear. For example, pipework for sprinkler systems in normal-use multi-storey buildings in the highest hazard zones of New Zealand should be designed for horizontal accelerations of up to $1.0g$. Such equivalent-static force values should be adequate unless equipment with low damping has a natural period of vibration close to one of the important periods of the building (see also the discussion relating to Table 11.1).

11.3 Seismic Protection of Equipment

11.3.1 Introduction

As discussed by Blackwell (1970), there are two main problems affecting the protection of many types of equipment. The first concerns movements, and the second energy dissipation. Both problems are worsened if resonance or quasi-resonance exists. Movements can be dealt with in either of the following ways:

(1) by preventing serious relative displacement during an earthquake by anchoring the components of the installation to the building structure;
(2) by accommodating the relative movements of components without fracture of pipelines, ducts, cables and other connections (these relative movements may result from movements of either the building fabric or the mechanical services components themselves).

Table 11.2 Seismic coefficients for equipment in buildings, as previously required in New Zealand (NZS 4219, 1983). For preliminary guidance only. (Reproduced by permission of Standards New Zealand)

Item	Part or portion of building	$C_{p_{max}}$	$C_{p_{min}}$
7	Towers not exceeding 10% of the mass of the building. Tanks and full contents, not included in item 8 or item 9; chimneys and smoke stacks and penthouses connected to or part of the building except where acting as vertical cantilevers:		
	(a) Single-storey buildings where the height to depth ratio of the horizontal force resisting system is:		
	(i) Less than or equal to 3	0.2	
	(ii) Greater than 3	0.3	
	(b) Multi-storey buildings where the height to depth ratio of the horizontal force resisting system is:		
	(i) Less than or equal to 3	0.3	
	(ii) Greater than 3	0.5	
8	Containers and full contents and their supporting structures; pipelines, and valves:		
	(a) For toxic liquids and gases, spirits, acids, alkalis, molten metal, or poisonous substances, liquid and gaseous fuels including containers for materials that could form dangerous gases if released:		
	(i) Single-storey buildings	0.6	0.5
	(ii) Multi-storey buildings	1.3	0.9
	(b) Fixed firefighting equipment including fire sprinklers, wet and dry riser installations, and hose reels:		
	(i) Single-storey buildings	0.5	0.3
	(ii) Multi-storey buildings	1.0	0.6
	(c) Other		
	(i) Single-storey buildings	0.3	0.2
	(ii) Multi-storey buildings	0.7	0.4
9	Furnaces, steam boilers, and other combustion devices, steam or other pressure vessels, hot liquid containers; transformers and switchgear; shelving for batteries and dangerous goods:		
	(i) Single-storey buildings	0.6	0.5
	(ii) Multi-storey buildings	1.3	0.9
10	Machinery; shelving not included in item 9; trestling, bins, hoppers, electrical equipment not specifically included in other item 8, 9 or 11, other fixtures:		
	(i) Single-storey buildings	0.3	0.2
	(ii) Multi-storey buildings	0.7	0.3
11	Lift machinery, guides, etc., emergency standby equipment	0.6	
12	Connections for items 8 to 11 inclusive shall be designed for the specified forces provided that the gravity effects of dead and live loads shall not be taken to reduce these forces		
13	Suspended ceilings including attached equipment, lighting and attached partitions, see clause 3.6.5	0.6	
14	Communications, detection or alarm equipment for use in fire or other emergency:		
	(i) Single-storey buildings	0.5	0.3
	(ii) Multi-storey buildings	1.0	0.6

The energy-absorption problem means dealing with the seismic stresses occurring in the equipment, its mountings, and its fastenings to the structure. The equipment may have to be strengthened, and damping devices may have to be fitted. Mountings should not be made too strong because, apart from the expense, this may cause the equipment to fail somewhere else. Also the resulting lower period of vibration sometimes leads to higher stresses. For example, it can be seen from Figure 11.2 that if equipment with a natural period of 1.4 s and 5% damping is stiffened so that its natural period decreases by 0.5 s the inertial force will have increased three times. It would generally be impracticable to make connections at positions of maximum sway on equipment whose natural period is above 1.0 s. Fitting limit stops might overcome this problem but such stops would have to be designed to limit shock loading (Section 11.3.3). Energy is absorbed by the deformation of fasteners, springs, and rubber mountings, but, if the materials have little natural damping, the deformation remains within the limits of elasticity, and dissipation of energy may be insufficient for earthquake protection. In this respect, springs and even rubber mountings may prove unsatisfactory. Hydraulic or friction dampers could be added to increase the energy absorption, but this would be expensive and require detailed design.

Simpler methods of absorbing energy are usually possible, including the plastic deformation of supports and holding-down bolts, and the frictional work done when units slide about on the floor. Once plastic deformation has taken place, bolts will be slack on the return movement, and this is when floor friction is useful. Floor friction is free from backlash and shock effects, apart from the deceleration at the end of the slide, and is generally free of costs; the unit of course must be designed not to tip over.

When fastenings are designed for plastic deformation, they should be proportioned and sized so that the stresses are evenly distributed throughout the whole volume of the material, because the amount of energy dissipated is directly proportional to the stress developed and to the volume of material developing stress. Fastenings should be free of weak links or stress concentrations, which would result in early fracture of the fastening without much dissipation of energy.

11.3.2 Rigidly mounted equipment

Boilers, calorifiers, control panels, batteries, air-conditioners, kitchen equipment, and hospital equipment

(1) The first requirement is to prevent the equipment sliding across the floor. The coefficient of friction rarely exceeds 0.3, and the effectiveness of friction can be greatly reduced by the upward component of the earthquake, so friction alone is unlikely to be sufficient. Mounting on bituminous felt or lead would increase the friction and may be sufficient for less important equipment. The use of a suitable glue with neoprene pads would also give increased security against sliding.
(2) The next requirement is to ensure stability against overturning. In the first instance a simple geometric calculation will show whether the equipment is inherently stable or not. This is a function of the base width and the height of centre of mass of the equipment (Figure 11.3). If the horizontal acceleration is $0.6g$ (the maximum formerly required for boilers in single-storey buildings in New Zealand; see Table 11.2), the holding-down force would be zero if the centre of gravity of the equipment is not higher than 0.84 times the width of the base.

Seismic Protection of Equipment

(3) Where overturning stability cannot be obtained from geometric considerations, the equipment will have to be fastened to the building structure in some way. If this is done by holding-down bolts fixed into the floor, the bolts should be the weakest part of the system so that they protect the equipment by yielding first. This is particularly desirable when the equipment itself is not very strong. Fine-thread bolts with a length not less than 10 times the diameter should be used, and the thread should be designed such that the ultimate strength of the bolt based on the thread root area exceeds the yield strength of the gross bolt area (see also Section 11.3.1). Restraint against overturning can in some cases be obtained by fastening the top of the equipment to walls or columns; but the walls in particular must be seen to be strong enough for this purpose.

(4) Pipework and electrical wiring connections are vulnerable and therefore must be strong. It would also be wise to allow some flexibility in the pipes and wires away from the equipment in case of relative seismic movement between the items on either side of the connections.

(5) Doors to control panels should be hinged to prevent them being dislodged in earthquakes; loose covers can fall against live contacts, shorting out the equipment.

(6) Mercury switches should be avoided, as should essential instruments that have heavy movable components likely to break away from their supports.

(7) Boilers with extensive brickwork are undesirable, as it is very difficult to reinforce the fire brick.

Chimneys

Chimneys should be subjected to a thorough seismic structural design. Lightweight double wall sheet-metal flues should be used where possible, and prefabricated stacks should be avoided or used with great care.

Tanks

As well as considerations of sliding and overturning as discussed above, the following points are peculiar to tanks:

(1) Corrugations of copper tanks are liable to collapse with subsequent failure of the bottom joint. This can be remedied by making a stronger joint and possibly reducing the number of corrugations. Alternatively, welded stainless steel tanks can be used to increase the tank strength while retaining corrosion resistance.

(2) Where there will be a possibility of a tank sliding, severance of the connections can be avoided by flexibility in the pipes (10 diameters on each side of a bend should be adequate) and by provision of strong connections between the pipes and the tank. The bottom connection can be strengthened by passing it right through the tank and welding it at each end. The top connection could be similarly treated unless a large arm ball valve were required, when extra strengthening at the connection would be satisfactory.

(3) Suspended tanks should be strapped to their larger systems, and provided with lateral bracing.

(4) Because of the build-up of surface waves in liquid during earthquakes, some protection against liquid spillage may be desirable. This may be either in the form of a lid, or a

spill tray with a drain under the tank. The effects of pressures on the tank due to the liquid oscillation may have to be taken into account in the design of larger tanks.
(5) In large tanks the walls need to be designed for various load cases, including buckling of the tank walls. Special analysis and design guidance is available from specialist references, such as New Zealand Society for Earthquake Engineering (in preparation).
(6) For a method of restraining small tanks, see Figure 13.5.

11.3.3 Equipment mounted on isolating or energy-dissipating devices

Introduction

Blackwell (1970) described flexible mountings as falling into groups relating to the predominant motions of short-period earthquakes, as follows:

Group 1. Mountings with a natural period less than the predominant earthquake period (i.e. below about 0.07 s). Felt, cork and most rubber mountings would usually come into this category. Provided the mountings will not permit sliding to occur (e.g. by gluing), and the connections to the equipment are reasonably flexible, no further mounting precautions should be necessary.

Group 2. Mountings with a natural period corresponding to the predominant category, such as those used for low-speed fans, engines, compressors, and possibly electric motors. As resonance is likely, some method must be provided which limits the movement and transfers the forces directly to the floor instead of through the mounts. Steel rods or angles would be suitable and should be designed to yield at the design acceleration. An example of such a fastener is shown in Figure 11.7. Note the covering round the rod to reduce shock loading. Alternatively, the rods could be replaced by multiple-strand steel wire. The flexibility normally provided in pipe, duct, and electric wire connections would be adequate for an earthquake.

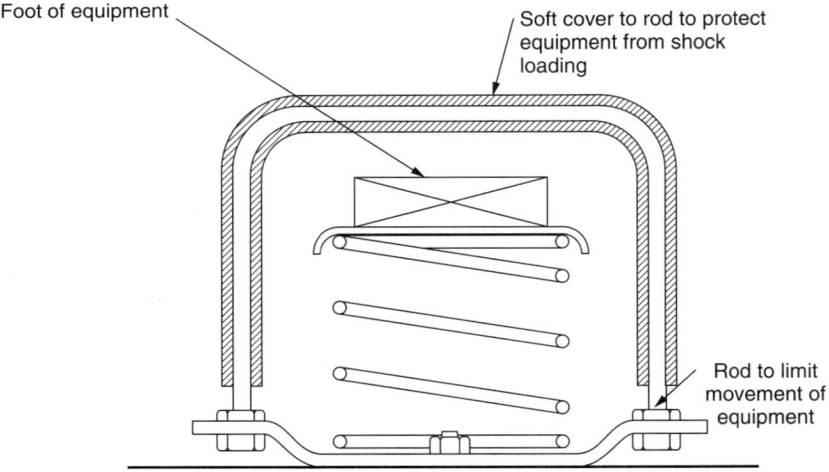

Figure 11.7 Detail of a flexible mounting with a resilient restraint against movements (after Blackwell, 1970). (Reproduced by permission of the Institution of Professional Engineers NZ)

Developments in isolation and energy-dissipation methods

The successful development of seismic isolation and energy dissipation devices for major structures, as described in Section 8.5, has been paralleled by similar applications for equipment. For example, brittle equipment can be protected by energy-dissipating supports (see page 461–2).

Subsequent developments of the principles of Section 8.5 include the use on large industrial boilers supported from the top on laterally flexible hangers, together with the use of energy absorbers (Hollings *et al.*, 1986; Lopez *et al.*, 1984). This permits the use in seismic regions of equipment that is not, in itself, designed for earthquakes. The same approach has also been used for seismic protection of nuclear installations such as reactor vessels, and much other equipment such as piping or computers (Chang, 1984; Schneider *et al.*, 1982; Skinner *et al.*, 1993; Spencer, 1980).

An example of the use of isolation systems for the seismic protection of a large piece of equipment comes from that provided for a three-storeys high 50 m long printing press in Wellington, New Zealand (Dowrick *et al.*, 1992). In this case the plant was protection by isolating the whole of the part of the building (the press hall) in which the brittle cast printing iron presses are located (Figure 11.8). Note the 0.46 m whole seismic movement gap required around the press hall. The seismic stresses were greatly reduced (to as little as a twelfth) from those that the unisolated press would experience in shaking twice that of the El Centro (1940) record (Figure 11.9).

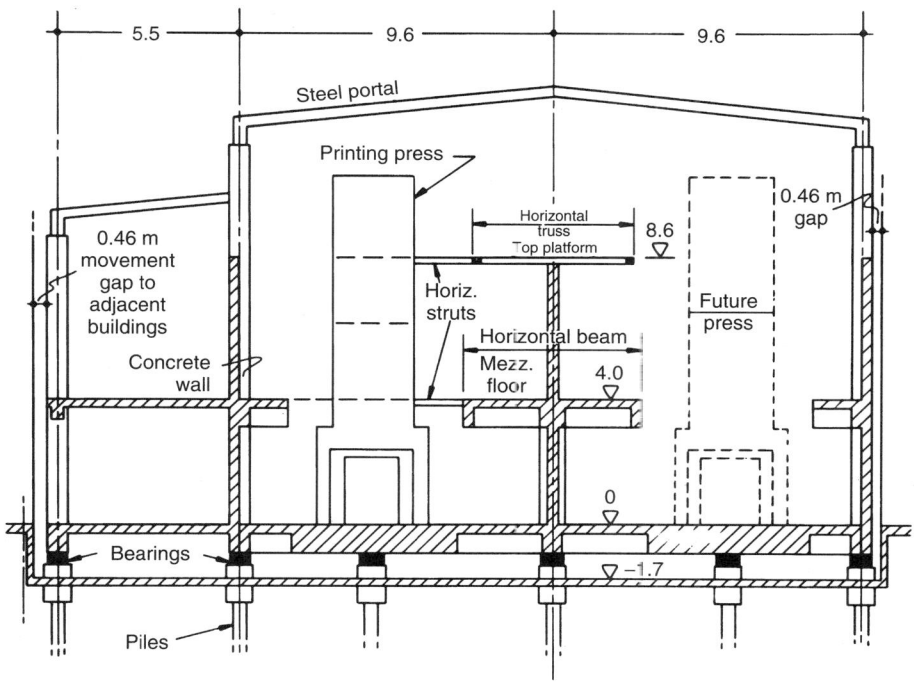

Figure 11.8 Section through the press hall on lead-rubber bearings at Wellington Newspapers plant (from Dowrick *et al.*, 1992)

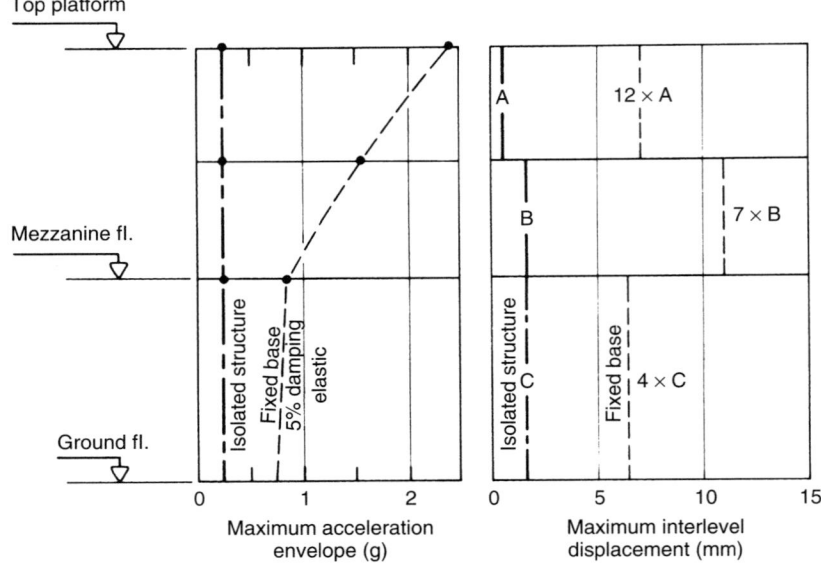

Figure 11.9 Comparison of isolated and fixed-base structures at Wellington New Zealand Papers (see Figure 11.8) under shaking twice that of El Centro (from Dowrick *et al.*, 1992)

At the other end of the size scale, for small items of equipment Belleville washers provide an inexpensive and simple means of damping seismic motions. As can be seen in Figure 11.10, these washers are conical spring washers, which are designed to achieve: a shift in the natural period of the equipment they support to outside the range where the ground acceleration is strongest; and an increase in the damping of the whole structure by the extra damping provided by the washers. This damping is obtained by the friction between the Belleville washers and the flat washers between them. This comes about as the Belleville washers are deflected by compression and decompression during the earthquake loading cycles.

Structures such as high-voltage electrical equipment do not have much damping (about 2%), but this can be increased up to 8% by mounting it on Belleville washers. In the Edgecumbe, New Zealand, earthquake at intensity MM9 there was extensive damage to basework, transformers and circuit breakers. One pole of a 220 kV circuit breaker was destroyed, but the Delle 220 kV circuit breakers supported on Belleville washers were undamaged (Pham and Hoby, 1991).

11.3.4 Light fittings

Pendant fittings can have a wide range of natural frequencies. Wire-supported fittings may not fail but could swing and smash if brittle. Heavy fittings and brittle materials for supports should be avoided, as should any combination of low damping and low fundamental periods (in the range 0.2–1.0 s). In many cases the seismic design of lighting will be related to that of suspended ceilings when the light fittings are attached to them (see also Table 11.2 and Section 12.4.2).

Seismic Protection of Equipment

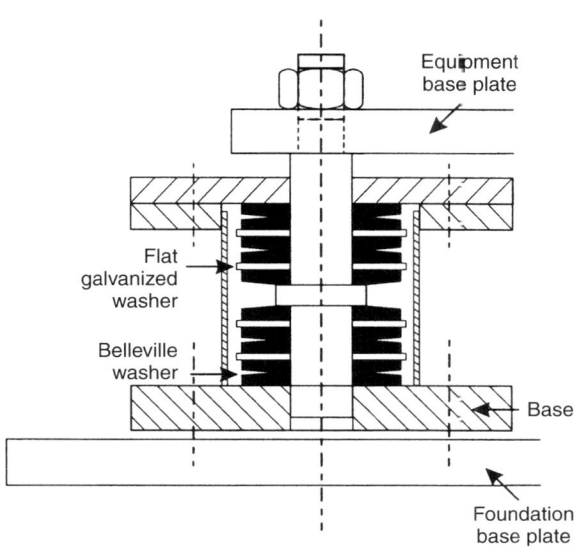

Figure 11.10 Belleville washer damper arrangement for the earthquake protection of equipment (from Pham and Hoby, 1991)

11.3.5 Ductwork

Ductwork is usually quite strong in itself, and despite relatively flexible hangers, it is usually susceptible to earthquake damage only where it crosses seismic movement gaps in buildings. At these points flexible joints should be provided which are long enough to take up the seismic movements. Canvas joints may be suitable (except where there is a fire risk), or lead-impregnated plastic if noise is a problem. Wherever possible, seismic movement gaps in buildings should not be crossed. It may be possible to locate fire walls at seismic movement gaps, and to design pipe and duct systems to be separate on each side of the gap, thus avoiding crossing the gap as well as keeping the number of systems down to a minimum.

The other most vulnerable position in ductwork is at its connection to machines (e.g. fans). At these positions flexible duct connections should be installed in a semifolded condition with enough material to allow for the expected differential deflection between the machines and the ductwork.

Duct openings and pipe sleeves through walls or floors should be large enough to allow for the anticipated movement of the pipes and ducts.

Beattie (2000) discusses shake table experiments aimed at identifying resonances between ductwork and structure, as follows:

> Both the fan coil unit and the duct system associated with it hung on 700 mm long threaded rods from a rigid frame supported on the shake table [Figure 11.11]. Because each item had a different mass from the other and the hanger length for each was different, they oscillated at different natural frequencies. Therefore their unbraced responses to the shake table motion were also different and often out of phase. Differential displacements reached as high as 55 mm at only $0.18g$ table acceleration. These items are connected by a canvas skirt. Also,

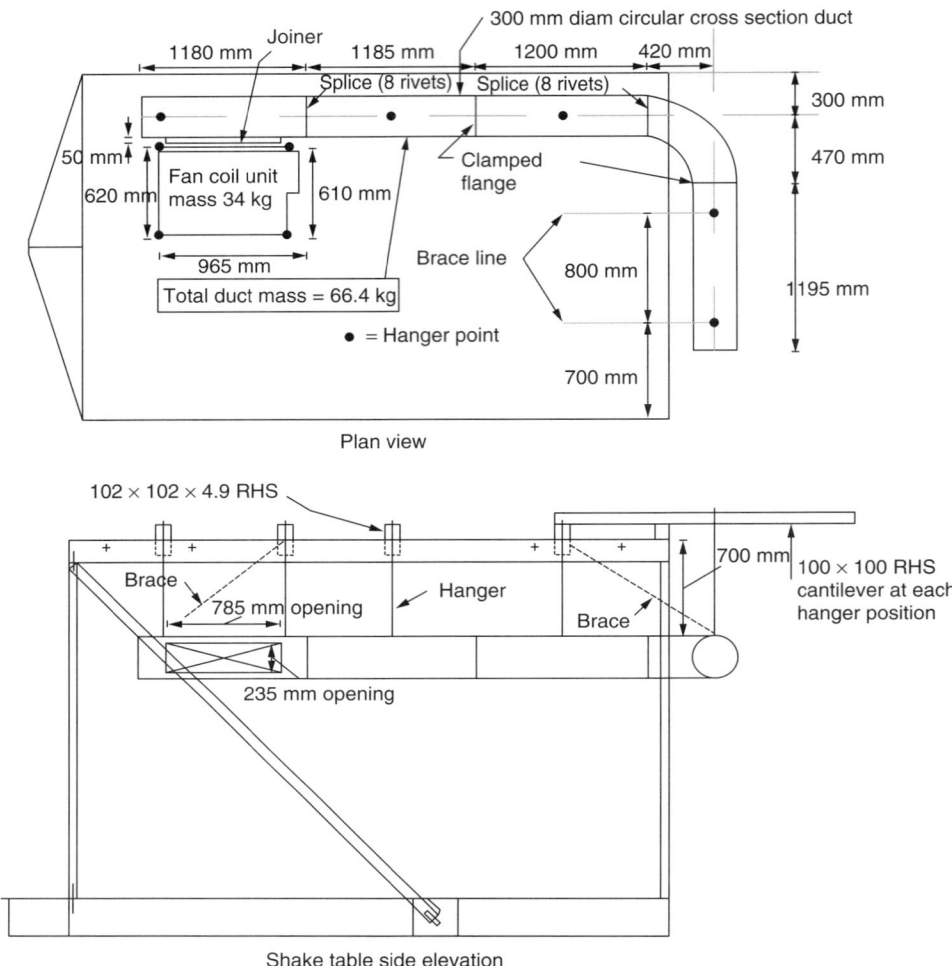

Figure 11.11 Duct system tested on shake table (from Beattie, 2000)

at the natural frequency of the individual units, the amplification of the table acceleration was between four and five times.

Diagonal braces were added to provide horizontal support to the elements. This addition resulted in a reduction in fan coil acceleration to about twice the table acceleration and to about the same level as the table in the case of the duct. In order to determine the forces in the braces, the table acceleration at any point in time was subtracted from the fan coil acceleration and the duct acceleration. At a table acceleration of approximately $0.75g$, the differential acceleration between the table and the fan coil unit was about $0.9g$. For the duct the figure was about $1.2g$.

The important point to note with such systems is that the skirt joints between units, provided to prevent the transfer of fan vibrations to the duct, must also be able to accommodate any differential movement between the components.

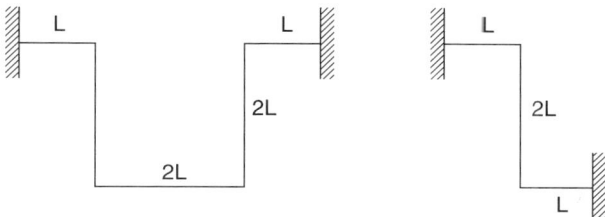

Figure 11.12 Suggested pipe arrangements for crossing movement gaps

11.3.6 Pipework

Flexibility requirements

Flexibility is required in pipework to allow for building and equipment movement. Seismic flexibility requirements are different from those for accommodating thermal expansion, as seismic movements take place in three dimensions. Sliding joints or bellows cannot be used, as they do not have the required flexibility and introduce a weakness which could cause early failure without making use of the ductility of the pipework as a whole. Accordingly, those expansion joints which are installed to accommodate thermal expansion must be fully protected from earthquake movements.

The movements should be taken up by bends, off-sets or loops which have no local stress concentrations and which are so arranged that if yielding occurs there will not be any local failure. Note that short-radius bends can cause stress concentrations. Anchors adjacent to loops must also be strong, and connections to equipment must be able to resist the pipe forces caused by earthquake movements. Connections using screwed nipples and some types of compression fittings should be avoided, unless they can be arranged so as to be unaffected by seismic movements.

U-bends and Z-bends, as shown in Figure 11.12, can be used to obtain flexibility. The dimensions L should be determined by calculation, so as to give safe stresses in the pipes and at the supports for the applied seismic movements.

Where the laying of pipes across seismic movement gaps in buildings cannot be avoided, details as shown in Figures 11.13, 11.14 or 11.15 can be used. Such crossings should be made at the lowest floor possible to minimize the amount of movement which has to be accommodated.

Pipework should be tied to only one structural system. Where structural systems change, the relative deflections are anticipated, and flexible joints should be provided in the pipework to allow for the same amount of movement. Suspended pipework systems should have consistent degrees of freedom throughout. For example, branch lines should not be anchored to structural elements if the main line is allowed to sway. If pipework is allowed to sway, flexible joints should be installed at equipment connections.

Methods of supporting pipework

Simple hangers will allow the pipe to swing like a pendulum. With usual support spacings, pipes will have a fundamental period of about 0.1 s if sideways movement is prevented at every support, and the period will increase to 0.2 s with twice this spacing, and to

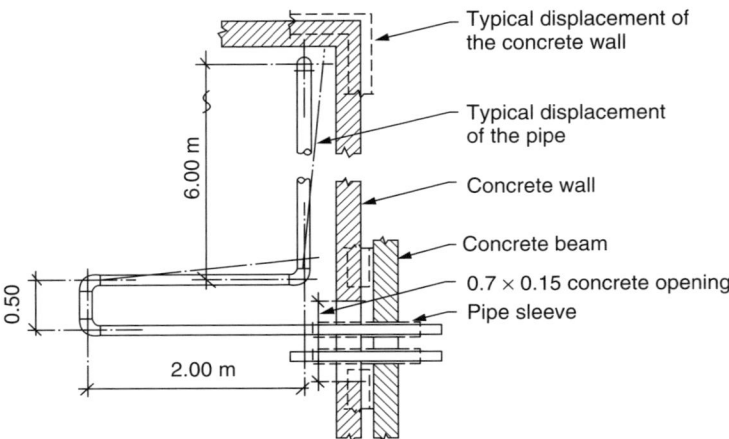

Figure 11.13 Plan view of pipework crossing a seismic movement gap (after Berry, 1972)

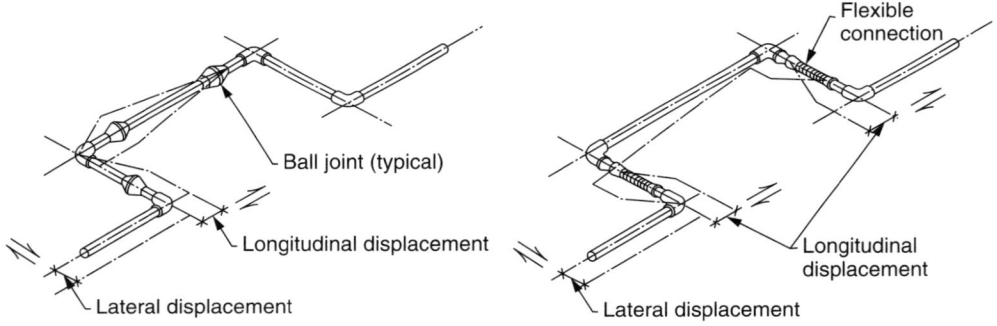

Figure 11.14 Pipework details for crossing seismic movement gaps where space limitations prevent the use of the pipe loop shown in Figure 11.13 (after Berry, 1972)

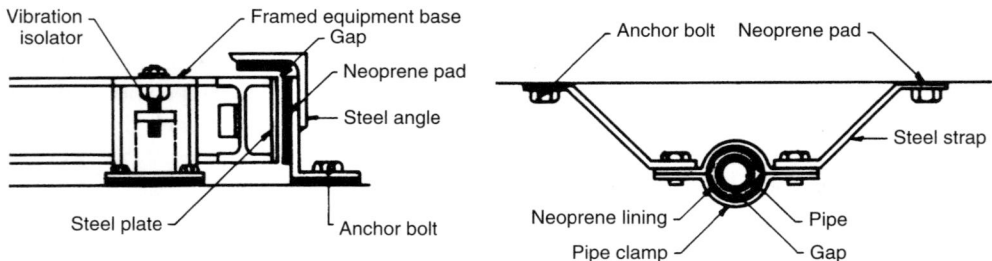

Figure 11.15 Combined earthquake mountings and vibration isolation for machine bases and pipework (after Berry, 1972)

about 1.0 s with three times the spacing. The latter two periods are very close to common building periods and the resulting resonance would cause large movements, considerable noise and possible failure. This can be avoided by the provision of horizontal restraints or by the use of two hangers in a V-formation.

As noted in Section 11.3.3, specially designed energy-dissipating supports for pipework are an aseismic design alternative that may have design advantages in some circumstances, especially in critical facilities.

References

Beattie GJ (2000) The design of building services for earthquake resistance. *Proc. 12th World Conf on Earthq Eng*, Auckland, Paper No. 2462 (CD-ROM).

Berry OR (1972) Architectural seismic detailing. State of the Art Report No. 3, Technical Committee No 12, Architectural–Structural Interaction IABSE-ASCE, Int Conf on Planning and Design of Tall Buildings, Lehigh University (Conf Preprints, Reports Vol. Ia-12).

Blackwell FN (1970) Earthquake protection for mechanical services. *NZ Engineering* **25**(10): 271–275.

Chandrasekeran AR and Singhal NC (1984) Behaviour of a 220 kV transformer under simulated earthquake conditions. *Proc. 8th World Conf on Earthq Eng*, San Francisco **VIII**: 149–156.

Chang IK (1984) Dynamic testing and analysis of improved computer/clean room raised floor system. *Proc. 8th World Conf on Earthq Eng*, San Francisco **V**: 1175–1180.

Dowrick DJ, Babor J, Cousins WJ and Skinner RI (1992) Seismic isolation of a printing press in Wellington, New Zealand. *Bull NZ Nat Soc Earthq Eng* **25**(3): 161–166.

Gilmour RM and Hitchcock HC (1971) Use of yield ratio response spectra to design yielding members for improving earthquake resistance of brittle structures. *Bull NZ Soc Earthq Eng* **4**(2): 285–393.

Hitchcock HC (1969) Electrical equipment and earthquakes. *NZ Engineering* **24**(1).

Hollings JP, Sharpe RD and Jury RD (1986) Earthquake performance of a large boiler. *Proc. 8th Europ Symp on Earthq Eng*, Portugal.

Housner GW and Jennings PC (1972) The San Fernando earthquake. *Earthq Eng Struc Dyn* **1**(1): 5–32.

Igusa T and Der Kiureghian A (1985) Generation of floor response spectra include oscillator-structure interaction. *Earthq Eng Struc Dyn* **13**(5): 661–676.

Lopez G, Makimoto Y, Mii T, Mitsuhashi K and Pankow B (1984) Analytical-experimental dynamic analysis of the El Centro power plant unit No. 4 steam generator. *Proc. 8th World Conf on Earthq Eng*, San Francisco **VII**: 189–196.

Makris N and Roussos Y (1998) Rocking response and overturning of equipment under horizontal pulse-type motions. Pacific Earthq Eng Res Center, University of California. PEER-98/05.

Makris N, Roussos Y and Zhang J (1999) Rocking response and overturning of electrical equipment under pulse-type motions. *PEER Center News* **2**(2): 1–6, 12.

NZS 4219 (1983) *Specification for Seismic Resistance of Engineering Systems in Buildings*. Standards Association of New Zealand, Wellington.

New Zealand Society for Earthquake Engineering (in preparation) *Seismic Design of Storage Tanks*: Recommendations of a Study Group of the NZSEE.

Pender MJ and Robertson TW (1987) Edgecumbe earthquake: reconnaissance report. *Bull NZ Nat Soc Earthq Eng* **20**(3): 201–49.

Pham T and Hoby P (1991) The design, testing and application of Belleville washers as a damping device for seismic protection of electrical equipment. *Proc. Pacific Conf on Earthq Eng*, Auckland **3**: 381–391.

Schiff AJ (1984) Seismic design practice for electrical power substations. *Proc. 8th World Conf on Earthq Eng*, San Francisco **VIII**: 181–188.

Schneider S, Lee HM and Godden WG (1982) Behaviour of a piping system under excitation: (etc). Report No. UCB/EERC-809/03, Earthquake Engineering Research Center, University of California, Berkeley.

Shibata H, Sata H and Shigeta T (1964) Aseismic design of machine structures. *Proc. 3rd World Conf on Earthq Eng*, New Zealand **2**, session II, 552–568.

Skinner RI, Robinson WH and McVerry GH (1993) *An Introduction to Seismic Isolation*. John Wiley & Sons, Ltd, Chichester.

Spencer P (1980) The design of steel energy-absorbing restrainers and their incorporation into nuclear power plant for enhanced safety (Vol 5): Summary report. Report No UCB/EERC – 80/34, Earthquake Engineering Research Center, University of California, Berkeley.

US Geological Survey (1971) *The San Fernando Earthquake of February 9th, 1971*, Geological Survey Professional Paper 733. US Government Printing Office, Washington, DC.

Winthrop D and Hitchcock HC (1971) Use of yield ratio response spectra to design yielding members for improving earthquake resistance of brittle structures. *Bull NZ Soc Earthq Eng* **4**(2): 294–300.

12

Architectural Design and Detailing for Earthquake Resistance

12.1 Introduction

Architects have a crucial role in both design and detailing to ensure the adequate earthquake performance of buildings. First, the choice of a suitable *structural form* is crucial, involving full collaboration at conceptual design stage between architects and engineers (for this aspect of design see Chapter 8, particularly Sections 8.3–8.6). Second, a large part of the damage done to buildings by earthquakes is non-structural. For instance, in the San Fernando, California, earthquake of February 1971, a total of $500 million worth of damage was done to the built environment, of which over half was non-structural. The importance of sound anti-seismic detailing in earthquake areas should need no further emphasizing.

Buildings in their entirety should be tailored to ride safely through an earthquake, and the appropriate relationship between structure and non-structure must be logically sought. For the effect of non-structure on the overall dynamic behaviour of a building see Section 8.3.8, where the question of full separation or integration of infill panels into the structure is discussed.

Architectural items such as partitions, doors, windows, cladding and finishes need proper seismic detailing; many non-seismic construction techniques do not survive strong earthquake motion as they do not provide for the right kinds or size of movements. Detailing for earthquake movements should, however, be considered in conjunction with details for the usual movements due to live loads, creep, shrinkage and temperature effects. As with so many other problems, it is worth saying that good planning can provide the right framework for practical seismic details.

An ironic example of the inadequacy of a non-structural item comes from the San Fernando earthquake; a modern fire station withstood the earthquake satisfactorily with regard to its structure, but the main doors were so badly jammed that all the fire engines were trapped inside. Arnold (1991) notes that engineers tend to emphasize structural damage in earthquakes, but in certain situations earthquake damage to non-structural components will greatly exceed the cost of structural damage. For example, in an analysis of a new 27-storey condominium building in Los Angeles, Shipp and Johnson (1990) estimated that in a maximum credible event the building would suffer structural damage of just over $1 million compared to non-structural damage of just under $6.7 million, relative to a total construction cost of $42.8 million. This estimated cost is for direct economic loss only, excluding indirect losses of revenue and building use. Moreover, costly damage to non-structural elements can occur in earthquakes of moderate intensities which would cause little or no structural damage.

In the last three decades, much useful work on architectural design and non-structural detailing has been carried out, such as that by Massey whose work is now incorporated in Charleson (2006), Arnold (2007) and Charleson (2008).

12.2 Non-structural Infill Panels and Partitions

12.2.1 Introduction

The recommendations of this section should be applied in conjunction with normal design considerations regarding creep, shrinkage and temperature effects which overlap, but are generally less exacting than the seismic design requirements for infill panels.

In earthquakes all buildings sway horizontally, producing differential movements of each floor relative to its neighbours. This is termed *inter-storey drift* (Figure 12.1), and is accompanied by vertical deformations which involve changes in the clear height h between floors and beams.

Any infill panel should be designed to deal with both these movements. This can be done by either (1) integrating the infill with the structure, or (2) separating the infill from the structure. A discussion of both systems of constructing infill panels follows, while further guidance on the seismic effectiveness of some types of partitions may be found in Rihal and Granneman (1984), while the need to avoid accidental formation of soft (weak) storeys in infilled walls is discussed by Dolsek and Fajfar (2001) and also in Section 8.3.8.

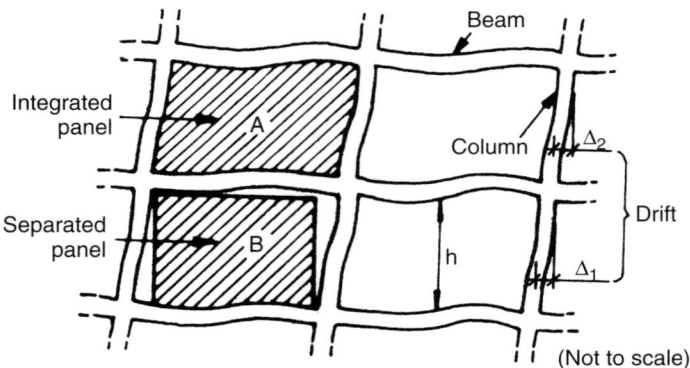

Figure 12.1 Diagrammatic elevation of structural frame and non-structural infill panels

12.2.2 Integrating infill panels with the structure

In this case, the panels will be in effective structural contact with the frame such that the frame and panels will have equal drift deformations (panel A in Figure 12.1). Such panels must be strong enough (or flexible enough) to absorb this deformation, and the forces and deformations should be computed properly. Where appreciably rigid materials are used the panels should be considered as *structural* elements in their own right, as discussed in Sections 5.4.6 and 10.4.5. Reinforcement of integrated rigid walls is usually necessary if seismic deformations are to be satisfactorily withstood.

Integration of infill and structure is most likely to be successful when very flexible partitions are combined with a very stiff structure (with many shear walls). Attention is drawn to the fact that partitions not located in the plane of a shear wall may be subjected to deformations substantially different from those of the shear wall. This is particularly true of upper-storey partitions.

Light partitions may be dealt with by detailing them to fail in controlled local areas, thus minimizing earthquake repairs to replaceable strips (Figure 12.2).

Finding suitable flexible construction for integral infill may not be easy, especially in beam and column frames of normal flexibility. These may experience an inter-storey drift of as much as 1% of the storey height in an earthquake.

12.2.3 Separating infill panels from the structure

For important structural reasons, this method of dealing with non-structural infill is likely to be preferable to integral construction when using flexible frames in strong earthquake regions (see Figure 12.1, panel B). The size of the gap between the infill panels and the structure is considerably greater than that required in non-seismic construction. In the absence of reliable computed structural movement, it is recommended that horizontal and vertical movements of between 20 mm and 40 mm should be allowed for. The appropriate amount will depend upon the stiffness of the structure, and the structural engineer's advice should be taken on this.

This type of construction has two inherent detailing problems which are not experienced to the same extent in non-seismic areas. First, awkward details may be required to ensure lateral stability of the elements against out-of-plane forces. Secondly, soundproofing and fireproofing of the separation gap is difficult. Moderate soundproofing of the movement

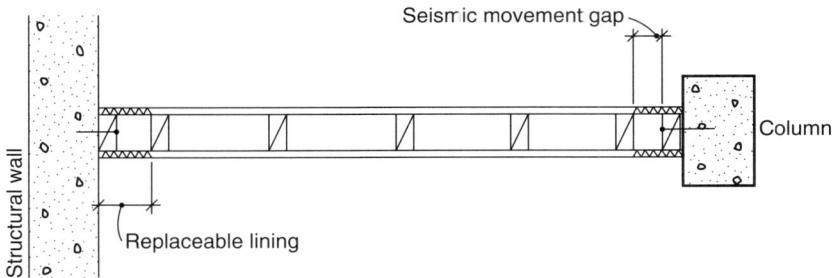

Figure 12.2 Lightweight partition detailed so that earthquake hammering by the structure will damage limited end strips only

gap can be achieved with cover plates or flexible sealants, but where stringent fireproofing and sound proofing requirements exist, the separation of infill panels from the structure is inappropriate. Designers should be careful in the choice of so-called 'flexible' materials in movement gaps; the material must be not only sufficiently soft but also permanently soft. Both polysulphide and foamed polyethylene are *not* flexible enough (or weak enough) in this situation.

It is in fact difficult to find a suitable material; Mono-Lasto-Meric is both permanently and sufficiently soft, but is not suitable for gap widths exceeding 20 mm. Foamed polyurethane is probably the best material from a flexibility point of view and will provide modest sound-insulation, but may have little fire resistance. A fire resistant possibility is Declon 156, a polyester/polyurethane foam which intumesces in fire conditions.

Figures 12.3–12.6 show some details used for separated infill panels. Note that great care has to be taken during both detailing and building to prevent the gaps being accidentally filled with mortar or plaster. Figure 12.6 shows a detail which helps prevent plaster bridging the gap. Further details suitable for small seismic movements may be found elsewhere (Arnold, 1984; Charleson, 2006).

12.2.4 Separating infill panels from intersecting services

Where ducts of any type penetrate a full-height partition, the ducts should not be tied to the partition for support. Support should occur on either side of the partition from the building structure above. If the opening is required to be sealed because of fire resistance or acoustics, the sealant should be of a resilient non-combustible type to permit motion of the duct without affecting the partition or duct. It is important for both seismic and acoustic considerations that the duct be independently supported by hangers and horizontal restraints from the building structure.

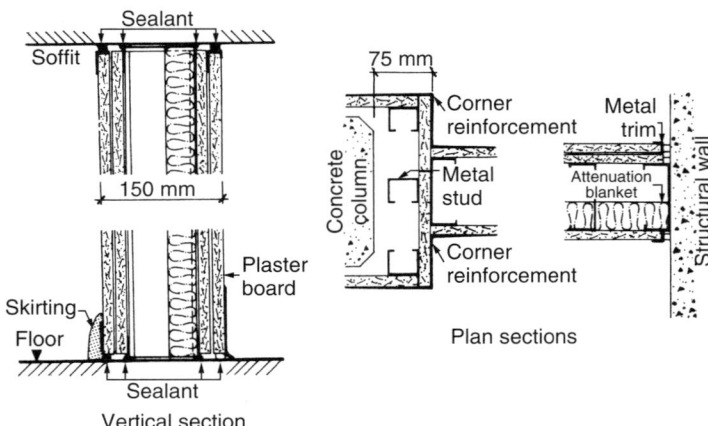

Figure 12.3 Light partition details for small seismic movements (i.e. suitable for stiff-framed buildings or small earthquakes)

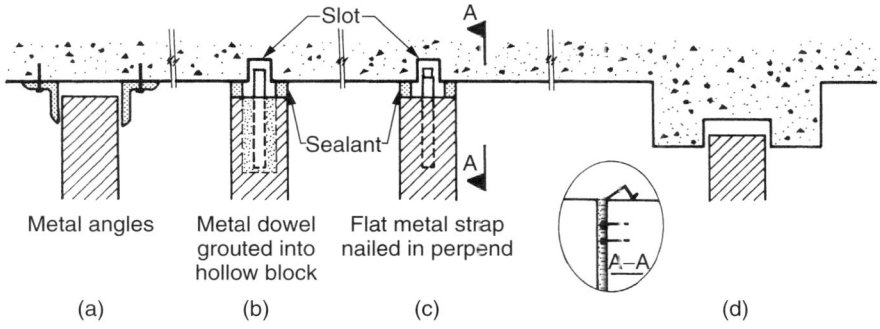

Figure 12.4 Separated stiff partitions: top details for lateral stability of brick or block walls (see Section 12.2.2)

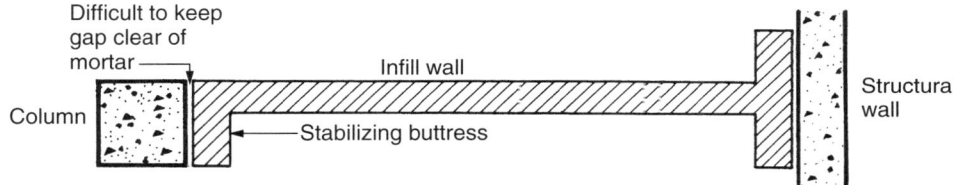

Figure 12.5 Separated stiff partition: plan view of stabilizing buttress systems

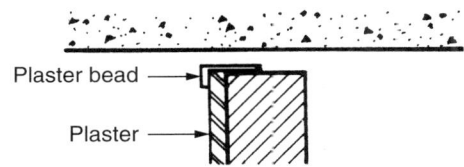

Figure 12.6 Plastering detail to ensure preservation of gap between partition and structure

Further discussion of ducts can be found in Section 11.3.6; for some remarks on the required properties of gap sealants around ducts, see the discussion on infill panels in Section 12.2.3.

12.3 Cladding, Wall Finishes, Windows and Doors

12.3.1 Introduction

The problems involved in providing earthquake proof details for these items are the same in principle as those for partitions, as discussed in the preceding section. Their in-plane stiffness renders them liable to damage during the horizontal drift of the building, and the techniques of integral or separated construction must again be logically applied.

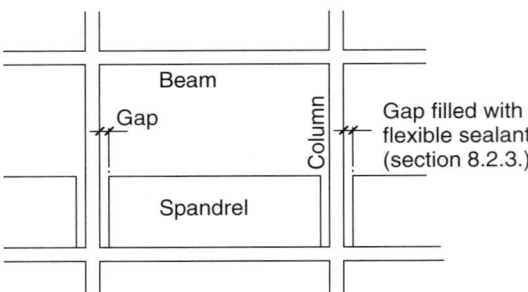

Figure 12.7 Detail of external frame showing separation of spandrel from columns to avoid unwanted interaction

12.3.2 Cladding and curtain walls

Precast concrete cladding is discussed in Section 10.3.12. Suffice it here to point out that in flexible buildings, non-structural precast concrete cladding should be mounted on specially designed fixings which ensure that it is fully separated from horizontal drift movements of the structure. Brick or other rigid cladding should be either fully integral and treated like infill walls (Section 5.4.6), or should be properly separated with details similar to those for rigid partitions (Figures 12.4 and 12.5) or for spandrels (Figure 12.7).

External curtain walling may well be best dealt with as fully framed prefabricated storey-height units mounted on specially designed fixings capable of dealing with seismic movements in a similar way to precast concrete cladding, as mentioned above (see also Section 12.3.5).

12.3.3 Weather seals

Weather seals that may be damaged in severe earthquakes should be accessible and suitable for replacement.

12.3.4 Wall finishes

Brittle or rigid finishes should be avoided or specially detailed on any walls subjected to shear deformations, i.e. drift as applied to panel A in Figure 12.1. This applies to materials such as stone facings or most plasters. In Japan it is recommended that stone facings should not be used on walls where the storey drift is likely to be more than 1/300.

Brittle veneers such as tiles, glass or stone should not be applied directly to the inside of stairwells, escalators or open wells. If they must be used, they should be mounted on separate stud walls or furrings. Preferably the stairwells should be free of material which may spall or fall off and thus clog the exit way or cause injury to persons using the area.

Heavy ornamentation such as marble veneers should be avoided in exit lobbies. If a veneer of this type must be used, it should be securely fastened to structural elements using appropriate structural fastenings to prevent the veneers from spalling off in the event of seismic disturbance.

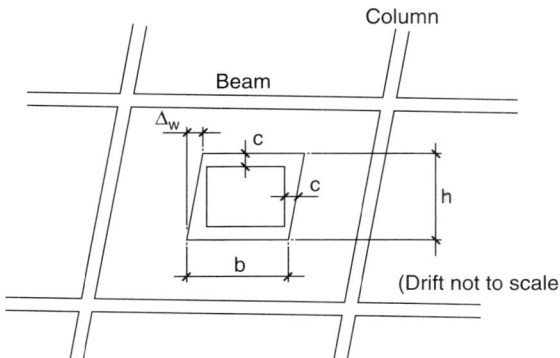

Figure 12.8 Detail of external frame with window glazing set in soft putty

Plaster on separated infill panels must be carefully detailed to prevent its bridging the gap between panel and structure (Figure 12.6) as this may defeat the purpose of the gap, resulting in damage to the plaster, the infill panel and the structure.

12.3.5 Windows and architectural glass panels

It is worth observing that in the San Fernando earthquake, which caused $500 million worth of damage, glass breakage cost more than any other single item.

Window sashes should be separated from frame action except where it can be shown that no glass breakage will result. If the drift is small, sufficient protection of the glass may be achieved by windows glazed in soft putty (Figure 12.8), where the minimum clearance c all round between glass and sash is such that

$$c > \frac{\Delta_w}{2[1 + (h/b)]}. \tag{12.1}$$

The failure mode of hard putty glazed windows tends to be of the explosive buckling type, and should be used only where sashes are fully separated from the structure, for example when glass is in a panel or frame which is mounted on rockers or rollers as described in Section 10.3.12. Further discussion of window behaviour in earthquakes may be found elsewhere (Osawa *et al.*, 1965).

Regarding curtain walls, dynamic racking tests have shown (Memari *et al.*, 2006) that glass cracking and fallout drift resistance of architectural glass panels used in curtain walls are significantly improved by modified corner geometries and edge finish conditions, and provisions for these have been incorporated in various American standards, such as the 2003 International Building Code.

12.3.6 Doors

Doors which are vital means of egress, particularly main doors of highly populated and emergency service buildings, should be specially designed to remain functional after a strong earthquake. For doors on rollers, the problem may not be simply a geometric one dealing with the frame drift Δ, but may also involve the dynamic behaviour of the door itself.

12.4 Miscellaneous Architectural Details

12.4.1 Exit requirements

Every consideration should be given to keeping the exit ways clear of obstructions or debris in the event of an earthquake. As well as the requirements for wall finishes and doors outlined in Sections 12.3.4 and 12.3.6, the following points should be considered.

Floor covers for seismic joints in corridors should be designed to take three-dimensional movements, i.e. lateral, vertical and longitudinal. Special attention should be given to the lateral movement of the joints.

Free-standing showcases or glass lay-in shelves should not be placed in public areas, especially near exit doors. Displays in wall-mounted or recessed showcases should be tied down so that they cannot come loose and break the glass front during an earthquake. Where this is impracticable, tempered or laminated safety glass should be used for greater strength.

Pendant-mounted light fixtures should not be used in exit ways. Recessed or surface-mounted independently supported lights are preferred.

12.4.2 Suspended ceilings

In seismic conditions ceilings become potentially lethal. Individual tiles or lumps of plaster may jar loose from the supports and fall. Ceiling-supported light fixtures may loosen and drop out, endangering persons below. Thus, alternatives to the standard ceiling construction procedures should be considered. A thorough review of the seismic hazard from suspended ceilings, along with detailing recommendations, has been given by Clarke and Glogau (1979), while studies on dynamic response behaviour have been made by Rihal and Granneman (1984).

The horizontal components of seismic forces to which a ceiling may be subjected can be allowed for in several ways. A dimensional allowance should be made at the ceiling perimeter for this motion so as to minimize damage to the ceiling where it abuts the walls: one way of doing this is to provide a gap and a sliding cover (Figure 12.9). Some ceiling suspension systems need additional horizontal restraints at columns and other structural elements, such as diagonal braces to the floor above, in order to minimize ceiling motion in relation to the structural frame. This will reduce hammering damage to the ceiling, and tiles will be less likely to fall out. The suspension system for the ceiling should also minimize vertical motion in relation to the structure.

Lighting fixtures which are dependent upon the ceiling system for support should be securely tied to the ceiling grid members. If such support is likely to be inadequate in earthquakes, the light fixtures should be supported independently of the building structure above. Diffuser grilles, if required for the air supply system, should also be hung independently.

In seismic areas, a lay-in T-bar system for ceiling construction should be avoided if at all possible, as its tiles and lighting fixtures drop out in earthquakes. In both the 1964 Alaska and the 1971 San Fernando earthquakes, the economical (and therefore

Miscellaneous Architectural Details 483

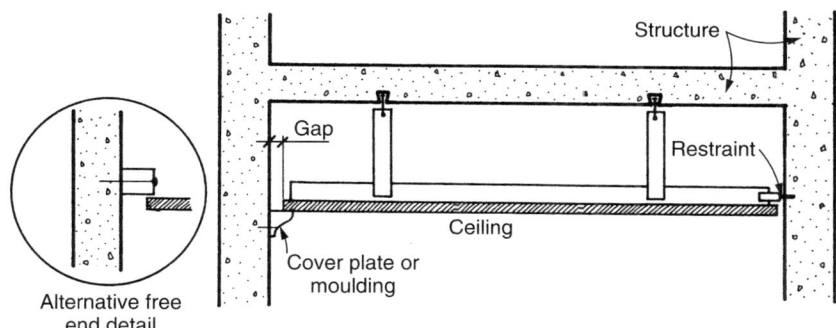

Figure 12.9 Details of periphery of suspended ceilings to prevent hammering and excessive movement

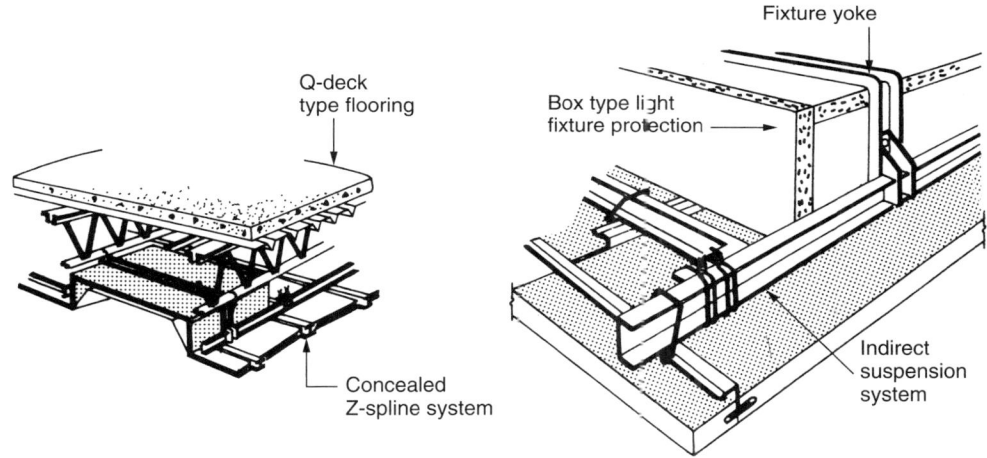

Figure 12.10 Two details of suspended ceiling construction providing movement restraint and secure tile fixing (after Berry, 1972)

popular) exposed tee grid suspended ceilings suffered the greatest damage. Evidently the differential movement between the partitions and the suspended ceilings damaged the suspension systems, and as the earthquake progressed the ceilings started to sway and were battered against the surrounding walls. This damage was aggravated when the ceilings supported lighting fixtures, and in many instances the suspension systems were so badly damaged that the lighting fixtures fell.

Damage to ceilings can also occur where sprinkler heads project below the ceiling tiles. One way of minimizing this problem is to mount the heads with a swivel joint connection so that the pipe may move with the ceiling. Figures 12.9 and 12.10 give suggestions for seismic detailing of suspended ceilings.

References

Arnold C (1984) *Non-structural Issues of Seismic Design and Construction*. Earthquake Engineering Research Institute, Oakland, CA.

Arnold C (1991) The seismic response of non-structural elements in buildings. *Bull NZ Nat Soc Earthq Eng* **24**(4): 306–316.

Arnold C (ed.) (2007) *Designing for Earthquakes: A Manual for Architects*. Earthquake Engineering Research Institute, Oakland, California. Available as a book or online from http://www.fema.gov/library/viewRecord.do?id=2418.

Berry DR (1972) Architectural seismic detailing. State of the Art Report No 3, Technical Committee No 12, Architectural-Structural Interaction. *IABSE-ASCE Int Conf on Planning and Design of Tall Buildings*, Lehigh University.

Charleson A (2006) *Architectural Design for Earthquakes – A Guide to the Design of Non-structural Elements*, 2nd edn. New Zealand Society for Earthquake Engineering. Free download available from http://www.nzsee.org.nz.

Charleson AW (2008) *Seismic Design for Architects: Outwitting the Quake*. Elsevier, Oxford.

Clarke WD and Glogau OA (1979) Suspended ceilings: the seismic hazard and damage problem and some practical solutions. *Bull NZ Nat Soc Earthq Eng* **12**(4): 292–304.

Dolsek M and Fajfar P (2001) Soft storey effects in uniformly infilled reinforced concrete frames. *J Earthq Eng* **5**(1): 1–12.

Memari AL, Kremer PA and Behr RA (2006) Architectural glass panels with rounded corners to mitigate earthquake damage. *Earthq Spectra* **22**(1): 129–150.

Osawa Y, Morishita T and Murakami M (1965) On the damage to window glass in reinforced concrete buildings during the earthquake of April 20, 1965. *Bull Earthq Res Institute, Univ. Tokyo* **43**: 819–827.

Rihal SS and Granneman G (1984) Experimental investigation of dynamic behaviour of building partitions and suspended ceilings during earthquakes. *Proc. 8th World Conf on Earthq Eng, San Francisco* **V**: 1135–1140.

Shipp JG and Johnson MW (1990) Seismic loss estimation for non-structural components in high-rise buildings. *Proc. 4th US Nat Conf on Earthq Eng*. EERI, Oakland, CA.

13

Retrofitting

13.1 Introduction

The previous parts of this book discuss the determination of earthquake hazard and how to improve measures for mitigating earthquake risk for new items of the built environment. Standards of design and construction for earthquake resistance are slowly improving, more so in wealthier parts of the world than elsewhere. As most of what we build lasts much longer than any given construction standard, this obviously means that at the time of any given earthquake most of the built environment is not as earthquake resistant as the latest standards could make them. Hence, there is a huge need to improve the existing built environment by replacing or making safer (i.e. retrofitting) parts of the existing built environment.

Calls for retrofitting have been made after successive earthquakes for many decades, indeed probably for centuries. The earliest cases of retrofitting actually being carried out known to the author are some brick buildings which were strengthened in the 1930s in New Zealand (Dowrick and Rhoades, 2002) after the 1934 $M_W = 7.4$ Pahiatua (renamed Waione in 2008) earthquake. However, there are no doubt earlier examples in other countries, e.g. in California after the 1906 San Francisco earthquake, or in other highly seismic regions such as Chile and Japan. Considering that formal attempts at earthquake resistant design started in about the 1920s, it is salutary to look at the historical incidence of deaths and damage costs worldwide in the twentieth century. Earthquake deaths worldwide, plotted in Figure 1.2, show no tendency for risk reduction over time. The casualties are dominated by those from earthquakes in developing (i.e. poor) countries and places where unreinforced masonry (URM) construction is the norm. The huge totals in the 1920s and 1970s result from three earthquakes in the world's most populous country China.

Next consider the damage costs plotted in Figure 1.3. Again, there is no tendency for global risk to reduce, rather it appears to be increasing with time. It is important to note that (ironically) the two worst years for monetary loss, 1994 and 1995, were caused by damage in two of the richest and most technically advanced countries in the world, in the Northridge (California) and the Hyogoken-Nanbu (Kobe, Japan) earthquakes. It is no wonder that there is great concern worldwide about the continuing growth of megacities in places of high seismic hazard, and often with low building standards.

Because of the vast variety of existing structures, the development of general rules of real use is difficult and to a large extent each structure must be approached as a strengthening problem on its own merits. Some of the factors which need consideration are as follows:

(1) the form of the structure and non-structure, and the need for change (e.g. to create symmetry);
(2) the materials used in the existing construction;
(3) the permissible visual and functional effect of the strengthening;
(4) the desired further design life;
(5) the desired seismic resistance.
(6) the acceptable damage to the existing fabric in the design event;
(7) the parts requiring strengthening and the problems of access thereto (e.g. piles);
(8) the degree to which ductile failure modes are required (significant ductility is not reliably achievable at reasonable cost in many older constructions, particularly of masonry, or may imply heavy damage to the existing fabric);
(9) the extent to which other components are to be upgraded as well as the strength (e.g. architectural features and building services);
(10) continuance of normal function during the strengthening works;
(11) costs.

Depending upon the above factors, significant seismic resistance can be obtained for most structures for only a small fraction (5–30%) of their replacement cost, while the long-term upgrading of historical buildings or monuments may exceed the cost of their replacement (where that is meaningful).

13.2 To Retrofit or Not?

Looking at Figures 1.2 and 1.3, it is evident that, in order to reduce earthquake risk worldwide, it will be necessary to speed up the rate at which retrofitting is being carried out, as we cannot rely on the renewal rate to win this race. Unfortunately, the incentives for voluntary retrofitting are not very compelling. The cost of retrofitting is balanced against its benefits, i.e. reduction in loss of life, reduced cost of material damage, reduced business interruptions, and possibly reduced earthquake insurance. In real terms the chances of substantial earthquake damage occurring to any given structure may not appear to be that great. For example, there is only a 2% chance of an earthquake with a 500-year return period occurring in the next 10 years (one-fifth of the design life of many structures). The issue is whether a 2% chance is acceptable or not.

However, the total loss in a 500-year event in a given city could be catastrophic for that city, as it was for Kobe. By way of example, consider the case of Wellington, New Zealand, the 500-year earthquake scenario for which is shown in Figure 7.1. The expected casualties for this event comprise hundreds of deaths and injuries (Table 7.3). In addition, the financial costs amount to billions of dollars. In studies carried out by the present author, it was found that the cost of damage to non-domestic buildings located within

Table 13.1 Estimated material damage losses due to shaking and fire for non-domestic buildings of three vulnerabilities in a Wellington fault earthquake, New Zealand (Figure 7.1)

Intensity zone	Replacement value ($m)	Material damage cost ($m)		
		Case (a) All bldgs URM	Case (b) Actual bldgs	Case (c) Lower bound
MM7	1700	50	12	1.4
MM8	400	76	16	1.6
MM9	1500	824	190	22.5
MM10	8400	8200	1752	294.0
Total	12000	9150	1970	319.5
	100%	75%	16%	2.6%

the intensity MM7 isoseismal of Figure 7.1 is likely to amount to about 16% of the total replacement value of those buildings.

The additional cost of business losses could well be two or three times the material damage cost. Thus, the total loss arising from this event may be equal to about half the replacement value of the built environment within the MM7 isoseismal. This large total arises despite the fact that relatively few of the buildings are highly vulnerable, i.e. of URM. A summary of the estimated damage costs are given in Table 13.1, for three cases: (a) assuming all buildings to be of URM; (b) the existing non-domestic buildings; and (c) assuming all buildings transformed to have lower-bound vulnerability (Figure 6.23). In case (a) the total direct losses of NZ$9.2 billion are about three-quarters of the total replacement value, excluding any costs of strengthening.

It follows from the above losses that there is a strong case for some degree of compulsion for retrofitting to considerably reduce vulnerability to the best standard possible, whenever intensity MM10 is reasonably likely to occur (e.g. once in 500 years). As resources will always be limited for retrofitting, a strategy of assigning priorities to what should be retrofitted first is likely to be adopted, e.g.:

(1) post-earthquake emergency facilities;
(2) lifelines;
(3) URM buildings;
(4) buildings which are cheap to retrofit;
(5) vulnerable buildings containing many people;
(6) cultural heritage property;
(7) other property.

In addition to the obvious cases of URM and pre-code construction, structures that may warrant retrofitting are those of the early earthquake resistant design eras which are made of what is now deemed *brittle* construction in concrete or steel, or those having undesirable structural features such as soft (weak) storeys or large eccentricities. The following list of principal weaknesses in reinforced concrete buildings is given by Coburn and Spence (2002):

(a) Insufficient lateral load resistance, as a result of designing for too small a lateral load.

(b) Inadequate ductility caused by insufficient confinement of longitudinal reinforcement, especially at beam column or slab column junctions.
(c) A tendency to local overstressing due to complex and irregular geometry in plan and elevation.
(d) Interaction between structure and non-structural walls resulting in unintended torsional forces and stress concentrations.
(e) Weak ground floor due to lack of shear walls.
(f) High flexibility combined with insufficient spacing between buildings, resulting in risk of neighbouring structures pounding each other during shaking.
(g) Poor-quality materials or work in the construction.

In different countries, local construction styles may give rise to other common weaknesses, such as changes in column construction at 'mid' height up multi-storey buildings in Kobe (Park et al., 1995). Unrepaired damage from previous earthquakes has also been found to be a significant cause of failure in some areas (Aguilar et al., 1989).

In endeavouring to find socially acceptable non-draconian criteria for retrofitting, one approach is to require retrofitting of only the most dangerous steel or concrete structures. For example, the New Zealand Society for Earthquake Engineering (NZSEE, 2006) recommends that the expected performance level should be set at as nearly as is reasonably practicable to New Building Standard [NBS]. Thus the initial target level for improvement should be 100%NBS. In many cases this will not be practicable and it will be necessary to establish a reasoned reduction in acceptable level. In any event NZSEE recommends that 67%NBS be regarded as a minimum to be achieved in the structural improvement measures notwithstanding the legal minimum is possibly only 34%.

NZSEE (2006) sets out a seismic safety evaluation process for existing buildings, a flowchart for which is shown in Figure 13.1. This process is driven by the territorial authority (TA), e.g. city or district council, which has responsibility for local construction standards. The process comprises two phases, an initial TA-funded appraisal followed if necessary by a detailed assessment paid for by the building's owner. Obviously, the initial appraisal of the complete pre-1976 building stock is a large task (even if restricted to the most vulnerable era, i.e. pre-1976 in New Zealand), and to make this task affordable, the initial appraisal of each building is such that it would take not more than four hours of the time of an experienced earthquake engineer.

One of the steps in the initial evaluation process is to evaluate the performance achievement ratio (PAR), taking six factors A–F into account, as set out in Table 13.2. Four of these factors are the critical structural weaknesses illustrated in Figure 13.2, of which the first three are structural form considerations discussed above in Section 8.3. If a building fails to get 34 points (out of 100) in the initial evaluation, a detailed assessment is required. If a building obtains 34–67 points a more detailed evaluation may be recommended.

For reinforced concrete buildings professional engineering opinion is divided over whether a force-based or a displacement-based procedure is the more satisfactory. For example, Park (1997) has proposed a force-based procedure (limited to reinforced concrete moment resisting frames), while Priestley et al. (2007) have proposed a displacement-based procedure. In the USA, a considerable commitment has been made to the force-based approach by the Applied Technology Council (1996) (see also Comartin et al., 2000) in its two-volume document ATC-40, which deals with reinforced concrete

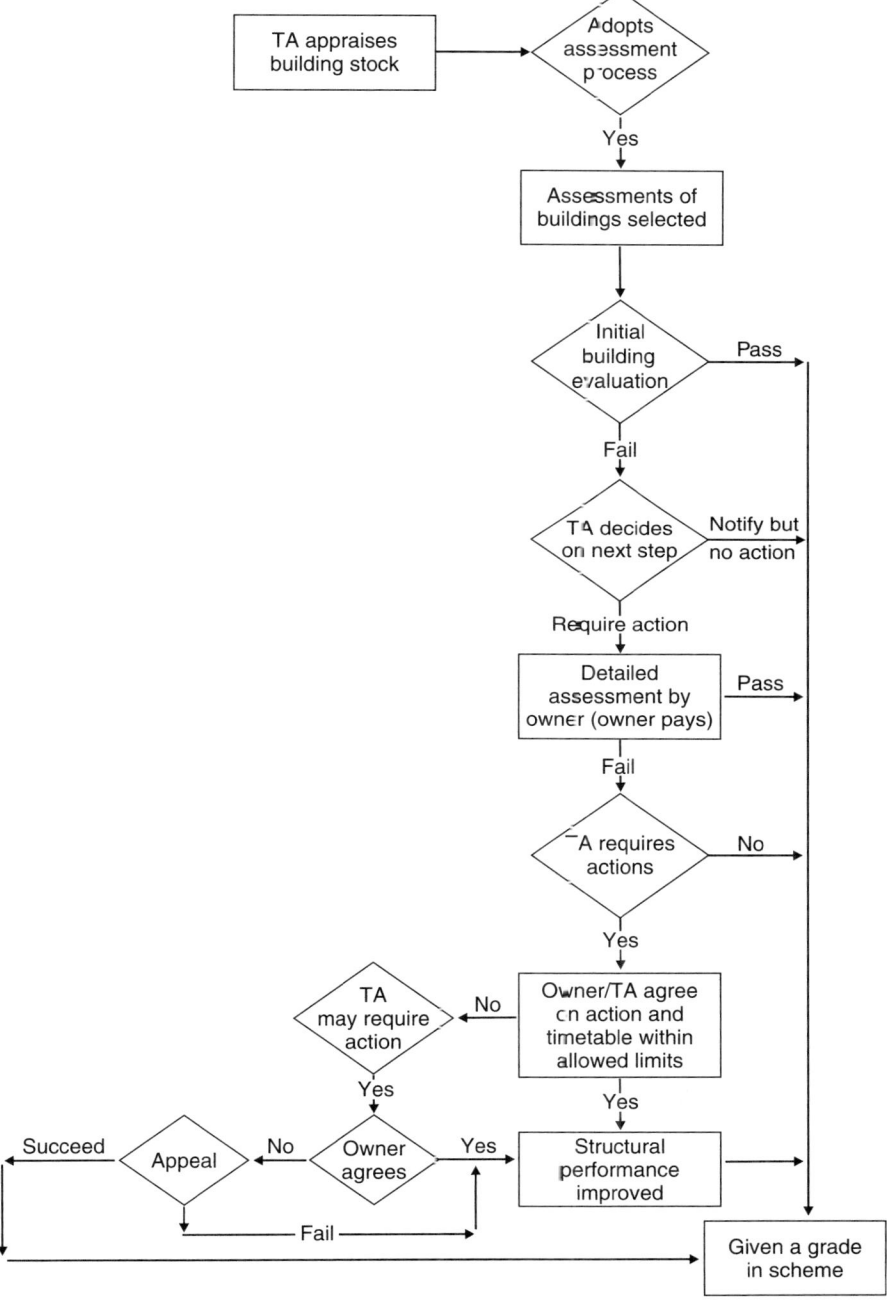

Figure 13.1 Flowchart for identifying, assessing and retrofitting of buildings, as proposed for New Zealand. TA = Territorial Authority (NZSEE, 2006)

Table 13.2 Step 3 of initial evaluation procedure of seismic safety of existing building (NZSEE, 2006): assessment of performance achievement ratio

Critical structural weakness			Effect on structural performance *(choose a value—do not interpolate)*		
3.1 Plan Irregularity					
Effect on Structural Performance			Severe	Significant	Insignificant
	Factor	A	0.4 max	0.7	1
Comment					
3.2 Vertical Irregularity					
Effect on Structural Performance			Severe	Significant	Insignificant
	Factor	B	0.4 max	0.7	1
Comment					
3.3 Short Columns					
Effect on Structural Performance			Severe	Significant	Insignificant
	Factor	C	0.4 max	0.7	1
Comment					

3.4 Pounding Potential
(Estimate D1 and D2 and set D = the lower of the two, or = 1.0 if no potential for pounding)
a) Factor D1: Pounding Effect
   Select appropriate value from Table

> **Note:**
> *Values given assume the building has a frame structure. For stiff buildings (e.g. with shear walls), the effect of pounding may be reduced by taking the coefficient to the right of the value applicable to frame buildings*

Factor D1 =

Table for Selection of Factory D1	Severe	Significant	Insignificant
Separation	0<Sep<.005H	.005<Sep<.01H	Sep>.01H
Alignment of Floors within 20% of Storey Height	0.7	0.8	1
Alignment of Floors not within 20% of Storey Height	0.4	0.7	0.8

b) Factor SD2: Height Difference Effect
   Select appropriate value from Table
   Factor D2 =

Table for Selection of Factor D2	Severe	Significant	Insignificant
	0<Sep<.005H	.005<Sep<.01H	Sep>.01H
Height Difference >4 Storeys	0.4	0.7	1
Height difference 2 to 4 Storeys	0.7	0.9	1
Height Difference <2 Storeys	1	1	1

	Factor	D	*(Set D = lesser of D1 and D2 or… set D = 1.0 if no prospect of pounding)*

3.5 Site Characteristics *(Stability, landslide threat, liquefactions etc.)*

Effect on Structural Performance			Severe	Significant	Insignificant
	Factor	E	0.5 max	0.7	1
3.6 Other Factors					
	Factor	F	For ≤3 storeys, maximum value 2.5 Otherwise, maximum value 1.5. No minimum		

3.7 Performance Achievement Ratio (PAR)
   (equals A × B × C × D × E × F)

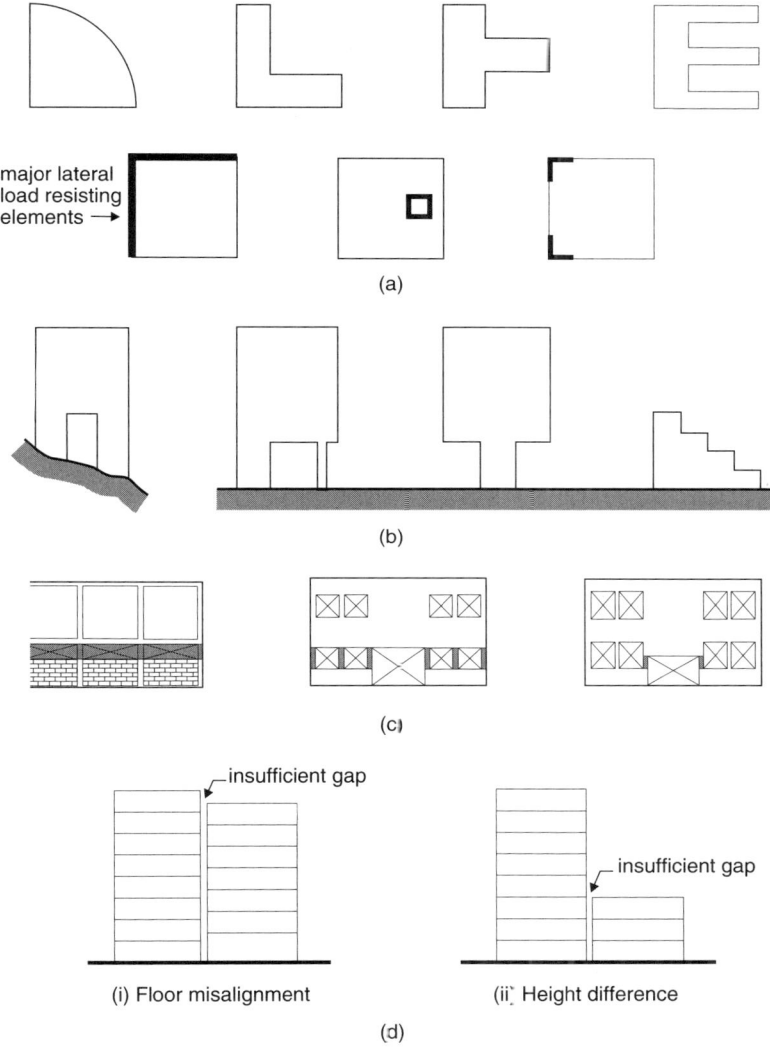

Figure 13.2 Four critical structural weaknesses used in evaluating the PAR of buildings (Table 13.2) in the assessment of the need to retrofit (from NZSEE, 2006)

structures. Ironically, steel-framed buildings in Los Angeles were among the most damaged in the 1994 Northridge earthquake. The damage arose mainly from brittle weld details, which required very expensive remedial work for both repair and upgrading (Bertero et al., 1994; Krawinkler, 1996). The displacement-based approach appears to be the better of the two methods, because displacements more directly reflect damage than does force-based analysis. NZSEE (2006) has resolved this difference in view of force-based and displacement-based approach by including them both in the detailed evaluation of reinforced concrete buildings.

Table 13.3 Statistics of damage states of all pre-1976 (i.e. brittle) reinforced concrete buildings subject to strong shaking in New Zealand earthquakes

Event date	M_W	MMI	PGA (g)	OK			Cracks			Collapse		
				\multicolumn{9}{c}{No. of bldgs per no. of storeys}				
				1	2	3+	1	2	3+	1	2	3+
1922 Dec 25*	6.4	8	0.3	1								
1929 Jun 16*	7.8	9	0.6	3			1	1				
		8	0.4–0.15	12	11	1	2	3	1			
1931 Feb 2*	7.8	10	0.8–0.5	9	27	9	12	18	4			1
		9	0.5	2	6							
		8	0.3	13	8	3	2	1				
1932 Sept 15*	6.8	8	0.5–0.3	9	3	3	2	2				
1934 March 5*	7.3	8	0.4				1					
1942 Jun 24*	7.1	8	0.5–0.4	56	16	2	5	8	2			
1946 Jun 26	6.4	7+	0.3				1					
1948 May 22	6.4	8	0.35	1	2							
		7?	0.25	2	2							
1968 May 23	7.2	9	0.6	15								
(prelim data)		8	0.5–0.2	26	2		20					
1987 Mar 2	6.5	9	0.4	86	5		44	16	2			
		8	0.3	6	6	2						
1994 Jun 18	6.7	9	0.5	1	1							
		8	0.4		1							
Totals				242	90	22	88	49	9			1

Notes: Tallest buildings 5 storeys.
 *All buildings pre-code in these six earthquakes.

Any assessment method should of course adequately reflect the inherent strength of structural walls. For example, as reported by Dowrick and Rhoades (2000), and quantified in Table 13.3, all 11 New Zealand earthquakes which have shaken brittle concrete buildings strongly (intensities MM8–MM10, and PGAs 0.2–0.8g), have shown that the 500 such buildings of one to three storeys have been collapse-free, except that one soft-storey building collapsed. The rest had walls of concrete or brick infill and many were asymmetric, including 'corner' buildings. Of the 500 buildings in Table 13.3, about 300 were of pre-code construction.

Other documents which give guidelines on the assessment of existing buildings have been produced by the American Society of Civil Engineers (ASCE, 2003), the Federal Emergency Management Agency (FEMA, 2000) and the state-of-the-art paper by Holmes (2000). Guidance for estimating costs of retrofitting in the USA has been given by FEMA (1994).

13.3 Benefit-Cost of Retrofitting

Benefit-cost studies are sometimes carried out to help decide whether to retrofit a given property or not. For a global view of this approach, consider the buildings involved in the Wellington fault earthquake, as given in Table 13.1. Let us assume (plausibly) that to retrofit all of the buildings from their existing vulnerability, case (b), to that of the lower bound, case (c), would cost on average 20% of their present replacement value. From Table 13.1, it can be seen that the reduction in material damage cost in the 500-year earthquake would be $(16 - 2.6) = 13.4\%$ of the replacement value. Next, if we assume that the business interruption losses are twice the material damage costs, then the total reduction in monetary losses would be $3 \times 13.4 = 40.2\%$ of the replacement value. Thus, considered only on the above basis, the cost of retrofitting (20% of replacement value) would be about half the monetary losses, so that retrofitting would more than pay for itself.

However, the full picture is much more complicated than in the simple argument given above, as account needs to be taken of all the costs and all savings (including casualties) over a long period of time. Also, the benefit-cost considerations for individual properties will vary widely from the average figures used above. For example, after the 1942 Wairarapa earthquakes in New Zealand, many owners of commercial URM buildings complained (with justification) that the income received from their buildings was insufficient to fund retrofitting.

As part of the Turkish Marmara Earthquake Emergency Reconstruction project, a substantial benefit-cost study was made of retrofitting 369 apartment buildings in Istanbul. To quote Hopkins *et al.* (2006):

> The study has provided valuable insights into the benefits and costs and relative merits of retrofitting. This applies to individual buildings used to the study and to many similar buildings in other parts of Istanbul. As such it should provide a valuable resource for owners, municipalities and the Turkish government in making decisions on retrofitting to reduce earthquake risk. The simple approach of using a scenario earthquake at different time in the future gives easy-to-understand results. The specially developed tabulations and graphs in the Individual Building Cost Reports provide a ready means of deciding on the best option.
>
> The analyses successfully identified buildings for which the replacement with a new building was the best option and gave some supporting data for this conclusion. More importantly, perhaps, the analysis and presentation of results highlights the critical need to retrofit or replace rather than do nothing.
>
> Even though challenges remain in order for retrofitting to be implanted by the private owners, including availability of funds and legal/planning constraints, the benefit-cost analysis showed that retrofitting was a good investment for most buildings.

13.4 Retrofitting Lifelines

The term 'lifelines' is used to cover all public utilities, i.e.:

- water supply;
- gas supply;

- electricity supply;
- sewerage systems;
- stormwater drainage;
- telecommunications;
- transportation (road, rail, sea, air);
- building services.

As most of the above utilities are networks covering whole regions, they are vulnerable to breakdown from earthquake actions at many points, often involving pipelines, cables or structures buried in poor ground. Since about 1990, efforts have been under way in various countries to improve the reliability of and minimize the damage to a range of lifelines by integrated multidisciplinary study groups. For example, the first such initiative in New Zealand was for the Wellington region, where an ongoing study group has issued a series of reports as first described by Hopkins *et al.* (1993). Lifelines have quickly become an important feature of earthquake engineering conferences, as typified by the state-of-the-art paper by Kameda (2000) at the 12th World Conference on Earthquake Engineering.

In the Wellington study the first stage involved identifying mitigation measures in outline. The main initial value of the project was to raise the awareness of all service providers and to highlight those parts of their services that were most at risk from seismic hazards. Most of the utility companies resolved to carry out more detailed reviews of their own installations and systems.

The initial list of recommended mitigation measures was extensive, with 18 of the more important noted by Hopkins *et al.* as listed below:

- Provide standby power plant at pumping stations as a matter of urgency.
- Installation of tie bolt couplings on the water main where it crosses the Wellington Fault.
- Installation of isolation valves.
- Review of alternative sources of water.
- Decommissioning of Karori's reservoir.
- Pinpointing of likely trouble spots in gas lines. Isolation and ready response is planned.
- Bracing of zone transformers and older models of switchgear.
- Review required spare parts and cables to ensure that reinstatement of electrical supply is not unnecessarily delayed.
- Pay closer attention to seismic issues for extensions and new equipment.
- Increase awareness and effectiveness of mitigation measures through staff training.
- Review key telephone exchanges for seismic integrity.
- Investigate more closely the usability of the Port of Wellington and the airport following an earthquake.
- Increase redundancy as roading system is developed.
- Review performance of important bridges.
- Make a closer and more detailed analysis of the Thorndon region and the motorway overbridges. Take steps to improve this vital area.
- Increase redundancy in the Cook Strait ferry service.
- Investigate the use of shut-off valves in buildings for gas.

- Encourage business to review the earthquake integrity of their operations, equipment and stock.

An important aspect of the study was to identify all the interdependencies of various lifelines for the first week after the earthquake. For example, railways, one of the least dependent services, are dependent to some extent on radio, roading, fuel supply and equipment, as set out in Table 13.4. To quote from Hopkins *et al.* (1993):

> Adding across the table gives a measure of importance, adding down the table gives a measure of dependency. The overall priority for attention can be gauged from the interdependence quotient.
> The analysis highlighted the fundamental importance of roading, equipment (of all kinds), standby power, fuel supply and telecommunications, together with the high dependency of building services, air transport and broadcasting on other lifelines.

Table 13.4 Interdependence of lifelines in the first week after a strong earthquake (from Hopkins *et al.*, 1993)

THESE ARE DEPENDENT ⇐ ON THESE ⇓	Water supply	Gas supply	Sanitary drainage	Storm drainage	Mains electricity	Standby electricity	VHF radio	Telephone systems	Roading	Railways	Sea transport	Air transport	Broadcasting	Fuel supply	Firefighting	Building services	Total importance
Water supply		•	1	•	•	•	•	•	•	•	•	•	•	•	3	3	7
Gas supply	•		•	•	•	•	•	•	•	•	•	•	•	•	•	2	2
Sanitary drainage	1	•		•	•	•	•	•	•	•	•	•	•	•	•	3	4
Storm drainage	•	•	2		•	•	•	•	•	•	•	•	•	•	•	1	3
Mains electricity	2	1	2	2		•	3	3	•	2	•	3	1	•	•	2	21
Standby electricity	3	1	2	2	•		3	3	•	•	•	3	2	2	•	3	24
VHF radio	3	3	3	2	3	•		3	2	2	2	2	2	•	3	•	30
Telephone systems	2	1	1	•	1	1	•		•	•	•	1	3	1	2	1	14
Roading	2	2	2	2	3	2	2	2		2	3	3	2	2	3	1	33
Railways	•	•	•	•	•	•	•	•	•		1	•	•	•	•	•	1
Sea transport	•	•	•	•	•	•	•	•	•	•		•	•	1	•	•	1
Air transport	1	1	•	•	1	•	•	•	•	•	•		1	•	•	•	4
Broadcasting	2	2	2	•	•	•	•	1	1	•	•	•		•	1	•	9
Fuel supply	3	1	1	1	•	3	1	1	3	2	•	1	1		3	1	22
Firefighting	•	2	•	•	•	•	•	1	•	•	•	2	•	1		1	7
Building services	•	•	•	•	•	•	•	2	•	•	•	1	1	•	2		6
Equipment	3	3	3	2	3	3	2	3	3	3	3	3	3	2	2	3	44
TOTAL DEPENDENCE	22	17	19	11	11	9	11	19	9	11	9	19	16	9	19	21	
PRIORITY FACTOR	29	19	23	14	32	33	41	33	42	12	10	23	25	31	26	27	

Note: 3 = High dependence, 2 = Moderate dependence, 1 = Low dependence, • = No dependence
Priority factor = Importance + Dependency

The methodologies developed to derive estimates of importance, dependency and rate of recovery can be applied to lifelines in other cities.

The analysis of interdependence [was] a new concept and the work done on the project is believed to be the first of its kind. It proved to be a most important technique for developing an understanding of the likely effects of earthquake. Many service providers were able to analyse their own vulnerability and assess consequences, but when the implications of the effects on others, on which they depended, were analysed and discussed, new and important factors came to light. For example, the dependency of radio and telecommunications remote sites on access roads to supply diesel to standby power generators. As a result, the need to increase the capacity of holding tanks was identified.

In evaluating the vulnerability of lifelines the concept of damage ratios, as discussed for buildings and equipment in Chapter 6, is valuable. In the case of pipelines, damage ratios may be estimated from the expected numbers of fractures per kilometre in a given class of ground. For lifelines, more reliable damage ratios than are currently available would improve the quality of decision-making in relation to upgrading or rerouting. In the case of the latter option, the choice of lower-risk routes for linear systems such as water supply, roads and railways can obviate the need to make special provisions for bad ground or fault lines.

Where existing lifelines such as pipelines or electricity power cables cross active faults, loops can sometimes easily be provided to accommodate several metres of shear displacement of the fault rupture. Such loops are similar in principle to arrangements for pipes crossing seismic movement gaps in buildings (Figures 11.13 and 11.14).

13.5 Retrofitting Structures

13.5.1 Strategies for improving structural performance

This section is a brief summary of a section of NZSEE (2006). Improving the structural performance of structures in earthquakes may be achieved by adopting one or more of the strategies given below.

Local modification of components

While some structures have substantial strength and stiffness, some of the components may be understrength or have inadequate deformation capacity. A strategy for this type of structure could be restricted largely to local improvements to those components (Figure 13.3).

Removal or lessening of irregularities and discontinuities

Stiffness, mass and strength irregularities are common causes of inadequate seismic performance of structures. Checking seismic displacements and forces often identifies high concentrations of forces at one level of a structure. Sometimes structural performance can be improved by deliberately weakening some elements.

Global structural strengthening and stiffening

Some flexible structures will have poor seismic performance because critical components or elements do not have adequate ductility to resist the large seismic deformations

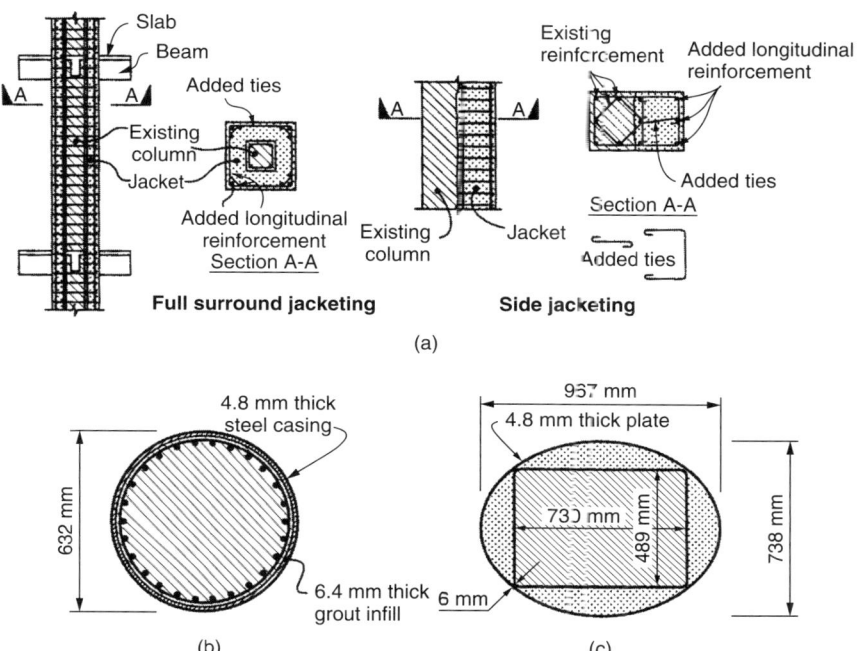

Figure 13.3 Three methods for retrofitting reinforced concrete columns: (a) reinforced concrete jackets; (b) grouted site-welded circular thin steel jacket; (c) site-welded elliptical thin steel jacket with concrete infill (from Park, 2001)

usually associated with that type of structure. For structures with many such elements a cost-effective way to improve performance is to stiffen the structure so as to reduce the ductility demand on those critical components. Construction of new braced frames, moment resisting frames or shear walls within an existing structure are effective methods for adding both additional stiffness and strength (see text related to Figure 13.7).

Seismic isolation and supplementary energy dissipation

Alternatives to strengthening a weak structure are to substantially isolate it from damaging ground motions, or to add energy dissipation devices (see Figures 13.4 and 13.6 and Section 8.5).

Removal of unnecessary seismic mass

Many older-style existing building have heavy non-structural components. Removal of some of these to reduce the seismic mass can often be beneficial.

Widening seismic joints or linking structures together across seismic joints

Seismic joints in existing structures are often too narrow or ineffective, causing damage from effects such as pounding or unexpected torsional responses. Substantial

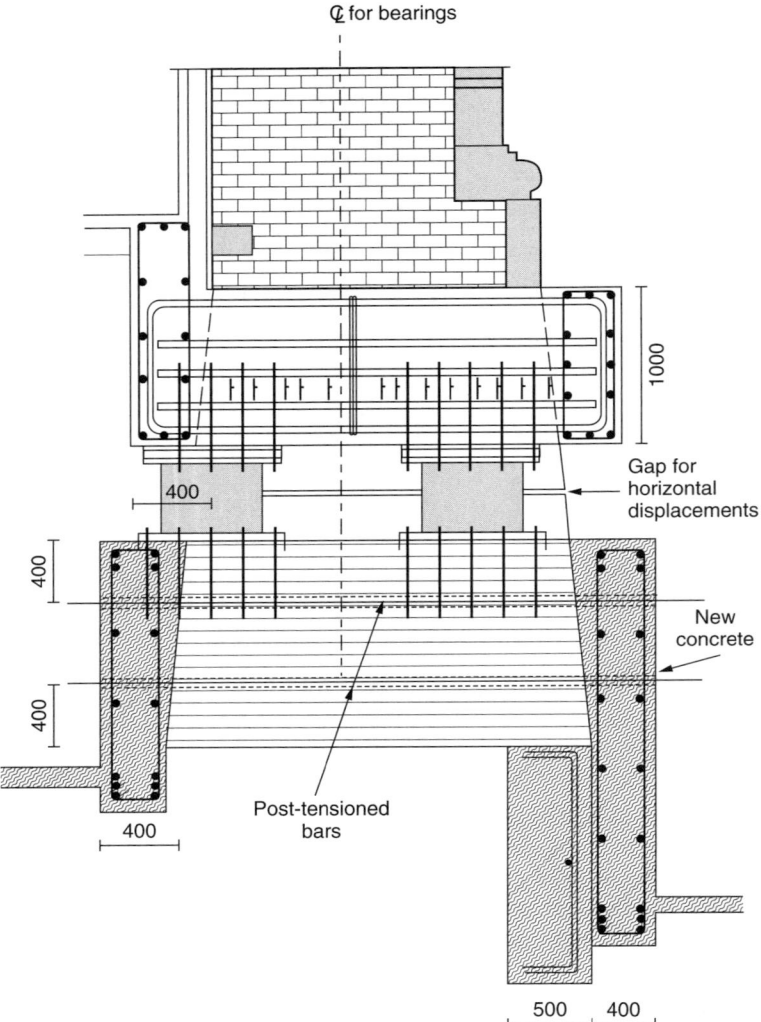

Figure 13.4 Section through the foundations of the URM New Zealand parliament building, showing retrofitting lead-rubber bearings (from Poole and Clendon, 1992)

improvements can be made, occasionally by widening joints, or more frequently by linking adjacent structures together across joints.

Seismic emergency gravity supports

Some columns in existing structures can be vulnerable to severe damage, leading to collapse in and earthquake. For example, where columns at a given level vary in height the shortest columns are likely to be overloaded prematurely and fail in shear. Rather than strengthen such columns, and alternative strategy is to install supplementary seismic emergency columns immediately adjacent to them.

Add stiffness or damping?

In giving advice on choosing between these two options, Kelly (2007) examined two common forms of hardware used to strengthen existing buildings, i.e. buckling restrained braces and viscous damping devices. Non-linear analyses were used to quantify the performance of two frame buildings strengthened with each type of device. It was shown that equivalent structural performance, in terms of overall deformations, can be achieved with both types of device, and generally for lower cost by buckling restrained braces if only moderate levels of drift reduction are required. However, when the total building performance is examined the viscous damping devices provide additional benefits in the form of reduced floor accelerations. The benefits of this may be sufficient to warrant the higher-cost solution.

13.5.2 Examples of retrofitting structures

The retrofitting of a structure involves improving its performance in earthquakes through one or more of:

- increasing its strength and/or stiffness;
- increasing its ductility;
- reducing the input seismic loads
- increasing its damping.

This may be done through modifications to one or more of:

- columns
- beams
- bracings
- walls
- foundations
- horizontal diaphragms
- joints between structural elements
- damping
- period of vibration.

By way of example, three ways of improving the performance of weak or brittle columns suggested by Park (2001) are illustrated in Figure 13.3. These comprise:

(a) adding reinforced concrete jackets, either as a full surround, or on one side;
(b) adding site-welded thin steel jackets with grouted infill;
(c) adding site-welded steel jackets with concrete infill.

In addition to increasing the load capacity of the original column, the jacket increases the ductility of the column by considerably enhancing the confinement of the core. Such techniques are used for bridge piers in Japan, New Zealand and the USA (Priestley *et al.*, 1996), as well as for columns in other structures.

Beam–column joints of reinforced concrete structures are inherently difficult to strengthen directly, because of the inaccessibility of the joint region. However, moderate strengthening of joints can be effected by encasing the top and bottom of the column immediately adjacent to the joint, as shown in laboratory tests carried out by Liu and Park (2001). They used fibreglass wrapping around the columns, the fibreglass having ultimate and design ultimate strength of 400 MPa) and 100 MPa, respectively. Their test on the retrofitted unit with zero axial column load showed that wrapping the column regions using fibre-glass jacketing adjacent to the joint core ensured the development of the postulated joint shear force path, leading to much improved stiffness and strength. The increase in the attained storey shear strength was up to 20%.

URM walls are generally retrofitted by casting or spraying concrete on one face of the existing wall. In Figure 13.4 the concrete strengthening layer is shown on the inner face of the wall so as not to spoil the outward appearance of the building, the New Zealand parliament building, which is a heritage building (Poole and Clendon, 1992). The wall is underpinned by a metre-deep concrete beam supported at intervals by lead-rubber bearings. The horizontal gap to allow for the expected seismic horizontal displacements of up to about a half a metre is shown half way up the bearings. Beneath the bearings the existing masonry foundation is encased by reinforced concrete which is post-tensioned to the existing footing.

Heritage structures need special sensitivity in treatment so as not to disfigure their appearance (Robinson and Bowman, 2000; Charleson *et al.*, 2001), and many such buildings have been successfully strengthened in various countries. Heritage buildings are often protected by statutory requirements, which make them harder to deal with, especially the most protected ones. In New Zealand, for example, the most precious buildings, designated as Category I by the NZ Historic Places Trust, comprise 'buildings of special or outstanding historical or cultural heritage significance or value'.

A theoretical scheme for protecting the brick fabric of Turnbull House in Wellington, a Category I building, is shown schematically in Figure 13.5 taken from Charleson *et al.* (2001). The use of several types of strengthening member was proposed: steel ribs and mullions, sheet steel shear walls and steel framing. The strengthening was designed to be expressed visually, but sympathetically to the existing fabric. Similar discrete strengthening has been done to the classical brick Shed 13 on Wellington's waterfront, using vertical prestessing of the walls and non-intrusive concrete columns (Cattanach *et al.*, 2008).

Older reinforced concrete buildings with masonry infill have often been recognized as being better than commonly believed; see Section 8.3.8 and Earthquake Engineering Research Institute (EERI, 2001). The evaluation of such buildings is not easy, as done (ideally) in non-linear time-history analysis (Sritharan and Dowrick,1994), therefore a much simpler equivalent strut approach for modelling the infill panels has been used by Bell and Davidson (2001).

Foundations of buildings may sometimes be inadequate. In some cases this may be due to the ground being liquefiable, and some means of ground improvement may be economic. Such a measure would depend on how vulnerable the building was to damage

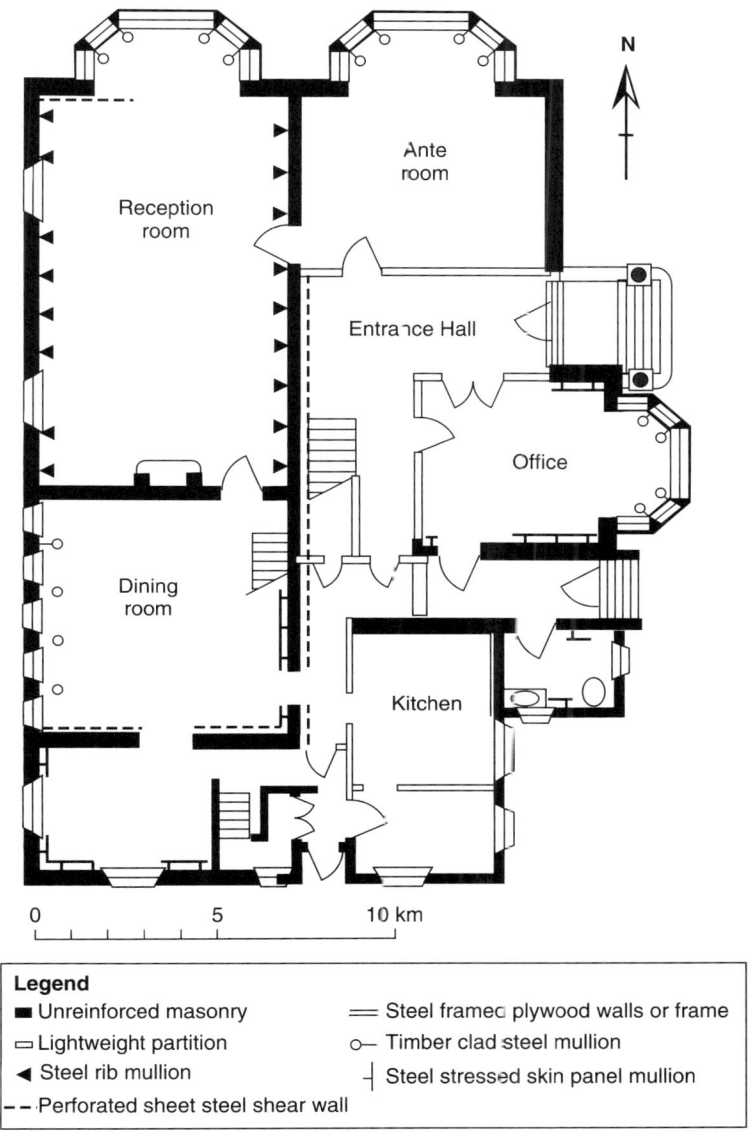

Figure 13.5 Ground-floor plan showing strengthening work of the URM Turnbull House, Wellington, New Zealand (from Charleson *et al.*, 2001). (Reproduced by permission of the Earthquake Engineering Research Institute)

from differential ground displacements due to liquefaction. A means of assessing this is given in Section 6.3.10.

Houses throughout the world vary widely in material used and in construction styles, and are often made of two or more materials, e.g. brick veneer. This often complicates the retrofitting of houses. Fortunately, a large international database of house styles and retrofitting techniques is hosted by EERI at http://www.eeri.org. Retrofitting schemes have been under development for many years, even the relatively well-performed timber framed house needing attention, as set out by Cooney (1982). As well as dealing with structural considerations, Cooney illustrates how to protect services in houses. An example is that of domestic hot water cylinders, retrofitted seismic restraints for which are illustrated in Figure 13.6.

An extremely vulnerable style of house is one that is all too common in Fiji. As shown in Figure 13.7, it comprises two storeys, the lower of which is open for parking cars and drying washing, with a staircase at one end. It is thus an eccentric soft-storey structure with reinforced concrete ground-floor columns, which are no doubt brittle. Buildings of this type, including non-domestic ones, also exist in many other countries, including Australia and New Zealand. Fortunately, such buildings can be easily and cheaply retrofitted by installing bracing or infill in one bay (or more) between adjacent columns, ideally on the three unbraced perimeter façades.

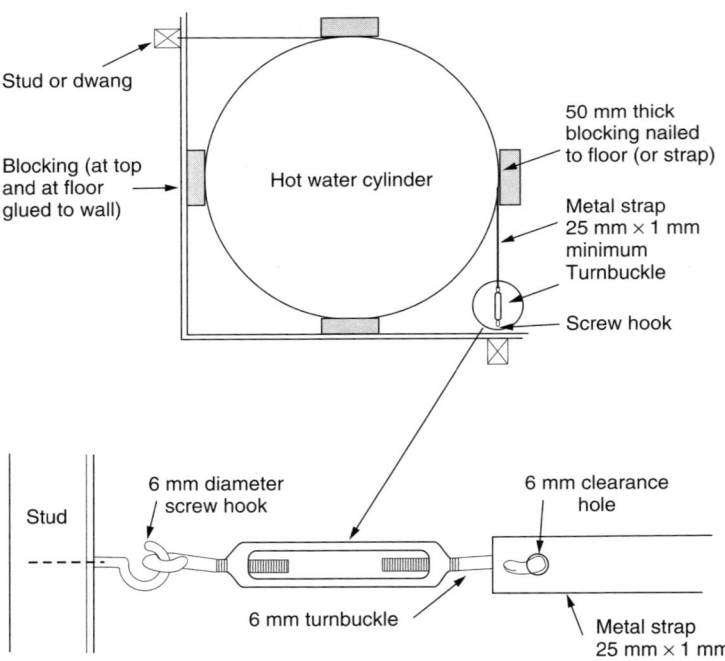

Figure 13.6 Methods of retrofitting a previously free-standing hot water cylinder in a dwelling (after Cooney, 1982). (Reproduced by permission of the Building Research Association of New Zealand)

Figure 13.7 Highly asymmetrical reinforced concrete house in Suva, Fiji, with weak bottom storey, which could be cheaply retrofitted with bracing to prevent collapse

Much work has been done in many other countries on retrofitting, using much the same principles as described above, but tuned to suit local needs, economics and construction styles. An excellent example concerns Greek experience as discussed by Dritsos (2005).

Further guidance on retrofitting of *buildings* is given in ASCE (2000), ASCE (2007) (concrete for concrete only), Holmes (2000) and NZSEE (2006). For further guidance on retrofitting of *bridges*, see Yashinsky and Karshenas (2003).

13.6 Retrofitting Equipment and Plant

Equipment and plant vary enormously in their inherent vulnerability in earthquakes. As discussed in Section 6.3.3, *fragile* equipment has been shown to be many times more vulnerable than *robust* equipment, so there is considerable potential for reducing damage by retrofitting *fragile* or *medium* equipment.

An example of such retrofitting comes from the Haywards HVDC Converter Station in the Wellington region of New Zealand. Here the AC filter capacitor banks, made essentially of brittle porcelain, were built in 1965. In 1988 their earthquake resistance was greatly enhanced by mounting them on low-stiffness elastomeric bearings (for isolation) and installing hysteretic steel dampers (T. Pham, personal communication, 1991); see Figure 13.8. The bearing shifted the natural periods of the capacitor banks from 0.2–0.5 s to 1.8 s, while the function of the dampers was to limit horizontal displacements in strong earthquake shaking and provide lateral restraint during weak ground shaking and wind loading. The specifications of the banks and retrofitting hardware are as follows:

Figure 13.8 Base of capacitor bank showing retrofitting with low-stiffness elastomeric bearing and a vertical steel cantilever damper (photo from Skinner *et al.*, 1993). (Reproduced by permission of John Wiley & Sons, Ltd.)

- *AC filter capacitor banks.* a total of 18 banks of three different types with individual masses varying from 20,000 kg to 32,000 kg. The heights of the banks vary from 6.6 m to 9.6 m.
- *Rubber bearings.* each bank has four to six bearings rated at 5000 kg each. Each bearing has 19 layers with a total height of 254 mm and a plan dimension of 400 × 400 mm. The shear stiffness is rated at 0.06 kN mm^{-1}.
- *Dampers.* each bank is provided with two circular tapered-steel dampers with a base diameter of 45 mm, a height of 500 mm and was designed for a yield force of 10.6 kN.

Perhaps the most common form of retrofitting is that done to robust items of equipment, such as pumps or transformers, which only need more protection against overturning. During the Edgecumbe, New Zealand, earthquake of 1987 ($M_W = 6.5$), much of the equipment was overturned in two electrical switchyards which were located in the near-source region. Ironically, this earthquake occurred not long before a long-planned upgrading of the holding-down arrangements at switchyards in the area around Edgecumbe. Thus, a large costly reinstatement was so nearly avoided by what would have been a low-cost seismic upgrade.

13.7 Retrofitting in Developing Countries

Many developing countries are located in highly seismic regions, and are in great need of economical methods of retrofitting all kinds of structures. The first international document to offer low-technology suggestions to help overcome this problem was that of the International Association for Earthquake Engineering (1986). This document on non-engineered construction contains practical ideas for new and existing buildings made of various materials.

Different countries often need to develop their own variations of the basic strengthening methods to suit their own construction styles and uses of materials. An example of this comes from a programme of strengthening of weak URM school buildings carried out in Nepal (Bothara *et al.* 2004). Most of the strengthening interventions were applied externally, aimed at improving the integrity of the building as a whole, as well as the shear strength of individual wall panels. No strength calculations were attempted. Figure 13.9 illustrates a method of providing a structural lintel over a window (called a bandage) and vertical reinforcement of a corner of a building (splint). These simple measures will no doubt increase the robustness of the building, thereby reducing earthquake damage and casualties.

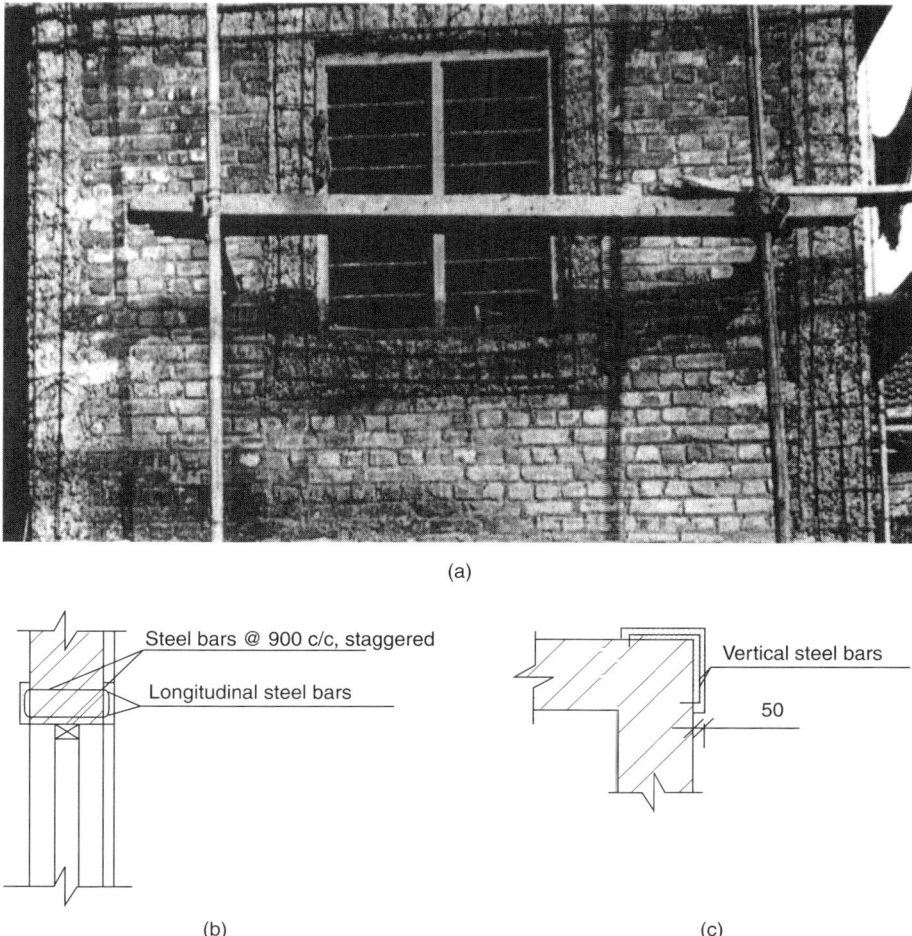

Figure 13.9 Strengthening of unreinforced masonry wall panel and window opening of a school in Nepal: (a) splint and bandage under implementation; (b) bandage: (c) splint (after Bothara *et al.*, 2004)

13.8 Performance of Retrofitted Property in Earthquakes

13.8.1 Introduction

The performance of various types of property that have been retrofitted is bound to be widely variable, because of the wide variety of structural types that are retrofitted, the variety of methods of retrofitting that are used, the degree of strengthening used, and the strength of ground motions to which the retrofitted structures are subjected. Thus, the effectiveness of retrofitting in withstanding real earthquakes, as distinct from theoretical analysis, needs careful interpretation on a case-by-case basis.

Since retrofitting began in the first half of the twentieth century, a now rapidly growing number of retrofitted buildings and other structures have been shaken by earthquakes. For example, in Los Angeles retrofitting of URM buildings began in 1981, and six years later approximately 1100 buildings had been strengthened and were subjected to the $M_W = 5.9$ Whittier Narrows earthquake (Deppe, 1988). This left 4500 URM buildings in Los Angeles, which would have been strengthened or demolished within the next several years.

13.8.2 Earthquake performance of retrofitted unreinforced masonry buildings

When the Northridge earthquake struck the Los Angeles area in 1994, about 5700 URM buildings there had been retrofitted. This had been done to various standards of earthquake protection, depending on the requirements of the local authority, and had not necessarily been aimed at eliminating death risk (EERI, 1996).

In a survey of damage to URM buildings in Los Angeles after the earthquake (Schmid, 1994), damage levels were assigned subjectively to 208 unstrengthened and 637 strengthened buildings, at intensities MM7–MM9. The probabilities of occurrence of the three damage levels are plotted in Figure 13.10, which shows that although the damage to the strengthened buildings is reduced to about half that of the unstrengthened buildings, much damage still occurred.

A similar degree of benefit was experienced by very early partially retrofitted URM buildings in the $M_W = 7.1$ Wairarapa, New Zealand, earthquake of 1942, at intensity MM8 (Dowrick and Rhoades, 2002). Here the mean damage ratio D_{rm} (Chapter 6) was 0.05 for the retrofitted buildings compared, with 0.17 for those that were unstrengthened.

It is noted that a greater degree of protection than that used in the examples given above can be provided by more comprehensive retrofitting techniques.

13.8.3 Earthquake performance of retrofitted reinforced concrete buildings

A study prepared for FEMA (Wiss, Janney, Elstner Associates, 1994) indicates that at least five brittle reinforced concrete buildings had been retrofitted and performed adequately in the Northridge earthquake. Of those buildings, two with concrete shear walls were in the area of strong shaking. One of the two latter buildings (Topanga Plaza) was a three-storey building built around 1964. In 1989 a structural evaluation found that it had a potential weak bottom storey, so two shotcrete concrete shear walls were added in

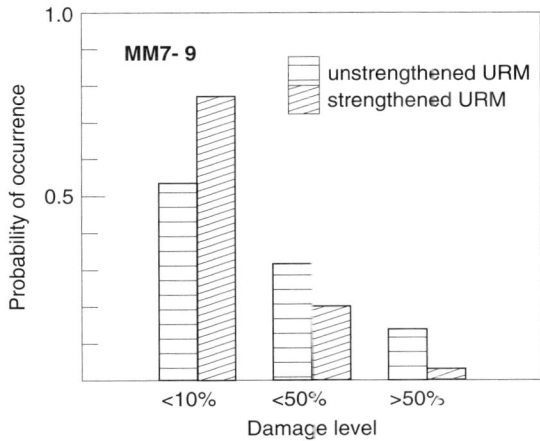

Figure 13.10 Histogram of probabilities of occurrence of three subjective damage levels to URM buildings, strengthened and unstrengthened, in the Northridge earthquake of 1994. Derived from data from Schmid (1994)

the short direction at ground level. The building suffered substantial cracking to both the original and added walls in the bottom storey, but the otherwise likely collapse was prevented.

The other strongly shaken building (the Van Nuys building, only 5 km from the epicentre) was an eight-storey concrete bearing wall structure built in 1953. It was retrofitted in 1989 to conform to safety levels of the then current code, by thickening selected existing walls and adding some newer ones. Following the 1994 earthquake, the only damage that was observed comprised some minor to moderate cracks in the shear walls (Seismic Safety Commission, 1994).

References

Aguilar J, Juares H, Ortega R and Iglesisas J (1989) Statistics of retrofitting techniques in reinforced concrete buildings. *Earthq Spectra* **5**(1): 145–152.

American Society of Civil Engineers (2000) *Prestandard and Commentary for the Seismic Rehabilitation of Buildings*, FEMA-356. Federal Emergency Management Agency, Washington, DC.

American Society of Civil Engineers (2003) *Seismic Evaluation of Existing Buildings*, ASCE/SEI 31-03. ASCE, Reston, VA.

American Society of Civil Engineers (2007) *Seismic Rehabilitation of Existing Buildings*, ASCE/SEI 41. American Society of Civil Engineers, Reston, VA.

Applied Technology Council (1996) *ATC-40: The Seismic Evaluation and Retrofit of Concrete Buildings* (2 volumes). Redwood City, CA.

Bell DK and Davidson BJ (2001) Evaluation of earthquake risk buildings with masonry infill panels. *Technical Conference of NZ Soc for Earthq Eng*, Wairakei, New Zealand, Paper No. 4.02.01.

Bertero VV, Anderson JC and Krawinkler H (1994) Performance of steel building structures during the Northridge earthquake. Report No UCB/EERC-94/09, Earthquake Engineering Research Center, University of California, Berkeley.

Bothara JK, Pandey B and Guragain R (2004) Seismic retrofitting of low strength unreinforced masonry non-engineered school buildings. *Bull NZ Soc Earthq Eng* **37**(1): 13–22.

Cattanach A, Alley G and Thornton A (2008) Appropriateness of seismic strengthening interventions in heritage building: a framework for appraisal. *NZ Soc Earthq Eng 2008 Conference*, Wairakei, New Zealand, Paper 30 (CD-ROM).

Charleson A, Preston J and Taylor M (2001) Architectural expression of seismic strengthening. *Earthq Spectra* **17**(3): 417–426.

Coburn A and Spence R (2002) *Earthquake Protection*, 2nd edn. John Wiley & Sons, Ltd, Chichester.

Comartin CD, Niewiarowski RW, Freeman SA and Turner FM (2000) Seismic evaluation and retrofit of concrete buildings: A practical overview of the ATC-40 document. *Earthq Spectra* **16**(1): 241–261.

Cooney R (1982) Strengthening houses against earthquakes: A handbook of remedial measures. Technical Paper P37. Building Research Association of New Zealand, Judgeford.

Deppe K (1988) The Whittier Narrows, California earthquake of October 1, 1987: Evaluation of strengthened and unstrengthened unreinforced masonry in Los Angeles city. *Earthq Spectra* **4**(1): 157–180.

Dowrick DJ and Rhoades DA (2000) Earthquake damage and risk experience and modelling in New Zealand earthquakes. *Proc. 12th World Conf on Earthq Eng*, Auckland, Paper No. 0403 (CD-ROM).

Dowrick DJ and Rhoades DA (2002) Damage ratios for low-rise non-domestic brick buildings in the magnitude 7.1 Wairarapa, New Zealand earthquake of 24 June 1942. *Bull NZ Soc Earthq Eng* **35**(3): 135–148.

Dritsos SE (2005) Seismic retrofit of buildings: A Greek perspective. *Bull NZ Soc Earthq Eng* **38**(3): 165–181.

EERI (1996) Northridge earthquake of January 17, 1994 reconnaissance report, Volume 2. *Earthq Spectra*, Supplement C, **11** Chapters 8 and 9.

EERI (2001) Preliminary observations on the origin and effects of the January 26, 2001 Bhuj (Gujarat, India) earthquake. EERI Special Earthquake Report. *Earthquake Engineering Research Institute Newsletter* **35**(4).

Federal Emergency Management Agency (1994) *Typical Costs for Seismic Rehabilitation of Existing Buildings*. FEMA-156, Volume 1, 2nd edn. FEMA, Washington DC.

Federal Emergency Management Agency (2000) *Recommended Seismic Evaluation and Upgrade Criteria for Existing Welded Steel Moment-Frame Buildings*, FEMA-351. FEMA, Washington DC.

Holmes WT (2000) Risk assessment and retrofit of existing buildings. *Proc. 12th World Conf on Earthq Eng*, Auckland, Paper No. 2826 (CD-ROM).

Hopkins DC, Lumsden JL and Norton JA (1993). Lifelines in earthquakes: A case study based on Wellington. *Bull NZ Nat Soc Earthq Eng* **26**(2): 208–21.

Hopkins DC, Sharpe R, Sucuoglu H and Kubin D (2006) Feasibility of retrofitting residential buildings in Istanbul. *Proc NZ Soc Earthq Eng Conf, Napier, Paper 7* (CD-ROM).

International Association for Earthquake Engineering (1986) *Guidelines for Earthquake Resistant Non-engineered Construction*. IAEE, Tokyo.

Kameda H (2000) Engineering management of lifeline systems under earthquake risk. *Proc. 12th World Conf on Earthq Eng*, Auckland, Paper 2827 (CD-ROM).

Kelly T (2007) Improving seismic performance: Add stiffness or damping? *New Zealand Society for Earthquake Engineering Conference 2007*, Palmerston North, Paper No 11 (CD-ROM).

Krawinkler H (1996) Earthquake design and performance of steel structures. *Bull NZ Nat Soc Earthq Eng* **29**(4): 229–241.

Liu A and Park R (2001) Seismic behaviour and retrofit of pre-1970's as-built exterior beam-column joints reinforced by plain round bars. *Bull NZ Soc Earthq Eng* **34**(1): 68–81.

New Zealand Society for Earthquake Engineering (2006) *Assessment and Improvement of the Structural Performance of Buildings in Earthquakes*. NZSEE, Wellington. http://www.nzsee.org.nz/PUBS/2006AISBEGUIDELINES_Corr_06a.pdf.

Park R (1997) A static force-based procedure for the seismic assessment of existing reinforced concrete moment resisting frames. *Bull NZ Nat Soc Earthq Eng* **30**(3): 213–226.

Park R (2001) Improving the resistance of structures to earthquakes. *Bull NZ Soc Earthq Eng* **34**(1): 1–39.

Park R, Billings IJ, Clifton GC et al. (1995) The Hyogo-ken Nanbu earthquake of 17 January 1995. *Bull NZ Nat Soc Earthq Eng* **28**(1): 1–98.

Poole RA and Clendon JE (1992) NZ parliament buildings seismic protection by base isolation. *Bull NZ Nat Soc Earthq Eng* **25**(3): 147–60.

Priestley MJN, Seible F and Calvi GM (1996) *Seismic Design and Retrofit of Bridges*. John Wiley & Sons, Inc., New York.

Priestley MJN, Calvi GM and Kowalsky MJ (2007) *Displacement-Based Design of Structures*. IUSS Press, Pavia, Italy.

Robinson L and Bowman I (2000) *Guidelines for Earthquake Strengthening*. New Zealand Historic Places Trust, Wellington.

Schmid B (1994) Report on URM buildings for the City of Los Angeles Task Force on building damage. September 27.

Seismic Safety Commission (1994) B18: Performance of retrofitted buildings. A compendium of background reports on the Northridge earthquake. Prepared under Executive Order W-79-94. California Seismic Safety Commission, November.

Skinner RI, Robinson WH and McVerry GH (1993) *An Introduction to Seismic Isolation*. John Wiley & Sons, Ltd., Chichester.

Sritharan S and Dowrick DJ (1994) Response of low-rise buildings to moderate earthquake shaking, particularly the May 1990 Weber earthquake. *Bull NZ Nat Soc Earthq Eng* **27**(3): 205–221.

Wiss, Janney, Elstner Associates, Inc. (1994) Performance of seismically rehabilitated buildings during the January 17, 1994 Northridge earthquake. Report for the Federal Emergency Management Agency, Washington DC.

Yashinsky M and Karshenas MJ (2003) *Fundamentals of Seismic Protection for Bridges*. MNO-9. EERI, Oakland, CA.

Appendix A

Modified Mercalli Intensity Scale (NZ 2007)

The following version of the MM scale is as used in New Zealand (Dowrick *et al*., 2008); items marked * in the scale are defined in the note following. Revisions in this version of the scale are shown in italics.

MM1   *People*
  Not felt except by a very few people under exceptionally favourable circumstances.
MM2   *People*
  Felt by persons at rest, on upper floors or favourably placed.
MM3   *People*
  Felt indoors; hanging objects may swing, vibration similar to passing of light trucks, duration may be estimated, may not be recognised as an earthquake.
  Fittings
  Liquids in large open containers may be disturbed (sometimes considerably) in large magnitude (long duration) earthquakes. Pendulum clocks may stop, start, or change rate (H*).
MM4   *People*
  Generally noticed indoors but not outside. Light sleepers may be awakened. Vibration may be likened to the passing of heavy traffic, or to the jolt of a heavy object falling or striking the building.
  Fittings
  Doors and windows rattle. Glassware and crockery rattle. Liquids in open vessels may be slightly disturbed *in small to medium-sized earthquakes*. Standing motorcars may rock.
  Structures
  Walls and frames of buildings, and partitions and suspended ceilings in commercial buildings, may be heard to creak.
MM5   *People*
  Generally felt outside, and by almost everyone indoors. Most sleepers awakened. A few people alarmed.

Fittings
Small unstable objects are displaced or upset. Some glassware and crockery may be broken. Hanging pictures knock against the wall. Open doors may swing. Cupboard doors secured by magnetic catches may open.
Structures
Some windows Type I* cracked. A few earthenware toilet fixtures cracked, *in timber buildings with inadequately braced piles*.
Environment
Loose boulders may occasionally be dislodged from steep slopes.

MM6 *People*
Felt by all.
People and animals alarmed.
Many run outside.*
Difficulty experienced in walking steadily.
Fittings
Objects fall from shelves.
Pictures fall from walls (H*).
Some furniture moved on smooth floors, some unsecured free-standing fireplaces moved.
Glassware and crockery broken.
Very unstable furniture overturned.
Small church and school bells ring (H*).
Appliances move on bench or table tops.
Filing cabinets or 'easy glide' drawers may open (or shut).
Structures
Slight damage to Buildings Type I*.
Some stucco or cement plaster falls.
Windows Type I* broken.
Damage to a few weak domestic chimneys, some may fall.
Environment
Trees and bushes shake, or are heard to rustle.
Loose material may be dislodged from sloping ground, e.g. existing slides, talus *and scree* slopes.
A few very small ($\leq 10^3$ m^3) soil and regolith slides and rock falls from steep banks and cuts.
A few minor cases of liquefaction (sand boil) in highly susceptible alluvial and estuarine deposits.

MM7 *People*
General alarm.
Difficulty experienced in standing.
Noticed by motorcar drivers who may stop.
Fittings
Large bells ring.
Furniture moves on smooth floors, may move on carpeted floors.
Substantial damage to fragile* contents of buildings.
Structures
Unreinforced stone and brick walls cracked.
Buildings Type I cracked some with minor masonry falls.
A few instances of damage to Buildings Type II.
Unbraced parapets, unbraced brick gables, and architectural ornaments fall.

Roofing tiles, especially ridge tiles may be dislodged.
Many unreinforced domestic chimneys damaged, often falling from roof-line.
Water tanks Type I* burst.
A few instances of damage to brick veneers and plaster or cement-based linings.
Unrestrained water cylinders (Water Tanks Type II*) may move and leak.
Some windows Type II* cracked. Suspended ceilings damaged.
Environment
Water made turbid by stirred up mud.
Small slides such as falls of sand and gravel banks, and small rock-falls from steep slopes and cuttings common.
Instances of settlement of unconsolidated, or wet, or weak soils.
A few instances of liquefaction (i.e. small water and sand ejections).
Very small ($\leq 10^3$ m^3) disrupted soil slides and falls of sand and gravel banks, and small rock falls from steep slopes and cuttings are common
Fine cracking on some slopes and ridge crests.
A few small to moderate landslides ($10^3 - 10^5$ m^3), mainly rock falls on steeper slopes ($>30°$) such as gorges, coastal cliffs, road cuts and excavations.
Small discontinuous areas of minor shallow sliding and mobilisation of scree slopes in places.
Minor to widespread small failures in road cuts in more susceptible materials.
A few instances of non-damaging liquefaction (small water and sand ejections) in alluvium.

MM8 *People*
Alarm may approach panic.
Steering of motorcars greatly affected.
Structures
Building Type I, heavily damaged, some collapse*.
Buildings Type II damaged, some with partial collapse*.
Buildings Type III damaged in some cases.
A few instances of damage to Structures Type IV.
Monuments and pre-1976 elevated tanks and factory stacks twisted or brought down.
Some pre-1965 infill masonry panels damaged.
A few post-1980 brick veneers damaged.
Decayed timber piles of houses damaged.
Houses not secured to foundations may move, *and damage to earthenware sanitary fittings may occur*.
Most unreinforced domestic chimneys damaged, some below roof-line, many brought down.
Environment
Cracks appear on steep slopes and in wet ground.
Significant landsliding likely in susceptible areas.
Small to moderate ($10^3 - 10^5$ m^3) slides widespread; many rock and disrupted soil falls on steeper slopes (steep banks, terrace edges, gorges, cliffs, cuts etc.).
Significant areas of shallow regolith landsliding, and some reactivation of scree slopes.
A few large ($10^5 - 10^6$ m^3) landslides from coastal cliffs, and possibly large to very large ($\geq 10^6$ m^3) rock slides and avalanches from steep mountain slopes.
Larger landslides in narrow valleys may form small temporary landslide-dammed lakes.
Roads damaged and blocked by small to moderate failures of cuts and slumping of road-edge fills.

| | Evidence of soil liquefaction common, with small sand boils and water ejections in alluvium, and localised lateral spreading (fissuring, sand and water ejections) and settlements along banks of rivers, lakes, and canals etc.
Increased instances of settlement of unconsolidated, or wet, or weak soils. |
|---|---|
| **MM9** | *Structures*
Many Buildings Type I destroyed*.
Buildings Type II heavily damaged, some collapse*.
Buildings Type III damaged, some with partial collapse*.
Structures Type IV damaged in some cases, some with flexible frames seriously damaged.
Damage or permanent distortion to some Structures Type V.
Houses not secured to foundations shifted off.
Brick veneers fall and expose frames.
Environment
Cracking of ground conspicuous.
Landsliding widespread and damaging in susceptible terrain, particularly on slopes steeper than 20°.
Extensive areas of shallow regolith failures and many rock falls and disrupted rock and soil slides on moderate and steep slopes (20°–35° or greater), cliffs, escarpments, gorges, and man-made cuts.
Many small to large (10^3–10^6 m^3) failures of regolith and bedrock, and some very large landslides (10^6 m^3 or greater) on steep susceptible slopes.
Very large failures on coastal cliffs and low-angle bedding planes in Tertiary rocks. Large rock/debris avalanches on steep mountain slopes in well-jointed greywacke and granitic rocks. Landslide-dammed lakes formed by large landslides in narrow valleys. Damage to road and rail infrastructure widespread with moderate to large failures of road cuts and slumping of road-edge fills. Small to large cut slope failures and rock falls in open mines and quarries.
Liquefaction effects widespread with numerous sand boils and water ejections on alluvial plains, and extensive, potentially damaging lateral spreading (fissuring and sand ejections) along banks of rivers, lakes, canals etc.). Spreading and settlements of river stop-banks likely. |
| **MM10** | *Structures*
Virtually all Buildings Type I destroyed*.
Most Buildings Type II destroyed*.
Buildings Type III $^\triangledown$ heavily damaged, some collapse*.
Structures Type IV $^\triangledown$ damaged, some with partial collapse*.
Structures Type V $^\triangledown$ moderately damaged, but few partial collapses.
A few instances of damage to Structures Type VI.
Some well-built* timber buildings moderately damaged (excluding damage from falling chimneys).
Environment
Landsliding very widespread in susceptible terrain.
Similar effects to MM9, but more intensive and severe, with very large rock masses displaced on steep mountain slopes and coastal cliffs. Landslide-dammed lakes formed. Many moderate to large failures of road and rail cuts and slumping of road-edge fills and embankments may cause great damage and closure of roads and railway lines. |

Liquefaction effects (as for MM9) widespread and severe. Lateral spreading and slumping may cause rents over large areas, causing extensive damage, particularly along river banks, and affecting bridges, wharfs, port facilities, and road and rail embankments on swampy, alluvial or estuarine areas.

MM11 *Structures*
All Buildings Type II $^\nabla$ destroyed *.
Many Buildings Type III $^\nabla$ destroyed *.
Structures Type IV $^\nabla$ heavily damaged, some collapse*.
Structures Type V $^\nabla$ damaged, some with partial collapse.
Structures Type VI suffer minor damage, a few moderately damaged.
Environment
Environmental response criteria have not been suggested for MM11 as that level of shaking has not been reported in New Zealand or (definitively) elsewhere. As discussed in the text, it is likely that the MM scale in fact saturates between MM10 and MM11.

Notes to the 2007 NZ MM Scale

Items marked * in the scale are defined below.

Construction types

Buildings Type I (Masonry D in the NZ 1965 mm scale)

Buildings with low standard of workmanship, poor mortar, or constructed of weak materials like mud brick or rammed earth soft storey structures (e.g. shops) made of masonry weak reinforced concrete or composite materials (e.g. some walls timber, some brick) not well tied together. Masonry buildings otherwise conforming to buildings Types I–III, but also having heavy unreinforced masonry towers. (Buildings constructed entirely of timber must be of extremely low quality to be Type I.)

Buildings Type II (Masonry C in the NZ 1966 MM scale)

Buildings of ordinary workmanship, with mortar of average quality. No extreme weakness, such as inadequate bonding of the corners, but neither designed nor reinforced to resist lateral forces. Such buildings not having heavy unreinforced masonry towers.

Buildings Type III (Masonry B in the NZ 1966 MM scale)

Reinforced masonry or concrete buildings of good workmanship and with sound mortar, but not formally designed to resist earthquake forces.

Structures Type IV (Masonry A in the NZ 1966 MM scale)

Buildings and bridges designed and built to resist earthquakes to normal use standards, i.e. no special collapse or damage limiting measures taken (mid-1930s to *c*. 1970 for concrete and to *c*. 1980 for other materials).

Structures Type V

Buildings and bridges, designed and built to normal use standards, i.e. no special damage limiting measures taken, other than code requirements, dating from since $c.$ 1970 for concrete and $c.$ 1980 for other materials.

Structures Type VI

Structures, dating from $c.$ 1980, with well-defined foundation behaviour, which have been specially designed for minimal damage, e.g. seismically isolated emergency facilities, some structures with dangerous or high contents, or new generation low damage structures.

Windows

Type I – Large display windows, especially shop windows.
Type II – Ordinary sash or casement windows.

Water Tanks

Type I – External, stand mounted, corrugated iron tanks.
Type II – Domestic hot-water cylinders unrestrained except by supply and delivery pipes.
H – (Historical) More likely to be used for historical events.

Other Comments

'Some' or 'a few' indicates that the threshold of a particular effect has just been reached at that intensity.

'Many run outside' (MM6) variable depending on mass behaviour, or conditioning by occurrence or absence of previous quakes, i.e. may occur at MM5 or not until MM7.

'Fragile Contents of Buildings'. Fragile contents include weak, brittle, unstable, unrestrained objects in any kind of building. 'Well-built timber buildings' have: wall openings not too large; robust piles or reinforced concrete strip foundations; superstructure tied to foundation.

∇ *Buildings Type III – V* at MM10 and greater intensities are more likely to exhibit the damage levels indicated for low-rise buildings on firm or stiff ground and for high-rise buildings on soft ground. By inference, lesser damage to low-rise buildings on soft ground and high-rise buildings on firm or stiff ground may indicate the same intensity. These effects are due to attenuation of short period vibrations and amplification of longer period vibrations in soft soils.

References

Dowrick DJ, Hancox GT, Perrin ND and Dellow GD (2008) The Modified Mercalli earthquake intensity scale: Revisions arising from New Zealand experience. *Bull NZ Soc Earthq Eng* **41**(3): 193–205.

Appendix B

Structural Steel Standards for Earthquake Resistant Structures

The following is an extract from the New Zealand *Steel Structures Standard* NZS 3404:Part 1:1997 (incorporating Amendment No 2 of October 2007), Standards New Zealand. (Reproduced from NZS 3404:1997 by permission of Standards New Zealand under Licence 000710.)

All structural steel coming within the scope of [NZS 3404 above] shall, before fabrication, comply with the requirements (a), (b), or (c) below.

(a) Australian or Joint Australian/New Zealand Standards:

AS 1163	Structural steel hollow sections
AS 1594	Hot-rolled steel flat products
AS/NZS 3678	Structural steel–Hot-rolled plates, floorplates and slabs
AS/NZS 3679	Structural steel
Part 1	Hot-rolled bars and sections
Part 2	Welded I-sections

(b) British Standards:

BS 4	Structural steel sections
Part 1	Specification for hot-rolled sections
BS 4848	Hot-rolled structural steel sections
Part 2	Specification for hot-finished hollow sections
Part 4	Equal and unequal angles
BS 7668	Weldable structural steels. Hot-finished structural hollow sections in weather resistant steels. Specification
BS EN 10 025	Hot-rolled products of non-alloy structural steel. Technical delivery conditions of Parts 1–6

Earthquake Resistant Design and Risk Reduction D. Dowrick
© 2009, John Wiley & Sons, Ltd

BS EN 10 029	Specification for tolerances on dimensions, shape and mass for hot-rolled steel plates 3 mm thick or above
BS EN 10 210	Hot-finished structural hollow sections of non-alloy and fine grain structural steels
Part 1	Technical delivery requirements
BS EN 10219	Cold formed welded structural hollow sections of non-alloy and fine grin steels
Part 2	Tolerances, dimensions and sectional properties

(c) Japanese Standards:

JIS G 3101	Rolled steel for general structure
JIS G 3106	Rolled steels for welded structure
JIS G 3114	Hot-rolled atmospheric corrosion resisting steels for welded structure
JIS G 3132	Hot-rolled carbon steel strip for pipes and tubes
JIS G 3136	Rolled steel for building structure
JIS G 3141	Cold reduced carbon steel sheets and strip
JIS G 3192	Dimensions, mass and permissible variations of hot-rolled steel sections
JIS G 3193	Dimensions, mass and permissible variations of hot-rolled steel plates, sheets and strip

Index

Accelerograms 105–16, 141, 196, 452
   as design earthquakes 114
   of real earthquakes 114
   and response spectra, sources of 111
   scaling of 112
   simulated 111
   synthetic 115–6
Accelerograph 72, 452
Acceptable risk 5, 105
Added mass of soil 155, 162
Adobe 416, 444
Advocacy 10
Aftershock
   epidemic-type (ETAS) 69
   power-law decay of 66
Alaska 1964 earthquake, Great 31, 147, 257
Allowable bearing pressure on soils 315
Alluvium *see* Soil
Amplification (in soil) 138, 147, 257
   See also Topograhical effects
Analysis, method of, for structures, selection of method of 193–6
   see also Seismic analysis
Anatolian fault zone 63
Architectural detailing 475–83
Arias intensity 74
Asperities 70–1
Attenuation
   of displacement 102
   of ground motions 91–5, 97–8
   in interplate regions 96–8

   in intraplate regions 99–100
   model 91, 119
   in volcanic regions 100–1
   within soil 144, 150
Available ductility for reinforced concrete members 320–7
Averaged response spectra 95

Banco Central, Managua 116
Bar bending 389
Base isolation 293
   see also Isolation, seismic
Basin edge effects 104
Basins (geological) 47
   see also Sedimentary basins
Bauschinger effect 172
Bay of Plenty, NZ, 1914 earthquake 62
Beam-column joints 304, 362–4, 397–9
Beam-hinging failure mechanisms 284–5
Beams
   reinforced concrete 397
   shear strength 369, 395–6
   steel 347–351
Bedrock
   depth to 28, 35, 140
   effective (equivalent) 35
   motion 141
Benioff zone 16
Bhuj, India, 2001 earthquake 287
Bilinear (hysteresis) model 177–8
Body wave magnitude 21
Bolting 343, 365

Borah Peak, Idaho, 1983 earthquake 31
Braced frames 289–91, 304–5, 356–60, 441
Branching renewal processes 69
Brick, pure 214–5, 217
Bridges 296, 300–1
Brittle materials 194, 416, 429, 455, 457–8, 480
Bucharest 1977 earthquake 62
Buckling of steel columns 354–5
Buffer plates (tectonic) 19
Buildability 274–5
Buller 1929 earthquake 87, 92
Business interruption 253–6
  reduction of 255–6
  upstream effects 253

Caissons 316
Cantilever walls (concrete) 290, 379–81
  bending strength 379–81
  ductility 381
  shear strength 384
Capacity design 285–342
  see also Failure mode control
Caracas 1967 earthquake 28, 39, 150, 378
Casualties 3–5, 237–40
  capacity design and 251
  daytime 250–1
  estimation by risk modelling 249–53
  night time 250
Ceilings, suspended 482–3
Chiba-ken Chubu, Japan, 1980 earthquake 143
Chilean earthquakes
  1985 275
  1960 62
Chimneys 210–2, 433, 465
Chinese earthquake catalogue 120
Cladding 406, 410, 480
Clay 30, 77–9, 315
  sensitive (quick) 27, 314
Cleats 365
Clyde Dam, New Zealand 117
Cohesionless soils 321, 324, 333
Cold formed sections 345

Columns
  reinforced concrete 393–6
  steel 352–6
Compaction
  degree of 36, 40
  dynamic 326
  grouting 326
  piles 327
Complete Quadratic Combination (CQC) 192
Composite construction 366
Compressive underthrust faults 50–1
Compressive overthrust faults 50–1
Concentrically braced frames 290, 304–5, 356–8
Concrete blocks, hollow 416–27
Concrete quality see Quality
Concrete structures 367–410
  available ductility 370–7
    doubly reinforced sections 372–3
    effect of confinement of ductility 374–7
    flexure and axial load 377
    singly reinforced sections 371–2
  in situ concrete design and detailing 387–92
  required ductility (robustness) 370
Cone penetrometer tests (CPT) 36
Configuration of construction see Strucural form
Confined masonry 444
Confinement
  effect on ductility 374–7
  reinforcing (columns) 393–5
  see also Transverse steel
Connections
  in precast concrete 399–410
  in steelwork 361–5
  in timber construction 441–3
Consequences of earthquakes 3–6
Construction materials, choice of 282–3
Corner frequency 72
Corruption, effect on losses 261, 306
Cost
  of construction 9, 10, 270, 298
  of damage 8, 206, 209

Index

directly due to ground shaking
246–7
due to earthquake-induced fires
247–8
using structural response parameters
248–9
of earthquake resistance 9, 10
of retrofitting 486
Coulomb theory 330–3
Coupled walls 384–5
Coupling beams 385–7
Critical structures 274
Crossing active faults 117–8, 496
Crustal strain 45–7
Crustal waveguide effects 103–4
Curtain walls 480
glass panels 481
Cyclic loading behaviour see Hysteresis
Cyclic mobility 147
Cyclic torsional shear test 43
Cyclic triaxial test 40–2

Damage Avoidance Design 224, 304–6
Damage costs see Cost of damage
Damage function, building 223–4
Damage models, ground motion measures
and 223–4
Damage ratio 206–13
for buildings and contents 213–20
damage costs directly due to ground
shaking 246–7
definition 206
mean 209–13
probability distribution 208–9, 211–2
Damage states 205–6
Damageability 177
Dampers 295–303
Damping 132, 135–8
adding 461, 499
effective 165
equivalent viscous 136, 142, 167–8
hysteretic 160, 295–6, 302–3
material 135
radiation 135–8, 165
ratios for structures 177–9
soil-structure system 165–6

timber 431–2
Dams, large 117
Deaths see Casualties
Deflection
control (in isolated structures) 295, 298
ductility factor 196, 298
see also Deformation control; Drift
Deformation control 281
see also Deflection; Drift
Densification of soils 326
See also Differential settlement
Density, mass 133
Design
aims 6
events 109–11
seismic, of prestressed concrete,
recommendations 411–2
Design brief 265–6
Design displacement spectra 340, 342
Design earthquakes 105–9
Design procedures using dynamic
analysis 456–7
using equivalent-static analysis 457–62
Developing countries 259–61, 306,
444–5, 504–5
earthquake risk management in
259–61
Development (anchorage of reinforcing)
389
Diagonal braces 356–60
Diaphragms, timber horizontal 437–8
Differential settlement 241–2, 288
Direct integration 192–3
Directivity see Near source directivity
Displacement spectra see Design
displacement spectra
Distributed seismicity 111, 112, 118
Distribution in time of damaging
earthquakes 63–70
Doors 481–2
Drift (inter-storey) 269, 282, 296, 476
see also Deflection control;
Deformation control; Lateral drift
Ductility
adequate 345, 367, 441
cantilever walls (concrete) 379–84

Ductility (*continued*)
  casualties and 251
  column (steel) 353–4
  demand 164, 370
  effect of axial load 415
  factors 338–9
  limited 194, 345
  notch (of steel) 346
  prestressed concrete 413
  ratio, section 370
  required 370
Duhamel integral 185–6
Ductwork 469–70
Duration
  effective 73–5
  of loading, effect (on timber) 429–32
  of strong motion 73–5
Dutch cone 36
Duty of care 13
Dynamic analysis 184–93
  soil models for 151–63
  of soil-structure systems 151
Dynamic compaction 326
Dynamic properties of soil 131–8
Dynamic site response analysis, two- and three-dimensional 144
Dynamic yield stress 344

Earthquake
  distribution in time and size 63–70
    EEPAS model 69
  forecasting 70
  prediction *see* Earthquake forecasting
  processes, model of 66–70
  source models 70–2
  see also Seismic response
Earthquake risk modelling
  business interruption 253–6
  casualty estimation 237–40, 249–53
  earthquake insurance *see* Insurance
  impediments to earthquake risk reduction 10–3, 261–2
  material damage costs 245–9
  planning for earthquakes 256–9
Earthquake risk reduction potential 236–9
Earthquakes and Megacities Initiative (EMI) 261

Eccentrically braced frames 290–1, 358–60
Economic consequences of earthquakes 7–9
Edgecumbe, New Zealand, 1987 earthquake 209, 214–23, 254
Effective length of steel columns 354
El Centro 1940 ground motions 171, 187–8, 196
Elastic continuum analysis of single piles 318
Elastic homogeneous half-space 158
Elastic response spectra 32, 80, 96–7, 110–3, 186–8
Elastic seismic response of structures 169–72
Elastoplastic hysteresis model 178
Electrical equipment *see* Equipment
Embedment
  effects in soil-structure systems 168
  of footings 162
Energy
  absorption 175
  dissipation 175, 180, 292, 304, 466–7
  dissipators 291–305
  isolating devices 293–301
  (strain) release 46, 62–4
Epicentre 60–1, 86, 110
Epidemic-type aftershock (ETAS) 69
Equipment 451–72
  base-mounted free-standing 458–62
  brittle 462
  code seismic coefficients 462–3
  design requirement 453–4
  material behaviour in 455–6
  mounted in buildings 454–5
  on suspended floors 460–1
  retrofitting of 503–4
  rigidly mounted 464–6
  seismic protection of 462–73
  vulnerability of 217–20
Equivalent radius (of footings) 137–8
Equivalent static force analysis 183–8
European Macroseismic Scale 81
Exit requirements (from buildings) 482
Extensional faults 50–1

Index 525

Failure modes 265–6, 281, 284–6, 344–5, 368, 390, 419, 439
  of piles 317
Failure mode control 284–8
  see also Capacity design
Faulting see Faults
Faults
  activity, degree of 51–7
  displacements, probability of 116–8
  earthquake magnitude and 57–9
  location of active 49–50
  movements, designing for 117–8, 496
  planning for 256–8
  rupture dimensions 57–9
  system, Nevis-Cardrona, New Zealand 52–5
  trace 51–2, 110
  types 51
  see also Focal mechanism
Felt intensity 21, 83
Field determination
  of fundamental period of soil 39
  of shear wave velocity 36–9
Field-tests for soil properties 34
Finite element analysis 162–3
Fire
  earthquake induced, cost of 247–8
  resistance of timber construction 443
Flow liquefaction see Liquefaction
Focal depth, effect of 62, 148
Focal mechanism 88, 98–9
Focus 20, 85–6
Folds (geological) 47
Foundation(s)
  column bases and pile caps 392
  concrete 392
  construction, damage to housing and 227–235
  damping 165–6
  dashpot 152–4, 157–62
  deep box 313–4
  in liquefiable ground 325–8
  microzoning and 225–35
  modelling, finite elements 162–3
  piled 288, 298–9
  reinforced concrete see Concrete

  shallow 312–3
  spring stiffness 153–7
  tie-beams 392
  timber structures 433–5
  see also Substructure
Fourier amplitude spectrum 72
Fragility functions 223–5
Framed tube structures 289
Franki method (ground improvement) 326–7
Free-field motion 150–1
Frequency
  content 79–81
  domain 143–4, 163
Friction angles for foundations 313
Fundamental period
  effective 164–5
  for soil deposits 140
  see also Period of vibration

Gediz, Turkey, 1970 earthquake 150
Geology 15, 228
  local 27–31
Glass panels 481
Grabens 47–8
Gravity retaining walls 333–4
Gravity walls 331–2
Ground classes
Ground motions
  attenuation 91, 99–100
  characteristics 72–84
  spatial patterns 84–91
Ground oscillations during liquefaction 147
Groundwater
  conditions 34, 36, 40
  discharge 31
Grouting
  of masonry 423
  of soil 326–7

Half-space
  elastic 136, 154
  layered 154, 161
  theory, limitations of 136–7
  viscoelastic 158–9

Hammering 278, 477, 482–3
  see also Pounding
Hanging wall effects 103
Hawke's Bay, NZ, 1931, earthquake 4, 85, 205–7, 225–7, 237, 253
Hazard
  definition 1
  functions 67–8
Heritage structures 500
Household contents, damage to 211, 216, 218, 232, 235, 238
Houses, retrofitting of 502
Hybrid structural systems 291
Hyogo-ken Nanbu 1995 earthqhuake see Kobe
Hypocentre 20, 85–6
Hysteresis 135–6, 176, 303, 340–1
  damping see Dampiong
  monotonic behaviour 347–53, 374, 429
  pinched loops 177, 432

Imperial Valley 1940 earthquake see El Centro 1940 ground motions
Inangahua New Zealand 1968 earthquake 208–18, 227–35
Inelastic response spectra 113–4
Infill, masonry, reinforced concrete buildings with 500
Infill panels
  effect on member forces 182–3
  effect on seismic response 180–2
  interaction of frames and 180–3
  non-structural 286–7, 476–9
Infill walls, structural 424
Inhomogeneous soil 160–2
Insurance (earthquake) 9, 259
Intensity
  definition 21
  see also Modified Mercalli intensity
Interaction of frames and infill panels 180–2
Inter-earthquake effects 240–1
Intermediate technology 306
International decade for Natural Disaster Reduction (IDNDR)

Inter-storey drift see Drift
Intraplate earthquakes 47, 57, 79–80, 99–100
Isolating devices, location of 293–4
Isolation
  from seismic motion 291–8
  of equipment 466–8
  using flexible piles 298–9
  using rocking 298, 300–1

Kobe 1995 earthquake 4, 253, 255, 279, 293
Kocaeli, Turkey, 1999 earthquake 279, 293
Kanto 1923 earthquake 7
Koyna dam, India 15

Laboratory tests for soils 34, 39–43
Lamellar tearing 345–7
Laminations in steel 346
Landslides (avalanches) 12, 27, 30–1, 251–2
Lateral drift see Drift
Lateral restraints, spacing of 354–5
Lateral spreading 147
Lead-rubber bearings 292, 295–7
Lifelines, retrofitting 493–6
Light fittings 468
Limit state design 268, 270, 312, 356–7
Limited ductility see Ductility
Line source models 85–7
Link beams in EBFs 360
Liquefaction
  damage due to 241–2
  flow 147
  level ground 147
  of saturated cohesionless soils 147–9
Liquefiable ground (foundations) 241–2, 325–6
Local magnitude 21–3
Lognormal distribution 209
Loma Prieta 1989 earthquake 103, 429
Longitudinal steel
  in r.c.beams 397
  in prestressed concrete members 413–5

Losses *see* Damage costs
Low damage structures *see* Damage avoidance design
Low seismic hazard regions *see* Masonry
Low-rise construction 182, 206, 226, 277, 283, 291
Lumped mass model for soil 142
Lysmer's analogue 159

Magnitude (earthquake) 21–4
Magnitude-frequency relationship 63–5
Managua 1972 earthquake 116, 117, 428
Masonry 416–27
   confined 444
   in low seismic hazards regions 427
   infill, reinforced concrete buildings with 424
   reinforced, design and construction details 421–4
   see Stone buildings
   unreinforced *see* Unreinfoced masonry
Mass densities 133
Material (internal) damping *see* Damping, material
Material behaviour 141–2, 192–7
Maximum Considered Earthquake (MCE) 268
Maximum magnitude (M_{max}) 46–9, 51, 53, 64–5, 117
Mechanical equipment *see* Equipment
Mexico City 28, 105, 140, 143
   Lake Zone of 28, 105, 140, 143
Mexico earthquakes
   1957 28
   1985 28, 105, 143
Michigan Basin 49
Microzones 31–4
   effects on vulnerability 225–36
      classification 228–9
      effects on damage to houses 229–35
      foundation effects 227–9
      house foundation type, household contents and 235
      risk assessment methodology 236
      very strong shaking 225–7
Mode shapes 190, Figure 5.21

Mode superposition 191
Modified Mercalli intensity scale 81–3, 204–5, 511–6
   upper bound on MMI 203–5
   uses of MM intensities 83
Modified Omori relation 69
Modulus of elasticity values for soils 135
Moment magnitude 22
Moment (seismic) release 44–7
Moment-resisting frames 280–2, 286, 289
Monetary Seismic Risk 3
Mononobe-Okabe equations 329–33
Monte Carlo process for synthetic earthquake catalogue 258
Movement gaps 278, 433, 469, 471–2
   See also Seismic joints
Multi-degree-of-freedom system 170, 183–92
Murchison, New Zealand 1929 earthquake *see* Buller

Napier, New Zealand 1931 205–6, 225–7
Near-source directivity effects 102–3
Near fault fling 296
Nevis-Cardrona fault *see* Faults
Next Generation Attenuation Models 96–7
Niigata, Japan 1964 earthquake 147
Non-linear dynamic analysis 192, 298
Non-linear seismic response of structures 172–7
Non-linear soil behaviour 160, 193
Non-structure and failure mode control 286–7
Normal faults *see* Fault types
Normal mode analysis 192
Normal risk construction 273–4
North Sea 30, 48
Northridge, California, 1994 earthquake 70–1, 84, 102, 104–5, 347, 418, 429, 506
   casualties 251–3
Notch ductility *see* Ductilty

One-dimensional site response analysis 140–3
Operational Basis Earthquake (OBE) 109, 274

Pacoima Dam 72, 79, 144
Pahiatua New Zealand 1934 earthquake 485
Panel zones, steel 361–4
Parkfield California 1966 earthquake 196
Particle
  board 431, 435, 438
  size distribution (of soil) 39–40
Partitions 181, 280, 281, 286
  see also Infill
P-delta effect 196–7, 281, 282
Peak Ground Acceleration (PGA) 81–3, 91, 93, 121, 140, 143–4, 225, 229
  versus Modified Mercalli intensity–81 83
Peak ground motions, upper bounds 76–9
Penetration resistance tests (soils) 36
Performance Achievement Ratio see Retrofitting
Performance-Based Seismic Design 267–70
Period of vibration 164–6, 169–72, 179–80
  effective 340–2
  of soil sites 139–40
Peruvian 1970 earthquake 4, 30, 286
Piers 316
Piles
  dynamic response 317–8
  elastic continuum analysis 319
  equivalent static lateral load analysis 318
  foundations 166, 288, 316–7
  house 229–35
  in cohesionless soils 324–5
  in cohesive soils 321–4
  in isolated structures 298–9
  lateral elastic displacements of a single 'long' pile 321–5
  non-linear lateral displacements of a single 'long' pile 320–1

Pipelines 116–7
Pipework 471–3
Planar source models 86–8
Planning
  aims 5
  for earthquakes 256–9
Plant 451–73
  retrofitting of 503–4
  vulnerability 214, 217–20
  see also Equipment
Plastic design method 345
Plastic hinge
  mechanisms 285
  rotation capacity 348–9
Plasticity 134, 172, 349
Platforms (geological) 47–8
Plumbing see Equipment; Pipework; Services
Plywood 435–9, 442
Point-source models 85–6
Poisson model, or process 68–70
Poisson's ratio 34, 41, 43
Porcelain 194–5
  see also Brittle materials
Post-yield behaviour 271–4
  see also Hysteresis
Pounding 286, 424
  see also Hammering
Precast concrete
  cladding 302, 406, 409–11, 480
  structural 399–406
Precursory scale increase 66, 69
PRESSS 304
Prestressed concrete 263, 304, 410–6
Probabilistic Seismic Hazard Assessment (PSHA) 120–2
Probability of occurrence, or exceedance 105, 110, 112
  of occurrence of fault displacements 116–7
Pseudo-acceleration 187

Quality assurance 266, 306
Quality
  of concrete 390
  of reinforcement 390–1

Index 527

of structural steel 345–7, 517–8
Quito Project 260

Radiation damping 135–8, 158–67
RADIUS Initiative 260
Ramberg-Osgood model 177–8
Rankine conditions 333–4
Rate
   of loading 343–4
   of strain energy release 62–3
Ray-theory method 115
Reduction factor, for loading 195–6
Reinforcement quality *see* Quality
Relative density
   of soil 146
   test 34, 36–7, 40–1
Reliability of performance 270, 274–6, 282, 284, 286
Reliable seismic behaviour
   concrete structures 367–77
   masonry 419–21
   steel structures 344–7
   timber structures 432–3
Repairability 223–4, 273–4, 285, 290
Repairable (structures) 244, 268, 273
Replacement Value 206, 208, 210, 219, 247, 487, 493
Resonance
   of soils 105, 113, 187, 229, 235
   of strucures 169–70
Resonant column test(for soils) 42–3
Response spectrum
   analysis 184–8, 191–2, 193
   averaged 95, 188
   design 114, 452–3
   elastic 111–3, 197
   inelastic 113–4
   of design earthquakes 105–6
   of real earthquakes 111
   of simulated earthquakes 111
   sources of 111
   special features of 112–3
Retaining walls *see* soil-retaining structures
Retrofitting
   benefit-cost 493

   benefits of 486
   equipment and plant 503–4
   in developing countries 504–5
   lifelines 493–6
   Performance Achievement Ratio (PAR) 488–9
   performance in earthquakes 506–7
   structures 496–503
   see also Upgrading
Reverse faults *see* Fault types
Rise time 71–2
Risk reduction 1–2, 9–13, 237, 261–2
   effect of corruption *see* Corruption
Robustness of concrete structures 370
Rocking structures 137–9, 153, 158–68, 298–301
Root-mean-square acceleration 75–6
Rupture (fault)
   area 57–9, 70–1
   displacement 53, 70, 116–8
   length 57–60, 70
   slip *see* Displacement
   surface 57, 87, 97
   velocity 71

Safe-Shutdown Earthquake (SSE)
Safety criteria 270–1
San Antonio, Chile 1985 earthquake 144, 276, 278
San Fernando, California 1971 earthquake 28, 72, 93, 277, 314, 454, 475
San Francisco Bay area 103, 140
San Francisco earthquakes
   1906 4, 7, 59, 248
   1957 144–5
Sedimentary basins, effects on ground motions 105
Seiches 31
Seismic analysis of structures, methods of 183–97
Seismic design
   criteria 451–6
   joints 482, 497
   procedures for equipment 456–62
Seismic gaps 52
   See also Movement gaps

Seismic hazard (definition) 1
Seismic isolation *see* Isolation, seismic
Seismic moment 22, 46, 86
Seismic response
    masonry 417–8
    prestressed concrete 412–3
    reinforced concrete 367
    of soil-structure systems 149–68
    soils 131–49
    steel structures 343–4
    structures
        elastic 169–72
        non-linear 172–7
    timber structures 429–32
Seismic risk (definition) 1
Seismic soil pressures 329–34
    in cohesionless soils containing water 333
    in cohesive soils or with irregular ground surface 333
    completely rigid walls 333–4
    in unsaturated cohesionless soils 331–2
Seismicity 1, 15–7, 19
Seismicity model 118
Seismology 21
Seismotectonics
    global 16–20
    regional 47–9
Serviceability criteria 270–2
Serviceability Limit State 268, 270
Services 451–73, 478
    see also Equipment; Lifelines
Settlement of dry sands 34, 40, 138, 146
Shaanxi 1556 earthquake 3
Shear beam 141–4, 146, 157
Shear in columns (steel) 356
Shear modulus 20, 34, 36–7, 40–2, 132–4, 147
Shear strain 20, 134, 136
    effect on damping and shear modulus 135–6, 139–40, 156
Shear strength
    of concrete columns 395–6
    of structural walls 384
Shear walls *see* Structural walls; Walls

Shear wave velocity 34–40
Sian 1556 earthquake 48
Simulated earthquakes 115–6
    see also Accelerograms
Single-degree-of freedom-systems 169–70, 184–7, 339–40
Site
    characteristics 27–42, 107, 490
    investigations 27, 34–9
    period 39, 139–40
    response to earthquakes 138–49
    soil conditions 27–34, 143–5, 163
Slip-predictable models 69
Soft storey concept 279, 487, 492, 502
    See also Weak storeys
Soil
    conditions *see* Site conditions
    improvement 326–7
    layers, effect on bedrock excitation 131–7, 145–51
    models 145–51
    reinforcement of 326–7
    retaining structures 328–34
    structure interacion 149–62
    tests 39–43
    types 133
Space geodesy 45–7
Spatial distribution of earthquakes 60–2
Spectra for different site conditions 31–2
Spectral acceleration, displacement, velocity 187–8
Splices (in reinforcement) 388–9, 393, 422
Spring and block models 66
Springs and dashpots 152–4, 157
Sprinkler systems 462–3, 483
Square-Root-of-Sum-of-Squares (SRSS) 192
Standard penetration resistance 36
Standards
    of construction 306
    reinforced concrete *see* Quality
    structural steel *see* Quality
Stationary random processes 115
Steel
    beams 347–51

design of 350–1
  under cyclic loading 350
  under monotonic loading 347–50
connections under cyclic loading 361
design forces for connections 364
quality *see* Quality
Stiff structures versus flexible 273, 281–2
Stiffness
  Appropriate, of structures 280–2
  degradation 172–5
  effective 269, 340–2
Stochastic methods 115
Stock, vulnerability of 211, 217–22, 238
Stone buildings 445
Stone columns 327–8
Strain-hardening 172
Strain-release map 174, 348–9, 361, 390
Strain-softening 172, 311
Strength of an earthquake 20–1
Strengthening structures *see* Retrofitting
Stress drop (on faults) 71, 76
Stress-release model 69
Strike-slip faults *see* Fault types
Structural form 275–80
  for masonry 419–21
  for steel 344
  for timber 429
Structural walls (concrete)
  flanged 380–2
  reinforced concrete 377–87
  see also Walls
Struts, steel, forces in 355
Subduction zone interface (underthrust) faults 50
Subduction zones 16–8, 93–4
Substructure 287–8
  see also Foundations
Supervision of construction
  in masonry 422, 427
  in timber 429
Surface wave magnitude 21–2, 24
Survivability 268, 274, 298
Symmetry of structures 276–7

Tanks 463, 465–6
  underground 314

Tectonic
  plates 16, 18–9
  provinces 47, 49
Tectonics *see* Seismotectonics
Tilting (of strata) 47, 49
Timber sheathed walls 431, 435–7
Time-domain analysis 163
Time-predictable models 69
Tokachi-Oki 1968 earthquake 30–1
Tokyo 1923 earthquake 248
Topographical effects 28, 30, 144–6
Torsional (effects) 276
Transcurrent faults 50
Trigger models 69
Tsunami 31

Uncertainty intervals 210
Unified magnitude 21
Uniform hazard response spectra 79–81, 112–3
Unreinforced masonry (URM) 4, 13, 204, 207, 281
  see also Brick, pure
  in developing countries 444, 505
  in low seismic hazard regions 427
  retrofitting, earthquake performance 501, 506

Value (of property) 1–3
Vertical shear beam model 141–3
  see also Shear beam
Vibro techniques 326
Viscoelastic half space 158
Viscous damping 135–6
Void ratio 40
  critical 146–7
Volcanic activity and earthquakes 15–6
Von Mises criterion 365
Vulnerability
  casualties 237–40
  contents of buildings 214–35
  definition 2–3
  of different classes of buildings 213–4
  household contents 216
  inter-earthquake effects 240–1
  qualitative measures of 203–7

Vulnerability (*continued*)
　quantitative measures of 206–41
　upper and lower bounds on 236–8

Wairarapa New Zealand 1942 earthquake
　206, 214–5, 493, 506
Wall finishes 480–1
Walls *see also* Structural walls
　apertures in 419–21
　curtain 480–1
　for reliable behaviour 257–60
Warping of strata 47, 49
Water
　content of soil 31, 33
　table 35, 315
　see also Groundwater; Liquefaction;
　　Tanks
Weak storeys 204
　see also Soft storey concept
Weaknesses in earthquake risk reduction
　processes 9–13
Weibull distribution 68
Weldability 346
Welding
　of reinforcing bars 391, 402–7
　of structural steel 347, 364
Western Montana 49
Whittier Narrows, California 1987
　earthquake 506
Windows 481
Winkler spring method 329
Wood–Anderson seismograph 21
Workmanship 274–5, 278, 286

Young's modulus 134, 319
　See also Modulus of elasticity